**EMIS Datareviews Series No. 26**

Series Advisor: Professor B. L. Weiss

# PROPERTIES, GROWTH AND APPLICATIONS OF

# Diamond

## ELECTRONIC MATERIALS INFORMATION SERVICE

**Other books in the EMIS Datareviews Series from INSPEC:**

No. 1 Properties of Amorphous Silicon (2nd Edition, 1989; out of print)

No. 2 Properties of Gallium Arsenide (2nd Edition, 1990)

No. 3 Properties of Mercury Cadmium Telluride (1987; out of print)

No. 4 Properties of Silicon (1988; out of print)

No. 5 Properties of Lithium Niobate (1989; out of print)

No. 6 Properties of Indium Phosphide (1991)

No. 7 Properties of Aluminium Gallium Arsenide (1993)

No. 8 Properties of Lattice-Matched and Strained Indium Gallium Arsenide (1993)

No. 9 Properties and Growth of Diamond (1994)

No. 10 Properties of Narrow Gap Cadmium-based Compounds (1994)

No. 11 Properties of Group III Nitrides (1994)

No. 12 Properties of Strained and Relaxed Silicon Germanium (1995)

No. 13 Properties of Silicon Carbide (1995)

No. 14 Properties of Metal Silicides (1995)

No. 15 Properties of III-V Quantum Wells and Superlattices (1996)

No. 16 Properties of Gallium Arsenide (3rd Edition, 1996)

No. 17 Properties of Wide Bandgap II-VI Semiconductors (1997)

No. 18 Properties of Porous Silicon (1997)

No. 19 Properties of Amorphous Silicon and its Alloys (1998)

No. 20 Properties of Crystalline Silicon (1999)

No. 21 Properties, Processing and Applications of Indium Phosphide (2000)

No. 22 Properties, Processing and Applications of Glass and Rare Earth-Doped Glasses for Optical Fibres (1998)

No. 23 Properties, Processing and Applications of Gallium Nitride and Related Semiconductors (1999)

No. 24 Properties of Silicon Germanium and SiGe:Carbon (2000)

No. 25 Physical Properties of Liquid Crystals: Nematics (2001)

Details of these and forthcoming books in the EMIS Datareviews Series are available at the following address on the World Wide Web:

http: //www.iee.org.uk/publish/books/emis.html

# PROPERTIES, GROWTH AND APPLICATIONS OF

# Diamond

Edited by

M. H. NAZARÉ and A. J. NEVES
University of Aveiro, Portugal

IEE

INSPEC
THE INSTITUTION OF ELECTRICAL ENGINEERS

Published by: INSPEC, The Institution of Electrical Engineers,
London, United Kingdom



**British Library Cataloguing in Publication Data**

A CIP catalogue record for this book
is available from the British Library

**ISBN 0 85296 785 3**

Printed in England by Short Run Press Ltd., Exeter

# Contents

# Foreword

Diamond has a reputation that is unique. Much of its popular image is associated with a small number of exceptional gemstones. Much of its market value relates to its effectiveness in cutting tools. For a metastable structure, the idea of a diamond being for ever gives only a small hint of its unusual properties. Yet, after centuries of being the most desirable of gemstones, another side to its character is emerging: diamond is showing signs of becoming a formidable 'carbon ceramic'. Underpinning this development are two major bodies of work. The first body of work, put simply, has reduced costs dramatically. Exceptional gemstones mean exceptional costs. Carbon ceramics, as working materials, have to be competitive in their perceived value for money. The second body of work is diamond science. It has helped to make the cost reductions possible: the two bodies of work support each other. Science has also opened up business options. These business opportunities, in turn, have revealed new scientific opportunities. That business should lead to new science should be no surprise: even J. Willard Gibbs was aware that thermodynamics probably owed more to the steam engine than the steam engine owed to thermodynamics. So, in this volume, we see science supporting the new technology of diamond film growth; we see this new technology creating isotopically pure diamonds which enable both extension of the science and improvements in thermal conductivity; we see scientific techniques which are new probes of diamond quality.

The first and obvious result of the developments in synthesis was the choice of form and content. Nature chose dopants and chose morphology; nature provided the obstacle of graphitisation. In their ways, both high temperature, high pressure growth and metastable growth processes widened this choice. No longer was it necessary to be limited by what nature provided or what one could make by careful and arduous treatments. The choice of film (albeit polycrystalline) or bulk existed. There was a choice of morphology. The film could even be persuaded to match a substrate shape. There was the possibility of controlling the surface, in both its tribological and electronic properties.

A second gain was defect science and engineering. Isotopically pure diamonds meant that the full power of spin resonance could be used at last. Selective doping made it easier to define impurity phenomena. This does not mean that puzzles are all resolved. The status of shallow states is still in question, despite some optimistic signs. But some of the long-standing puzzles seem resolved. The self-interstitial is giving up its mystery; the vacancy and its various charge and excited states seem to offer explanations for the many transitions beautifully characterised in careful spectroscopies. These developments do not mean, yet, that diamond can be made even more resistant to radiation, or that a diamond laser will be possible, but they open up rational approaches.

Thirdly, the wealth of past experiments can be exploited in new contexts. Reproducibility has become more precise, so that special optical, thermal, mechanical or electronic properties can be managed systematically. Cross-links are gradually becoming possible. Historically, the science of diamond has often been limited by its range, since those who work on tribological properties (such as friction, wear, polishing) might not talk as much as they should to those whose concern is with thermal properties (exploiting the huge thermal conductivity), electronic properties (realising that negative electron affinity and electron emission can have tribological effects) or phase changes (the role of graphite and ill-defined and impure phases in polishing). These links do not concern diamond alone, of course, since key processes may involve graphite and the amorphous hydrogenated diamond-like carbons.

Fourthly, some striking new areas have emerged, one being electrochemical applications. No-one would have taken electrochemical uses seriously before thin films were available at acceptable cost; now, such films look seriously promising. This electrochemical example is one of a group of applications which relies on the versatility of diamond. The factors of robustness and some sort of transparency are powerful together. The optical bandgap is important in electrochemistry, just as it is in photonic applications. The robustness is important for electrodes in an aggressive environment, just as a different form of robustness (and indeed a different form of transparency, largely associated with low atomic number) is needed for X-ray windows.

The future is as unpredictable as ever. Yet some themes are sure to continue. The first is the broadening of ways to assess quality for the many applications The second is the achievement of chosen quality at acceptable cost. This implies a third theme, the need for understanding. It is this quantitative understanding of diamond phenomena which opens the way to new fields of diamond science and technology.

Marshall Stoneham
University College London, UK
June 2000

# Diamond Materials Reviewed in this Volume

## Natural diamond

Nitrogen is the dominant impurity in the majority of diamonds. At the beginning of the twentieth century it was discovered that most natural diamonds have optical absorption in the infrared, in the 6.5–15 μm range, related to the impurity content. Diamonds exhibiting those optical features were classified as type I and those without them as type II. Later, it became known that the extrinsic absorption is due to the presence of nitrogen in type I diamond whereas type II diamonds are relatively free of nitrogen.

Type I diamonds can be further classified into type Ia if nitrogen is present in aggregated form or as type Ib if nitrogen is present as single substitutional atoms. Most natural diamonds belong to the subdivision type Ia and the nitrogen concentration can be as high as 3000 ppm. Only a small fraction of natural diamonds (less than 0.1%) are type Ib: typical nitrogen concentrations are 40 ppm. Both the isolated substitutional nitrogen in type Ib diamonds and the A-aggregate (a nearest-neighbour pair of substitutional nitrogen atoms ) in type Ia diamonds act as donors, with ionization energies of 1.7 and 4.0 eV, respectively, meaning that the crystals are electrically insulating at room temperature.

Some, very rare in nature, type II diamonds are highly conductive, exhibiting resistivities as low as 25 Ω cm; those were classified as type IIb, while the nitrogen-free high resistivity specimens, typically $10^{16}$ Ω cm, were designated as type IIa. Type IIb diamonds are p-type semiconductors with typical boron concentrations less than 1 ppm and an acceptor ionization energy of 0.37 meV.

## Synthetic diamond

Large single crystal diamonds can be grown by the high-temperature high-pressure technique (see Datareview B1.1). Again, nitrogen is a common impurity, being incorporated predominantly as single substitutional atoms with concentrations that can be as high as 800 ppm. To produce type II diamond nitrogen has to be eliminated from the growth environment, using nitrogen getters. Other atomic species, which can be incorporated into the diamond lattice during growth, are B, P, Ni, Co, Si and H. Hydrogen plays a crucial role in the low-pressure growth of diamond. Low-pressure growth from the gaseous phase using chemical vapour deposition (CVD) produces mainly polycrystalline diamond (see Section B1). Typically in diamond CVD, gas mixtures consisting of 1 vol% $CH_4$ in 99 vol% $H_2$ have been used in which atomic hydrogen is generated either by thermal decomposition or by collisional processes in a plasma. The diamond films so produced consist of micrometre-sized crystallites usually displaying columnar morphology. If diamond films are grown from hydrogen-poor (argon) plasma, they exhibit nanocrystalline structure, with randomly oriented nanocrystallites (see Section B2).

Naturally occurring diamond consists of 98.9% $^{12}C$ and 1.1% $^{13}C$, reflecting the natural abundance of the two stable isotopes of carbon. Single crystal diamond of any desired $^{12}C$:$^{13}C$ ratio can be produced combining the CVD and the HPHT techniques.

M.H. Nazaré
University of Aveiro, Portugal

# Contributing Authors


**A.B. Anderson**
A8.2
Department of Chemistry
Case Western Reserve University
Cleveland, OH 44106-7217, USA

**J.C. Angus**
B1.2
Chemical Engineering Department
Case Western Reserve University
Cleveland, OH 44106-7217, USA

**J.M. Baker**
A7.4, A7.5
Clarendon Laboratory, Department of Physics
Parks Road, Oxford OX1 3PU, UK

**E.J. Brookes**
A5.5
School of Engineering
University of Hull, Cottingham Road,
Hull, HU6 7RK, UK

**J.E. Butler**
B1.2
Gas/Surface Dynamics Section
Code 6174, Naval Research Laboratory
Washington DC 20375, USA

**G.Z. Cao**
B3.4
Department of Materials Science and Engineering
University of Washington
Roberts Hall 302M, Box 352120
Seattle, WA 98195, USA

**P.R. Chalker**
B3.3
Materials Science & Engineering
Department of Engineering
The University of Liverpool
Brownlow Hill, Liverpool L69 3GH, UK

**J. Chevallier**
A8.3
Laboratoire de Physique des Solides de Bellevue
1, Place Aristide Briand 92195 Meudon, France

**F.M. Costa**
B1.6
Departamento de Fisica
Universidade de Aveiro
3810 Aveiro, Portugal

**A. Crossley**
B3.3
AEA Technology
329 Harwell, Didcot
Oxfordshire OX11 0RA, UK

**R.D. Daniel**
A5.5
School of Engineering
University of Hull, Cottingham Road
Hull, HU6 7RK, UK

**G. Davies**
A7.3
Physics Department
King's College London
The Strand, London WC2R 2LS, UK

**D.L. Dreifus**
C1.4

Technology Division
General Electric Company
1975 Noble Rd., Nela Park
Cleveland, OH 44112-6300, USA

**W.P.J. van Enckevort**
A4.1, A4.6, A5.6

Department of Solid State Chemistry
Faculty of Science, University of Nijmegen
Toernooiveld 1, 6525 ED Nijmegen, The Netherlands

**H. Fan**
B1.6

Departamento de Fisica
Universidade de Aveiro
3810 Aveiro, Portugal

**A.J.S. Fernandes**
B1.6

Centre for Astrophysics
Porto University, Rua D. Manuel II
4003 Oporto Codex, Portugal

**J.E. Field, FRS**
A4.2, A4.3, A4.4, A4.5

Cavendish Laboratory
University of Cambridge
Madingley Road, Cambridge, CB3 0HE, UK

**I.J. Ford**
B1.4

Department of Physics and Astronomy
University College London
Gower Street, London WC1E 7JE, UK

**B.A. Fox**
C1.4

Technology Division
General Electric Company
1975 Noble Rd., Nela Park
Cleveland, OH 44112-6300, USA

**D.G. Goodwin**
B1.2

Department of Mechanical Engineering and Applied Physics
California Institute of Technology
Pasadena, CA 91125, USA

**J. Goss**
A5.1

School of Physics
University of Exeter
Exeter, Devon, EX4 4QL, UK

**M. Grimsditch**
A1.4

Materials Science Division
Argonne National Laboratory
Argonne, IL 60439, USA

**D.M. Gruen**
B2.1, B2.2, B2.3, B2.4

Materials Science and Chemistry Division
Argonne National Laboratory
Argonne, IL 60439, USA

**J. Hartmann**
A1.5, A2.1

Physikalish-Technishe Bundesanstalt
Fachlabor 7.21 Temperaturstrahlung
Abbestrasse 2-12, 10587 Berlin, Germany

**R. Haubner**
B1.5, C2.1

Institute for Chemical Technology of Inorganic Materials
University of Technology Vienna
Getreidemarkt 9/161, A-1060 Vienna, Austria

**J.J. Hren**
C1.3

Department of Materials & Engineering
North Carolina State University, Box 7907
Raleigh, NC 27695-7907, USA

**R.B. Jackman**
C1.6

Department of Electronic and Electrical Engineering
University College London
Gower Street, London WC1E 6BT, UK

**C. Johnston**
B3.3

AEA Technology
329 Harwell, Didcot
Oxfordshire OX11 0RA, UK

**R. Jones**
A5.1

School of Physics
University of Exeter
Exeter, Devon, EX4 4QL, UK

**H. Kanda**
B1.1

National Institute for Research in Inorganic Materials (NIRIM)
1-1 Namiki, Ibaraki 305-0044, Japan

**R. Kalish**
B3.1

Solid State Institute and Physics Department
Technion Haifa 32000, Israel

**I. Kiflawi**
A5.2, A5.3, A6.2, A6.3

Department of Physics
University of Reading, Reading, Berkshire RG6 6AH, UK

**P. Koidl**
C1.1

Fraunhofer Institut Angewandte Festkörperphysik
Tullastraβe 72, D-79108 Freiburg, Germany

**S.C. Lawson**
A5.2, A5.3, A6.2, A6.3

DTC Research Centre
Maidenhead, UK

**J.C. Lopes**
A6.1

Departamento de Fisica
Universidade de Aveiro
3810 Aveiro, Portugal

**J.E. Lowther**
A7.1

University of the Witwatersrand
P.O. Wits 2050, Johannesburg, South Africa

**B. Lux**
B1.5, C2.1

Institute for Chemical Technology of Inorganic Materials
University of Technology Vienna
Getreidemarkt 9/161, A-1060 Vienna, Austria

**A. Mainwood**
A7.2

Physics Department
King's College London
The Strand, London WC2R 2LS, UK

**C. Manfredotti**
C1.5
Dipartimento di Fisica Sperimentale
Università degli Studio di Torino
Via Pietro Giuria, 1-10125 Torino, Italy

**V.A. Nadolinny**
A7.5
Department of Physics, Clarendon Laboratory
Parks Road, Oxford OX1 3PU, UK

**C. Nebel**
A1.6
Walter Schottky Institut
TU – München, D 85748 Garching, Germany

**A.J. Neves**
A6.1
Departamento de Fisica
Universidade de Aveiro
3810 Aveiro, Portugal

**M.E. Newton**
A5.4, A5.8, A7.4, A7.5
Department of Physics, Clarendon Laboratory
Parks Road, Oxford OX1 3PU, UK

**E. Pereira**
A8.4
Departamento de Fisica
Universidade de Aveiro
3810 Aveiro, Portugal

**D. Pickard**
A2.2
National Medical Laser Centre
Department of Surgery
University College London
Charles Bell House, 67-73 Riding House Street
London WC1 7NP, UK

**J.F. Prins**
B3.2
Department of Physics
University of Pretoria, Pretoria 0002
Gauteng, South Africa

**A.K. Ramdas**
A1.1, A1.2, A1.3
Purdue University
Physics Building, West Lafayette
IN 47907-1396, USA

**M. Reichling**
A1.5, A2.1
Department Chemie
Universität München Butenandtstraβe 5-13
81377 München, Germany

**J. Ristein**
A3.1
Institut fur Technische Physik
Universität Erlanger-Nürnberg
Erwin-Rommel-str. 1
D-91058 Erlanger, Germany

**R. Sauer**
A8.1
Abteilung Halbleiterphysik
Universitaet Ulm
Albert Einstein-Alle 45, D 89069 Ulm, Germany

**R. Schlesser**
A3.2
Materials Science and Engineering
North Carolina State University
1001 Capability Drive, RB#1 Raleigh, NC 27695-7919, USA

**T. Sekine**
B1.1

National Institute for Research in Inorganic Materials (NIRIM)
1-1 Namiki, Ibaraki 305-0044, Japan

**A. Shih**
A3.3, A3.4

Naval Research Laboratory, Code 6844
4555 Overlook Avenue, S.W.
Washington, D.C. 20375-5320, USA

**R.F. Silva**
B1.6

Department of Quimica
Sao Paulo University
Sao Carlos, Brazil

**Z. Sitar**
A3.2

Materials Science and Engineering
North Carolina State University
1001 Capability Drive, RB#1
Raleigh, NC 27695-7919, USA

**M. Stoneham, FRS**
Foreword

Centre for Materials Research
Department of Physics and Astronomy
University College London
Gower Street, London WC1E 6BT, UK

**M. Stutzmann**
A1.6

Walter Schottky Institut
TU – München, D 85748 Garching, Germany

**F.K. de Theije**
A4.6, A5.6

Department of Solid State Chemistry
Faculty of Science, University of Nijmegen
Toernooiveld 1, 6525 ED Nijmegen, The Netherlands

**K. Thonke**
A8.1

Abteilung Halbleiterphysik
Universitaet Ulm
Albert Einstein-Alle 45, D 89069 Ulm, Germany

**D.J. Twitchen**
A7.4, A7.5

Clarendon Laboratory, Department of Physics
Parks Road, Oxford OX1 3PU, UK

**C. Uzan-Saguy**
B3.1

Solid State Institute and Physics Department
Technion Haifa 32000, Israel

**E. Vittone**
C1.5

Dipartimento di Fisica Sperimentale
Università degli Studio di Torino
Via Pietro Giuria, 1-10125 Torino, Italy

**M. Werner**
B3.3

Deutsche Bank AG, CORE
Microtechnology Innovation Team
Unter den Linden 13/15
10117 Berlin, Germany

**M.D. Whitfield**
B1.4

Department of Electronic and Electrical Engineering
University College London
Gower Street, London WC1E 7JE, UK

**C. Wild**
C1.1

Fraunhofer Institut Angewandte Festkörperphysik
Tullastraβe 72, D-79108 Freiburg, Germany

**H. Windischmann**
C1.2, C2.2

Norton Diamond Film
Northboro, MA 01532, USA

**J.E. Yater**
A3.3, A3.4

Naval Research Laboratory, Code 6844
4555 Overlook Avenue, S.W.
Washington, D.C. 20375-5320, USA

**A.M. Zaitsev**
A5.7

Institut für Mineralogie NA 031167
Ruhr-Universitaet Bochum, Universitaetsstr. 150
D-44801, Bochum, Germany

**V.V. Zhirnov**
C1.3

Department of Materials & Engineering
North Carolina State University, Box 7907
Raleigh, NC 27695-7907, USA

# ABBREVIATIONS

| | |
|---|---|
| AFM | atomic force microscopy |
| ASED | atom superposition and electron delocalization |
| | |
| BDT | brittle ductile transition |
| BE | bound exciton |
| BEN | bias enhanced nucleation |
| | |
| CAMPP | chemically assisted polishing and planing |
| CBM | conduction band minimum |
| CIRA | cold implantation rapid annealing |
| CL | cathodoluminescence |
| CNDO | complete neglect of differential overlap |
| CVD | chemical vapour deposition |
| | |
| D | donor |
| DFT | density functional theory |
| DLC | diamond-like carbon |
| DOS | density of states |
| | |
| EBD | electron beam device |
| ECR | electron cyclotron resonance |
| EDC | energy distribution curve |
| EELS | electron energy loss spectroscopy |
| EH | extended Hückel |
| EMF | electromotive force |
| ENDOR | electron nuclear double resonance |
| EPR | electron paramagnetic resonance |
| ERD | elastic recoil detection |
| ESEEM | electron spin echo envelope modulation |
| ESR | electron spin resonance |
| | |
| FE | free exciton |
| FEA | finite element analysis |
| FEED | field emission energy distribution |
| FET | field effect transistor |
| FTC | frequency temperature coefficient |
| FTIR | Fourier transform infrared |
| FWHM | full width half maximum |
| | |
| HFCVD | hot filament chemical vapour deposition |
| HFI | hyperfine interactions |
| HOD | highly oriented diamond |
| HPHT | high pressure high temperature |
| HPMS | high pressure microwave source |
| HREELS | high resolution electron energy loss spectroscopy |
| HRTEM | high resolution transmission electron microscopy |
| HT | high temperature |
| HTHP | high temperature high pressure |

| | |
|---|---|
| IC | integrated circuit |
| IC | ionisation chamber |
| IDT | interdigitated transducer |
| IR | infrared |
| | |
| KMC | kinetic Monte Carlo |
| | |
| LADS | large area deposition system |
| LDF | local density functional |
| LDLT | low dose low temperature |
| LEA | low electron affinity |
| LEED | low energy electron diffraction |
| LO | longitudinal optical |
| LODDI | low damage drive in |
| | |
| MAC | mass absorption coefficient |
| MBE | molecular beam epitaxy |
| MCD | microcrystalline diamond |
| MCDA | magnetic circular dichroism of the absorption |
| MEMS | microelectromechanical system |
| MEMS-MMA | microelectromechanical system moving mechanical assembly |
| MESFET | metal semiconductor field effect transistor |
| MNDO | medium neglect of differential overlap |
| MO | molecular orbital |
| MPCVD | microwave plasma-assisted chemical vapour deposition |
| MPDR | microwave plasma disk reactor |
| MSM | metal-semiconductor-metal |
| | |
| NEA | negative electron affinity |
| NN | nearest neighbour |
| NNN | next nearest neighbour |
| NP | no-phonon |
| NRA | nuclear reaction analysis |
| | |
| ODMR | optically detected magnetic resonance |
| Op-FET | optically activated field effect transistor |
| | |
| PACVD | plasma-assisted chemical vapour deposition |
| PBC | periodic bond chains |
| PCD | polycrystalline diamond composites |
| PL | photoluminescence |
| | |
| QA | quality assessment |
| QL | quasi-longitudinal |
| QPY | quantum photoyield |
| QT | quasi-transverse |
| | |
| RF | radio frequency |
| RIE | reactive ion etching |

| | |
|---|---|
| RMS | root mean square |
| RT | room temperature |
| RTA | rapid thermal annealing |
| RTD | resistive temperature detector |
| SAW | surface acoustic waves |
| SCL | space charge layer |
| SE | secondary emission |
| SEM | scanning electron microscopy |
| SIMS | secondary ion mass spectrometry |
| STM | scanning tunnelling microscopy |
| TA | transverse acoustic |
| TC | thermal conductivity |
| TCD | temperature coefficient of delay |
| TCR | temperature coefficient of resistance |
| TEM | transmission electron microscopy |
| TL | thermoluminescence |
| TM | transition metals |
| TO | transverse optical |
| TSC | thermoconductive |
| UDD | ultradispersed diamond |
| UNCD | ultra-nanocrystalline diamond |
| UV | ultraviolet |
| VL | vacuum level |
| XRD | X-ray diffraction |
| ZPL | zero phonon line |

CHAPTER A1

# OPTICAL, MECHANICAL AND TRANSPORT PROPERTIES

# A1.1 Raman, Brillouin and infrared spectroscopy of phonons in conventional diamond[1]

A.K. Ramdas

July 2000

## A INTRODUCTION

Diamond, the cubic modification of crystalline carbon, belongs to the space group $O_h^7 (F4_1/d\bar{3}2/m)$ with two atoms per primitive (Bravais) cell. The structure can be viewed as two interpenetrating face centred cubic (Bravais) lattices shifted along the body diagonal by (¼, ¼, ¼)a where a is the dimension of the cubic (mineralogical) unit cell (FIGURE 1).

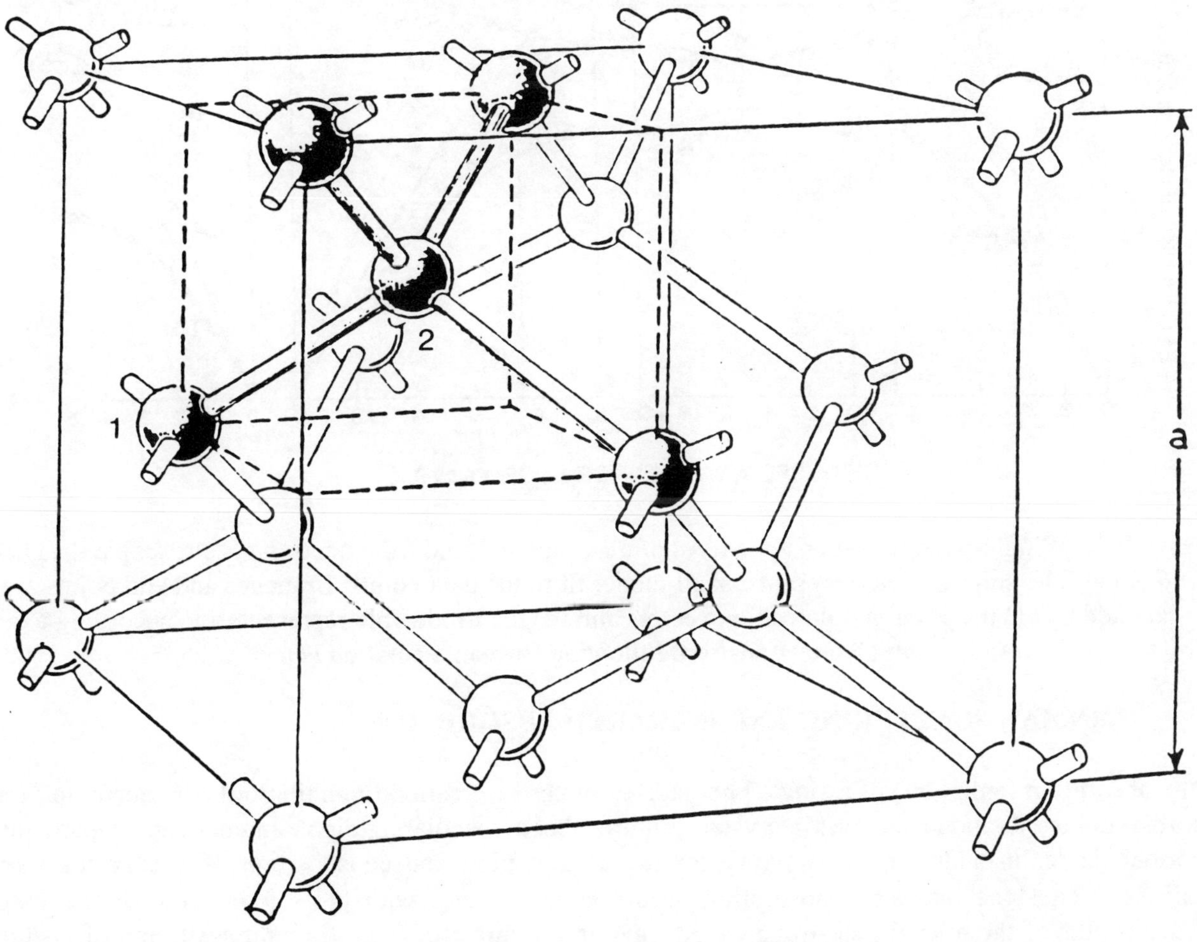


FIGURE 1 The diamond structure showing the tetrahedral coordination and the two interpenetrating face centred cubic (fcc) Bravais lattices the carbon atoms occupy. The mineralogical (cubic) unit cell of dimensions a × a × a and the two inequivalent atoms (1 and 2) of the smallest unit cell are indicated. (After W. Shockley [ *Electrons and Holes in Semiconductors* (Van Nostrand, Princeton, 1950) ].)

---

[1] Conventional (naturally occurring) diamond consists of 98.9% $^{12}C$ and 1.1% $^{13}C$.

The dispersion curves for the optical and acoustic branches of the vibrational spectrum of diamond, i.e. frequency (ω) of the ith branch as a function of wave vector (**q**), $\omega_{\mathbf{q}}^{i}$ versus **q**, with **q**'s spanning the entire Brillouin zone, have been determined by Warren et al [1] on the basis of inelastic slow neutron scattering. These are reproduced in FIGURE 2. The labelling of the special **q**'s corresponds to the standard convention [2,3] used in the deposition of special points of the Brillouin zone of diamond (FIGURE 3). Inelastic neutron scattering in diamond has been revisited by Kulda et al [4] who discovered an overbending in the dispersion curve for the longitudinal optical mode along [100], i.e. in the LO Δ-branch, with a maximum above the zone centre optical phonon. A high resolution inelastic X-ray scattering study [5] with synchrotron radiation has confirmed this overbending along Δ.

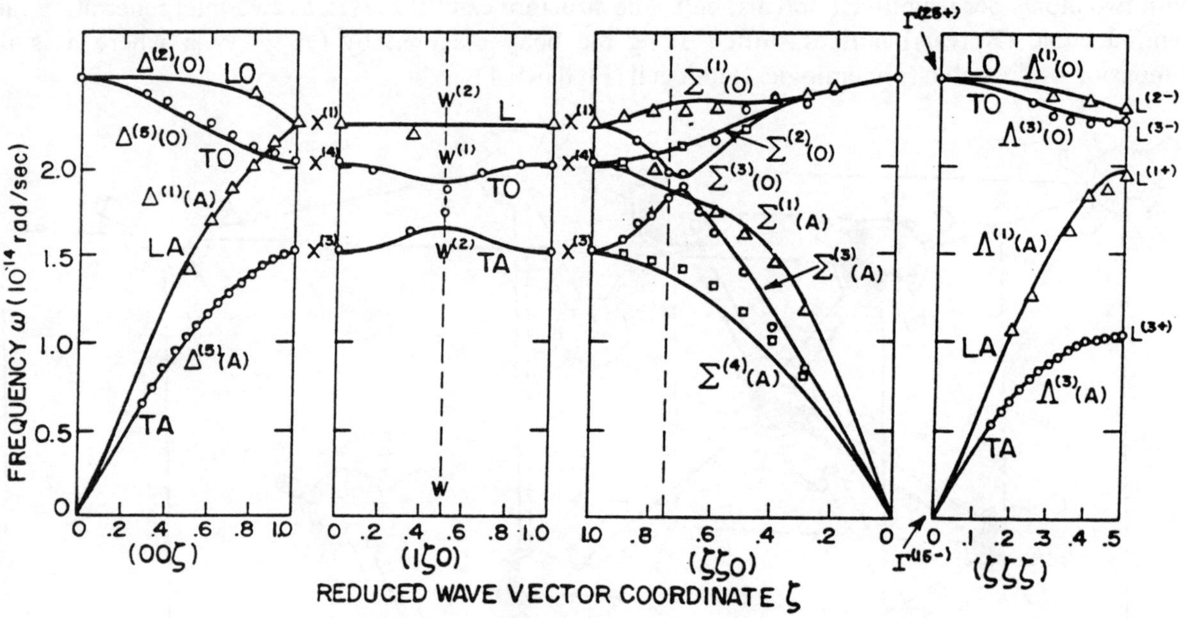


FIGURE 2 Single-phonon dispersion curves of diamond deduced from neutron spectroscopic data at 296 K [1]. The solid curves represent a shell-model fit to the data points. Branches and end points are labelled by (1) the phonon polarisation vectors and (2) the irreducible representation according to which they transform following Birman's notation [3].

## B RAMAN SCATTERING AND INFRARED ABSORPTION

The absorption (emission) of a single phonon ($\hbar\omega$) with the creation (annihilation) of a quantum of a normal mode vibration (phonon) in crystals requires that an oscillating dipole moment accompany the normal mode; in addition, the wave-vector (k) conservation and conservation of energy must be satisfied. Thus the resonance absorption occurs at $\hbar\omega = \hbar\omega_{\mathbf{q}}$, with $|\mathbf{q}| \sim 0$ in view of the long wavelengths of the relevant electromagnetic wave in the infrared. Inelastic light scattering of visible monochromatic radiation (wavelength $\lambda_L$, angular frequency $\omega_L$ and wave vector in the medium, $\mathbf{k}_L$) by a crystal, free of imperfections and hence with perfect translational symmetry, again conforms to k-conservation and energy conservation. The latter dictates that frequency shifts with respect to $\omega_L$ be $\pm\omega_{\mathbf{q}}$, + for phonon emission and – for absorption. The wave-vector of the phonon participating in the scattering process satisfies

$$|\mathbf{q}| = 2|\mathbf{k}_L|\sin(\theta/2) \tag{1}$$

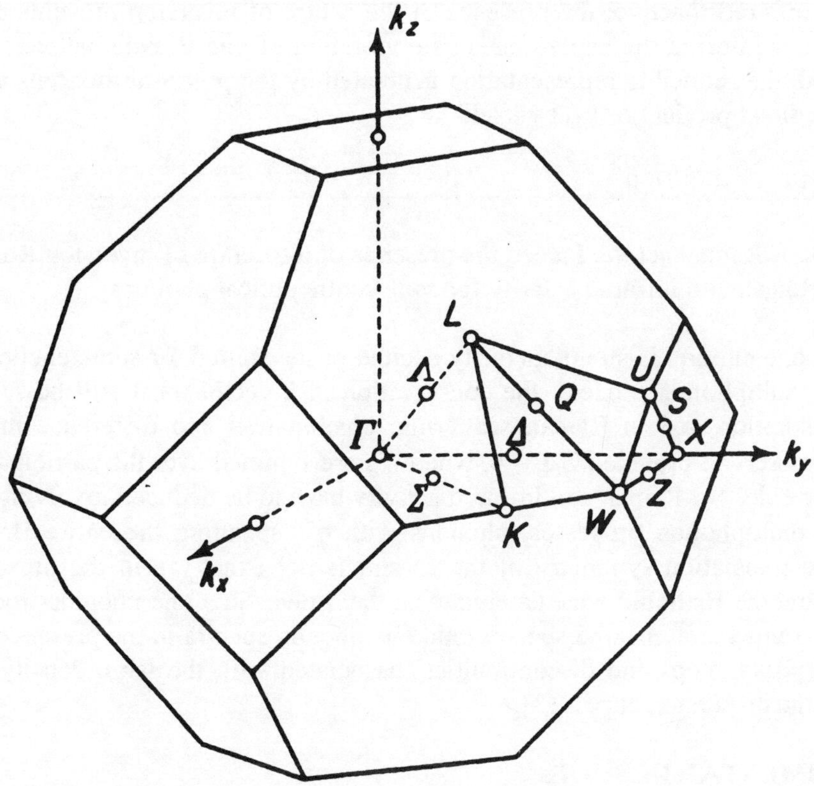


FIGURE 3 Brillouin zone for the face centred cubic translation group [2].

where θ is the scattering angle. For visible light, **q** is near the centre of the Brillouin zone, i.e. $|\mathbf{q}| \sim 0$. Inelastic light scattering (Raman scattering) associated with the optical phonon is allowed only if a non-vanishing oscillating polarisability tensor accompanies it. Thus both infrared absorption and inelastic light scattering (Raman scattering) associated with optical phonons can occur only for $|\mathbf{q}| \sim 0$, i.e. zone-centre optical phonons, if only a single quantum is absorbed or emitted and provided the selection rules are satisfied. It is for this reason that one is justified in classifying these 'zone centre' excitations in terms of the irreducible representations of the point-group of the crystal, $O_h$ in the case of diamond [6].

## C SELECTION RULES: INFRARED AND RAMAN

Applying group theoretical techniques [6], one can show that the reducible representation $\Gamma(Q_i)$ generated by the zone-centre excitations ($Q_i$) of diamond is decomposed as

$$\Gamma(Q_i) = F_{2g} + F_{1u} \tag{2}$$

It can be shown that the $F_{2g}$ ($\Gamma_{25}^{+}$ in the notation of [2]) corresponds precisely to the triply degenerate excitation in which the two interpenetrating Bravais lattices rigidly vibrate against each other; the $F_{1u}$ mode is the rigid translation of the crystal as a whole and corresponds to the polar representation. The dispersion curves in FIGURE 2 show $F_{2g}$ at the highest frequency at $|\mathbf{q}| = 0$ and the $F_{1u}$ mode with zero frequency, evolving into the acoustic modes (the long wavelength elastic or sound waves) whose ω's are linear in $|\mathbf{q}|$.

The $F_{2g}$ mode is infrared inactive; the presence of the centre of inversion prevents a dipole moment from being developed during the triply degenerate vibration of one Bravais lattice against the other. On the other hand, the reducible representation generated by the polarisability tensor for $O_h$ is given by the symmetric direct product of $F_{1u}$ by itself:

$$[F_{1u} \times F_{1u}] = A_{1g} + E_g + F_{2g} \tag{3}$$

Thus the $F_{2g}$ mode is Raman active. Indeed the presence of the centre of inversion leads to this mutual exclusion of the Raman and infrared activity for zone centre optical phonons.

When more than one phonon is simultaneously excited or de-excited, or some excited and others de-excited, i.e. in a multiphonon process, the conservation of k-vector must still be fulfilled. Hence in both infrared absorption and in Raman scattering, combination and difference modes as well as overtones can be observed provided $\Sigma_i \mathbf{q}_i \sim 0$, where i's are summed over the participating phonons. In addition, selection rules for Raman and infrared activity have to be deduced involving the space group [3,7,8]. In these multiphonon processes, phonons with **q**'s spanning the entire Brillouin zone can participate. If the translation symmetry of the crystal is not exact (as in the presence of defects), phonons throughout the Brillouin zone can manifest themselves in a one phonon process. The second and higher order Raman and infrared spectra (and one phonon spectra in the presence of defects) then exhibit features (peaks, steps and discontinuities) associated with the large density of states of the critical points in the dispersion curves [3].

## D EXPERIMENTAL RESULTS

### D1 Infrared and Raman Spectra

Diamond, when free from nitrogen impurities (the so-called type II diamond), is transparent throughout the near ultraviolet and the visible range of the electromagnetic spectrum; except for the 3.7 - 6 micron range where absorption due to two phonon processes occurs, it is transparent in the entire near and far infrared [9]. The two phonon infrared spectrum has been studied by many workers [8,10,11]; we refer here to Vogelgesang et al [8] who exploited the high S/N and resolution of an FTIR spectrometer. For the first and second order Raman spectra of diamond we refer to Solin and Ramdas [7] who used laser excitation, a scanning grating spectrometer and photoelectric detection with photon counting electronics, and to Vogelgesang et al [8] who also used laser excitation but a CCD-based triple grating spectrometer. The first and second order Raman and the second order infrared spectra recorded by Vogelgesang et al [8] are displayed in FIGURES 4 and 5. On the basis of space group selection rules for two phonon processes in the infrared and Raman scattering [3,5,8] on the one hand and the inelastic neutron scattering data [1] on the other, Solin and Ramdas [7] and Vogelgesang et al [8] arrived at the phonon energies (frequencies) at the high symmetry points of the Brillouin zone; in TABLE 1 the values obtained by Solin and Ramdas, and Vogelgesang et al are listed.

### D2 Brillouin Scattering

Near the zone-centre, for the acoustic branch of the vibration spectrum, ω is linear in **q**, the slope being given by the velocity of sound characteristic of sound waves travelling along **q**. The energies of the phonons involved in Brillouin scattering studied with visible radiation are usually much smaller than the thermal energy; also, the wavelengths of these phonons are much longer than the interatomic distances. A continuum classical description of the phenomenon is therefore generally adequate for measurements at room temperature as is well described in the literature [12]. From the classical point of view the scattering is a result of a Bragg reflection from the optical stratifications produced by the long wavelength acoustic phonons (elastic waves) in the medium; the Doppler shifts associated with these 'moving mirrors' produce the Brillouin components, inelastic light scattering from the elastic waves being referred to as Brillouin scattering. For given incident (i) and scattered (s) directions, the frequency shifts $\Delta\omega$ of the Brillouin components in a cubic crystal are given by

$$\Delta\omega = \pm\, 2\omega_L n \frac{v_s}{c} \sin(\theta/2) \tag{4}$$

Here $\omega_L$ is the excitation frequency, n the refractive index, c the velocity of light in vacuum, $\theta$ the scattering angle, and $v_s$ the velocity of the appropriate sound wave responsible for the Bragg reflection. For a general direction of propagation, there are three values for $v_s$ corresponding to two quasi-transverse (QT) and one quasi-longitudinal (QL) sound waves; $v_s$ in turn can be written as $(X/\rho)^{1/2}$ where X is an appropriate combination of the elastic moduli ($c_{ij}$) and $\rho$ is the density [12]. The Brillouin components for $\mathbf{q} \parallel [1\bar{1}0]$ are displayed in FIGURE 6 [13].

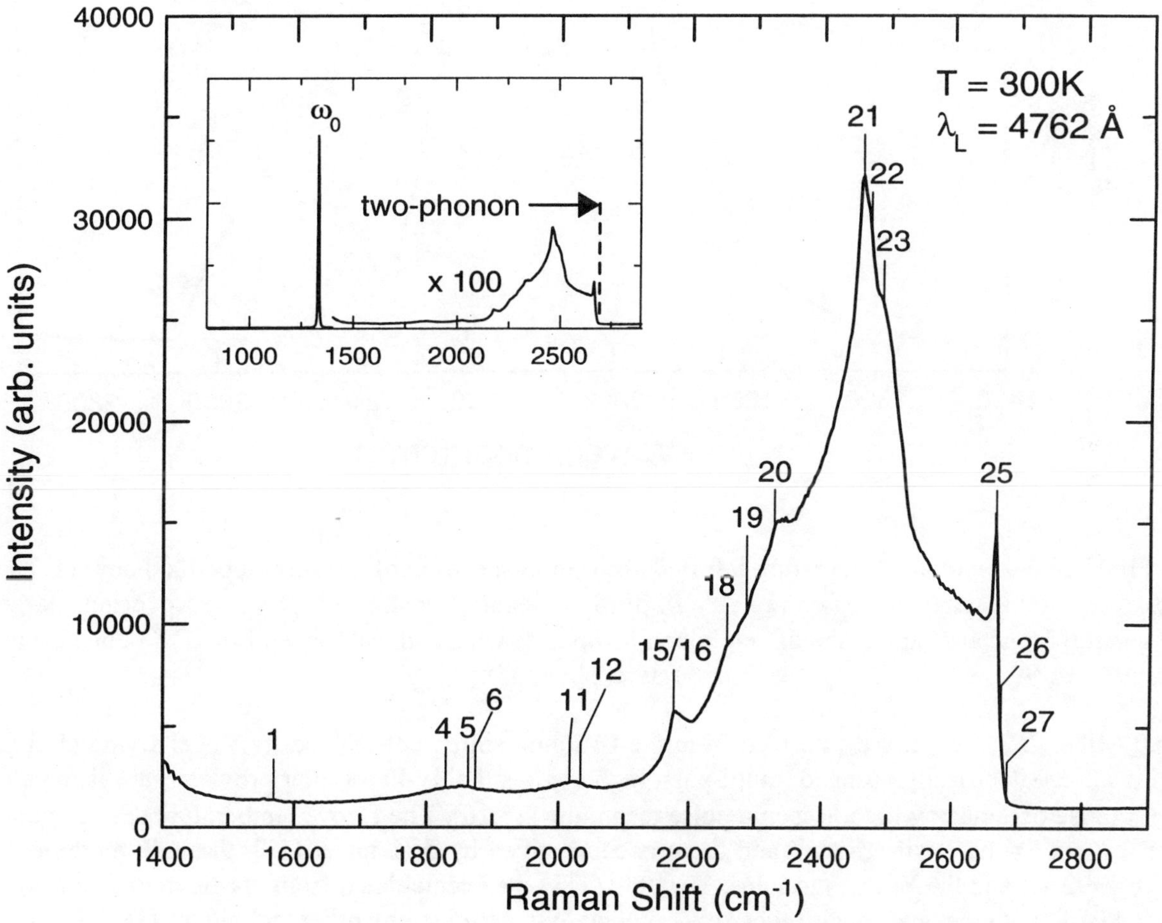


FIGURE 4 The Raman spectrum of a natural type IIb diamond, recorded in the backscattering geometry along [111] with the incident light vertically polarised in the (111) plane and the scattered light unanalysed. The polarisation configuration allows the simultaneous observation of the features characterised by the $\Gamma^{(1+)}$, $\Gamma^{(12+)}$ and $\Gamma^{(25+)}$ polarisability tensors. The two-phonon features identified in Table III of [8] are explicitly labelled. The inset underscores the relative intensities of the strong $\Gamma^{(25+)}$ zone centre optical phonon and the significantly weaker two-phonon features [8].

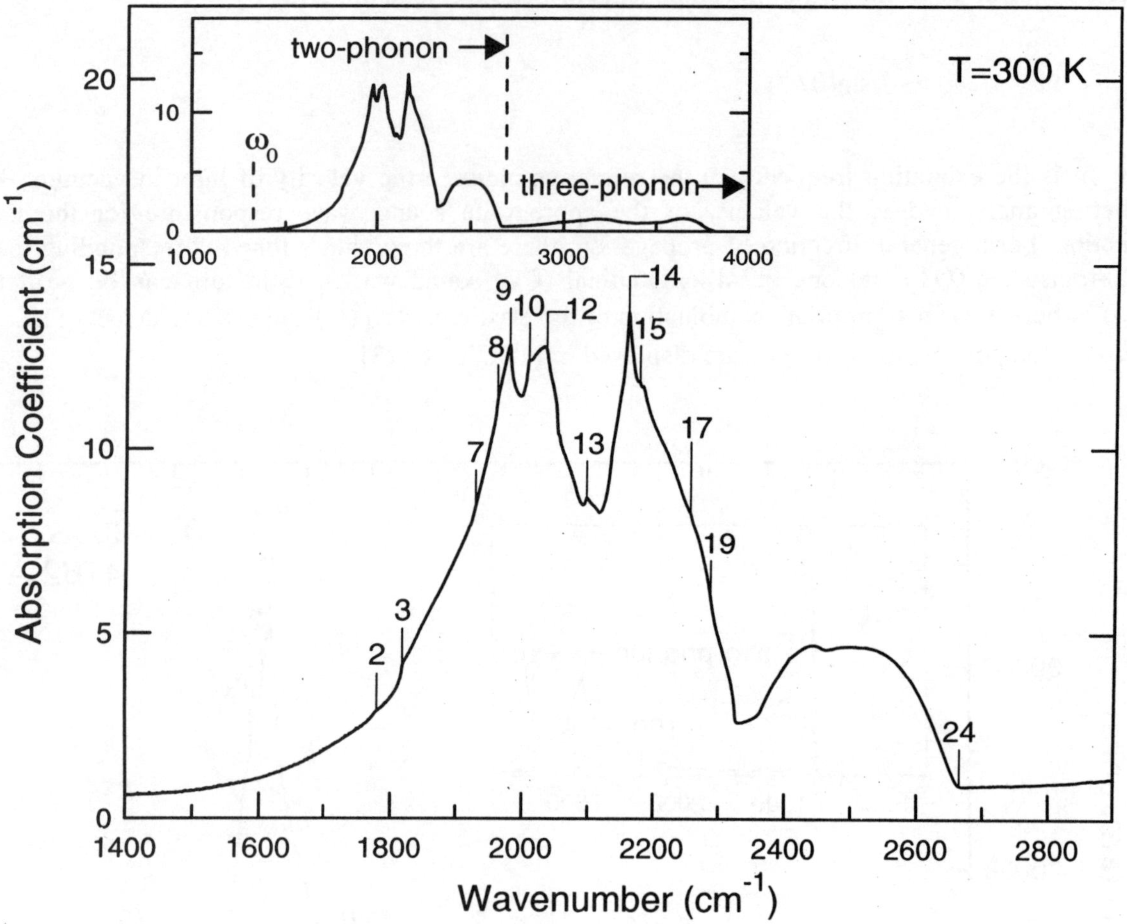


FIGURE 5 The room temperature infrared absorption spectrum of a natural type IIa diamond. The two-phonon features identified in Table III of [8] are explicitly labelled. The inset covering a larger spectral range also shows the three-phonon features as well as the absence of the $\Gamma^{(25+)}$ zone centre optical phonon [8].

In TABLE 2, $c_{11}$, $c_{12}$ and $c_{44}$ derived from the Brillouin shifts, determined by Vogelgesang et al [14] with a piezoelectrically scanned, multi-pass, high contrast Fabry-Perot interferometer, are listed along with those obtained by the ultrasonic pulse technique [15,16]. The Debye temperature, $\Theta_D$, calculated for 0 K using the elastic moduli and the procedure given by de Launay [17], the bulk modulus $B = {}^1/_3(c_{11} + 2c_{12})$ and the Young modulus, Y, in the (111) plane calculated from elastic moduli are shown in TABLE 3 and compared with those experimentally determined by other techniques [18–22].

TABLE 1 Phonon frequencies at critical and high symmetry points of the phonon dispersion curves. The representations of the associated phonon(s) are indicated in the notation of Birman [3] along with their nature (O = optical, A = acoustic, L = longitudinal, T = transverse). The phonon frequencies as determined in the present study are given in the last column with errors of ±2 $cm^{-1}$, except where stated differently [8].

| **q** (2π/a) | Representation | Frequency ($cm^{-1}$) | | |
|---|---|---|---|---|
| | | Neutron[1] | [7] | [8] |
| Γ: (0,0,0) | $\Gamma^{(15-)}$ (A) | | | 0.0 ± 0.0 |
| | $\Gamma^{(25+)}$ (O) | | 1332 | 1332.4 ± 0.01 |
| L: (½,½,½) | $L^{(3-)}$ (TO) | 1210 ± 37 | 1206 | 1208 |
| | $L^{(2-)}$ (LO) | 1242 ± 37 | 1252 | 1245 |
| | $L^{(1+)}$ (LA) | 1035 ± 32 | 1006 | 1009 |
| | $L^{(3+)}$ (TA) | 552 ± 16 | 563 | 572 |
| X: (1,0,0) | $X^{(1)}$ (L) | 1184 ± 21 | 1185 | 1170 |
| | $X^{(3)}$ (TA) | 807 ± 32 | 807 | 786 |
| | $X^{(4)}$ (TO) | 1072 ± 26 | 1069 | 1088 |
| W: (1,½,0) | $W^{(1)}$ (TO) | 993 ± 53 | 999 | 1012 |
| | $W^{(2)}$ (L) | 1168 ± 53 | 1179 | 1164 |
| | $W^{(2)}$ (TA) | 918 ± 11 | 908 | 915 |
| Σ: (ξ,ξ,0) | $\Sigma^{(1)}$ (O) | 1231 ± 32 | 1230 | 1236 |
| | $\Sigma^{(2)}$ (O) | 1120 ± 21 | 1109 | 1112 |
| ξ ~ 0.72 | $\Sigma^{(3)}$ (O) | 1046 ± 21 | 1045 | 1051 |
| | $\Sigma^{(1)}$ (A) | 982 ± 11 | 988 | 986 |
| | $\Sigma^{(3)}$ (A) | 993 ± 16 | 980 | 982 |
| | $\Sigma^{(4)}$ (A) | 748 ± 16 | | 748 |

TABLE 2 Elastic moduli of diamond (units: $10^{12}$ dyn/$cm^2$).

| Method | $c_{11}$ | $c_{12}$ | $c_{44}$ |
|---|---|---|---|
| Brillouin scattering [13] (~23°C) | 10.764 ± 0.002 | 1.252 ± 0.023 | 5.774 ± 0.014 |
| Brillouin scattering [14] (~23°C) | 10.804 (5) | 1.270 (10) | 5.766 (5) |
| Ultrasonic pulse technique [15] | 10.76 | 1.25 | 5.76 |
| Ultrasonic pulse technique [16] (25°C) | 10.79 ± 0.05 | 1.24 ± 0.05 | 5.78 ± 0.02 |

TABLE 3 Constants derived from elastic moduli (independent experimental values obtained from other techniques are presented in the 3rd column).

| | | |
|---|---|---|
| $\Theta_D$, Debye temperature (K) | 2246 ± 5 | 2246 ± 15 [18]<br>2219 ± 20 [19] |
| B = $^1/_3(c_{11} + 2c_{12})$, bulk modulus ($10^{12}$ dyn/$cm^2$) | 4.448 (8) | 6.3 [20]<br>5.6 [21] |
| Y, Young modulus in (111) plane ($10^{12}$ dyn/$cm^2$) | 11.64 | 5.5 [22] |

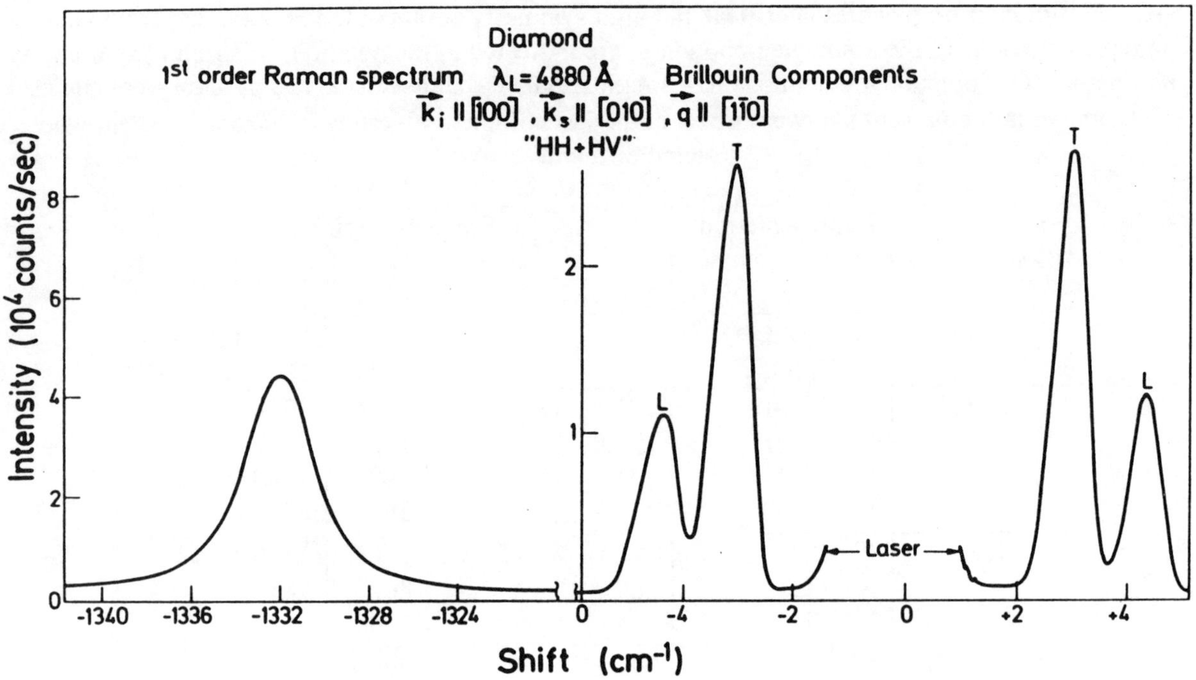


FIGURE 6 First order Raman spectrum and Brillouin components of diamond recorded with $\lambda_L$ = 4880 Å and T = 300 K. Incident light along $\mathbf{k}_i \parallel [100]$ and scattered along $\mathbf{k}_s \parallel [010]$. The incident light was polarised in the horizontal (scattering) plane, (H), and, when unanalysed, yielded the configuration HH + HV, where V is vertical (Grimsditch and Ramdas, unpublished [13]).

Under external uniaxial stress the $F_{2g}$ ($\Gamma_{25}^{+}$) triply degenerate zone-centre optical mode exhibits shifts and splittings [23,24]; with compressive force **F** applied along [001] or [111] it splits into a singlet and a doublet identified as such on the basis of their polarisation features. The splittings as a function of stress are $\Delta\omega_{[111]}$ = 0.22 ± 0.02 cm$^{-1}$/kbar, and $\Delta\omega_{[001]}$ = 0.073 ± 0.010 cm$^{-1}$/kbar. The hydrostatic component of the uniaxial stress yielded $\Delta\omega_H$ = 0.32 ± 0.02 cm$^{-1}$/kbar.

A number of workers have studied the $F_{2g}$ zone-centre Raman line as a function of hydrostatic stress [25–30] employing a variety of hydrostatic stress apparatus. The data are summarised in TABLE 4. Parsons measured both the first and second order Raman spectrum as a function of hydrostatic stress. Following the assignments for the features of the second order spectrum in [7], he deduced the pressure coefficients of the frequencies of many combinations and overtones of the phonons with wavevectors of the critical points.

TABLE 4 Hydrostatic shift of the zone-centre $F_{2g}$ optical phonon frequency ($\omega_0$).

| Range of pressure (GPa) | $d\omega_0/dP$ (cm$^{-1}$/GPa) | Ref |
|---|---|---|
| 0–1.0 | 2.8 ± 0.03 | [25] |
| 0–2.3 | 2.96 ± 0.11 | [26] |
| 0–2.4 | 3.6 ± 0.3 | [27] |
| 0–40.0 | 2.9 ± 0.05 | [28] |
| 0–27 | 2.87 ± 0.1 | [20] |
| 0–72 | 2.7 | [30] |
| 0–10 | 3.2 ± 0.2 | [23] |

## D3 Raman Spectra: Temperature Effects

The shift of the $F_{2g}$ phonon frequency as a function of temperature ($d\omega_0/dT$) has been studied by Zouboulis and Grimsditch [31] from 294 to 1900 K and by Herchen and Capelli [32] from 293 to 1850 K. Both sets of data for high temperatures yield mutually consistent values for ($d\omega_0/dT$), i.e. $3.84 \times 10^{-2}$ ($cm^{-1}/K$), and $3.6 \times 10^{-2}$ ($cm^{-1}/K$) respectively. Zouboulis and Grimsditch have deduced $(d\omega_0/dT)_V$, i.e. the 'pure' temperature dependence at constant volume using experimentally known thermal expansion coefficients as a function of temperature; they obtain $(d\omega_0/dT)_V = (1.21 \times 10^{-2})$ $cm^{-1}/K$. Herchen and Capelli have represented their entire data by $\omega_0 = a_1T^2 + a_2T + a_3$ with $a_1 = -1.075 \times 10^{-5}$ ($cm^{-1}/K^2$), $a_2 = -0.00777$ ($cm^{-1}/K$) and $a_3 = 1334.5\ cm^{-1}$. Earlier data, in particular below 300 K, are reviewed by Herchen and Capelli.

## E CONCLUSION

Spectroscopic studies of diamond have revealed numerous defect related features in luminescence spectra. Many display a zero-phonon line accompanied by vibronic side bands [33]. Local modes associated with impurities, collective vibrational modes, electron-phonon coupling and the Jahn–Teller effect - all of these physical phenomena play a role in these spectra.

In the intrinsic electronic transitions associated with absorption and recombination in diamond, the indirect nature of the fundamental gap is manifested: the valence band maximum of diamond occurs at the zone-centre (Γ point) and its conduction band minima lie at wavevectors $k_{min} = 0.76 \pm 0.002$ of the <001> zone boundary. The indirect transitions are accompanied by the absorption or emission of transverse acoustic, transverse optic and longitudinal optic phonons with energies of $\hbar\omega_{TA} = 87 \pm 2$, $\hbar\omega_{TO} = 141 \pm 2$ and $\hbar\omega_{LO} = 163 \pm 1$ meV (1 meV = 8.0645 $cm^{-1}$) [34].

Raman and infrared spectroscopy in conjunction with inelastic neutron scattering of diamond have provided a comprehensive picture of the phonon dispersion curves in diamond. The simplicity of its structure has enabled a thorough theoretical analysis of its lattice dynamics [35–40] based on different microscopic models. Some of the additional experimental studies which address novel aspects of the phonon spectroscopy of diamond are: the observation of the $F_{2g}$ zone centre optical phonon in infrared absorption under an external electric field [41,42] and the appearance of the 'second order Raman spectrum' as an electric field induced infrared absorption [43]; the absolute scattering cross section of the $F_{2g}$ mode from an intercomparison of the intensities of the Brillouin and Raman lines [12]; three wave mixing experiments [44]; and Raman-induced Kerr effect spectroscopy [45].

## REFERENCES


[1] J.L. Warren, J.L. Yarnell, G. Dolling, R.A. Cowley [ *Phys. Rev. (USA)* vol.158 (1967) p.805 ]. See also G. Peckham [ *Solid State Commun. (USA)* vol.5 (1967) p.311 ]

[2] The labelling of the symmetry points in the Brillouin zone of the face centred cubic lattice relevant for the diamond structure follows L.P. Bouckaert, R. Smoluchowski, E. Wigner [ *Phys. Rev. (USA)* vol.50 (1936) p.58 ]

[3] J.L. Birman [ *Theory of Crystal Space Groups and Lattice Dynamics* (Springer-Verlag, Berlin, 1984) ]

[4] J. Kulda et al [ *Solid State Commun. (USA)* vol.99 (1996) p.799 ]

[5] M. Schwoerer-Böhnig, A.T. Macrander, D.A. Rams [ *Phys. Rev. Lett. (USA)* vol.80 (1998) p.5572 ]

[6] S. Bhagavantam, T. Venkatarayudu [ *Proc. Indian Acad. Sci. A (India)* vol.8 (1938) p.101 and vol.9 (1939) p.224 ]. We use the notation in S. Bhagavantam, T. Venkatarayudu [ *Theory of Groups and its Applications to Physical Problems* (Academic Press, New York, 1969) ]. See also T. Venkatarayudu [ *Proc. Indian Acad. Sci. A (India)* vol.8 (1938) p.349 ]

[7] S.A. Solin, A.K. Ramdas [ *Phys. Rev. B (USA)* vol.1 (1970) p.1687 ]

[8] R. Vogelgesang et al [ *Phys. Rev. B (USA)* vol.58 (1998) p.5408 ]

[9] W. Kaiser, W.L. Bond [ *Phys. Rev. (USA)* vol.115 (1959) p.857 ]

[10] J.R. Hardy, S.D. Smith [ *Philos. Mag. (UK)* vol.6 (1961) p.1163 ]

[11] R. Wehner, H. Borik, W. Kress, A.R. Goodwin, J.D. Smith [ *Solid State Commun. (USA)* vol.5 (1967) p.307 ]

[12] M.H. Grimsditch, A.K. Ramdas [ *Phys. Rev. B (USA)* vol.11 (1975) p.3139 ]

[13] M.H. Grimsditch, A.K. Ramdas [ unpublished ]

[14] R. Vogelgesang, A.K. Ramdas, S. Rodriguez, M. Grimsditch, T.R. Anthony [ *Phys. Rev. B (USA)* vol.54 (1996) p.3989 ]; the values of $c_{ij}$'s quoted here supersede those in A.K. Ramdas, S. Rodriguez, M. Grimsditch, T.R. Anthony, W.F. Banholzer [ *Phys. Rev. Lett. (USA)* vol.71 (1993) p.189 ]

[15] H.J. McSkimin, W.L. Bond [ *Phys. Rev. (USA)* vol.105 (1957) p.116 ]

[16] H.J. McSkimin, P. Andreatch [ *J. Appl. Phys. (USA)* vol.43 (1972) p.2944 ]

[17] J. de Launay [ *J. Chem. Phys. (USA)* vol.24 (1956) p.1071 ]

[18] D.L. Burk, S.A. Friedberg [ *Phys. Rev. (USA)* vol.111 (1958) p.1275 ]

[19] J.E. Desnoyers, J.A. Morrison [ *Philos. Mag. (UK)* vol.3 (1958) p.42 ]

[20] L.H. Adams [ *J. Wash. Acad. Sci. (USA)* vol.11 (1921) p.45 ]

[21] E.D. Williamson [ *J. Franklin Inst. (UK)* vol.193 (1922) p.491 ]

[22] P.R. Pisharoty [ *Proc. Indian Acad. Sci. A (India)* vol.12 (1940) p.208 ]

[23] M.H. Grimsditch, E. Anastassakis, M. Cardona [ *Phys. Rev. B (USA)* vol.18 (1978) p.901 ]

[24] E. Anastassakis, A. Cantarero, M. Cardona [ *Phys. Rev. B (USA)* vol.41 (1990) p.7529 ]

[25] S.S. Mitra, O. Brafman, W.B. Daniels, R.K. Crawford [ *Phys. Rev. (USA)* vol.186 (1969) p.942 ]

[26] E. Whalley, A. Lavergne, P.T.T. Wong [ *Rev. Sci. Instrum. (USA)* vol.47 (1976) p.845 ]

[27] B.J. Parsons [ *Proc. R. Soc. Lond. A (UK)* vol.352 (1977) p.397 ]

[28] M. Hanfland, K. Syassen, S. Fahy, S.G. Louie, M.L. Cohen [ *Phys. Rev. B (USA)* vol.31 (1985) p.6896 ]

[29] H. Boppart, J. van Straaten, I.F. Silvera [ *Phys. Rev. B (USA)* vol.32 (1985) p.1423 ]

[30] A.F. Goncharov, I.N. Makarenko, S.M. Stishov [ *JETP Lett. (USA)* vol.41 (1985) p.184 ]

[31] E.S. Zouboulis, M. Grimsditch [ *Phys. Rev. B (USA)* vol.43 (1991) p.12490 ]

[32] H. Herchen, M.A. Capelli [ *Phys. Rev. B (USA)* vol.43 (1991) p.11740 ]

[33] See, for example, C.D. Clark, E.W.J. Mitchell, B.J. Parsons [ in *The Properties of Diamond* Ed. J.E. Field (Academic, New York, 1979) p.23-77 ]; G. Davies [ in *The Properties of Diamond* Ed. J.E. Field (Academic, New York, 1979) p.165-81 ]

[34] P.J. Dean, E.C. Lightowlers, D.R. Wight [ *Phys. Rev. (USA)* vol.140 (1965) p.A352 ]

[35] M.H. Cohen, J. Ruvalds [ *Phys. Rev. Lett. (USA)* vol.23 (1969) p.1378 ]

[36] S. Go, H. Bilz, M. Cardona [ *Phys. Rev. Lett. (USA)* vol.34 (1975) p.580 ]

[37] R. Tubino, J.L. Birman [ *Phys. Rev. B (USA)* vol.15 (1977) p.5843 ]

[38] D. Vanderbilt, S.G. Luie, M.L. Cohen [ *Phys. Rev. Lett. (USA)* vol.53 (1984) p.1477 and *Phys. Rev. B (USA)* vol.33 (1986) p.8740 ]

[39] K.C. Hass, M.A. Tamor, T.R. Anthony, W.F. Banholzer [ *Phys. Rev. B (USA)* vol.45 (1992) p.7171 ]

[40] P. Pavone et al [ *Phys. Rev. B (USA)* vol.48 (1993) p.3156 ]; W. Windl et al [ *Phys. Rev. B (USA)* vol.48 (1993) p.3164 ]

[41] E. Anastassakis, E. Burstein [ *Phys. Rev. B (USA)* vol.2 (1970) p.1952 ]

[42] J.F. Angress, C. Cooke, A.J. Maiden [ *J. Phys. C (UK)* vol.1 (1968) p.1769 ]

[43] J.F. Angress, A.J. Maiden [ *J. Phys. C (UK)* vol.4 (1971) p.235 ]
[44] M.D. Levenson, C. Flytzanis, N. Bloembergen [ *Phys. Rev. B (USA)* vol.6 (1972) p.3962 ]
[45] G.L. Eesely, M.D. Levenson [ *Opt. Lett. (USA)* vol.3 (1978) p.178 ]

# A1.2 Modifications to $^{12}$C-diamond by the $^{13}$C-isotope: Raman, Brillouin and infrared spectroscopy of phonons

A.K. Ramdas

July 2000

## A INTRODUCTION

Conventional (naturally occurring) diamond consists of 98.9% $^{12}$C and 1.1% $^{13}$C, reflecting the natural abundance of the two stable isotopes of carbon. As is well-known, diamond can be synthesised either by the high temperature-high pressure technique (HPHT) [1] or by low pressure growth from the gaseous phase using chemical vapour deposition (CVD) [2]. Diamonds thus produced have the isotopic composition of the starting material. While large single crystal diamonds several carats in size can be produced by HPHT, CVD diamond is polycrystalline. However it is possible to combine the two techniques; one can first grow diamond in the polycrystalline form with CVD and, using that as the starting material, grow carat size single crystals with HPHT [3]. In this manner one can now produce single crystal diamond of any desired $^{12}$C:$^{13}$C ratio by starting with methane of the requisite isotopic composition for CVD.

## B RAMAN SPECTRA

Raman spectroscopy has proved to be the preferred characterisation technique for establishing the successful growth of CVD diamond. The Raman line at $\omega_0$ associated with the zone centre ($F_{2g} \equiv \Gamma_{25}^{+}$) optical phonon frequency is the unmistakable signature that the material deposited on a suitable substrate (a Si single crystal wafer, for example) is indeed diamond. The high sensitivity of the Raman technique (thanks to micro-Raman features including multichannel detection using charge coupled devices) and the large frequency shift of $\omega_0$ make it possible to detect the occurrence of diamond of micron dimensions and discriminate it, with confidence, from the signature of graphitic carbon [4]. It is also feasible to grow CVD diamond of any desired $^{12}$C:$^{13}$C ratio. Chrenko [5] has reported shifts in the $F_{2g}$ mode as this ratio was altered; these measurements were made on gem quality diamond made by CVD followed with HPHT. Particularly comprehensive studies of the lattice dynamics and Raman spectra of isotopically mixed diamond have been published by Hass et al [6] and by Vogelgesang et al [7]. These authors have experimentally investigated both the first and the second order Raman spectra as a function of isotopic constitution of a series of gem quality single crystals; the isotopic constitution of the crystals was specifically established by burning the diamonds in oxygen and a subsequent mass spectrometric analysis of CO [3].

Vogelgesang et al [7] have interpreted their results for $^{13}C_x{}^{12}C_{1-x}$ with x ranging from 0.003 to 0.992 in terms of a valence-force-field model in which the effects of zero-point motion and the anharmonicity of the lattice vibration are explicitly incorporated [7]. Expressing the strain energy in terms of the changes in the bond lengths and bond angles, they introduce two harmonic parameters, $k_1$ and $k_2$, associated with the bond length and bond angle, respectively, and two corresponding anharmonic parameters, $g_1$ and $g_2$. It transpires that only $g_1$ is of significance in the present context. One can compute the zero temperature bulk modulus (B) and the frequency of the $F_{2g}$ zone centre optical phonon $\Omega_0$ in the harmonic approximation. A variational calculation based on the zero-point motion and anharmonicity parameter enables one to deduce the renormalised force constant (K), bulk modulus (B) as well as the lattice parameter (a) in terms of $g_1$, $k_1$ and $M = xM_{13} + (1-x)M_{12}$, where $M_{12}$ and $M_{13}$ are the masses of $^{12}$C and $^{13}$C, respectively. In TABLE 1, the expressions deduced in the harmonic approximation (first row) are compared with their renormalised counterparts (second row). In TABLE 1, $K_1 = k_1\left(1 - \frac{\hbar g_1^2}{8\sqrt{2}k_1^{5/2}M^{1/2}}\right)$ and $a_\infty$ is the lattice parameter of a diamond made up of 'infinitely' massive carbon atoms.

TABLE 1 Effects of zero-point motion and anharmonicity.

| $F_{2g}$ frequency | Lattice parameter | Bulk modulus |
|---|---|---|
| $\Omega_0 = \left[\frac{8(k_1 + 4k_2)}{3M}\right]^{1/2}$ | $a_\infty$ | $B = (k_1/3a)$ |
| $\omega_0 = \Omega_0\left(1 - \frac{32\hbar g_1^2}{81M^3\Omega_0^5}\right)$ | $a = a_\infty + \frac{\hbar g_1}{\sqrt{6}k_1^{3/2}M^{1/2}}$ | $B = (K_1/3a)$ |

The experimental values of $\omega_0(x)$ from the Raman spectra reported by Chrenko [5], Hass et al [6], Vogelgesang et al [7], and Spitzer et al [8], are well represented by the polynomial $\omega_0(x) = 1332.82 - 34.77x - 16.98x^2$ (FIGURE 1(a)). Theory predicts

$$\omega_0(x) = \omega_0(0)(M_{12}/M)^{1/2}\left[1 + \frac{32\hbar g_1^2}{81M_{12}^3\omega_0(0)^5}\left\{1 - \left(\frac{M_{12}}{M}\right)^{1/2}\right\}\right] \quad (1)$$

keeping corrections to second power of $g_1$ only. Isotopic disorder produces phonon lifetimes proportional to $x(1-x)$ accompanied by shifts in the real part of the frequency of the form $Cx(1-x)$ as shown by Hass et al [6] and Spitzer et al [8]. The anharmonicity constant $g_1$ is deduced from $a(x)$ determined experimentally by Holloway et al [9]. Thus EQN (1), along with the addition of $Cx(1-x)$ on the right hand side, is in excellent agreement with the polynomial fit to $\omega_0(x)$. Since the disorder induced term vanishes at $x = 0$ and 1, one can deduce the contribution of zero-point motion and anharmonicity combined, over and above the $M^{-1/2}$ dependence; $\omega_0(^{13}C)$ predicted to be 1280.9 $cm^{-1}$, instead of 1280.6 $cm^{-1}$ expected on the basis of $M^{-1/2}$ dependence only, is in reasonable agreement with 1281.07 $cm^{-1}$ deduced from the polynomial fit.

## C BRILLOUIN SCATTERING

The elastic moduli $c_{ij}$ are deduced from the Brillouin shift, $\Delta\omega = \pm 2\omega_L n(v_s/c)\sin(\theta/2)$, where $\omega_L$ is the frequency of the incident laser radiation, n the refractive index, c the speed of light in vacuum, $\theta$ the scattering angle and $v_s$ the velocity of the sound wave responsible for the Doppler shift producing the Brillouin component. The appropriate combination of the elastic moduli (X) is given by $X = v_s^2\rho$ where $\rho$, the density, is a function of x. This immediately underscores the isotopic effects one might expect. The density can be determined from a(x) reported by Holloway et al [9]. Vogelgesang et al [7] have investigated Brillouin scattering in the backscattering geometry using a high contrast, piezoelectrically scanned, multipassed Fabry–Pérot interferometer; FIGURE 1(b) shows their results for **q** || [111]. Backscattering by acoustic phonons with wave vectors **q** along [001], [110] and [111] by longitudinal acoustic (LA) phonons and by transverse acoustic (TA) phonons for **q** along [111] yielded: $c_{11}$, $\frac{1}{2}(c_{11} + c_{12} + 2c_{44})$, $\frac{1}{3}(c_{11} + 2c_{12} + 4c_{44})$, and $\frac{1}{3}(c_{11} - c_{12} + c_{44})$, respectively.

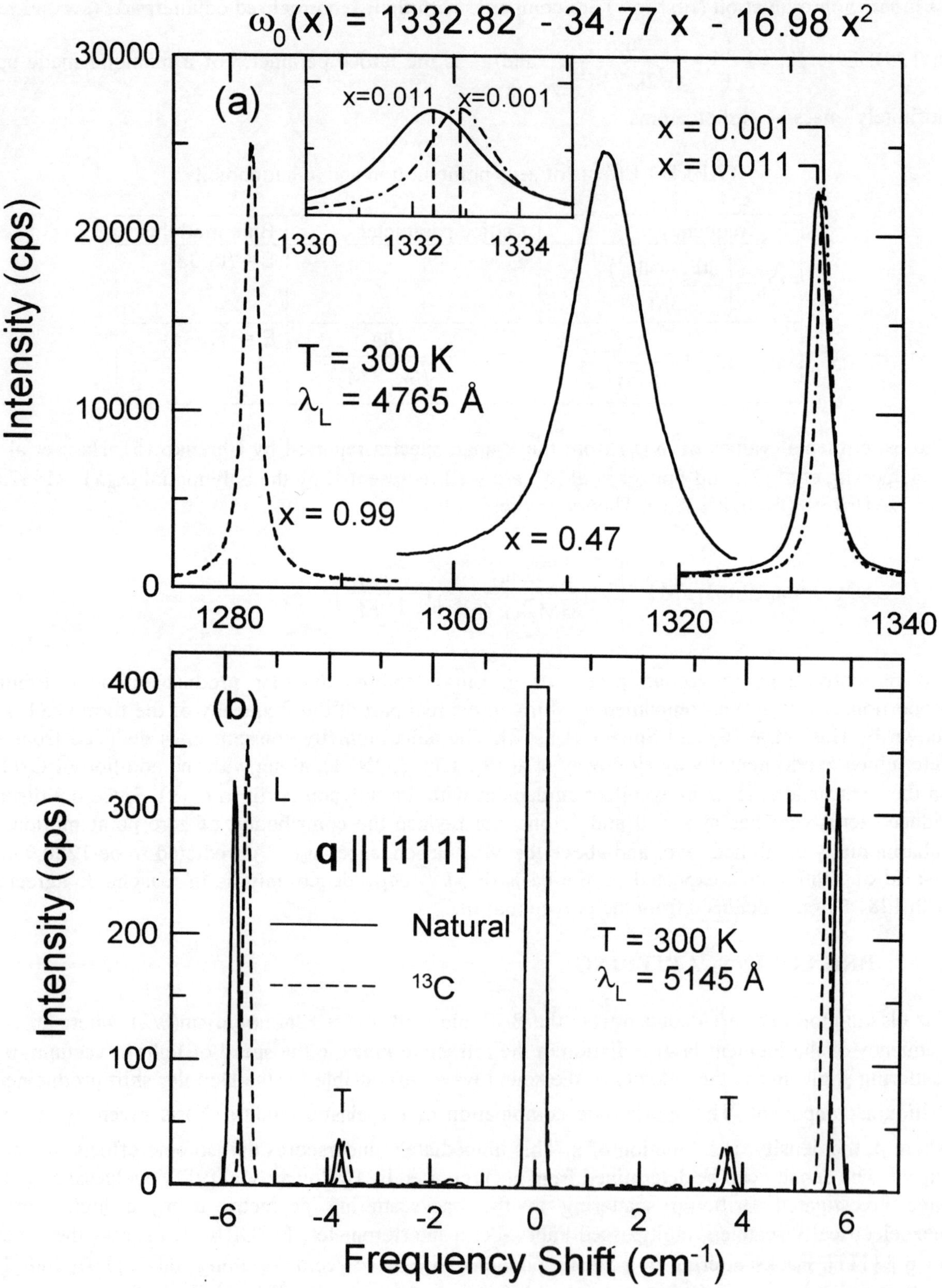


FIGURE 1 (a) Raman and (b) Brillouin spectra of type IIa $^{12}C_{1-x}{}^{13}C_x$ diamonds. The Brillouin spectra obtained with a large free spectral range of a tandem Fabry–Pérot interferometer clearly show the signatures of longitudinal (L) and transverse (T) components [7].

Results for **q** || [001] and [111] are displayed in FIGURE 2 as $v_s^2$ versus $\rho^{-1}(x)$. The elastic moduli $c_{11}$, $c_{12}$ and $c_{44}$ deduced and the renormalised bulk modulus B for $^{12}$C, natural, and $^{13}$C diamond are presented in TABLE 2. Theory predicts that

$$\frac{B(x)}{B(0)} = \frac{a(0)}{a(x)}\left[1 + \frac{\hbar g_1^2}{8\left(2k_1^5 M_{12}\right)^{\frac{1}{2}}}\left\{1 - \left(\frac{M_{12}}{M}\right)^{1/2}\right\}\right] \simeq 1 + 0.0012x \quad (2)$$

with the values of $k_1$ and $g_1$ deduced from $\omega_0(x)$ and a(x), respectively. The ~0.1% increase in B expected from theory for B($^{12}$C)/B($^{13}$C) is just below the accuracy of the present Brillouin measurements [7].

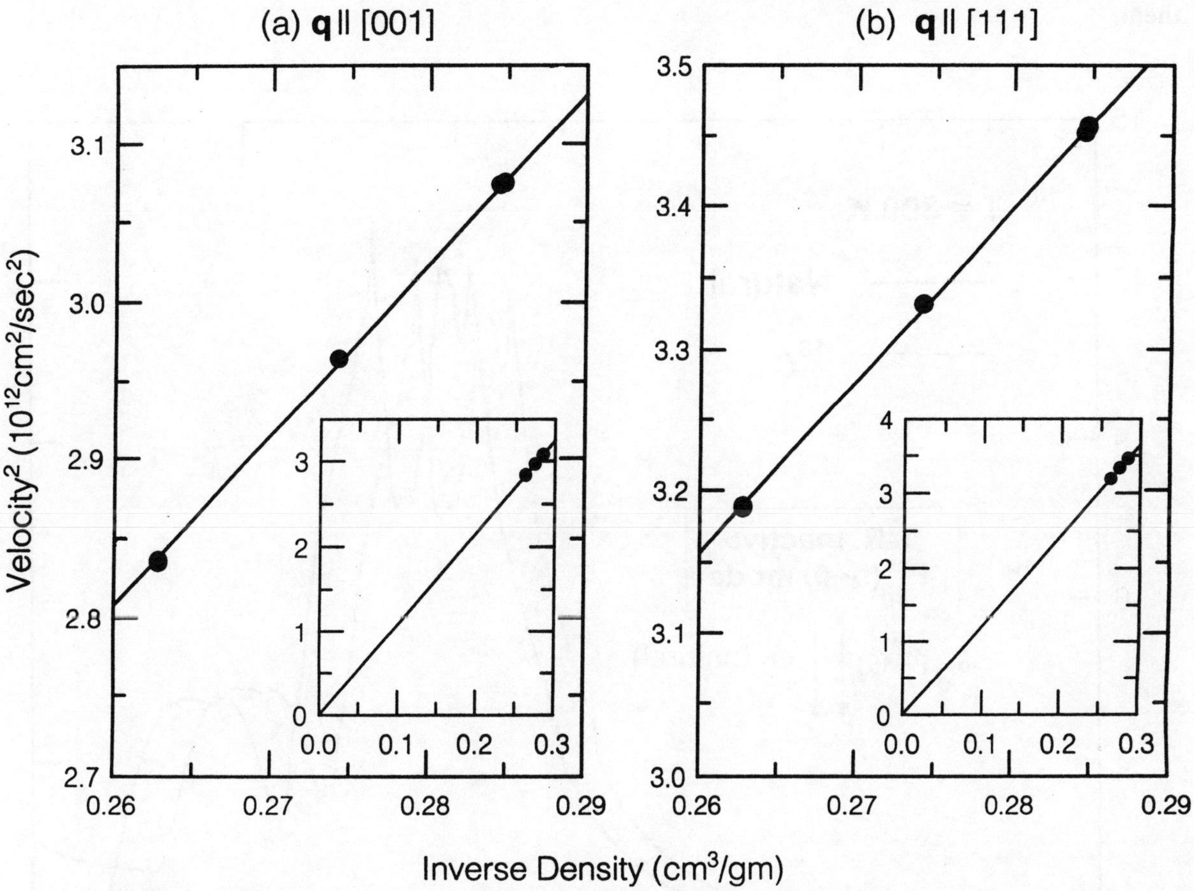


FIGURE 2 $v_s^2$ versus $\rho^{-1}(x)$ for longitudinal acoustic phonons travelling along [001] and [111] in diamond specimens of varying isotopic composition. The data points are indicated by full circles and the solid line represents a linear least squares fit passing through the origin ($\rho^{-1}(x) = 0$) [7].

TABLE 2 Elastic moduli $c_{ij}$ and bulk modulus B of diamond (in units of $10^{12}$ dyn/cm$^2$).

| x | $c_{11}$ | $c_{12}$ | $c_{44}$ | B |
|---|---|---|---|---|
| 0.0 | 10.799(5) | 1.248(10) | 5.783(5) | 4.432(8) |
| 0.01105 | 10.804(5) | 1.270(10) | 5.766(5) | 4.448(8) |
| 0.992 | 10.792(7) | 1.248(14) | 5.776(7) | 4.429(12) |

## D MULTIPHONON INFRARED AND RAMAN SPECTRA

The multiphonon infrared and Raman spectra of a $^{13}$C diamond are compared in FIGURES 3 and 4 with those of a natural diamond ($^{13}C_{0.11}{}^{12}C_{0.89}$) [10]. They display features—peaks and slope discontinuities—which are interpreted as combinations and overtones of critical points revealed in the phonon dispersion curves determined by inelastic neutron scattering [11]. The occurrence of identical spectroscopic features in both $^{13}$C and natural diamonds, with their frequencies essentially scaling as $M^{-1/2}$ as expected from the virtual crystal approximation, proved invaluable in arriving at assignments for them.

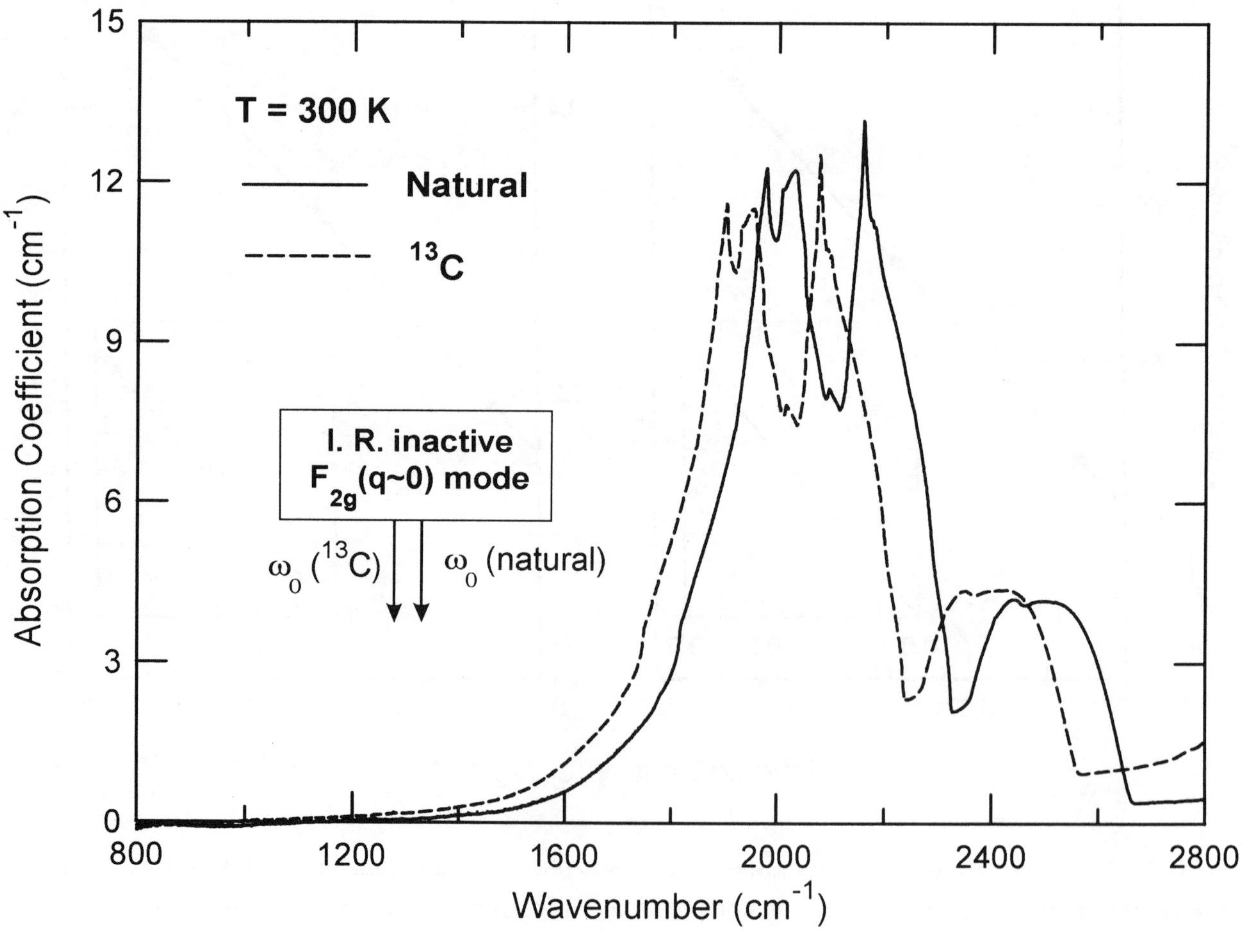


FIGURE 3 The multiphonon features of a natural and a $^{13}$C diamond in infrared spectra. Note the absence of the first-order line [10].

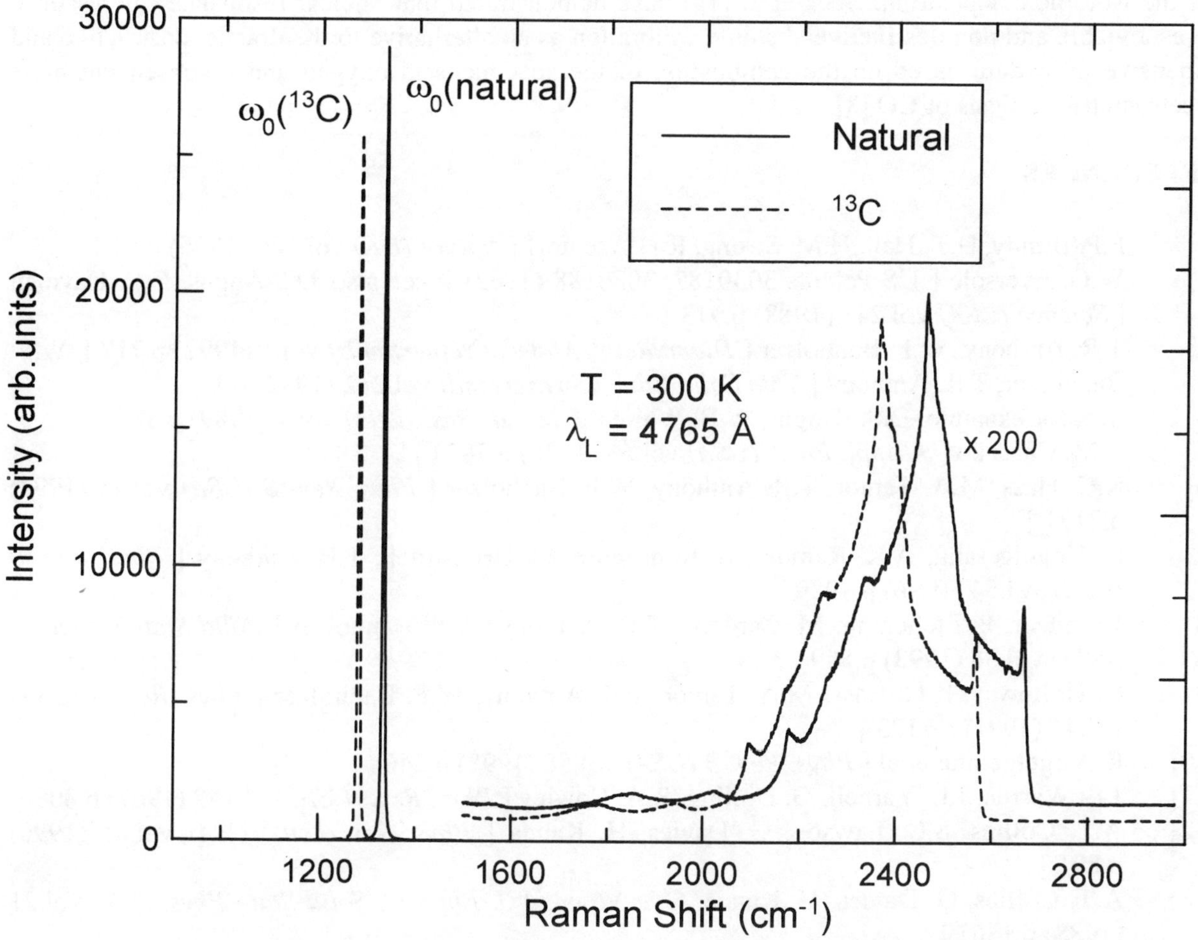


FIGURE 4 The Raman spectra of a natural and a $^{13}$C diamond. The spectra show the dominant first-order, Raman active $F_{2g}$ line, and the significantly weaker, quasi-continuous multiphonon features [10].

## E ABSORPTION (UV, VISIBLE AND INFRARED) AND LUMINESCENCE SPECTRA

From a study of the indirect transitions associated with the valence band maximum at the zone-centre (Γ-point) and conduction band minima at wavevectors $k_{min} = 0.76 \pm 0.02$ of the <001> zone-boundary, Collins et al [12] have deduced $\hbar\omega_{TO} = 135.4 \pm 0.6$ meV and $\hbar\omega_{LO} = 155 \pm 0.7$ meV for $^{13}$C; they report corresponding values of $141.2 \pm 0.6$ meV and $163 \pm 0.7$ meV for $^{12}$C diamond. The ratios of the corresponding phonon energies for $^{13}$C and $^{12}$C are very close: $(12/13)^{1/2} = 0.961$. Collins et al [13] have carried out a comprehensive study of the spectroscopy of $^{13}$C diamond in which they report the isotope related effects in the localised vibrational spectra in the infrared and in the vibronic spectra in the visible.

## F CONCLUSION

Spectroscopic techniques have disclosed many aspects of the vibrational and vibronic excitations in diamond which involve isotope-related effects. Besides providing valuable insights at the microscopic level, they offer precise methods for characterising the isotopic constitution. For example, Raman lines represented by $\omega_0(x) = 1332.82–34.77x–16.98x^2$ provide precisely such secondary calibration for the isotopic composition. Beck et al [14] have demonstrated that nuclear resonance fluorescence gives a viable and non-destructive absolute calibration as an alternative to the drastic, destructive and expensive procedure based on the combustion of the specimens in oxygen and a subsequent mass spectrometric analysis of CO [3].

**REFERENCES**

[1] F.P. Bundy, H.T. Hall, H.M. Strong, R.H. Wentorf [ *Nature (UK)* vol.176 (1955) p.51 ]
[2] W.G. Eversole [ US Patents 3030187, 3030188 (1962) ]; see also J.C. Angus, C.C. Hayman [ *Science (USA)* vol.241 (1988) p.913 ]
[3] T.R. Anthony, W.F. Banholzer [ *Diam. Relat. Mater. (Netherlands)* vol.1 (1992) p.717 ]; W.F. Banholzer, T.R. Anthony [ *Thin Solid Films (Switzerland)* vol.212 (1992) p.1 ]
[4] See, for example, D.S. Knight, W.B. White [ *J. Mater. Res. (USA)* vol.4 (1989) p.385 ]
[5] R.M. Chrenko [ *J. Appl. Phys. (USA)* vol.63 (1988) p.5873 ]
[6] K.C. Hass, M.A. Tamor, T.R. Anthony, W.F. Banholzer [ *Phys. Rev. B (USA)* vol.45 (1992) p.7171 ]
[7] R. Vogelgesang, A.K. Ramdas, S. Rodriguez, M. Grimsditch, T.R. Anthony [ *Phys. Rev. B (USA)* vol.54 (1996) p.3989 ]
[8] J. Spitzer, P. Etchegoin, M. Cardona, T.R. Anthony, W.F. Banholzer [ *Solid State Commun. (USA)* vol.88 (1993) p.509 ]
[9] H. Holloway, K.C. Hass, M.A. Tamor, T.R. Anthony, W.F. Banholzer [ *Phys. Rev. B (USA)* vol.44 (1991) p.7123 ]
[10] R. Vogelgesang et al [ *Phys. Rev. B (USA)* vol.58 (1998) p.5408 ]
[11] J.L. Warren, J.L. Yarnell, G. Dolling, R.A. Cowley [ *Phys. Rev. (USA)* vol.158 (1967) p.805 ]
[12] A.T. Collins, S.C. Lawson, G. Davies, H. Kanda [ *Phys. Rev. Lett. (USA)* vol.65 (1990) p.891 ]
[13] A.T. Collins, G. Davies, H. Kanda, G.S. Woods [ *J. Phys. C, Solid State Phys. (UK)* vol.21 (1988) p.1363 ]
[14] O. Beck et al [ *J. Appl. Phys. (USA)* vol.83 (1998) p.5484 ]

# A1.3 Electronic excitations in isotopically controlled diamonds: infrared and Raman spectroscopy of acceptor-bound holes

A.K. Ramdas

July 2000

## A INTRODUCTION

The isotopic composition of naturally occurring diamonds corresponds to the natural abundance of carbon, i.e. 98.9% $^{12}C$ and 1.1% $^{13}C$. Diamonds containing nitrogen as a dominant impurity are classified as type I. The relatively rare, nitrogen-free diamond is designated as type IIa. Further, nitrogen-free but semiconducting diamond (p-type due to substitutional boron acceptors) is called type IIb [1]. Type IIb diamonds are extremely rare and exhibit remarkably low resistivities (5 to $10^5$ Ω cm), whereas typical type I and type IIa diamonds have resistivities well in excess of $10^{18}$ Ω cm. Diamond can be viewed as an elemental group IV 'semiconductor' with a bandgap ~5.5 eV. Type IIa diamond is transparent from the ultraviolet to well into the microwave, barring only the 2–6 μm range of multiphonon absorption bands.

The ability to grow single crystals free of imperfections - be they lattice defects or chemical impurities - followed by the controlled introduction of a desired imperfection is the prerequisite for a semiconductor to be significant in solid state electronics. While the HPHT and CVD techniques of diamond synthesis [2] are milestones in this respect, the incorporation of shallow impurities (group III acceptors or group V donors) has been successful to date only for boron acceptors [3]. The study of the bound states of donors and acceptors in semiconductors can be performed with extraordinary detail using infrared spectroscopy [4]. The Lyman spectrum and the associated Zeeman and piezo-spectroscopic effects of an acceptor or a donor yield the binding energies and the symmetries of the group and excited states. To the extent that the bound states are described in terms of the parameters of the band extremum with which they are associated, such a study is of clear value in the context of the properties of the host. Cyclotron resonance experiments [5] as well as transport measurements on holes [6] have provided the Luttinger parameters and the spin-orbit splitting ($\Delta \approx 6$ meV) [7] characterising the valence band maximum of diamond at the zone centre. The zone centre valence band maxima of diamond, the acceptor-bound electronic levels, and the Lyman transitions are shown schematically in FIGURE 1. Kim et al investigated the Lyman spectrum of boron acceptors in isotopically controlled diamonds using Fourier transform infrared [8] as well as Raman spectroscopy [9,10]. The selection rules for infrared and Raman transitions being different, the two techniques can be complementary.

## B LYMAN SPECTRUM: INFRARED

The absorption spectra of natural type IIa and boron-doped type IIb diamonds in the spectral range 1800–6000 $cm^{-1}$ recorded at 5.0 K are shown in FIGURE 2. While both show identical intrinsic multiphonon features, the spectrum of the type IIb diamond clearly shows the characteristic sharp lines associated with the transitions of the acceptor bound hole from the ground to the various excited states [8]. The onset of the photoionisation (~3000 $cm^{-1}$) persists well into the red and endows the boron-doped diamond with its typical blue colour, the intensely blue Hope diamond being a striking example. FIGURE 3 shows the comparison of the Lyman spectrum of a man-made $^{13}C$ boron-doped diamond and that of a natural type IIb specimen. While the spectra are strikingly similar even in minutest detail, the spectral lines of the acceptors in the $^{13}C$ diamond are unambiguously shifted by

3.1 to 11.8 $cm^{-1}$ to higher energies. Since the acceptor in both natural and $^{13}C$ diamonds is boron, the small shifts strongly indicate that the acceptor states experience shifts associated with electron-phonon interaction arising from self-energy effects, larger in $^{13}C$ diamond than in the natural diamond. We thus have here a remarkable example of self energy correction shifting the spectra for the same substitutional acceptor but located in a host differing merely in its isotopic composition.

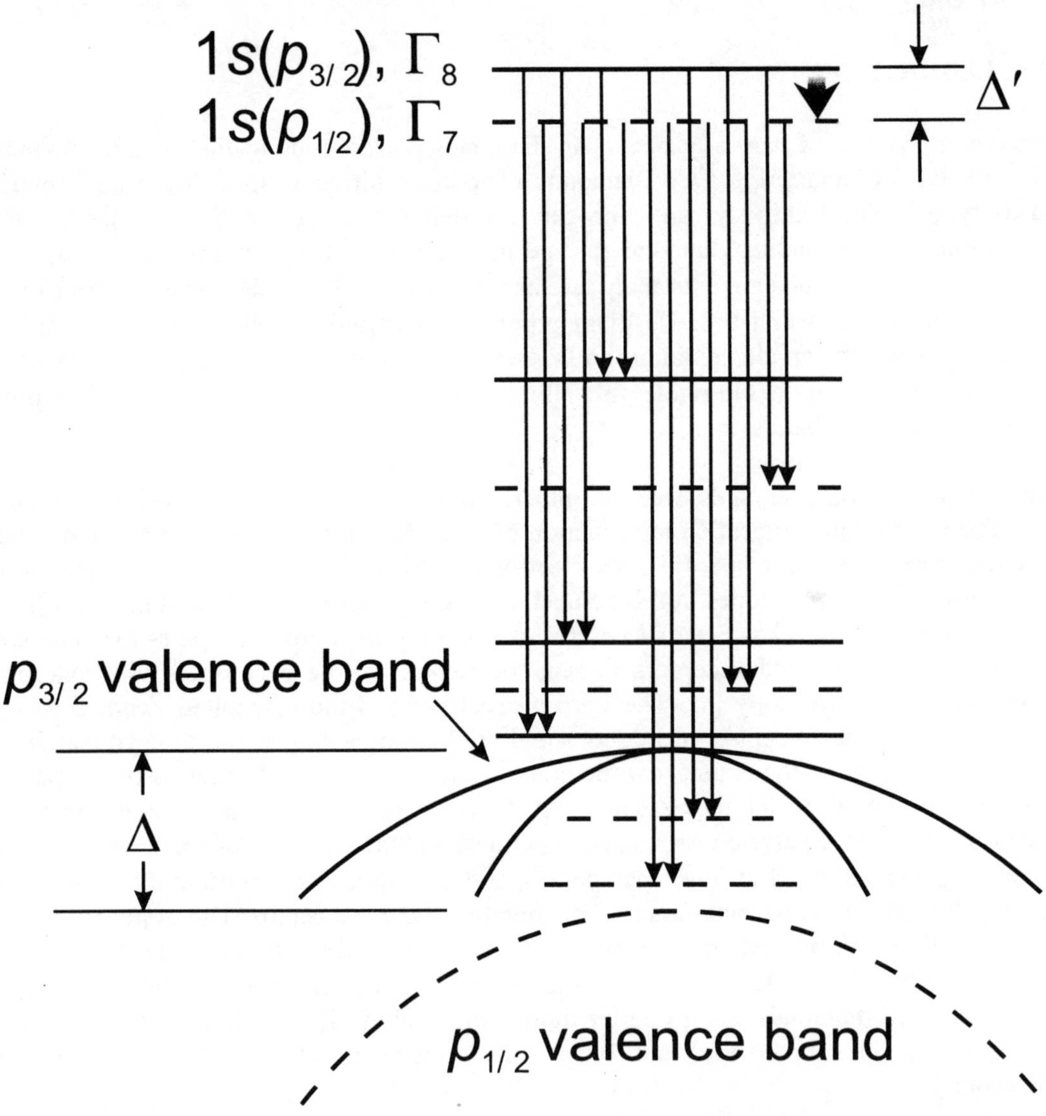


FIGURE 1 Energy levels of a hole bound to a substitutional boron acceptor shown schematically. The diagram also shows the Lyman transition from the ground state associated with the $p_{3/2}$ valence band (solid line) as well as those associated with the $p_{1/2}$ valence band (dashed line). $\Delta$ is the difference between the maxima of the two bands and $\Delta'$ that between the two ground states, associated with the $p_{3/2}$ and the $p_{1/2}$ valence bands, respectively [10].

When the infrared Lyman spectrum of the boron acceptors is studied as a function of temperature, new transitions appear at higher temperatures. The new lines can be ascribed to transitions originating from a higher lying, thermally populated, ground state ~2 meV above the lower ground state responsible for the Lyman spectrum in FIGURE 2 [10,11]. Indeed, the 1s ground state of the acceptor bound hole is expected to show a spin-orbit splitting, $\Delta'$, into $1s(p_{3/2})$ and $1s(p_{1/2})$ corresponding to $\Delta$

of the valence band [12,13]. It can be shown that the $1s(p_{3/2}) \rightarrow 1s(p_{1/2})$ transition is Raman active. Motivated by this prediction, Kim et al investigated the Raman spectrum of the boron-doped, isotopically controlled diamonds [9,10].

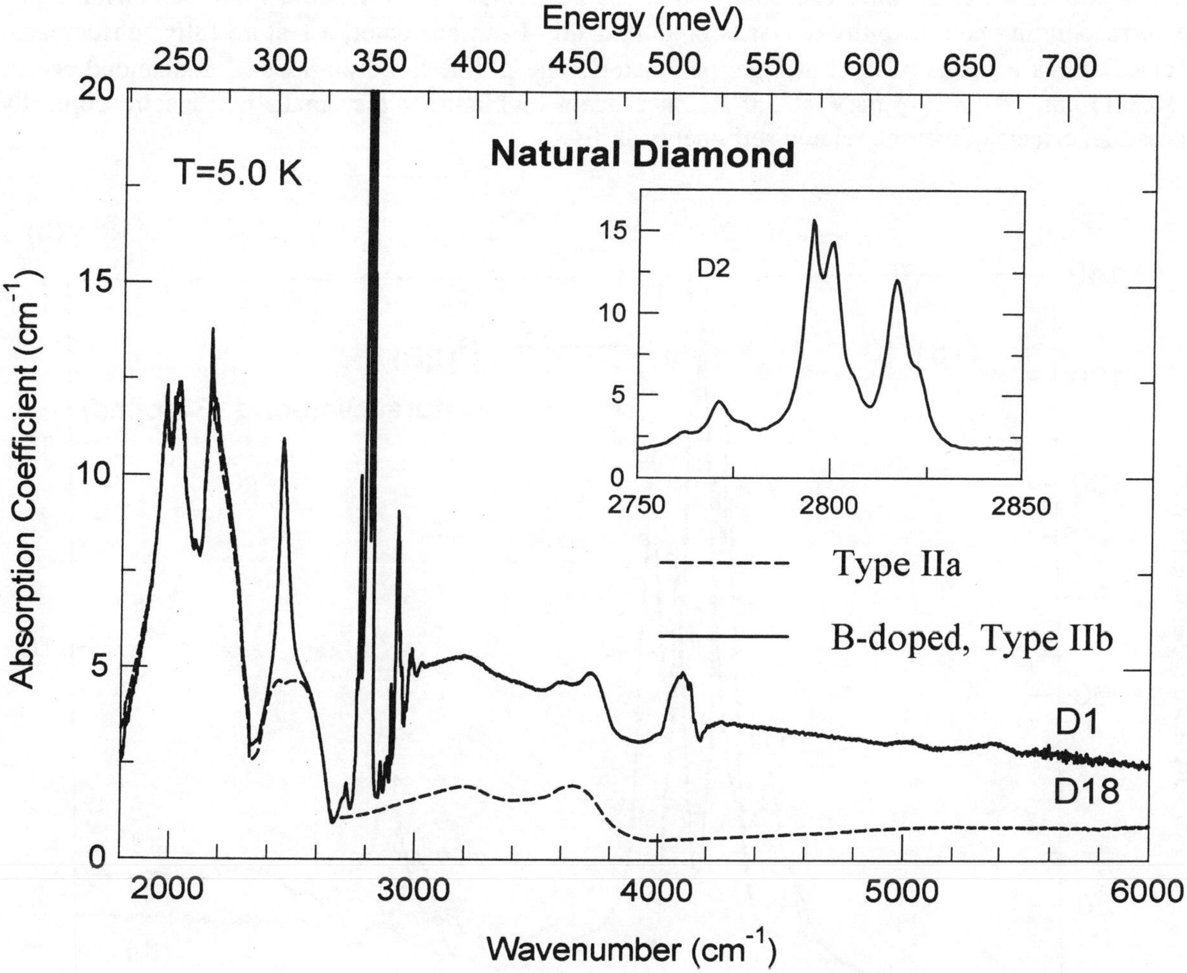


FIGURE 2 The Lyman spectra of natural type IIa (D18) and type IIb (D1) diamonds at temperature T = 5.0 K. The inset shows the most intense electronic transitions in the range 2750–2850 $cm^{-1}$ for a natural type IIb (D2) specimen having a smaller boron concentration [10].

The $p_{3/2} : \Gamma_8^+$ valence band of diamond with its maximum at the zone centre, the $p_{1/2} : \Gamma_7^+$ spin-orbit split valence band separated from the $p_{3/2}$ band by $\Delta$, the electronic states of the hole bound to group III acceptor (e.g. boron) with its $1s(p_{3/2})$:$\Gamma_8$ and the spin-orbit $1s(p_{1/2})$:$\Gamma_7$ state separated from it by $\Delta'$, and the Lyman transitions are depicted schematically in FIGURE 1. The figure explicitly recognises that while $\Delta$ and $\Delta'$ are related, they are not necessarily the same in magnitude.

## C LYMAN SPECTRUM: RAMAN

The electronic Raman spectrum of a man-made, boron-doped diamond of natural composition studied as a function of temperature shows a new Raman line emerging as the temperature is decreased below ~150 K as can be seen in FIGURE 4(a). The position of the line, 16.7(1) $cm^{-1}$ = 2.07(1) meV at 5.0 K, is entirely in agreement with that expected for the Raman active Δ′ transition. Its disappearance at the elevated temperatures is obviously due to the thermal ionisation of the boron acceptors, combined with its increased linewidth. Occurrence of the Δ′ line as a Stokes/anti-Stokes pair (FIGURES 4(b) and 4(c)) and its appearance with the same shift when excited with different laser lines established that it is indeed a Raman line. The polarisation characteristics of the Δ′ Raman line demonstrate that the corresponding polarisability tensor is predominantly $\Gamma_5$ in character, a feature fully substantiated by calculations in terms of the Luttinger parameters. The Δ′ line in boron-doped $^{13}$C diamond occurs at 16.2(1) $cm^{-1}$ = 2.01(1) meV at 5.0 K, once again highlighting the small, but spectroscopically accessible, effects of isotope related self-energy shifts.

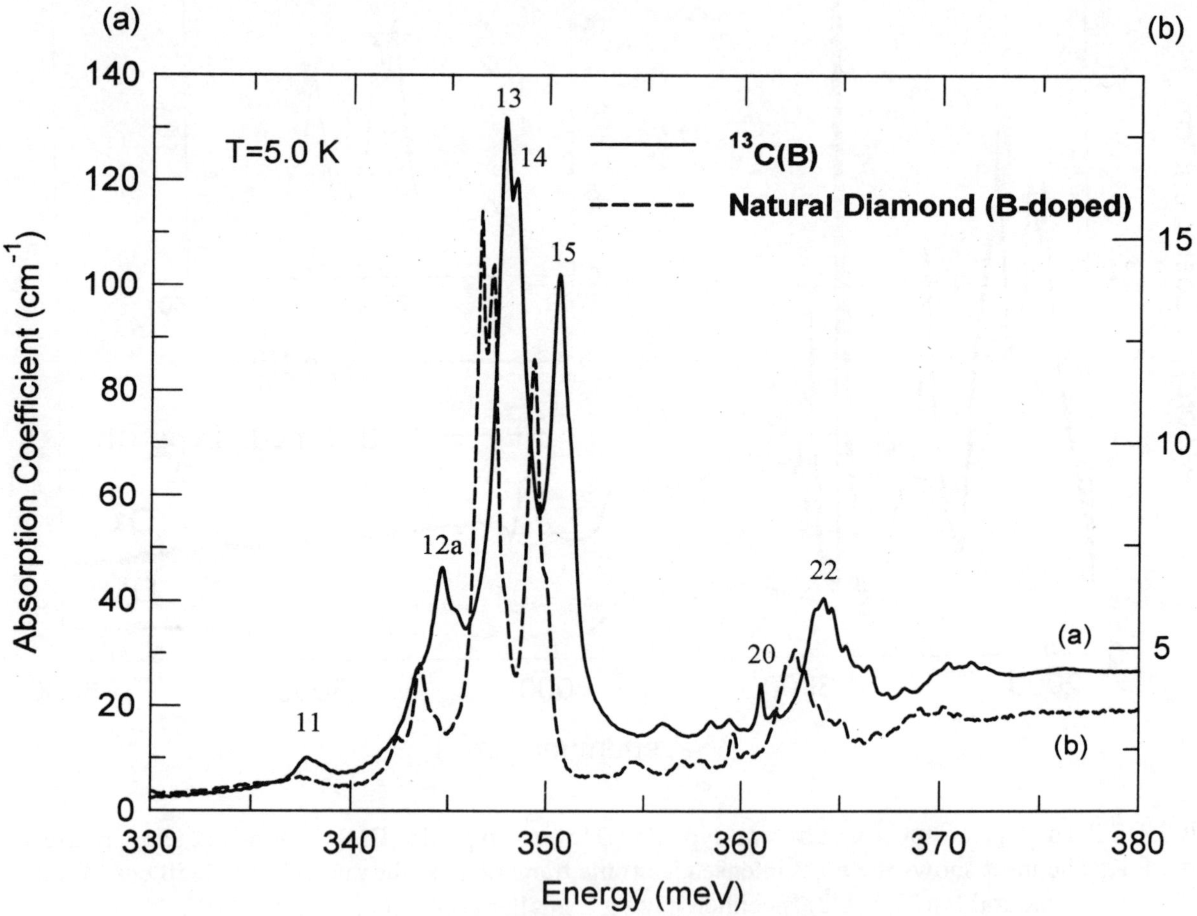


FIGURE 3 Comparison of the excitation spectrum of a natural type IIb and a $^{13}$C boron-doped diamond in the range 330 to 380 meV [10].

Kim et al [14,15] carried out a comprehensive experimental and theoretical investigation of the magneto-spectroscopy of the $\Delta'{:}1s(p_{3/2}) \rightarrow 1s(p_{1/2})$ Raman transition. The Zeeman effect of Δ′, observed in diverse polarisation configurations, has allowed the level ordering of the four $1s(p_{3/2}){:}\Gamma_8$ and the two $1s(p_{1/2}){:}\Gamma_7$ Zeeman sublevels and hence the determination of the sign and magnitude of $g_1$ and $g_2$, the orbital and spin g-factors of the acceptor-bound hole. High resolution Zeeman measurements with a multi-passed, tandem Fabry-Pérot interferometer led to the discovery of the Raman transitions within the $\Gamma_8$ multiplet (the Raman-EPR lines) [14,15]. Observations at zero magnetic field revealed that the Δ′ line is a doublet separated by (0.81 ± 0.15) $cm^{-1}$, indicative of a spontaneous symmetry breaking of the four-fold degenerate ground state [16].

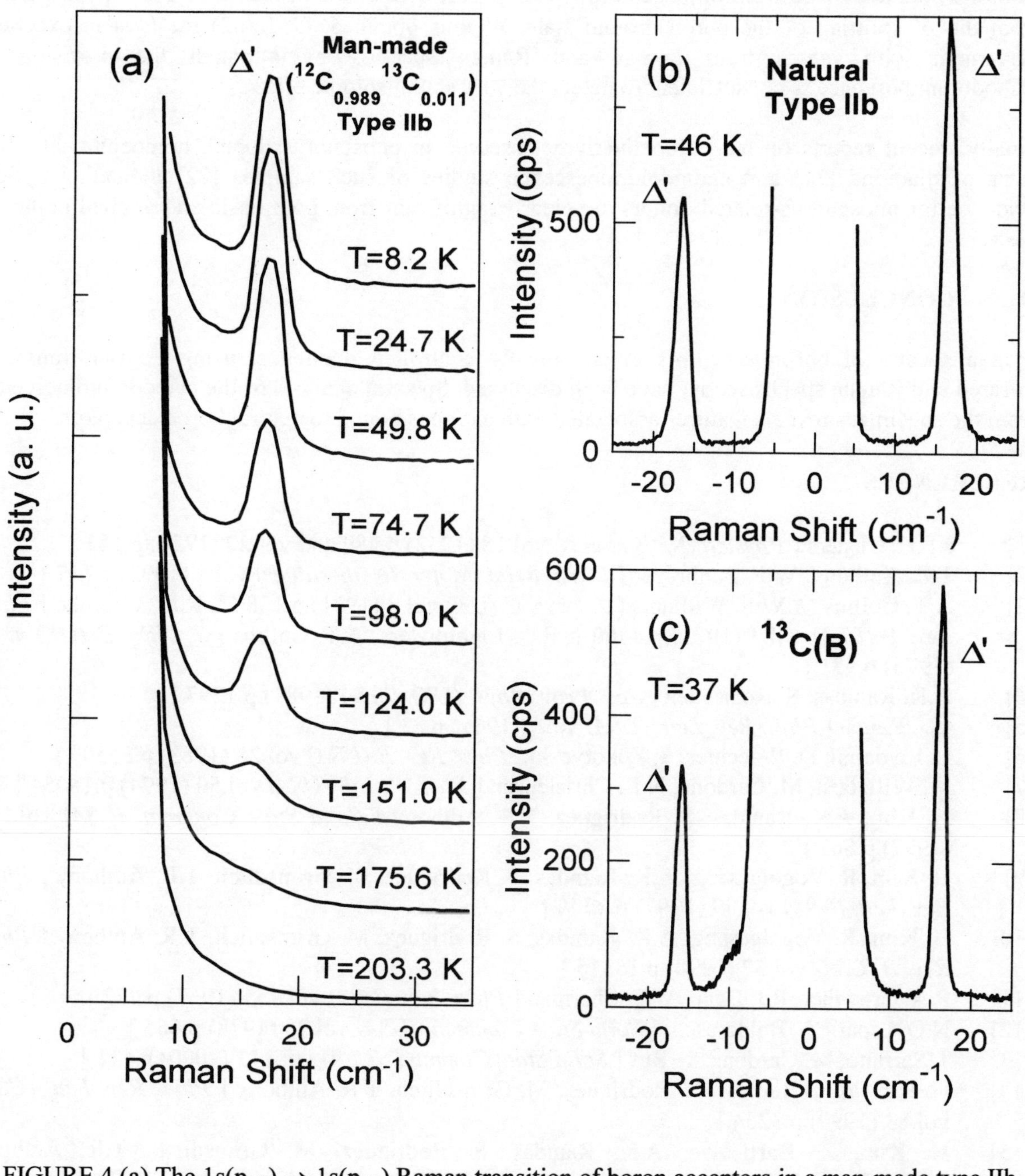


FIGURE 4 (a) The $1s(p_{3/2}) \rightarrow 1s(p_{1/2})$ Raman transition of boron acceptors in a man-made type IIb diamond with natural composition labelled here as Δ′, as a function of temperature. The spectra were excited with the 5309 Å $Kr^+$ laser line, and recorded in backscattering along [001]. (b) The Stokes and anti-Stokes components for Δ′ in natural type IIb diamond at 46 K, excited with the 4765 Å $Ar^+$ laser line in the right-angle geometry $x'(zz + zx')y'$, $x' \parallel [110]$, $y' \parallel [1\bar{1}0]$, and $z \parallel [001]$. (c) The Stokes and anti-Stokes lines in a $^{13}C$ diamond at 37 K, excited with the same laser line as in (b) but in backscattering.

In addition to Lyman and Raman spectroscopy of boron acceptors in diamond discussed above, the spectral analysis of cathodoluminescence from such specimens reveals features associated with excitons bound to neutral boron acceptors [17–20]. Part of the fine structure in such spectra arises from the $\Delta'$ splitting of the boron ground state: $\Delta'$ thus obtained, $(2 \pm 0.2)$ meV, is in excellent agreement with values from Lyman and Raman spectroscopy although the precision in cathodoluminescence is not sufficient to detect the very small isotopic effect.

Finally, recent reports on n-type conductivity observed in phosphorus doped, homoepitaxial CVD films of diamond [21] and cathodoluminescence studies of such samples [22] indicating optical evidence for phosphorus-related donors are clearly significant from both basic and applied points of view.

## D CONCLUSION

Lyman spectra of boron acceptors in isotopically controlled diamonds, using Fourier transform infrared and Raman spectroscopy, have been discussed. Spectral analysis of the cathodoluminescence from the specimens reveals features associated with excitons bound to neutral boron acceptors.

## REFERENCES


[1] J.F.H. Custers [ *Physica (Netherlands)* vol.18 (1952) p.489 and vol.20 (1954) p.183 ]

[2] T.R. Anthony, W.F. Banholzer [ *Diam. Relat. Mater. (Netherlands)* vol.1 (1992) p.717 ]

[3] A.T. Collins, A.W.S. Williams [ *J. Phys. C (UK)* vol.4 (1971) p.1789 ]; R.M. Chrenko [ *Phys. Rev. B (USA)* vol.7 (1973) p.4560 ]; E.C. Lightowlers, A.T. Collins [ *J. Phys. D (UK)* vol.9 (1976) p.951 ]

[4] A.K. Ramdas, S. Rodriguez [ *Rep. Prog. Phys. (UK)* vol.44 (1981) p.1297 ]

[5] C.J. Rauch [ *Phys. Rev. Lett. (USA)* vol.7 (1961) p.83 ]

[6] L. Regianni, D. Waechter, S. Zukotynski [ *Phys. Rev. B (USA)* vol.28 (1983) p.3550 ]

[7] M. Willatzen, M. Cardona, N.E. Christensen [ *Phys. Rev. B (USA)* vol.50 (1994) p.18054 ]

[8] H. Kim, A.K. Ramdas, S. Rodriguez, T.R. Anthony [ *Solid State Commun. (USA)* vol.102 (1997) p.861 ]

[9] H. Kim, R. Vogelgesang, A.K. Ramdas, S. Rodriguez, M. Grimsditch, T.R. Anthony [ *Phys. Rev. Lett. (USA)* vol.79 (1997) p.1706 ]

[10] H. Kim, R. Vogelgesang, A.K. Ramdas, S. Rodriguez, M. Grimsditch, T.R. Anthony [ *Phys. Rev. B (USA)* vol.57 (1998) p.15315 ]

[11] P.A. Crowther, P.J. Dean, W.F. Sherman [ *Phys. Rev. (USA)* vol.154 (1967) p.772 ]

[12] N.O. Lipari, A. Baldereschi [ *Solid State Commun. (USA)* vol.25 (1978) p.665 ]

[13] J. Serrano, M. Cardona, T. Ruf [ *Solid State Commun. (USA)* vol.113 (2000) p.411 ]

[14] H. Kim, A.K. Ramdas, S. Rodriguez, M. Grimsditch, T.R. Anthony [ *Phys. Rev. Lett. (USA)* vol.83 (1999) p.3254 ]

[15] H. Kim, Z. Barticevic, A.K. Ramdas, S. Rodriguez, M. Grimsditch, T.R. Anthony [ unpublished ]

[16] H. Kim, A.K. Ramdas, S. Rodriguez, M. Grimsditch, T.R. Anthony [ *Phys. Rev. Lett. (USA)* vol.83 (1999) p.4140 ]

[17] S.J. Sharp, A.T. Collins, G. Davies, G.S. Joyce [ *J. Phys., Condens. Matter (UK)* vol.9 (1997) p.L451 ]

[18] H. Sternschulte et al [ *Mater. Sci. Forum (Switzerland)* vol.258-263 (1997) p.757 ]

[19] T. Ruf et al [ *Solid State Commun. (USA)* vol.105 (1998) p.311 ]

[20] R. Sauer, H. Sternschulte, S. Wahl, K. Thonke, T.R. Anthony [ *Phys. Rev. Lett. (USA)* vol.84 (2000) p.4172 ]

[21] S. Koizumi, M. Kamo, Y. Sato, H. Ozaki, T. Inuzuka [ *Appl. Phys. Lett. (USA)* vol.71 (1997) p.1065 ]
[22] H. Sternschulte, K. Thonke, R. Sauer, S. Koizumi [ *Phys. Rev. B (USA)* vol.59 (1999) p.12924 ]

# A1.4 Elastic moduli of diamond

M. Grimsditch

March 2000

## A INTRODUCTION

The elastic constants of diamond are the largest of any material and have been determined by a variety of techniques over a period of over half a century. This Datareview outlines these studies and also discusses the temperature dependence of these constants.

## B DATA

The elastic constants of diamond are the largest of any known material reflecting the very strong interatomic carbon-carbon atomic bond, which is also responsible for the reputed hardness of this material. The determination of elastic constants is usually associated with ultrasonic techniques. A somewhat less well known technique is Brillouin scattering, the inelastic scattering of light from acoustic phonons. The first determination of the elastic constants of diamond, dating back to 1946, was accomplished using ultrasonics [1]. This was followed by an X-ray diffraction determination in 1953 [2] and a Brillouin scattering measurement in 1958 [3]. These early determinations are summarised in TABLE 1 where it can be see that $c_{11}$ and $c_{44}$ agreed to within ≅10% while the spread in $c_{12}$ was a factor of 2.

TABLE 1 Elastic moduli of diamond (GPa).

| $c_{11}$ | $c_{12}$ | $c_{44}$ | Method | Refs |
|---|---|---|---|---|
| 950 | 390 | 430 | Ultrasonics | [1] |
| 1100±110 | 330±30 | 440±40 | X-rays | [2] |
| 949 | 151 | 521 | Brillouin | [3] |
| 1079±5 | 124±5 | 578±2 | Ultrasonics | [4,5] |
| 1080.4±0.5 | 127.0±0.7 | 576.6±0.7 | Brillouin | [6,7] |

Subsequent improvements in techniques led to accurate determinations of the $c_{ij}$ using both ultrasonic [4,5] and Brillouin scattering techniques [6,7]. These later measurements, as can be seen in TABLE 1, produced agreement between the values obtained using different techniques. Interesting to note is that a 22-carat diamond was used for the ultrasonic experiments while ≅1/3-carat samples were typically used for the Brillouin experiments. The relative accuracy of the Brillouin and ultrasonic determinations is unusual. Typically ultrasonic experiments, in which sample size is not a concern, are more accurate than Brillouin determinations. Diamond is unique in this sense and the accuracy of the Brillouin determination is superior to that of ultrasonics due to a combination of factors: transparency, high sound velocities, etc.

The temperature dependence of the $c_{ij}$ has also been determined using Brillouin scattering [8]. For this purpose the technique is uniquely well suited since it allows the sample to be contained in a non-reactive environment (viz. argon) to delay the onset of graphitisation, and also it avoids the complications of the buffer rods used in ultrasonic experiments. FIGURE 1 shows the temperature dependence of four combinations of $c_{ij}$ measured in the backscattering geometry. The lines are obtained by fitting each of the three independent $c_{ij}$ to an expression of the form

$$c_{ij} = C_0 + C_2 (T^2 - 300^2) \tag{1}$$

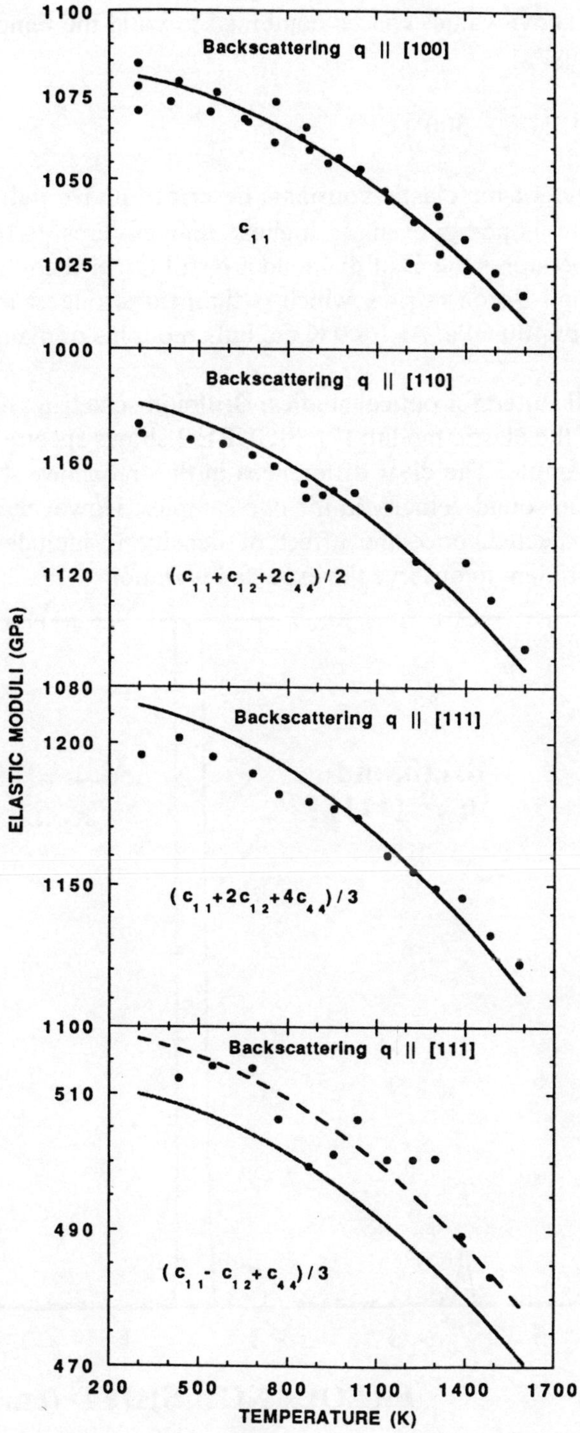


FIGURE 1 Temperature dependence of four combinations of elastic moduli of diamond [8]. The full lines are fits to the data discussed in the text and in [8].

where the $C_0$, the room temperature values, are obtained from TABLE 1, and the $C_2$ are adjusted to produce the best fit. The full and dashed lines in FIGURE 1 correspond to fits in which different sources of error have been included in the Brillouin analysis. A full description of the sources of error is given in [8] but, in particular, the dashed line in FIGURE 1 includes a 2° misalignment of the crystal. The $C_2$ values of EQN (1) corresponding to $c_{11}$, $c_{12}$ and $c_{44}$ are –29±8, –3±18 and –22±7 $10^{-6}$ GPa/K$^2$, respectively. The above values can be combined to yield the temperature dependence of the bulk modulus $\{B = (c_{11} + 2c_{12})/3\}$

$$B = 444.8 - 0.000012\,(T^2 - 300^2)\ \text{GPa} \tag{2}$$

The temperature dependence of the elastic constants described above indicates that diamond retains its very strong interatomic bonding even at highest temperatures (≅1600 K) reached in these experiments. At these temperatures the $c_{ij}$ of diamond are still larger than the room temperature values of any other known material. Boron nitride, which is the next strongest material [9,10], has a room temperature bulk modulus of 400 GPa. At 1600 K the bulk modulus of diamond is still 415 GPa.

Because diamond is so well suited for optical studies, Brillouin scattering has been used to investigate the isotopic dependence of the elastic moduli [7]. FIGURE 2 shows spectra taken from a natural and a $^{13}C$ isotopically enriched sample. The clear differences in the frequency shifts observed in the figure correspond to the changes in sound velocity in the two samples. However, although small differences in elastic constants are expected, once the effect of density is included, the accuracy of the $c_{ij}$ determinations was not sufficient to observe the expected variation.

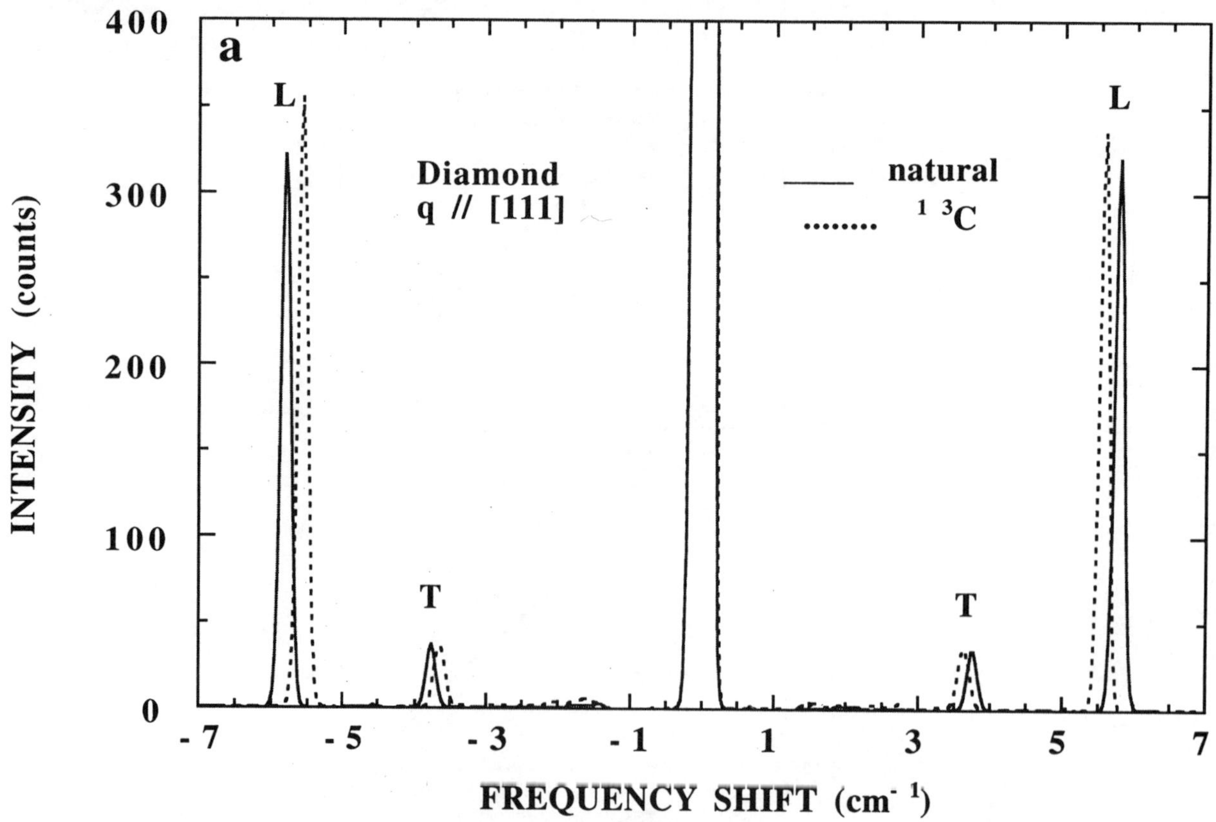


FIGURE 2 Brillouin spectra from natural and $^{13}C$ diamond samples. The observed changes in peak positions, which reflect changes in sound velocity, are mainly due to the density difference between the two samples.

## C CONCLUSION

This Datareview has summarised the elastic moduli of diamond as determined by a variety of techniques. The temperature dependence of the moduli has been established using Brillouin scattering and these data are also presented. Diamond retains its strong bonding even at ~1600 K.

## ACKNOWLEDGEMENT

Work supported by the US Department of Energy, BES Materials Sciences Grant W-31-109-ENG-38.

## REFERENCES

[1] S. Bhagavabtam, J. Bhimasenachar [ *Proc. R. Soc. Lond. A (UK)* vol.187 (1946) p.381 ]
[2] E. Prince, W. Wooster [ *Acta Crystallogr. (Denmark)* vol.6 (1953) p.450 ]
[3] R.S. Krishnan, V. Chandrasekharan, E.S. Rajagopal [ *Nature (UK)* vol.182 (1958) p.518 ]
[4] H.J. McSkimin, W.L. Bond [ *Phys. Rev. (USA)* vol.105 (1957) p.116 ]
[5] H.J. McSkimin, P. Andreatch [ *J. Appl. Phys. (USA)* vol.43 (1972) p.2944 ]
[6] M. Grimsditch, A.K. Ramdas [ *Phys. Rev. B (USA)* vol.11 (1975) p.3139 ]
[7] R. Vogelgesang, A.K. Ramdas, S. Rodriguez, M. Grimsditch, T. Anthony [ *Phys. Rev. B (USA)* vol.54 (1996) p.3989 ]
[8] E.S. Zouboulis, M. Grimsditch, A.K. Ramdas, S. Rodriguez [ *Phys. Rev. B (USA)* vol.57 (1998) p.2889 ]
[9] M. Grimsditch, E.S. Zouboulis, A. Polian [ *J. Appl. Phys. (USA)* vol.76 (1994) p.832-4 ]
[10] M. Grimsditch [ *Proc. 16th Int. Conf on Raman Spectroscopy* Ed. A. Heyns (Wiley, NY, 1998) p.85 ]

# A1.5 Thermal transport in diamond

J. Hartmann and M. Reichling

February 2000

## A INTRODUCTION

One of the most outstanding physical properties of diamond is its extremely high thermal conductivity. This is exploited in advanced micro- and optoelectronic systems where diamond is integrated as a thermal management material, an application of growing importance since diamond is now available in the form of high quality thin films.

Room temperature thermal conductivity values up to 25 W/cm K have been reported for the highest quality single crystals of type IIa. This is the highest value of all known materials and exceeds that of copper, which is commonly regarded as an excellent thermal conductor, by a factor of more than six. In contrast to a metal, where thermal conductivity is provided by the mobility of conduction band electrons, heat transfer in the insulator diamond is solely carried by lattice vibrations, i.e. phonons. The reason for the outstanding thermal conductivity and the high Debye-temperature of 2000 K of diamond is the stiffness of the $sp^3$ bonds forming its rigid structure together with the low mass of the carbon atoms. In most applications, the temperature is well below the Debye temperature and, hence, phonon–phonon scattering is small, resulting in little impedance for the phonon-mediated heat transport in a large high purity crystal.

Extrinsic phonon scattering mechanisms are the main source of thermal resistance in less pure material. For single crystal diamond, scattering at the sample's boundaries, at impurities and vacancies are the main contributions. For polycrystalline material, additional contributions arise from scattering at grain boundaries, dislocations and extended defects. All contributions influencing thermal conductivity depend on the wavelength of the contributing phonons and, therefore, on the temperature of the sample. Hence, for understanding phonon scattering mechanisms and their relative contributions, measurements of the temperature dependent thermal conductivity are of basic importance.

## B THERMAL TRANSPORT AND PHONON SCATTERING MECHANISMS

According to Fourier's empirical law for the conduction of heat [1], the heat flux vector q, i.e. the directed thermal energy per unit time and area, is related to the gradient of the temperature $\nabla T$ in the sample by the relation

$$q = -\kappa \nabla T$$

For an isotropic medium, the thermal conductivity $\kappa$ is the constant of proportionality between the absolute values of the temperature gradient and the heat flux. In the general case of an anisotropic material, however, the conductivity is a tensor. For describing thin films, the tensor properties can often be reduced to specifying a lateral thermal conductivity describing thermal transport parallel to the plane of the film and a perpendicular conductivity, characterising out-of-plane heat flow.

In the presence of a highly localised thermal barrier with resistance $R_{th}$, the temperature drop across the barrier $\Delta T$ is related to the heat flux by

$$\Delta T = q \cdot R_{th}$$

where q is the component of the heat flux perpendicular to the boundary plane.

Thermal conductivity and thermal resistance are macroscopic quantities that are related to microscopic properties by the gas kinetic definition of thermal conductivity based on phonons as quantised lattice vibrations [2]. In this physical picture, thermal conductivity κ can be described by

$$\kappa = \frac{1}{3} C v^2 \tau$$

where C is the phonon contribution to the heat capacity per unit volume, v the velocity of the phonons, and $\tau^{-1}$ the phonon scattering rate. As phonons of different wavelengths, ranging over a temperature dependent spectrum, contribute to the thermal transport, a more complete description of κ according to the Klemens–Callaway model [3,4] can be given by integration over the temperature-dependent Debye phonon wavelength spectrum.

In the following, scattering mechanisms relevant for heat transport in diamond and their dependence on temperature and phonon properties are briefly introduced. A more detailed general discussion can be found in [2–4] while [5] provides a survey on phonon scattering in diamond thin films.

In isotopically pure, defect-free diamond crystals of sufficient thickness, phonon–phonon scattering is the only mechanism generating thermal resistance and at elevated temperatures, i.e. well above the Debye temperature, this is the dominant contribution also in less pure material. Due to the low mass of the carbon atoms, changes in the isotopical content of diamond significantly affect its thermal conductivity, as the phonons are scattered due to the difference in mass of the different isotopes of carbon. Similar scattering also occurs at vacancies and impurity sites where the scattering rate depends on the wavelength of the phonons. For extended defects like clusters or aggregates of foreign or disordered atoms with a size much smaller than the wavelength of the lattice vibrations, scattering is similar to that at point defects. For phonon wavelengths smaller than the size of the extended defects, the scattering rate becomes independent of the phonon wavelength. Dislocations are another source of thermal resistance, since the phonons are scattered in the strain field in the vicinity of dislocations. Scattering at boundaries is most important at lowest temperatures where the phonon wavelength is comparable to or larger than the geometrical dimensions of the crystal. The type of reflection, i.e. specular or diffuse, critically depends on the microstructure of the boundaries and determines the impact on thermal resistance.

## C THERMAL CONDUCTIVITY DATA OF SINGLE CRYSTAL DIAMOND

In high quality material, the thermal conductivity is determined by the purity of the crystal. The most important impurities for lowering thermal conductivity are nitrogen, hydrogen, and the $^{13}C$ isotope of carbon. The purest type IIa material has the highest thermal conductivity while in type Ia and Ib material having nitrogen impurities, thermal conductivity is significantly lower. Several groups measured the thermal conductivity of natural single crystal diamond as a function of temperature. Typical results for type Ia and type IIa crystals are compiled in FIGURE 1 and TABLE 1.

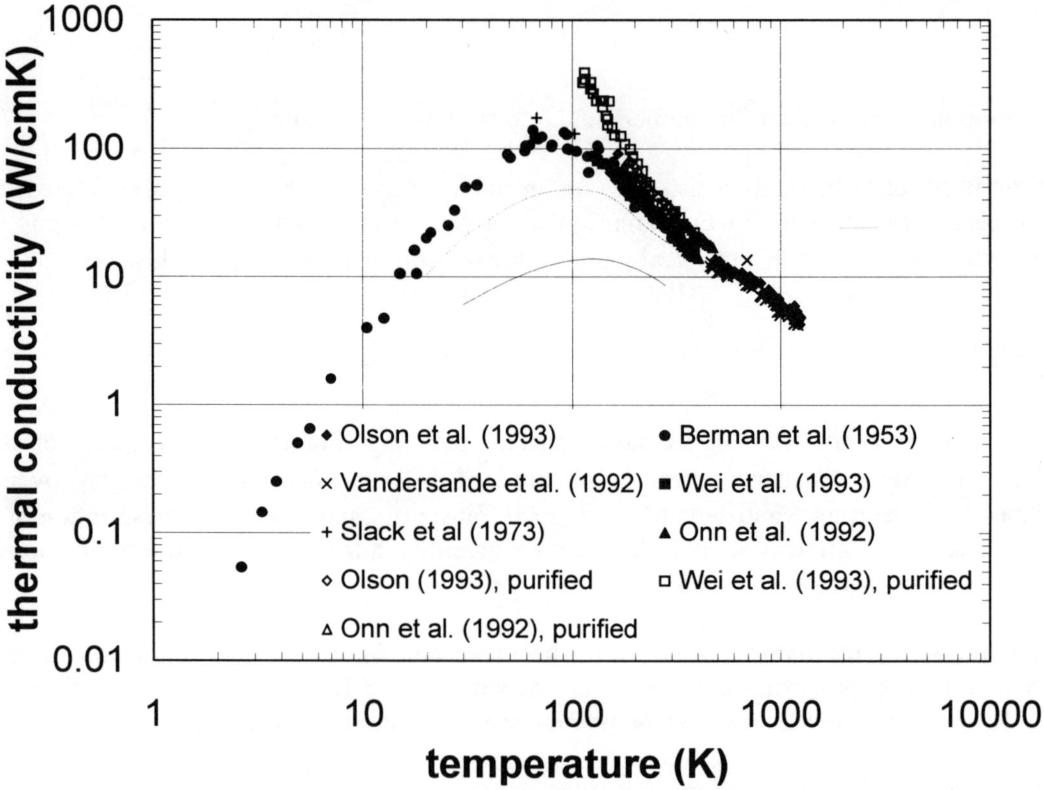


FIGURE 1 Thermal conductivity versus temperature for natural and isotopically purified single crystal diamond of type IIa. The lines show values for type Ia diamond crystals with a nitrogen content of $1.8 \times 10^{20}$ cm$^{-3}$ (solid line) and $1.0 \times 10^{20}$ cm$^{-3}$ (dotted line) [6–14].

The content of the heavier carbon isotope $^{13}$C has a tremendous impact on the phonon scattering rate and the highest thermal conductivity is expected for isotopically pure $^{12}$C diamond. A model calculation [10] predicts values up to 2000 W/cm K at temperatures around 80 K. The influence of the concentration of $^{13}$C on thermal conductivity has systematically been explored experimentally and theoretically in [10]. It was found that thermal conductivity varies by a factor of three when varying the $^{13}$C content from 0.001% to 50%, while it varies by nearly two orders of magnitude at 80 K.

## D THERMAL CONDUCTIVITY OF POLYCRYSTALLINE DIAMOND FILMS

Polycrystalline diamond films can be fabricated using several industrial processes like chemical vapour deposition (CVD), microwave or laser-assisted deposition (see Chapter B1 of this book), that yield polycrystalline films with different microstructure and thickness. Thermal properties are often determined by the size of the crystallites that depends on the deposition rate, the film thickness and the process gas mixture. Generally, crystallites have a columnar structure and are separated from each other by grain boundaries that act as local thermal barriers.

Measurements of thermal conductivity inside single crystallites of high quality CVD films reveal thermal conductivities up to 24 W/cm K, with some variation inside the grains (see Datareview A2.1) [15,16]. The overall conductivity of such a film, however, is much smaller due to grain boundaries that have been shown to be a source of highly localized thermal resistance [17]. The columnar structure also introduces an anisotropy in thermal conduction where the perpendicular conductivity is higher than the lateral conductivity (see FIGURE 2) since in the latter case the heat crosses more grain boundaries.

TABLE 1 Averaged values of thermal conductivity for natural and isotopically purified type IIa and type Ia diamond as a function of temperature. The errors given for type IIa material were determined from the scatter in available literature data and they demonstrate that such data can be measured with high accuracy and sample-to-sample scatter is small. For type Ia material, however, the conductivity strongly depends on the specific impurity level and we specify a range of values compiled from the literature.

| κ (W/cm K) / T (K) | Type Ia diamond | Type IIa diamond | Type IIa diamond, purified |
|---|---|---|---|
| 10 | | 3.5±2.2 | |
| 50 | 9.4 – 35.2 | 83.9±2.1 | |
| 70 | 11.7 – 44.6 | 132.9±32.4 | |
| 100 | 13.4 – 48.1 | 116.7±20.5 | |
| 150 | 13.5 – 42.7 | 70.2±11.1 | 184.0 |
| 200 | 11.6 – 29.5 | 44.8±7.0 | 77.7±7.9 |
| 250 | 9.7 – 21.6 | 29.8±4.2 | 44.5±1.5 |
| 300 | | 21.9±1.7 | 33.2±0.5 |
| 350 | | 17.6±1.4 | 24.8±0.2 |
| 400 | | 13.7±1.4 | 20.4 |
| 450 | | 12.4±1.2 | 17.9 |
| 500 | | 11.2±0.3 | |
| 550 | | 10.6±0.3 | |
| 600 | | 10.1±0.2 | |
| 650 | | 9.6±0.3 | |
| 700 | | 9.5±1.0 | |
| 750 | | 7.9±0.3 | |
| 800 | | 7.1±0.0 | |
| 850 | | 6.7±0.0 | |
| 900 | | 6.6±0.1 | |
| 950 | | 6.0±0.1 | |
| 1000 | | 5.1±0.1 | |
| 1050 | | 4.7±0.3 | |
| 1100 | | 4.6±0.4 | |
| 1150 | | 4.7±0.2 | |
| 1200 | | 4.7±0.5 | |

FIGURE 2 shows typical data for thin films and provides clear evidence that local thermal conductivity of diamond films, and especially lateral conductivity, depends on film thickness. Very thick films yield values close to that of crystalline material. The thickness dependence is a consequence of the film microstructure. The nucleation layer has a high content of amorphous graphite and small crystallites of pure diamond resulting in a strongly reduced thermal conductivity. With increasing film thickness, crystallites grow, finally forming a columnar structure, where the crystallite size increases with film thickness. FIGURE 2 also demonstrates the expected difference between lateral and perpendicular heat flow.

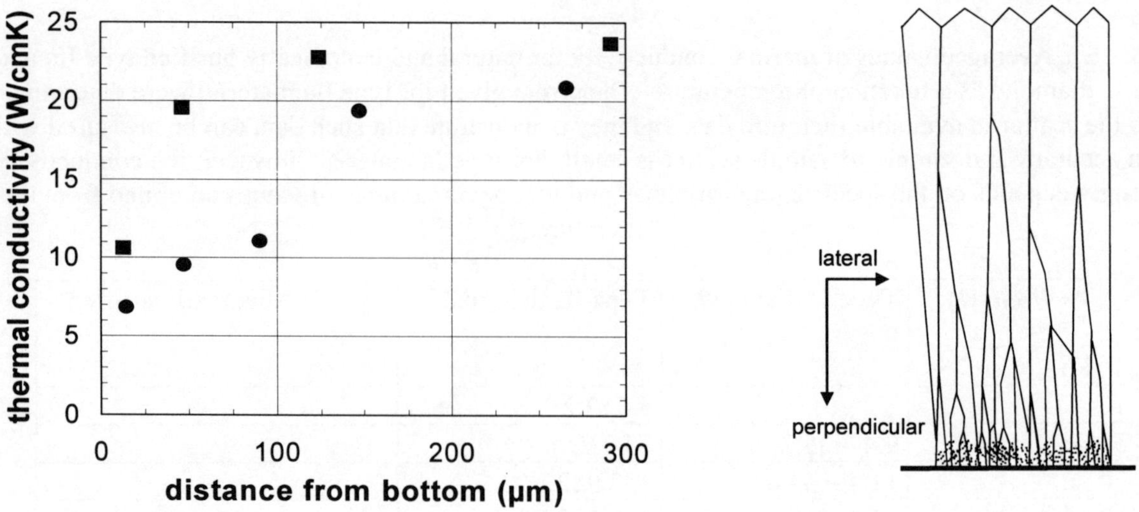


FIGURE 2 Measured local perpendicular (■) and lateral (•) thermal conductivity of polycrystalline diamond films versus distance from the bottom of the polycrystalline diamond films [18]. A schematic view of the microstructure of a polycrystalline diamond film and the directions of perpendicular and lateral heat flow are shown on the right side (after [19]).

Thermal resistances between single diamond crystallites have been measured to be of the order of between $10^{-5}$ and $10^{-4}$ $cm^2$ K/W [17,20]. For most applications of diamond films, thermal transport in the lateral direction is the most important one. For one-dimensional lateral heat flow over a length much larger than the average diameter of the grains s, the thermal conductivity $\kappa_{1D}$ of a diamond film can be estimated by the thermal conductivity $\kappa_0$ inside the grains and the thermal resistance $R_{th}$ located at the grain boundaries using the relation [17]

$$\kappa_{1D} = \left( \frac{1}{\kappa_0} + \frac{R_{th}}{s} \right)^{-1}$$

According to this equation, the lateral thermal conductivity of high quality diamond films, with a $\kappa_0$ of about 20 W/cm K, differs by only 1% from the thermal conductivity $\kappa_0$ inside the crystallites for an average crystallite diameter larger than 300 μm and a thermal resistance $R_{th}$ of $2 \times 10^{-5}$ $cm^2$ K/W.

Due to the peculiarities of different deposition processes, thermal conductivity varies significantly from one sample to another and results presented in the literature are spread over a large range. FIGURE 3 is a collection of temperature dependent lateral thermal conductivity values obtained by several groups on a variety of samples covering a range of preparation techniques.

The wide spread reported in thin film thermal data is due to the critical dependence of the conductivity on grain size, columnar structure and preparation related impurities. For example, the content of $CH_4$ in the CVD process influences thermal conductivity values as shown in FIGURE 4, where highest values are obtained for the lowest content of $CH_4$.

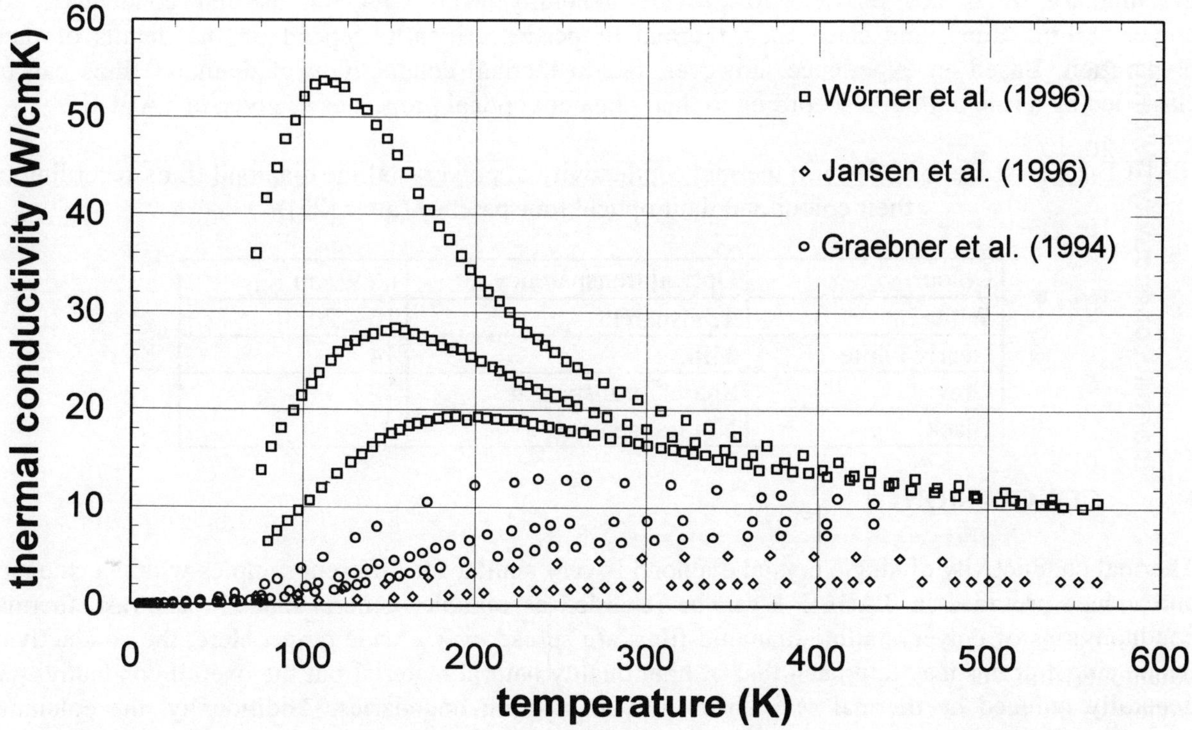


FIGURE 3 Thermal conductivity versus temperature for various polycrystalline diamond samples [21–24].

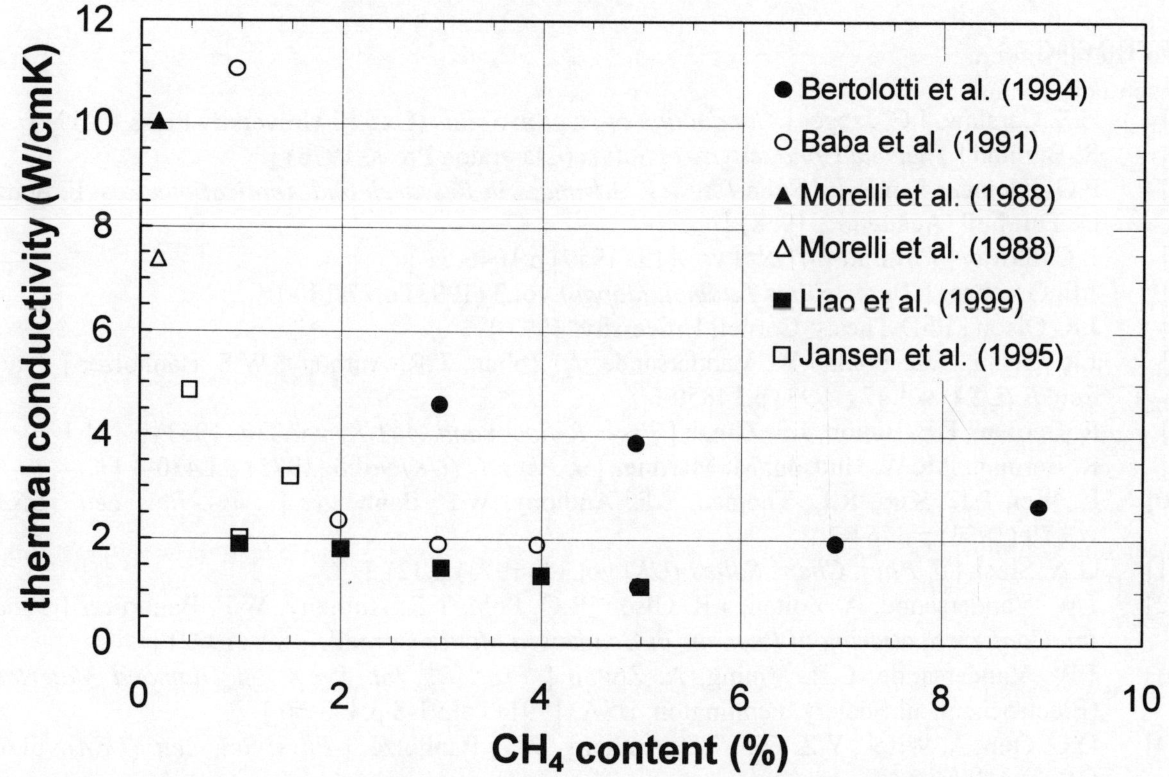


FIGURE 4 Variation of room-temperature lateral thermal conductivity of polycrystalline diamond films as a function of the volume fraction of $CH_4$ during the deposition process.

In summary, it is not possible to provide generally valid data for thermal conductivity of polycrystalline thin films since their thermal properties critically depend on the details of their preparation. Based on experience, however, lateral thermal conductivity of diamond films can be classified by a rule of thumb according to their apparent optical properties as given in TABLE 2.

TABLE 2 Classification of lateral thermal conductivity of polycrystalline diamond films according to their colour and their optical transparency (after [25]).

| Colour | Optical transparency | κ (W/cm K) |
|---|---|---|
| White | Transparent | 16 – 20 |
| Nearly white | Milky | 14 |
| Grey | Sparely transparent | 12 |
| Black | Not transparent | 9 |

## E CONCLUSION

Thermal conductivity of single crystal diamond is very similar for different samples within a category and values provided in TABLE 1 can be regarded as reliable, general data. In contrast, thermal conductivities of polycrystalline diamond films are spread over a wide range. Here, the conductivity inside single grains may approach that of high quality natural material but the overall conductivity is generally reduced by thermal resistances located at grain boundaries. Additionally, the columnar structure of thin films introduces a significant anisotropy in thermal properties. Accurate data can only be provided for a specific sample and general trends have always to be discussed in relation to the preparation technique.

## REFERENCES

[1] H.S. Carslaw, J.C. Jaeger [ *Conduction of Heat in Solids* (Oxford University Press, 1959) ]
[2] R. Berman [ *Thermal Conductivity in Solids* (Clarendon Press, 1976) ]
[3] P.G. Klemens [ in *Solid State Physics: Advances in Research and Applications* Eds. F. Seitz, D. Turnbull (Academic, 1958) ]
[4] J. Callaway [ *Phys. Rev. (USA)* vol.113 (1959) p.1046-51 ]
[5] J.E. Graebner [ *Diam. Films Technol. (Japan)* vol.3 (1993) p.77-130 ]
[6] J.R. Olson [ PhD Thesis, Cornell University, 1993 ]
[7] J.R. Olson, R.O. Pohl, J.W. Vandersande, A. Zoltan, T.R. Anthony, W.F. Banholzer [ *Phys. Rev. B (USA)* vol.47 (1993) p.14850-6 ]
[8] R. Berman, F.E. Simon, J.S. Ziman [ *Proc. R. Soc. Lond. A (UK)* vol.220 (1953) p.171 ]
[9] R. Berman, P.R.W. Hudson, M. Martinez [ *J. Phys. C (UK)* vol.8 (1975) p.L430-4 ]
[10] L. Wei, P.K. Kuo, R.L. Thomas, T.R. Anthony, W.F. Banholzer [ *Phys. Rev. Lett. (USA)* vol.70 (1993) p.3764-7 ]
[11] G.A. Slack [ *J. Phys. Chem. Solids (UK)* vol.34 (1973) p.321 ]
[12] J.W. Vandersande, A. Zoltan, J.R. Olson, R.O. Pohl, T.R. Anthony, W.F. Banholzer [ *Proc. Int. Conf. on Scattering of Phonons in Condensed Matter* Cornell, USA (1992) ]
[13] J.W. Vandersande, C.B. Vining, A. Zoltan [ *Proc. 2nd Int. Symp. on Diamond Materials* (Electrochemical Society, Pennington, USA, 1991) vol.91-8 p.443-50 ]
[14] D.G. Onn, A. Witek, Y.Z. Qiu, T.R. Anthony, W.F. Banholzer [ *Phys. Rev. Lett. (USA)* vol.68 (1992) p.2806-9 ]
[15] J. Hartmann, P.Voigt, M. Reichling [ *J. Appl. Phys. (USA)* vol.81 (1997) p.2966–77 ]
[16] M. Reichling, T. Klotzbücher, J. Hartmann [ *Appl. Phys. Lett (USA)* vol.73 (1998) p.756-8 ]
[17] J. Hartmann, M. Costello, M. Reichling [ *Phys. Rev. Lett. (USA)* vol.80 (1998) p.117-20 ]

[18] J.E. Graebner, J.A. Mucha, F.A. Baiochi [ *Diam. Relat. Mater. (Switzerland)* vol.5 (1996) p.682-7 ]
[19] J.E. Graebner, S. Jin, G.W. Kammlott, J.A. Herb, C.F. Gardinier [ *Nature (USA)* vol.359 (1992) p.401-3 ]
[20] J. Hartmann, M. Reichling [ *AIP Conf. Proc. (USA)* vol.463 (1999) p.295-9 ]
[21] E. Wörner, J. Wagner, W. Müller-Sebert, C. Wild, P.Koidl [ *Appl. Phys. Lett. (USA)* vol.68 (1996) p.1482-4 ]
[22] E. Jansen, E. Obermeier [ *Phys. Status Solidi A (Germany)* vol.154 (1996) p.395-402 ]
[23] D.T. Morelli, C.P. Beetz, T.A. Perry [ *J. Appl. Phys. (USA)* vol.64 (1988) p.3063-6 ]
[24] J.E Graebner, S. Jin, J.A. Herb, C.F. Gardinier [ *J. Appl. Phys. (USA)* vol.76 (1994) p.1552-6 ]
[25] J. Hartmann [ *Quantitative Bestimmung lokaler thermischer Eigenschaften in polykristallinem Diamant* (Dr. Köster, 1997) ]

# A1.6 Transport properties of diamond: carrier mobility and resistivity

C.E. Nebel and M. Stutzmann

May 2000

## A INTRODUCTION

Transport properties of diamond are determined by intrinsic defects, and extrinsic contaminants or dopants, most prominently nitrogen and boron, as well as different structural morphologies (e.g. polycrystallinity). Boron causes p-type conductivity and is an effective acceptor with a ground state level 370 meV above the valence band edge. Among natural diamond, boron doped crystals (named IIb) are very rare and contain less than 1 ppm boron. Synthetic high pressure high temperature diamond (HPHT) can contain up to 1000 ppm of boron. CVD-diamond has been doped in the regime $10^{16}$ cm$^{-3}$ to $10^{21}$ cm$^{-3}$ by adding gaseous compounds of boron to the plasma. Nitrogen is the dominant extrinsic impurity in diamond, and isolated substitutional nitrogen acts as a deep donor with an ionization energy of 1.7 eV. Recently, phosphorus doped CVD-diamond has been grown by adding phosphine ($PH_3$) to the plasma. At doping densities of $10^{19}$ cm$^{-3}$, n-type conductivity is achieved with an activation energy of 550 meV. Based on photoconductivity experiments it is assumed that phosphorus is a hydrogen-like dopant in diamond.

Natural and synthetic diamonds are monocrystalline wide bandgap semiconductors, while CVD-diamond grown on substrates such as silicon (heteroepitaxy) is polycrystalline. In these films, bulk properties and grain boundary effects have to be taken into account for an understanding of electronic properties.

Transport properties of diamond have been investigated ever since diamond was found to be a reasonably conducting solid. The primary characterization techniques for transport properties are temperature-dependent dark conductivity-, Hall-effect-, time-of-flight- and spectrally resolved photoconductivity measurements. In the following, we summarize the known properties concerning conductivity and carrier densities in Sections B and C and carrier mobilities in Section D.

In general, no truly intrinsic diamond with conductivity activation energies of about 2.5 eV expected for the large bandgap of 5.5 eV is available. At room temperature the carrier density of intrinsic diamond, $n = N_c \exp[-(E_c - E_F)/k_B T]$, would be much less than one electron per km$^3$ based on the effective density of states of the conduction band, $N_c \cong 2 \times 10^{19}$ cm$^{-3}$, and $E_c - E_F = 2.5$ eV. The electronic properties, therefore, always will be governed by defects, extrinsic contaminants or surface effects.

## B ELECTRICAL CONDUCTIVITY IN p-TYPE DIAMOND

For a p-type semiconductor, the current density j due to an externally applied electric field F, is given by:

$$j = \sigma F = \frac{1}{\rho} F = e p \mu F \quad (1)$$

where σ is the conductivity, ρ is the resistivity, e is the elementary charge, p is the free hole density in the valence band and μ is the hole mobility. For a non-degenerate p-type semiconductor (or equivalently for n-type material) containing a density $N_D$ of compensating donors, the density of holes p at a given temperature T can be calculated using [1]:

$$\frac{p(p+N_D)}{(N_A-N_D-p)}=\frac{2}{\beta}\left\{\frac{2\pi m^* k_B T}{h^2}\right\}^{3/2}\exp\left[-\frac{E_A}{k_B T}\right] \tag{2}$$

where $N_A$ and $N_D$ are the acceptor and donor densities, β is the spin degeneracy of acceptors and $m^*$ is the density of states effective hole mass given by [1]:

$$m_h^* = \left[m_h^{3/2} + m_l^{3/2}\right]^{2/3} \tag{3}$$

where $m_h$ and $m_l$ are the heavy and light hole masses, assuming that the split-off-band is not degenerate. For n-type doping, the density of states effective electron mass is given by:

$$m_e^* = \left[m_L m_{T1} m_{T2}\right]^{1/3} \tag{4}$$

where $m_L$ is the longitudinal and $m_{T1}$ and $m_{T2}$ are the transverse electron masses. Established values for the different effective masses will be discussed below.

In EQN (2), $E_A$ is the acceptor ionization energy and $k_B$ is the Boltzmann constant. At high temperatures, p approaches the saturation value $N_A$–$N_D$ and at low temperatures, where $p \ll N_A, N_D$, EQN (2) gives approximately [2]:

$$p \approx \left(\frac{N_A}{N_D}-1\right)\frac{2}{\beta}\left\{\frac{2\pi m^* k_B T}{h^2}\right\}^{3/2}\exp\left[-\frac{E_A}{k_B T}\right] \tag{5}$$

When an impurity level is created by splitting of states from the conduction or valence band with multiple or degenerate extrema, the impurity level spin degeneracy β will be larger than two. For acceptor levels introduced by group III impurities in Si and Ge, β = 4, because the heavy hole and light hole band are degenerate at **k** = 0 and the spin-orbit splitting energy Δ = 44 meV in Si and 295 meV in Ge is much larger than $k_BT$, so that in general the split-off band is well separated and thermally occupied only to a negligible degree [1,3]. In diamond however, spin-orbit coupling is weak; Δ = 6 meV [4,5] so that a threefold band degeneracy with β = 6 at temperatures above about 70 K can be expected. Below 70 K, a twofold band degeneracy with β = 4 is a better approximation.

The concentration p of holes is generally obtained from Hall effect data as:

$$p=\frac{r_H}{Re} \tag{6}$$

where R is the Hall coefficient, $r_H$ is the Hall factor and e is the elementary charge. For acoustic deformation potential scattering $r_H$ = 1.18 while for ionized impurity scattering $r_H$ = 1.93 [6]. Most Hall data in the literature have been analysed by assuming β = 2 and $r_H$ = 1.18. The discrepancy with the correct values discussed above makes relatively little difference for a quantitative analysis since effectively $(m^*)^{3/2}/\beta$ in EQNS (2) and (5) is used as an adjustable parameter to obtain the best fits to the experimental data. FIGURE 1 shows a plot of the hole density versus reciprocal temperature for a natural type IIb diamond [7], showing the predicted features given by EQNS (2) and (5). The evaluation gives a boron density of $N_A \cong 7 \times 10^{16}$ cm$^{-3}$, a compensating donor density of $N_D \cong 10^{15}$ cm$^{-3}$ and a boron ionization energy of $E_A \cong 368$ meV.

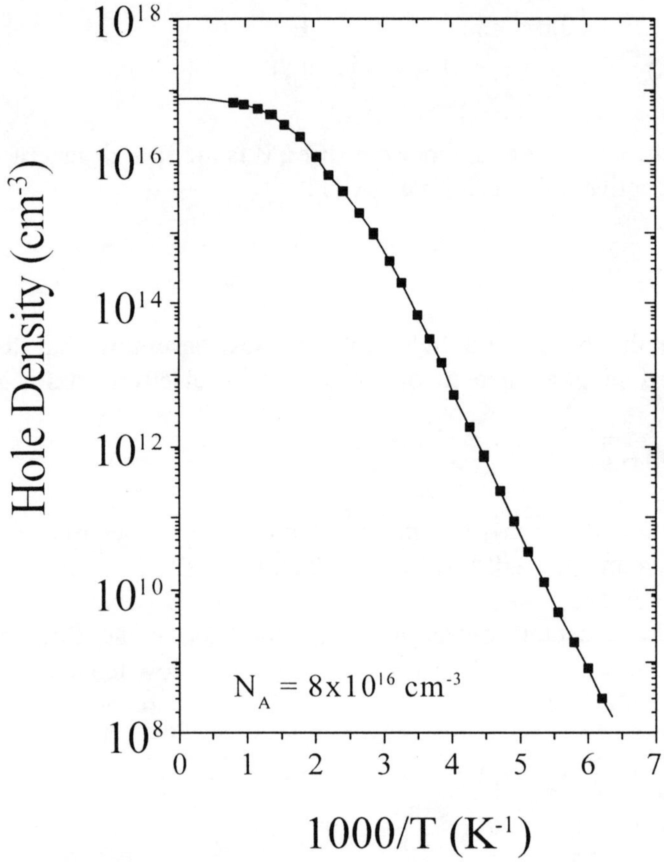


FIGURE 1 Hole concentration in the valence band as a function of reciprocal temperature [7]. The acceptor density is $8 \times 10^{16}$ cm$^{-3}$ and the compensating donor density $10^{15}$ cm$^{-3}$.

If the acceptor density $N_A$ is significantly larger than the compensating donor density $N_D$ ('nearly uncompensated semiconductor') and for the case that the free hole density in the valence band $p \gg N_D$ (which is the case at sufficiently high temperatures), the temperature dependent variation of p is given by [1]:

$$p(T) \cong \left\{ \frac{N_V}{\beta}(N_A - N_D) \right\}^{1/2} \exp\left( -\frac{E_A}{2k_B T} \right) \tag{7}$$

where $N_V$ is the effective density of valence band states, which varies with temperature as $N_V = N_{Vo}T^{3/2}$.

The typical dependence of the conductivity on reciprocal temperature for different boron concentrations is shown in FIGURE 2 [8]. The measured electrical conductivity can be described phenomenologically by a combination of various transport mechanisms:

$$\sigma = \sum^{i} \sigma_i \tag{8}$$

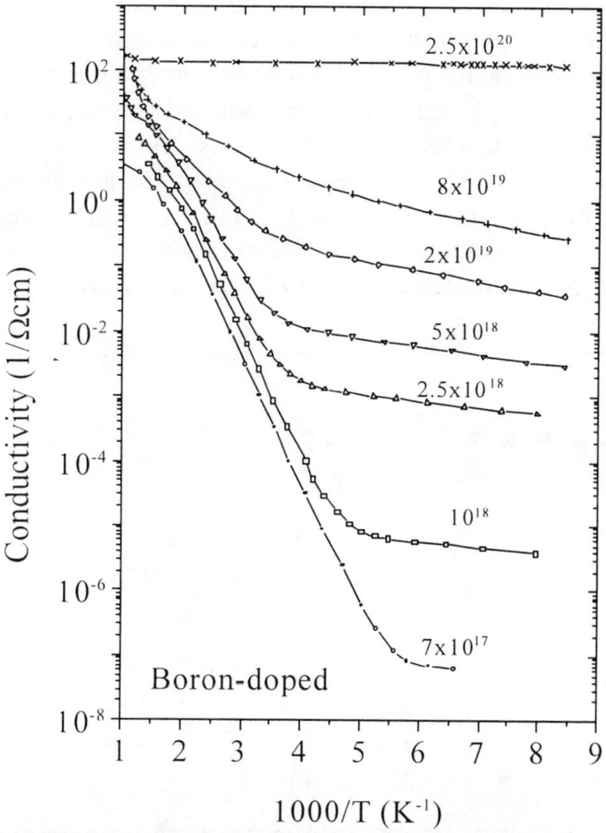


FIGURE 2 Conductivity data of boron doped diamond [8]. The acceptor density is indicated in the plot in units of $cm^{-3}$. The hopping regime at low temperature is followed by a thermally activated regime where $E_{act}$ = 370 meV. With increasing doping the onset of hopping shifts towards higher temperatures.

In the high temperature regime, holes are thermally excited from acceptor states into the valence band. In this regime, σ is thermally activated with $E_{act}$ = 370 meV, the acceptor ionization energy. At lower temperatures or higher doping densities, holes propagate via variable range hopping first discussed by Mott [9] where:

$$\sigma = \sigma_{hop} \exp\left\{ -\left( \frac{T_o}{T} \right)^{1/4} \right\} \tag{9}$$

with $T_o$ given by:

$$T_o = \frac{18.1}{k_B D(E_F) \alpha^3} \tag{10}$$

$D(E_F)$ is the density-of-states at the Fermi level $E_F$ and α is the effective Bohr radius of boron acceptor levels. The conductivity pre-factor $\sigma_{hop}$ can be calculated by:

$$\sigma_{hop} = e^2 D(E_F) R^2 \nu_{ph} \tag{11}$$

where R is the average hopping distance and $\nu_{ph}$ the corresponding exchange frequency. Based on these equations, reasonable fits have been achieved for the low temperature data (see for example [8,10,11]) which result in a Bohr radius $\alpha = (0.3 - 0.8)$ nm. At room temperature, hopping dominates transport at doping levels $\geq 5 \times 10^{19}$ cm$^{-3}$. Here the activation energy drops rapidly towards zero (FIGURE 3), in conjunction with a decrease of the resistivity over three orders of magnitude (FIGURE 4). Also shown in FIGURE 3 is the evolution predicted by the model of Pearson and Bardeen [12]. At acceptor densities exceeding $3 \times 10^{20}$ cm$^{-3}$, a metal-insulator transition takes place [8,13]. Such heavily doped samples show an increasing resistivity with increasing temperature. The metallic resistivity in boron doped diamond is about $10^{-3}$ to $10^{-2}$ Ω cm.

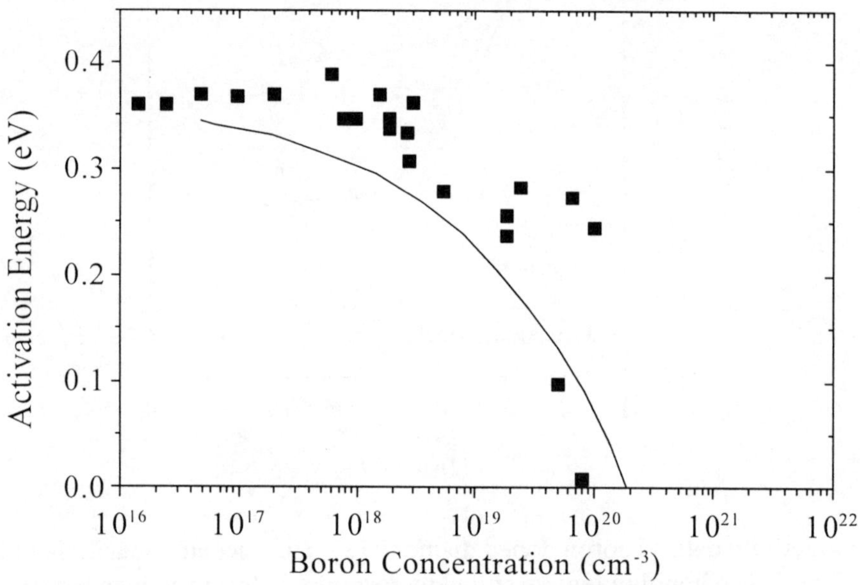


FIGURE 3 Conductivity activation energies at 300 K as a function of boron content. At about [B] > 3 × $10^{18}$ cm$^{-3}$ the activation energy starts to decrease and finally vanishes for [B] > 2 × $10^{20}$ cm$^{-3}$. The theoretical dependence according to the model of Pearson and Bardeen (solid curve) is given for comparison (solid line) [12].

## C CONDUCTIVITY IN n-TYPE DIAMOND

Type Ib diamonds have an optical absorption threshold at about 2 eV [14] which is attributed to the photoionization of substitutional nitrogen acting as a deep donor [15]. Based on the photoconductivity properties it is generally assumed that the dark conductivity in Ib-diamond is also n-type. Typical temperature dependent variations of the resistivity in several Ib-diamond samples are shown in FIGURE 5 [8,16]. At sufficiently high temperatures, > 500 K, the resistivity is activated with 1.7 eV, the thermal ionization energy of substitutional nitrogen. The resistivity in type IIa diamond is about three orders of magnitude higher, but shows the same activation energy of 1.7 eV (see FIGURE 5). Although nitrogen in natural diamond is mostly present in aggregated forms, the presence of small concentrations of isolated substitutional nitrogen, which acts as a deep donor, can be detected by electron spin resonance and photoluminescence [17]. Obviously, a small number of donors is sufficient to dominate the conductivity in type Ib and IIa diamonds. This result, however, can not be generalized. Redfield [18] for example detected by photo-Hall experiments in several highly resistive layers (comparable to type IIa diamond) a positive Hall-voltage which indicates that at least the photoresponse of such diamonds can be dominated by holes.

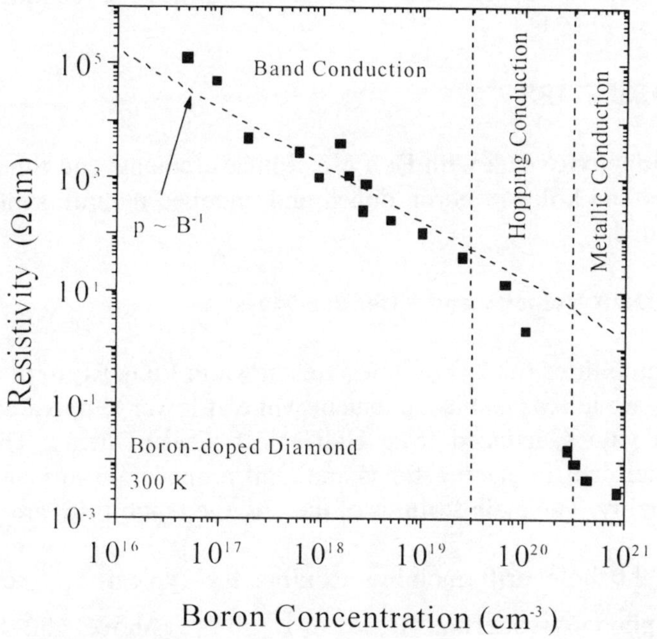


FIGURE 4 Variation of the resistivity of boron doped diamond at 300 K with boron doping levels. Below about $2 \times 10^{19}$ cm$^{-3}$, the decrease is inversely proportional to the acceptor density. In the range $2 \times 10^{19}$ to $3 \times 10^{20}$ cm$^{-3}$ hopping becomes dominant. For higher concentrations metallic conductivity is observed.

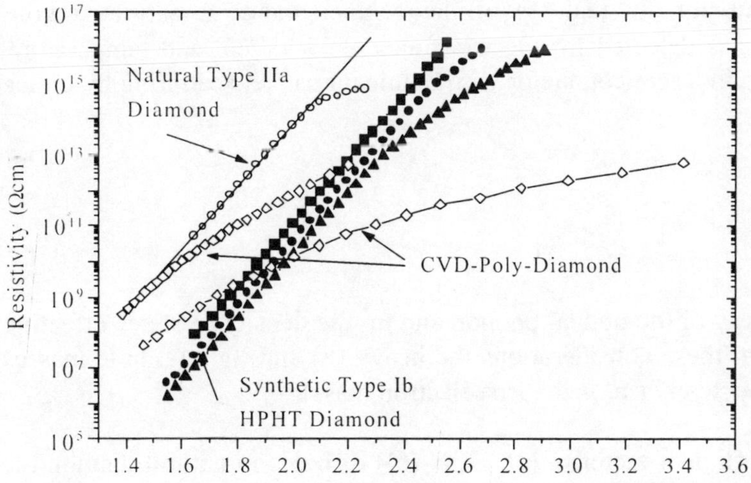


FIGURE 5 Resistivity of n-type diamond. Natural type IIa samples show the highest resistivity which is activated with 1.7 eV [16]. Several Ib-synthetic HPHT diamonds also show 1.7 eV activation energies but they are about three orders of magnitude lower in resistivity [8]. Transport in polycrystalline CVD-diamond is activated with energies increasing with increasing temperature [19].

FIGURE 5 also shows the dark conductivities of typical high resistive intrinsic CVD-polycrystalline diamond layers [19]. No singly activated behaviour is observed. At T > 650 K, the activation energy $E_{act}$ is about 1.5 eV. The different conductivity properties of poly-diamond are attributed to conduction in graphitic states at grain boundaries, which have a continuous density of states distribution in the bandgap [20,21].

## D CARRIER MOBILITIES

By combining basic conductivity data with Hall effect measurements and time-of-flight experiments, the mobility of electrons and holes in boron doped and undoped natural, synthetic HPHT and CVD diamond can be determined.

### D1 Hole Mobility, Drift Velocity and Effective Mass

The drift [22] and Hall mobilities [23,24] of holes are shown in FIGURE 6. In the temperature regime > 400 K both mobilities are in reasonable agreement while at lower temperatures the drift mobility is about 40% higher than values deduced from Hall effect measurements. This is attributed to the difference between carrier drift in electric fields and Hall propagation in magnetic fields or to some differences in sample quality. The main features of the observed mobilities are:

1) Below 400 K, the hole drift mobility exhibits the typical $T^{-3/2}$ temperature dependence expected for acoustic phonon scattering: $\mu_a \propto (m^*)^{-5/2} T^{-3/2}$. Above 400 K, the slope becomes steeper, showing a $T^{-s}$ dependence with s = 2.7 – 2.9 [25]. This is interpreted by the onset of optical phonon scattering [22].

2) For intrinsic diamond, the hole mobility at 290 K is in the range 1200 – 2000 $cm^2/V$ s [7,18,26]. With increasing electric field strength the drift velocity of holes exhibits a sublinear increase [22] which is characteristic for covalent semiconductors (FIGURE 7) [27]. For T ≤ 300 K an anisotropic drift velocity is observed, with $v_{d<100>} \geq v_{d<110>}$. Reggiani et al [28] attribute this result to the warped and nonparabolic features of the valence band. A saturated hole drift velocity of 1.1 (±0.1) × $10^7$ cm/s is detected for electric fields ≥ 25 kV/cm and temperatures T < 300 K. The limiting drift velocity in a semiconductor with dominant carrier scattering by optical phonons is given by [29]:

$$v_{sat} = \left[\frac{8E_{opt}}{3\pi m^*}\right]^{1/2} \qquad (12)$$

where $E_{opt}$ is the energy of the optical phonon and $m^*$ the density of states effective mass. Reggiani et al [22] calculate from these considerations the heavy (h) and light (l) hole masses to be $m_h$ =1.1 $m_o$ and $m_l$ = 0.3 $m_o$, respectively ($m_o$ is the free electron mass).

3) The Hall coefficient factor $r_H$ (cf. EQN (6)) of holes in natural diamond is less than unity in the temperature range 100 K < T < 1000 K [30] .

In comparison to the mentioned drift mobility data, Reggiani et al [30] applied pseudo-potential band structure calculations to determine the effective hole masses in diamond. The results are: $m_h$ = 1.08 $m_o$, $m_l$ = 0.36 $m_o$ and $m_{so}$ = 0.15 $m_o$ for the heavy (h), light (l) and split-off (so) band holes. Rauch [31] measured the effective hole masses in diamond by cyclotron resonance experiments. He was able to differentiate between two valence bands with masses of 0.7 $m_o$ (light hole) and about 2.1 $m_o$ (heavy hole) and a split-off mass of 1.06 $m_0$. So far, these results have not been verified by other workers although several attempts were made with no success. Thus, a large unexplained difference remains between theoretical predictions and experimental results.

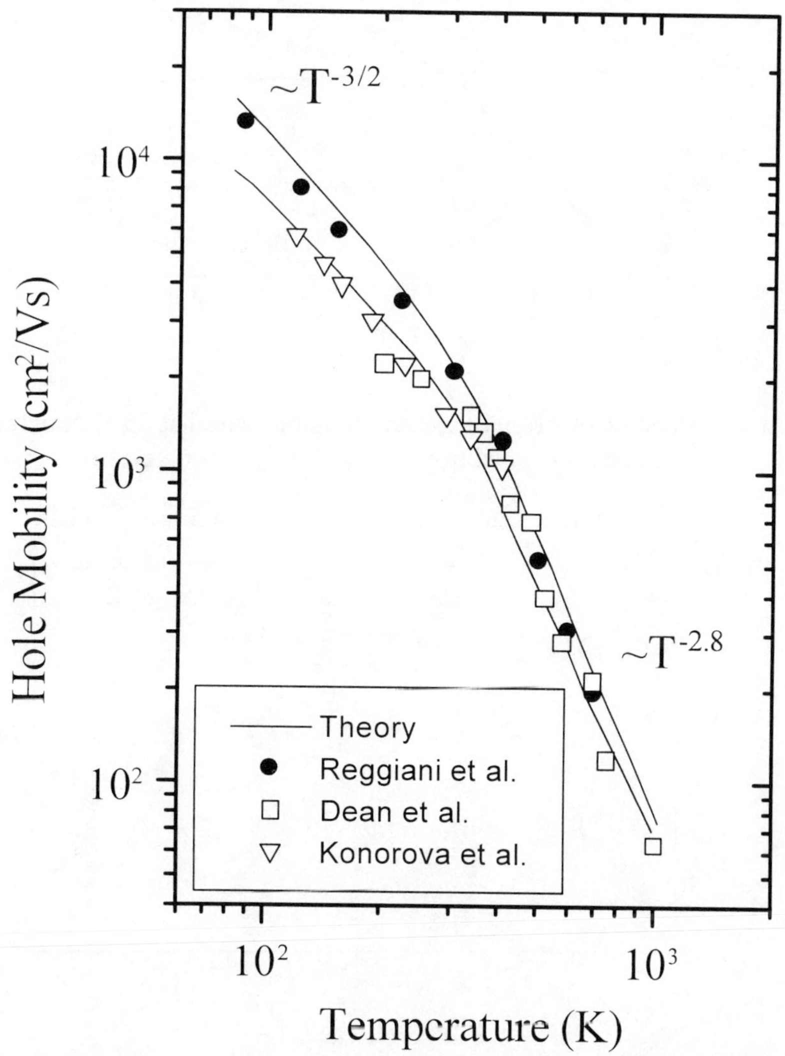


FIGURE 6 Hole mobilities as a function of temperature in natural diamond. The data have been determined by Hall effect (Dean [23], Konorova [24]) and time-of-flight experiments (Reggiani et al [22], drift mobility). The continuous curve refers to theoretical calculations by Reggiani et al [22].

Hole mobility data versus hole concentrations are shown in FIGURE 8 (data from [32–35]). As expected, $\mu_h$ decreases monotonically with increasing doping from about 1000 cm$^2$/V s (hole density: $10^{13}$ cm$^{-3}$) to 8 cm$^2$/V s (hole density: $7 \times 10^{16}$ cm$^{-3}$). The mobility resulting from a dominant ionized impurity scattering $\mu_i$ is [36]:

$$\mu_i \propto \left(m^*\right)^{-1/2} \frac{T^{3/2}}{N_i} \qquad (13)$$

where $N_i$ is the ionized impurity density. At room temperature and hole densities larger than $8 \times 10^{15}$ cm$^{-3}$ impurity scattering $\mu \propto 1/N_i$ is dominant as shown in FIGURE 8 (dashed line).

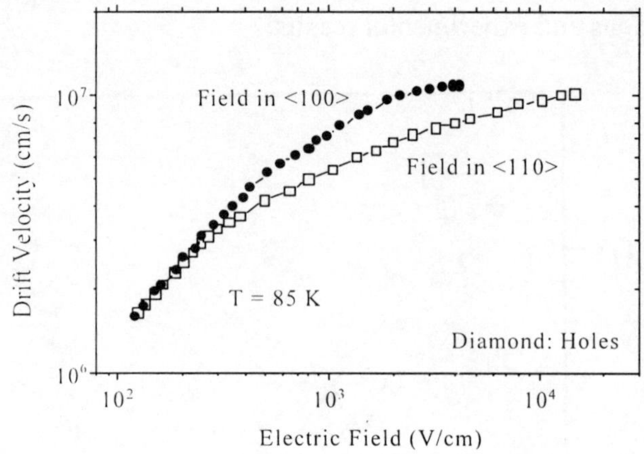


FIGURE 7 Hole drift velocity as a function of electric field measured at 85 K in natural diamond [22]. A characteristic anisotropy $v_{<100>} > v_{<110>}$ is detected.

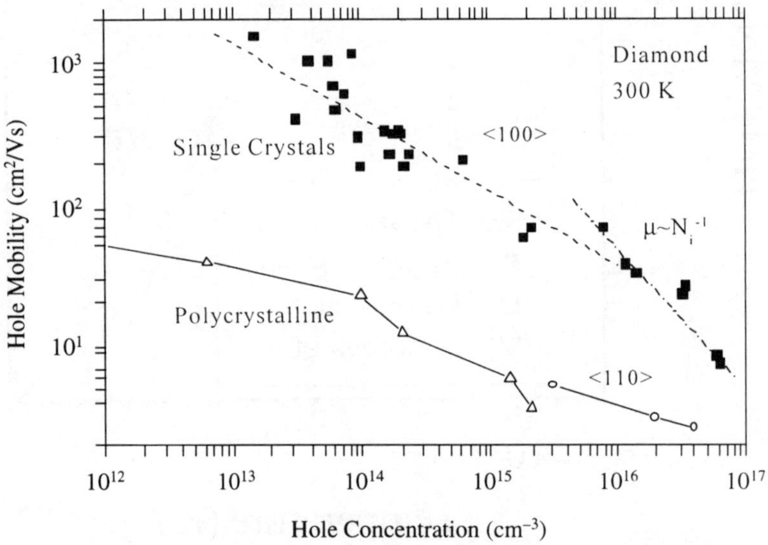


FIGURE 8 Relation between hole mobilities and hole concentration in boron doped monocrystalline diamond grown on <100> and <110> diamond substrates. The hole mobility in polycrystalline diamond is significantly smaller than in monocrystalline layers. The data are from [31–34].

The quality of boron-doped homoepitaxially grown CVD-layers strongly depends on the orientation of the used diamond substrate. The growth on <100> oriented substrates produces a higher quality (higher hole mobility) than on <110> layers, where a high density of stacking faults is present which accounts for the very low mobilities measured in these films (see FIGURE 8) [37].

In polycrystalline films the hole mobility is in the range (1 – 30) $cm^2/V$ s for T = 300 K [38], whereas in highly oriented <100> polydiamond films mobilities up to 229 $cm^2/V$ s have been reported [39]. The properties of carrier transport in polycrystalline diamond are affected by grain boundaries, where band bending generates energy barriers of height $E_B$. The mobility is then given by [40]:

$$\mu = \mu_{bulk} \exp\left[-\frac{E_B}{k_B T}\right] \qquad (14)$$

where $\mu_{bulk}$ is the mobility in the grain. Depending on the electronic properties of the grains and the energy barrier at the boundaries, mobilities can be expected which vary between zero (high energy barrier, low bulk mobility) and values detected in monocrystalline diamond of comparable doping (negligible energy barrier due to degenerate doping).

## D2 Electron Mobility, Drift Velocity and Effective Mass

Electron mobilities determined by time-of-flight experiments in natural diamond of high resistivity ($\rho > 10^{14}$ $\Omega$ cm) with less than $10^{19}$ $cm^{-3}$ nitrogen are shown in FIGURE 9 [41]. For comparison Hall-mobilities are also shown [18,24]. The main features of electron transport have been discussed by Nava et al [41] and can be summarized as follows:

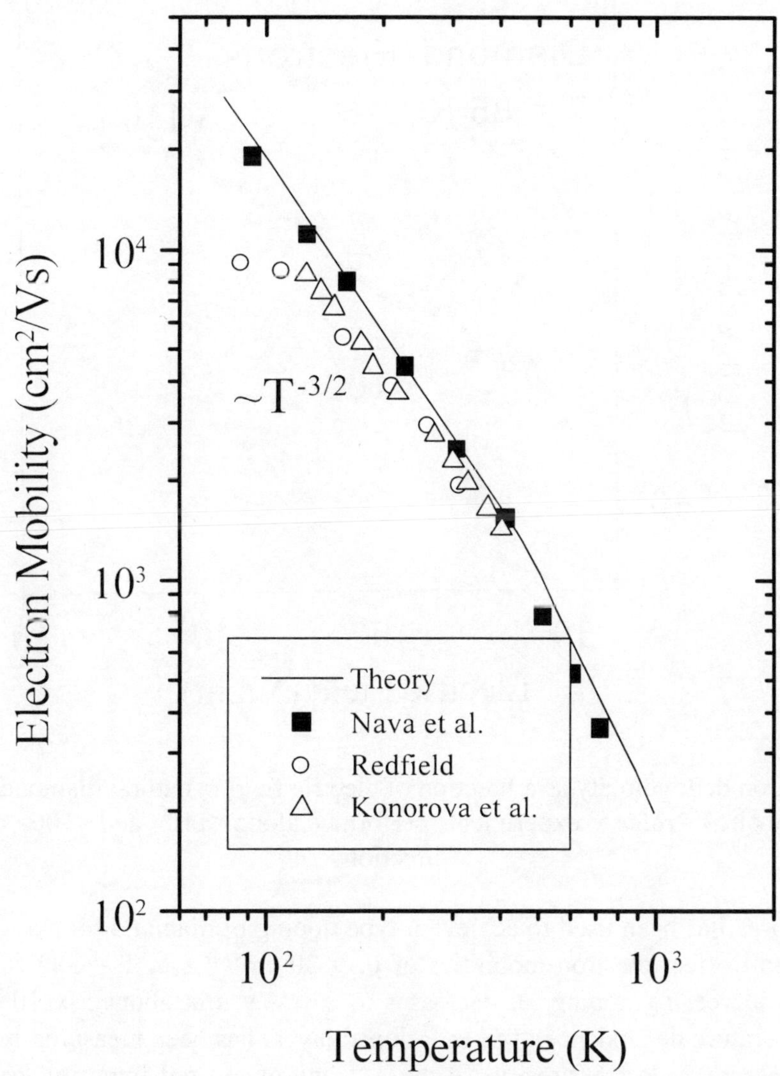


FIGURE 9 Electron mobilities as a function of temperature in natural diamond. The data are from Redfield [18] (Hall effect), Konorova [24] (Hall effect) and Nava [41] (time-of-flight). The solid line refers to theoretical calculations of Nava et al.

1) At 300 K, the electron mobility is about 2000 $cm^2/V$ s. As for holes, at temperatures < 400 K a typical $T^{-3/2}$ dependence is found (acoustic phonon scattering), while above 400 K the temperature dependence also becomes stronger. This is attributed to intervalley scattering.

2) The electron drift velocity versus electric field shows an anisotropy with $v_{<100>}$ larger than $v_{<110>}$ (FIGURE 10). This anisotropy is characteristic for a multi-valley band structure with ellipsoidal constant energy surfaces elongated along equivalent <100> directions. The saturation drift velocity is $1.5(\pm0.1) \times 10^7$ cm/s for all temperatures below 300 K.

3) Based on drift mobility data, Nava et al have estimated the transverse and longitudinal effective mass of electrons in diamond to be $m_t = 0.36\ m_o$ and $m_l = 1.4\ m_o$, which is in reasonable agreement with results from band structure calculations [42,43].

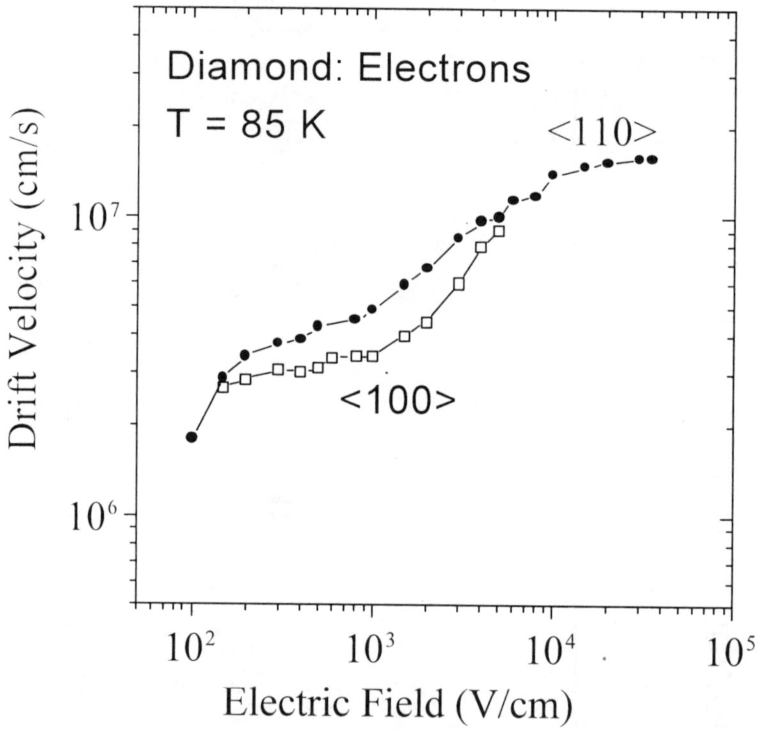


FIGURE 10 Electron drift velocity as a function of electric field in natural diamond [41] at 85 K. The closed and open circles refer to experiments performed along <110> and <100> crystallographic directions.

Recently, phosphorus has been used to achieve n-type doping of diamond [44,45]. Phosphorus doped diamond shows Hall effect electron mobilities of $\mu_e$ = 30 $cm^2/V$ s at T = 300 K for $2 \times 10^{18}$ $cm^{-3}$ phosphorus. With increasing doping, $\mu_e$ decreases to 2 $cm^2/V$ s at about $5 \times 10^{19}$ $cm^{-3}$ phosphorus content. The temperature dependence of $\mu_e$ in P-doped layers has been measured in the regime 300 to 400 K. The increase of $\mu_e$ is too steep to fit the $T^{3/2}$ law of ionized impurity scattering. Obviously more data have to be accumulated before a detailed transport model in phosphorus doped diamond can be developed.

## E CONCLUSION

This Datareview has reviewed the transport properties of diamond. These are determined by intrinsic defects, by contaminants or dopants (nitrogen, boron, phosphorus) and by structure (e.g. polycrystallinity). Boron is an acceptor while phosphorus is a donor impurity. Nitrogen, in isolated substitutional sites, acts as a deep donor. Data are presented for n- and p-type conductivities, mobilities, drift velocities and effective masses.

## REFERENCES


[1] J.S. Blakemore [ *Semiconductor Statistics* (Pergamon, Oxford, UK 1962) ]

[2] A.T. Collins [ *Ceram. Int. (UK)* vol.22 (1996) p.321 ]

[3] S.M. Sze [ *The Physics of Semiconductor Devices* 2nd edition (Wiley, New York, USA, 1981) ]

[4] C.J. Rauch [ *Phys. Rev. Lett. (USA)* vol.7 (1961) p.83 ]

[5] C.J. Rauch [ *The Physics of Semiconductors* (The Institute of Physics and Physical Society, London, 1962) p.276 ]

[6] K. Seeger [ *Semiconductor Physics, An Introduction* (Springer, Berlin, 1973) ]

[7] A.T. Collins, A.W.S. Williams [ *J. Phys. C, Solid State Phys. (UK)* vol.4 (1971) p.1789 ]

[8] T.H. Borst, O. Weis [ *Phys. Status Solidi A (Germany)* vol.154 (1996) p.423 ]

[9] N.F. Mott [ *Philos. Mag. (UK)* vol.19 (1969) p.835 ]

[10] B. Massarani, J.C. Bourgoin, R.M. Chrenko [ *Phys. Rev. B (USA)* vol.17 (1978) p.1758 ]

[11] Y.F. Tsay, K.P. Ananthanarayanan, P.J. Gielisse, S.S. Mitra [ *J. Appl. Phys. (USA)* vol.43 (1972) p.3677 ]

[12] G.L. Pearson, J. Bardeen [ *Phys. Rev. (USA)* vol.75 (1949) p.865 ]

[13] A.W.S. Williams, E.C. Lightowlers, A.T. Collins [ *J. Phys. C, Solid State Phys. (UK)* vol.3 (1970) p.1727 ]

[14] H.B. Dyer, F.A. Raal, L. du Preez, J.H.N. Loubser [ *Philos. Mag. (UK)* vol.11 (1965) p.763 ]

[15] R.G. Farrer [ *Solid State Commun. (USA)* vol.7 (1969) p.685 ]

[16] J.W. Vandersande, L.D. Zoltan [ *Surf. Coat. Technol. (Switzerland)* vol.47 (1991) p.392 ]

[17] A.T. Collins [ *Ind. Diam. Rev. (UK)* vol.34 (1974) p.131 ]

[18] A.G. Redfield [ *Phys. Rev. (USA)* vol.94 (1954) p.526 ]

[19] C.E. Nebel [ unpublished ]

[20] M. Nesladek, M. Vanecek, L.M. Stals [ *Phys. Status Solidi A (Germany)* vol.154 (1996) p.283 ]

[21] E. Rohrer et al [ *Diam. Relat. Mater. (Switzerland)* vol.7 (1998) p.879 ]

[22] L. Reggiani, S. Bosi, C. Canali, F. Nava, S.F. Kozlov [ *Phys. Rev. B (USA)* vol.23 (1981) p.3050 ]

[23] P.J. Dean, E.C. Lightowlers, D.R. Wight [ *Phys. Rev. A (USA)* vol.140 (1965) p.352 ]

[24] E.A. Konorova, S.A. Schevchenko [ *Fiz. Tekh. Poluprovodn. (USSR)* vol.1 (1967) p.299 ]

[25] A.T. Collins [ *Semicond. Sci. Technol. (UK)* vol.4 (1989) p.605 ]

[26] A.T. Collins, E.C. Lightowlers [ in *The Properties of Diamond* Ed. J.E. Field (Academic Press, London, 1979) p.79 ]

[27] C. Jacoboni, L. Reggiani [ *Adv. Phys. (UK)* vol.28 (1979) p.493 ]

[28] L. Reggiani, D. Waechter, S. Zukotynski [ *Phys. Rev. B (USA)* vol.28 (1983) p.3550 ]

[29] J.L. Moll [ in *Physics of Semiconductors* (Wiley, New York, 1964) ]

[30] L. Reggiani, D. Weechter, S. Zukotynski [ *Phys. Rev. B (USA)* vol.28 (1983) p.3550 ]

[31] C.J. Rauch [ in *Proc. 6th Int. Conf. on the Physics of Semiconductors* Ed. A.C. Stickland (The Institute of Physics and the Physical Society, London, 1962) p.276 ]

[32] N. Fujimori, H. Nakahata, T. Imai [ *Jpn. J. Appl. Phys. (Japan)* vol.29 (1990) p.824 ]

[33] E.P. Visser, G.J. Bauhi, G. Janssen, W. Vollenberg, W.J.P. van Enckevort, L.J. Gilling [ *J. Phys. Condens. Matter (UK)* vol.4 (1992) p.7365 ]
[34] T.H. Borst, O. Weis [ *Diam. Relat. Mater. (Switzerland)* vol.4 (1995) p.948 ]
[35] J.A. von Windheim, V. Venkatesan, D.M. Malta, K. Das [ *J. Electron. Mater. (USA)* vol.22 (1993) p.391 ]
[36] E. Conwell, V.F. Weisskopf [ *Phys. Rev. (USA)* vol.77 (1950) p.388 ]
[37] G.S. Gildenblat, S.A. Grot, A. Bazian [ *Proc. IEEE (USA)* vol.79 (1991) p.647 ]
[38] J.T. Huang, W.H. Guo, J. Hwang [ *Appl. Phys. Lett. (USA)* vol.68 (1996) p.3784 ]
[39] B.A. Fox, M.L. Harsell, D.M. Malta, H.A. Wynands, G.J. Tessmer, D.L. Dreifus [ *Mater. Res. Soc. Symp. Proc. (USA)* vol.416 (1996) p.319 ]
[40] C.H. Seager [ *Annu. Rev. Mater. Sci. (USA)* vol.15 (1985) p.271 ]
[41] F. Nava, C. Canali, C. Jacoboni, L. Reggiani, S.F. Kozlov [ *Solid State Commun. (USA)* vol.33 (1980) p.475 ]
[42] W. van Haeringen, H.G. Junginger [ *Solid State Commun. (USA)* vol.7 (1969) p.1135 ]
[43] J.A. van Vechten [ *Bull. Am. Phys. Soc. (USA)* vol.237 (1972) ]
[44] S. Koizumi, M. Kamo, Y. Sato, H. Ozaki, T. Inuzuka [ *Appl. Phys. Lett. (USA)* vol.71 (1997) p.1065 ]
[45] S. Koizumi [ *Diam. Relat. Mater. (Switzerland)* vol.7 (1998) p.540 ]

CHAPTER A2

# ASSESSING DIAMOND 'QUALITY'

# A2.1 Thermal characterisation of diamond

M. Reichling and J. Hartmann

February 2000

## A INTRODUCTION

Thermal characterisation of diamond is one of the key issues in assessing material quality since thermal parameters are strongly related to other features like isotopic constitution, defect content and microstructure (see Datareview A1.5). Hence, the macroscopic thermal conductivity can often be taken as a figure-of-merit for diamond quality relevant for industrial applications. Determining thermal conductivity of diamond materials does not require, in principle, techniques other than those used for thermal assessment of more common materials and the standard methods as well as their application to diamond have been described in detail in rather complete review articles [1,2]. Recent advancements in methodology have mainly been focused on thermal microscopy for measuring conductivity in extremely small volumes and finally single grains of polycrystalline material [3] and to quantitatively determine resistances across thermal barriers related to grain boundaries in thin films [4]. In this survey we summarise the main features of the standard techniques and briefly review the technology of quantitative thermal microscopy.

It is difficult to precisely control and measure heat flow and radiation and, hence, thermal measurements generally yield only a rather limited precision; this fully applies also to measurements on diamond materials. An accuracy of a few per cent has to be regarded as excellent while the precision of most measurements is at a 5 to 10% level or below. As demonstrated by a recent round robin test [5], this especially applies to the characterisation of polycrystalline thin films where measurements may be strongly affected by the presence of a substrate, sample inhomogeneities, microstructure and subtle experimental details. This test also revealed significant differences between results obtained on the same sample. Hence, thermal characterisation of diamond materials is presently well developed for a rough quality assessment; however, the precise determination of thermal parameters by standardised routine procedures is not yet available.

The choice of the method for thermal characterisation of a specific sample is, therefore, mostly not motivated by the obtainable level of accuracy but by other criteria related to the characteristics of the specimen, the desired information, the temperature range and practical or economic considerations (for detailed selection tables see [1]). Steady state techniques as described in Section B mostly do not require sophisticated equipment and analysis and are normally applied to large bulk samples. Dynamic methods described in Section C allow the restriction of the measured heat flow to a small volume and are, therefore, well suited for small samples and thin films. Heating of the sample can conveniently be provided by pulsed or modulated laser light which led to the development of photothermal techniques treated in Section D. Their strength is the non-contact determination of thermal parameters. The best resolution is obtained by photothermal microscopy as reviewed in Section E.

## B STEADY STATE TECHNIQUES

The physical property measured with steady state techniques is directly the thermal conductivity κ that is according to Fourier's empirical law for the conduction of heat [6] defined by the equation:

$$\mathbf{q} = -\kappa \nabla T$$

where **q** denotes the heat flux vector, i.e. the directed thermal energy per unit time and area, and $\nabla T$ the related temperature gradient. Heat flow is generally three-dimensional and $\kappa$ a tensor with components that have all to be determined for a complete characterisation. For practical purposes, however, the problem is reduced to one or two dimensions by a proper choice of experimental conditions and only one or two tensor components are determined. For a plate-like sample or a thin film these are the lateral thermal conductivity describing thermal transport parallel to the plane of the film and the perpendicular thermal conductivity, characterising out-of-plane heat flow.

The most straightforward approach to measuring thermal conductivity in a one-dimensional arrangement is the heated bar method [7], sketched in FIGURE 1. The sample in the form of a rectangular plate is attached to a thermal reservoir held at constant temperature $T_0$ on one side and heated by a line heater H1 on the other side. The temperature gradient is determined by several thermocouples attached along the sample. By operating the device in vacuum and using very thin leads conductive heat losses to the surrounding atmosphere and the attached devices can be reduced to a minimum. Distortions due to radiative heat loss that become an issue for measurements at temperatures above room temperature can be compensated by operating a second heater H2 and extending data analysis accordingly [7]. With these precautions, the method is well suited for the determination of conductivity in bulk samples and the lateral conductivity component in free standing thick films.

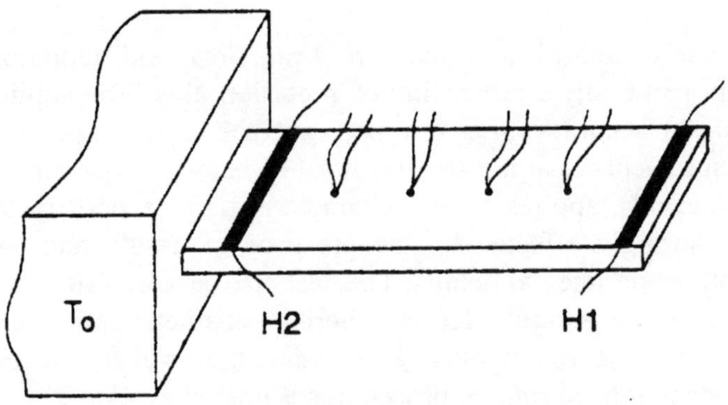


FIGURE 1 Set-up for the heated bar method. The rectangular sample is heated by line elements H1 and H2 and the temperature profile is determined via thermocouples attached along the specimen (after [7]).

The effects of radiation loss can effectively be minimised also by mounting the sample on a highly thermally insulating plate with a high coefficient of thermal reflection [8]. One end of the sample is connected to a thermal reservoir by a good connector like copper while the other end is heated. A further advantage of this insulated plate technique is that heaters and sensors can be integrated into the plate that is thermally attached by grease, reducing sample preparation to a minimum. For irregularly shaped samples it is advantageous to choose an arrangement yielding radial heat flow as shown in the schematic view of FIGURE 2 where the heater element H is formed as a ring, the sample is thermally grounded in the centre and the temperature gradient is monitored by thermocouples DS and SS.

Another way of minimising radiation effects is to reduce the heated area by integrating the sample into a micro device where thin-film heaters and thin-film thermocouples are evaporated directly onto the specimen as illustrated in FIGURE 3. This method is most conveniently applied to thin films on a silicon substrate that is etched away in a small area where the components of the device are attached [9].

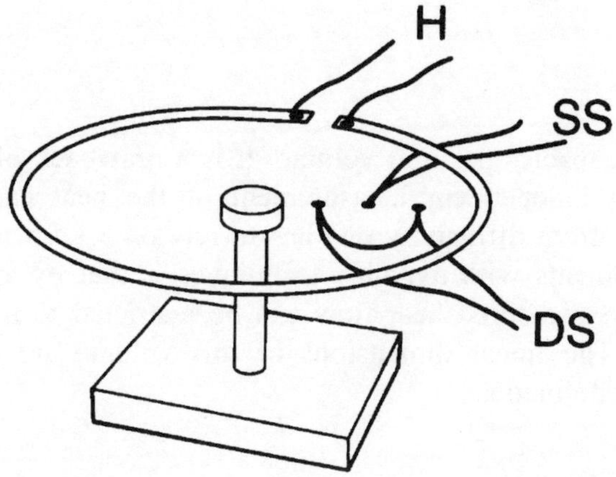


FIGURE 2 Set-up for the insulated plate technique. A sample of arbitrary shape is heated by a circular element H and the radial temperature profile is determined via thermocouples DS and SS. The supporting plate carrying the components is not shown. (After [8].)

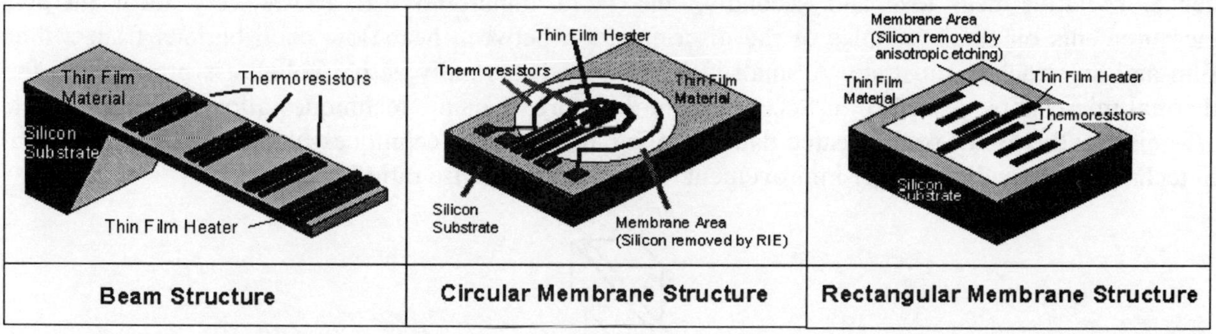


FIGURE 3 Examples of the arrangement of heater elements and thermosensors in micro devices used for thermal characterisation of thin diamond films on silicon substrates (after [9]).

In a geometry where the dominant mechanism of heat loss is radiation, the radiative energy transfer can, on the other hand, be utilised for the measurement of conductivity. Obviously, such a method is best suited for measurements at elevated temperatures [10]. The best geometry for this type of measurement is the radiating bar arrangement illustrated in FIGURE 4. A long, thin sample is clamped at its ends by two posts held at some fixed temperature above the temperature of the environment. Heat loss by radiation from the specimen is replaced by conduction through the bar and thermal conductivity can be determined from the temperature profile along the bar that is measured by a radiation thermometer D moved along the sample. Results may be improved by coating the specimen with a thermally thin film of a known, high emissivity. This technique was found to be especially useful for free-standing thin films of relatively poor thermal conductivity.

## C DYNAMIC TECHNIQUES

Dynamic techniques for measuring thermal parameters are based on periodic thermal excitation in the form of either a series of heat pulses or a harmonic oscillation. The resulting temperature fields varying in time and space are often referred to as thermal waves. The dynamics of thermal wave propagation is not only governed by the conduction properties of the material under investigation but also by its ability to store thermal energy. Therefore, the thermal parameter determined in dynamic measurements is not directly the thermal conductivity κ but the thermal diffusivity D defined as

$$D = \frac{\kappa}{C}$$

where C denotes the heat capacity per unit volume. It is a drawback of dynamic methods that they mostly do not provide an independent measurement of the heat capacity and, therefore, when deducing heat conductivity from diffusivity one has to rely on a known value for the heat capacity. The advantage of measurements with dynamic techniques is that by applying short pulses or high modulation frequencies, the measured heat flow can be restricted to a small volume, the so-called thermally active volume. The linear dimensions of this volume are approximately given by the thermal diffusion length $L_{th}$ defined as

$$L_{th} = \sqrt{2D\tau} \quad \text{and} \quad L_{th} = \sqrt{\frac{2D}{\omega}}$$

for excitation with pulses of duration τ or modulated excitation with a heating frequency ω. A small thermally active volume is of two-fold advantage. First, it can reduce the heated surface area and, hence, radiative heat loss and secondly, and more importantly, it allows very local thermal measurements on small samples or the discrimination between heat flow contributions from a thin film and an attached substrate. A small well controlled thermally active volume is prerequisite for thermal microscopy described in Section E. Furthermore, dynamic techniques allow one to diminish 1/f-noise by the use of sophisticated data sampling and analysis techniques like the box car and lock-in techniques providing further improvement of the signal-to-noise ratio [11].

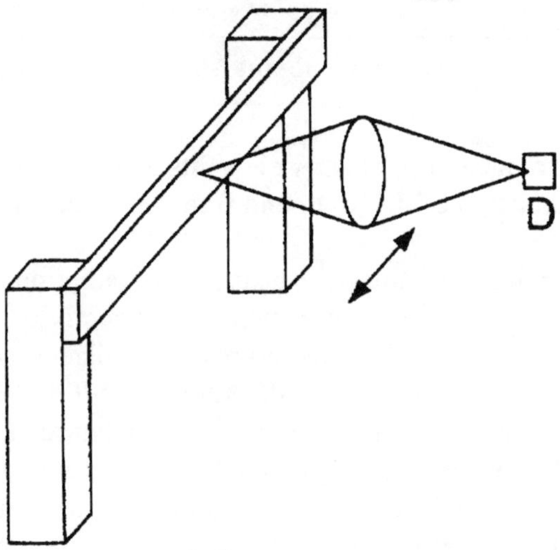


FIGURE 4 Set-up for the radiating bar technique where the temperature profile along a long thin sample is determined by a movable radiometric detector D (after [10]).

The measurement geometries used for dynamic measurements mostly do not differ significantly from those used for static measurements. A one-dimensional geometry of cylindrical heat flow into a bulk sample [12] is often used in conjunction with the 3ω-technique. This is a special method where the same resistive element is used for heating and temperature detection. A temperature field oscillating with frequency 2ω is created by an alternating current through the heater with frequency ω. The oscillating temperature causes a periodic change in resistance of the heater due to the non-zero temperature coefficient of its resistance. This, in turn, influences the electrical power dissipated into the material. Due to the mixing of the frequencies of electrical and thermal oscillation, the voltage across the heater is modulated at frequency 3ω where the amplitude and phase of this oscillation are related to the thermal characteristics of the sample. This technique has successfully been applied to both diamond bulk and film samples [13,14].

## D PHOTOTHERMAL TECHNIQUES

Most methods of dynamic thermal characterisation, however, are not operated with heater elements firmly attached to the sample but heating is provided by absorption of pulsed or modulated laser light. This offers the advantage of a non-contact measurement since temperature detection can also be done without contact to the sample, namely by a thermal radiation detector or a second probing laser beam. Techniques based on these principles are generally referred to as photothermal methods. They often require expensive equipment and sophisticated data analysis; however, besides providing non-contact characterisation, their major advantages can be high spatial resolution, high sensitivity and accuracy, high speed or simple sample handling for routine measurements.

The most classic of pulsed photothermal methods is the laser flash technique [15] with a rather simple set-up schematically depicted in the inset of FIGURE 5. The sample S is heated over a large area on one side by a pulsed laser beam. By covering the specimen with a highly absorptive coating, care is taken that the laser pulse energy is effectively transferred to the sample but only the surface is heated. The time-delayed transient temperature rise on the other side of the sample is monitored by a radiation detector probing a small spot in the centre of the heated area where thc detection sensitivity is increased by covering the back side of the sample with a thermally thin but highly emissive coating. With known sample thickness and heat capacity, the perpendicular component of thermal conductivity can be determined from an analysis of the temperature transient as shown in the main part of FIGURE 5. This is a simple method that is well applicable to large plate-like samples.

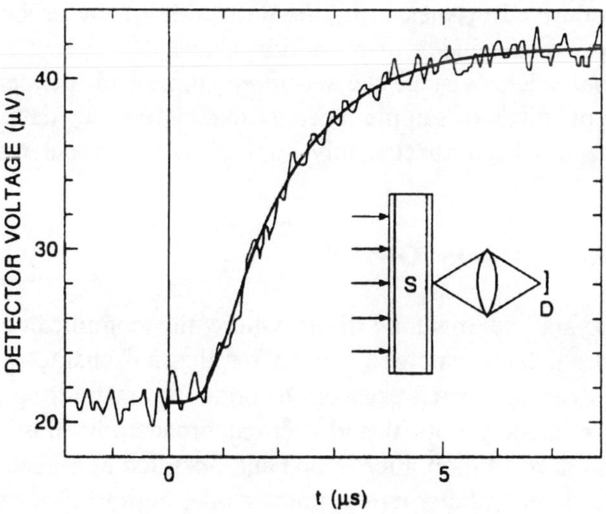


FIGURE 5 Temperature rise (proportional to detector voltage) at the back side of a sample as a function of time after heating the front side by a laser pulse. The inset shows the set-up for the laser flash technique with a coated sample S and the thermal radiation detector D (after [15]).

A sophistication of the laser flash method is the thermal grating technique measuring directional lateral heat flow in small volumes [16,17]. There, the heating laser beam is split into two parts that are reunited in a focal spot on the surface to form an interference pattern and, hence, an array of heated lines. The transient temperature profile created by lateral heat flow perpendicular to these lines is measured by a probe laser beam via thermoreflectance, a phenomenon described in more detail in Section E. The power of this technique is that it is very fast, measures directed heat flow on a ten-micrometre length scale and is potentially useful for detecting local anisotropies in thermal conduction. Prerequisite for obtaining the high resolution, however, is the use of nanosecond or shorter laser pulses.

For modulated photothermal measurements, the sample is mostly heated by a sinusoidally modulated or chopped laser beam focused into a single spot [18]. As for the pulsed radiometric methods, the temperature profile can be measured by infrared radiometry recording temperature variations at a modulation frequency of typically some tens of Hz. However, in this case, amplitude and phase of the temperature variation relative to the excitation are measured as a function of the distance to the heated centre by lock-in detection. A typical set-up involving an infrared scanner camera is shown in the upper part of FIGURE 6. The plots of amplitude and phase of the modulated photothermal signal shown in the lower part directly visualise the effects of lateral heat flow. The temperature amplitude and absolute phase decrease as a function of distance from the centre of excitation. The rapid decrease in amplitude shows the limited range of diffusion during the chopper period while the linear decrease in absolute phase reflects the increasing phase lag between excitation and response at a fixed distance from the centre of excitation and is a direct measure for the diffusivity. The resulting profiles are broader for diamond than for the metal due to the smaller diffusivity of the latter. With spot heating, the heat flow geometry is generally three-dimensional and needs the correspondingly complicated data analysis. However, by a proper choice of sample geometry, laser spot size and modulation frequency, the problem may be reduced to one of one-dimensional heat flow and either lateral or perpendicular conductivity can be extracted.

An alternative way of measuring modulated surface temperature is the mirage effect where the heat flow into the surrounding atmosphere or inside a transparent sample is sensed by a probe beam running parallel to the surface for a small distance or bouncing off the surface as shown in FIGURE 7. Quantitative information is derived by measuring the deflection of the probe beam in the modulated temperature-induced gradient of the index of refraction above the surface. The mirage technique is very popular in photothermal sciences since the set-up is simple and inexpensive and measurements mostly do not require sophisticated sample preparation; however, data analysis is sometimes complicated. The technique has been successfully applied to measurements on bulk and thin film samples [19,20].

## E PHOTOTHERMAL MICROSCOPY

With the advent of polycrystalline diamond films where the complicated microstructure heavily influences thermal conduction, there emerged a need for thermal characterisation on a micrometre length scale. To obtain this resolution with respect to both factors limiting the spatial resolution of photothermal measurements, namely optical and thermal broadening, the photothermal experiment has to be integrated into a high resolution microscope and operated at highest modulation frequencies up to more than 10 MHz [21]. In a photothermal microscope, both excitation and probing beams are focused almost to the diffraction limit by the microscope objective and the temperature measurement is done via thermoreflectance, i.e. based on the fact that the optical reflectivity of most materials depends on temperature. For small amplitudes there is a linear relation between the temperature rise and the relative change in reflectivity of the sample. To exploit this for measurements on diamond, the sample has to be coated with a very thin metallic film providing a reasonably high temperature coefficient for the reflectivity. This film also serves as an absorber for the heating laser light. It should be noted that the thermoreflectance technique requires a specular reflection of the probe laser beam from the sample surface and is, therefore, only applicable to samples with a sufficiently well polished surface.

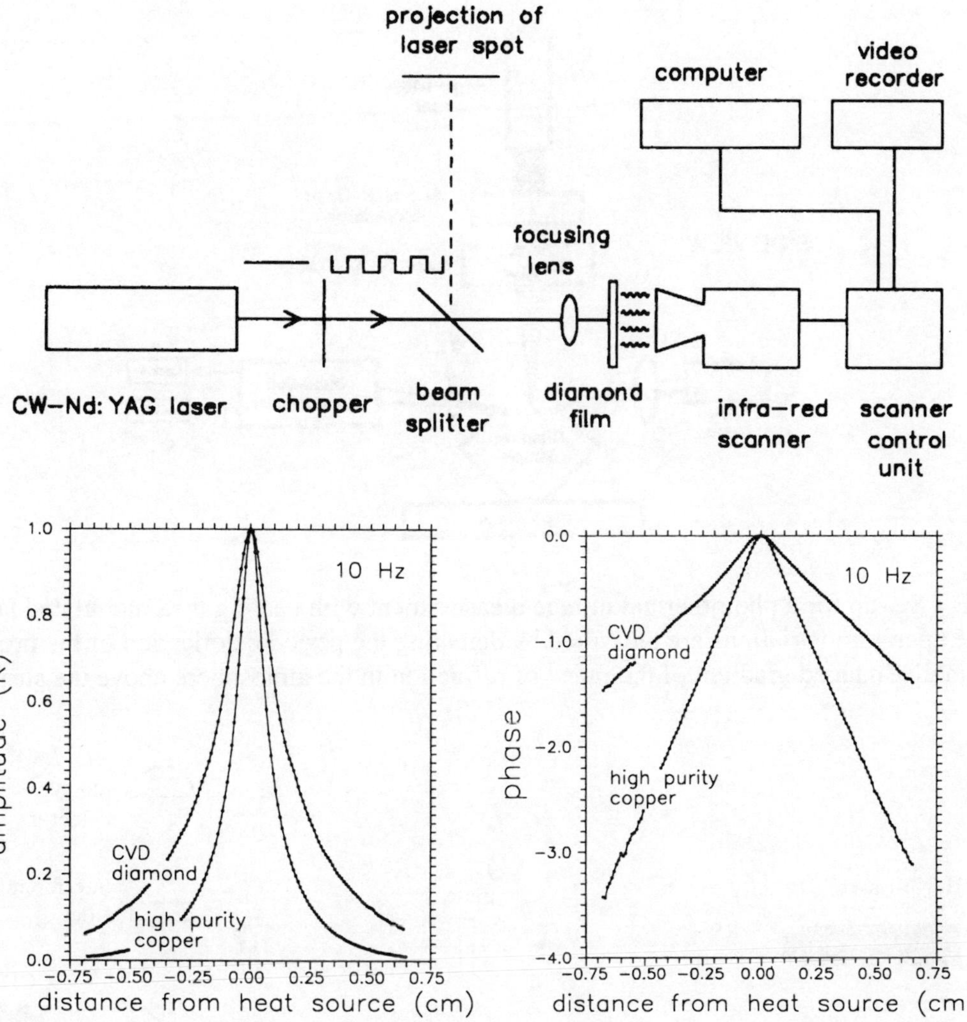


FIGURE 6 Set-up for a thermal diffusivity measurement via modulated thermal radiometry. An infrared scanner with lock-in detector determines amplitude and phase of the thermal waves radially travelling from the excitation in the centre. A comparison of typical thermal profiles between a diamond film and a metal are shown in the lower graphs. (After [18].)

The schematic view of a typical photothermal microscope [21] is shown in FIGURE 8. The heating laser beam supplied by an $Ar^+$-laser is modulated at a high frequency by an acousto-optical modulator and guided into the microscope through a dichroic mirror and a polarising beam splitter and focused on the sample surface by the objective. The probing HeNe-laser beam is united with the pump beam by the dichroic mirror and takes the same path to the sample for the rest of the optical path. The foci of both beams can be offset with respect to each other in a controlled way by moving a lens in the path of the heating beam, while the sample is positioned by another translation stage. Variations in the intensity of the reflected light are measured by a photodetector and processed by a high-frequency lock-in detector. The probe laser light is separated from the reflected pump laser light by an interference filter. The separation of the incoming and reflected probe beams is achieved by the $\lambda/4$ plate in conjunction with the polarising beam splitter. By modulating the pump laser beam also at a low frequency and analysing with a second lock-in detector, the photothermal signal can easily be separated from high-frequency electromagnetic stray fields picked up by the sensitive electronics. The integrated illumination and CCD camera system facilitates sample positioning that is crucial for measurements on polycrystalline material with strong sample inhomogeneities.

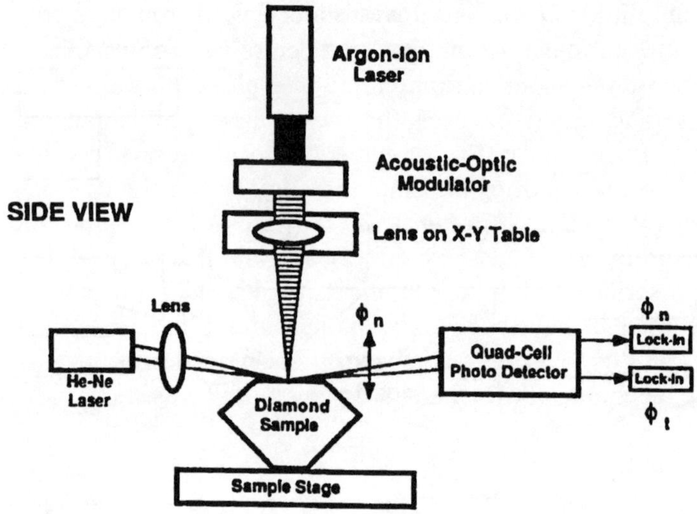


FIGURE 7 Set-up for a photothermal mirage measurement with heating by a modulated laser beam. Surface temperature variations are measured by detecting the periodic deflection of the probe beam in the thermally induced gradient of the index of refraction in the atmosphere above the surface (after [19]).

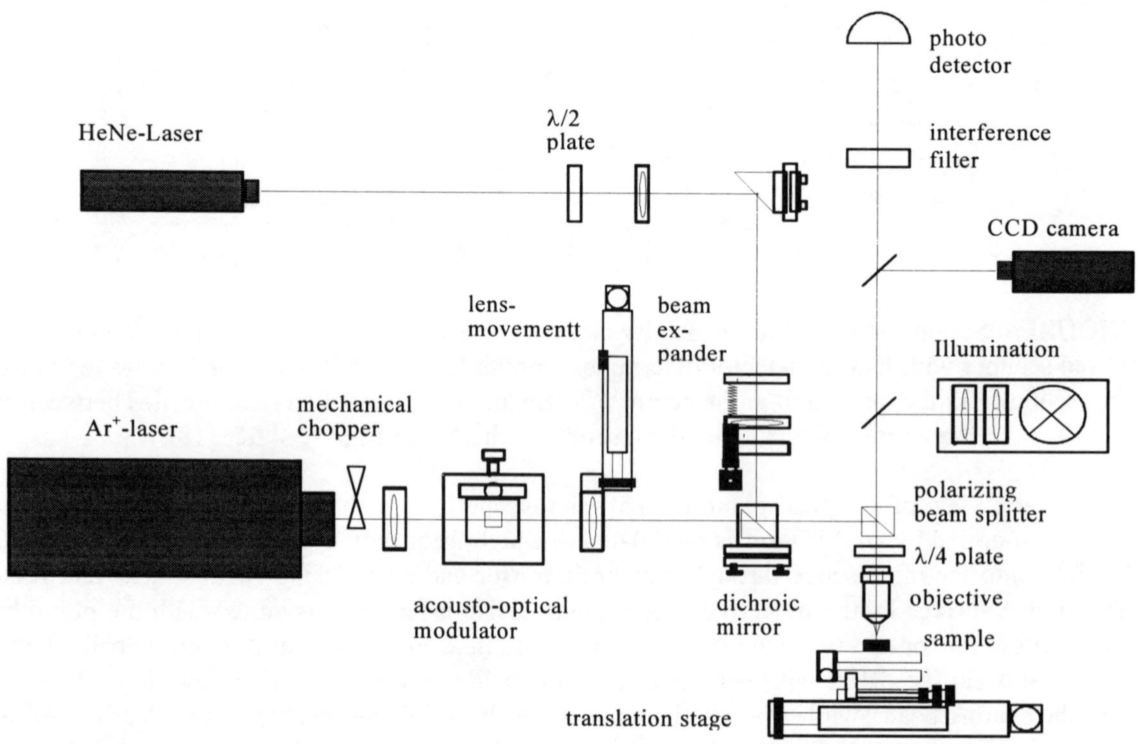


FIGURE 8 Set-up for photothermal microscopy based on thermoreflectance detection. Heating and probing laser beams are guided into a metallurgical microscope allowing focusing to the diffraction limit, precise sample positioning and observation during the measurement (after [21]).

The resolution power of such instrumentation is demonstrated in FIGURE 9. This graph shows amplitude and phase of the photothermal response for a one-dimensional scan of the pump beam with respect to the fixed probe beam in the centre of a grain of a polycrystalline sample. It can be seen that the resulting thermal amplitude profile has a width of a few micrometres only while the phase can be monitored over a distance of about ±5 μm from the centre before vanishing in the noise floor. If the beam is scanned in two dimensions, an amplitude and phase map can be drawn that reflects two-dimensional heat flow and might be used for the detection of lateral anisotropies [22]. From a quantitative analysis of such data one can obtain the local thermal conductivity with a resolution roughly corresponding to the width of the amplitude profile. As the size of grains in polycrystalline samples is often several tens of micrometres, this technique allows the determination of the conductivity inside a grain, unaffected by effects at grain boundaries. The image in FIGURE 10 displays results from a series of such measurements together with their respective positions inside a large grain. The measurements have been taken in the centre of the circles where their radius indicates the lateral resolution. At this level of spatial resolution, measurements can be performed in a dense network and, thus, the technique allows thermal conductivity mapping of individual grains [3].

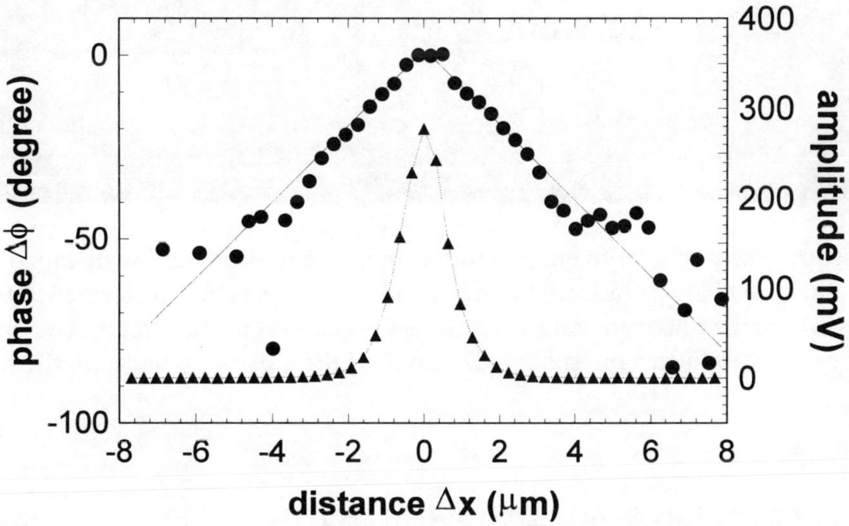


FIGURE 9 Photothermal amplitude and phase profiles obtained in the centre of a diamond grain of a 600 μm thick film sample demonstrating the resolution power of the photothermal microscope (after [21]).

## F CONCLUSION

There is a rich choice of methods for thermal characterisation of diamond in both its single crystal and polycrystalline form suited for a wide range of requirements with respect to sample size, shape and preparation and the desired spatial resolution and measured conductivity component. No general recommendation for a specific technique or a class of techniques can be given but instead a careful trade-off between several physical, technical and economic aspects using the criteria outlined in this Datareview is required to find the best method for a specific situation. However, without excessive experimental effort one cannot expect results with a precision better than 5 to 10% from an analysis with any of the available techniques.

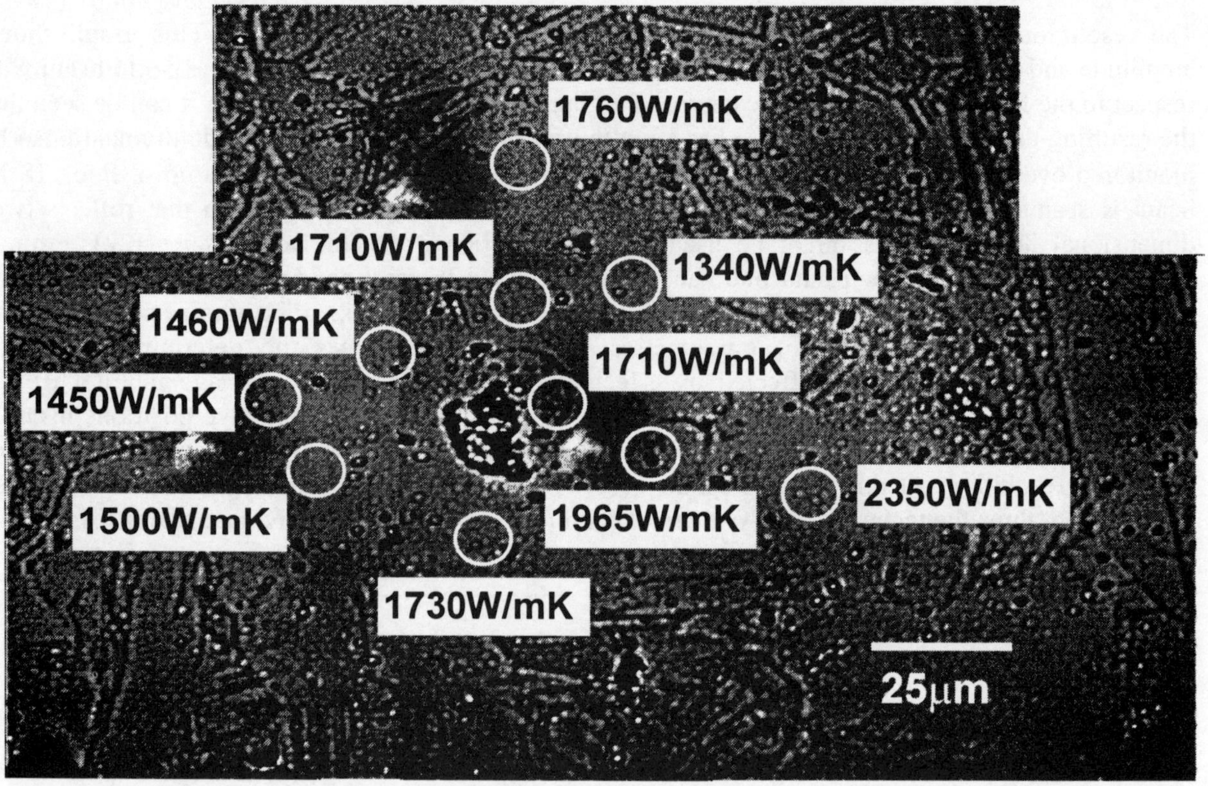


FIGURE 10 Microscopy image of a large grain at the surface of a polycrystalline diamond sample. A microscopic photothermal analysis of local thermal conductivity has been performed at positions indicated by circles where the obtained conductivities are given next to the circles. The diameter of the circles roughly indicates the spatial resolution of the measurement. (After [3].)

## REFERENCES

[1] J.E. Graebner [ *Diam. Films Technol. (Japan)* vol.3 (1993) p.77-130 ]
[2] K. Plamann, D. Fournier [ *Phys. Status Solidi A (Germany)* vol.154 (1996) p.351-69 ]
[3] M. Reichling, T. Klotzbücher, J. Hartmann [ *Appl. Phys. Lett. (USA)* vol.73 (1998) p.756-8 ]
[4] J. Hartmann, M. Costello, M. Reichling [ *Phys. Rev. Lett. (USA)* vol.80 (1998) p.117-20 ]
[5] J.E. Graebner et al [ *Diam. Relat. Mater. (Switzerland)* vol.7 (1998) p.1589-604 ]
[6] H.S. Carslaw, J.C. Jaeger [ *Conduction of Heat in Solids* (Oxford University Press, 1959) ]
[7] J.E. Graebner, J.A. Herb [ *Diam. Films Technol. (Japan)* vol.1 (1992) p.155-64 ]
[8] J.E. Graebner [ *Rev. Sci. Instrum. (USA)* vol.64 (1993) p.3245-7 ]
[9] E. Jansen, E. Obermeier [ *Phys. Status Solidi A (Germany)* vol.154 (1996) p.395-402
[10] A. Ono, T. Baba, H. Funamoto, A. Nishikawa [ *Jpn. J. Appl. Phys. (Japan)* vol.25 (1986) p.L808-13 ]
[11] Z. Chen, A. Mandelis [ *Phys. Rev. B (USA)* vol.46 (1992) p.13526-38 ]
[12] D.G. Cahill [ *Rev. Sci. Instrum. (USA)* vol.61 (1990) p.802-8 ]
[13] J.R. Olson, R.O. Pohl, J.W. Vandersande, A. Zoltan, T.R. Anthony, W.F. Banholzer [ *J. Appl. Phys. (USA)* vol.47 (1993) p.14850-6 ]
[14] T.R. Anthony, J.L Fleischer, J.R. Olson, D.G. Cahill [ *J. Appl. Phys. (USA)* vol.69 (1991) p.8122-5 ]
[15] J.E. Graebner [ *Diam. Relat. Mater. (Switzerland)* vol.4 (1995) p.1196-9 ]

[16] O.W. Käding, H. Skurk, A.A. Maznev, E. Matthias [ *Appl. Phys. A (Germany)* vol.61 (1995) p.253-61 ]

[17] A. Tokamakoff, W.F. Banholzer, M.D. Fayer [ *Appl. Phys. A (Germany)* vol.56 (1993) p.87-90 ]

[18] E.P. Visser, E.H. Vesteegen, W.J.P. van Enckevort [ *J. Appl. Phys. (USA)* vol.71 (1992) p.3238-48

[19] T.R. Anthony et al [ *Phys. Rev. B (USA)* vol.42 (1990) p.1104-11 ]

[20] M. Bertolotti et al [ *J. Appl. Phys. (USA)* vol.75 (1994) p.7795-8 ]

[21] J. Hartmann, P. Voigt, M. Reichling [ *J. Appl. Phys. (USA)* vol.81 (1997) p.2966-77 ]

[22] K. Plamann, D. Fournier, B.C. Forget, A.C. Boccara [ *Diam. Relat. Mater. (Switzerland)* vol.5 (1996) p.699-705 ]

# A2.2 Crystalline quality assessment in diamond by micro-Raman spectroscopy

D. Pickard

January 2000

## A INTRODUCTION

Details of the first and second order Raman spectra of diamond have been given in [1]. These represent an average or macroscopic spectrum from bulk natural diamond. Analysis of the differences from this normal spectrum can be shown to reveal details that relate to the 'quality' of the crystal. Quality is here defined in terms of stresses and strains in the crystal, defect or inclusion density and crystallite size and non-diamond carbon inclusions in chemical vapour deposition (CVD) diamond films. These factors can be related to other quality measures such as optical transmission, mechanical strength, carrier diffusion lengths, carrier lifetimes and macroscopic heat and electrical conduction.

The relationship of each of the observed features in the first order Raman spectrum, line position, line shape and width and the relative intensity of non-diamond Raman peaks, are examined separately below. Micro-Raman spectroscopy is the acquisition of spectra with a spectrometer attached to a confocal microscope. This allows observation of the crystalline quality at a micron level and with the addition of a mapping stage allows the variation in various attributes to be measured throughout a large crystal or CVD film [2].

Raman spectroscopy is a popular and relatively straightforward method of analysis for diamond and diamond films. Many good introductions exist to Raman spectroscopy [3–5] and these contain the details of Raman spectroscopy's dependence on symmetry in the scattering medium, scattering geometry and polarisation of the incident and detected light.

The first order Stokes Raman spectrum of bulk diamond is characterised by a single sharp phonon line which is due to a triply degenerate, $T_{2g}$ symmetrical, zone centre Raman active optical phonon. In unstressed natural diamond at room temperature the relative wavenumber position of this phonon mode has been accurately measured as 1332.4(1) $cm^{-1}$ with a measured line width of between 1.2 and 2.9 $cm^{-1}$ [6–13].

## B LINE POSITION

Although the position of the first order Raman line in natural diamond has been measured reasonably accurately this measurement is very sensitive to systematic errors arising from the spectrometer. It is therefore advisable to calibrate the instrument prior to any and every measurement by taking a reading from a 'reference' sample of natural diamond. However if only the relative shifts of Raman lines from different parts of the crystal are considered then the need for this exact calibration is effectively removed.

The frequency of the first order phonon bands and therefore the relative position of the first order Raman line is related to the mass of the atoms and the length and stiffness of the bonds in the crystal. Therefore the position of the first order line will be sensitive to anything that changes these parameters. The position of the first order band in specially grown isotopically controlled stones has been measured and has been shown to obey a non-linear relationship to isotopic ratio [8,14,15]. The temperature dependence of the line position has also been measured and well characterised in the range 10–1900 K [7,8,16,17].

Strains in the diamond lattice will also cause shifts in the diamond Raman line position. This strain may be the result of external pressure on the crystal or internal stresses caused by a variety of defects. In diamond films these stresses may arise from grain boundaries, non-diamond inclusions or mismatch strains at the substrate interface; in natural diamonds stresses may also be observed around growth sector boundaries and inclusions or indentations.

As mentioned above, the first order Raman line of diamond is triply degenerate and non-symmetrical stresses in the crystal will cause this degeneracy to be broken. Exactly which of the split lines are observed will depend on the symmetry of the stresses, the polarisation of the incident and scattered light and the scattering geometry. The presence of complex stress states implies inhomogeneities in the crystalline properties. Multiple lines may also be observed if different stress states exist within small areas of the crystal, or within adjacent crystals in a micro-crystalline matrix. In either case the appearance of multiple Raman lines will imply poor crystal quality.

The shifts in the diamond line will be related to the stresses by the lattice theory of morphic effects in diamond [18]. In general the larger the shift of the peak the greater the stress in the crystal, with shifts to higher wavenumbers implying compressive stresses and to lower wavenumbers tension. However, the possibilities of complex stress states as mentioned above should be noted in which the degeneracy of the first order line is broken and both upward and downward shifts may be observed. In those cases where the splitting is small it may be interpreted as a broadening of the line whereupon the interpretation of poor crystalline quality would be similar. The possibilities of resolving the stress state of the crystal have been investigated by multiple Raman measurements involving different polarisations [19]. These approaches involve finding the eigenvalues of the secular representation of the dynamical equation and solving for particular non-general cases.

## C LINE SHAPE AND WIDTH

The natural spectral line shape of Raman scattered light is Lorentzian, and has been shown to be so in diamond [20], with a full width (in $cm^{-1}$) inversely proportional to the lifetime (in picoseconds) of the scattered phonons (with a proportionality constant of $(2\pi c)^{-1}$). The natural width is defined as the width found from the intrinsic lifetime of the phonons, without the sources of decay described below. The intrinsic lifetime of the phonons can be calculated from a consideration of the cross-sections for the spontaneous decay processes, into acoustic phonons, using the anharmonic elastic constants of the crystal. Using perturbation theory and considering only the simplest, three-phonon processes the intrinsic lifetime of the optic phonon in diamond has been calculated as 5.2 ps [21], equivalent to a natural width of 1.01 $cm^{-1}$.

The observed width, however, is determined by a number of factors, which include, to a greater or lesser extent, the spectral width of the illumination source, the instrumental broadening due to the spectrometer and factors intrinsic to the scattering material. Although only the intrinsic factors are relevant to a discussion of diamond quality some description of the characteristics of the extrinsic factors may aid in their understanding and possible removal. The spectral width of the incident light is typically far narrower than the 'natural' width of the Raman lines (except perhaps for stimulation by diode or dye laser sources) and approximately Gaussian in spectral profile. The type and extent of broadening associated with the spectrometer will vary with the design of the instrument and the exact manner in which it is configured. The broadening from both these extrinsic factors will be convoluted into the observed shape of Raman lines. Therefore comparisons of crystalline quality based upon line width may be flawed if taken on different instruments or at different times. If it is possible to measure the extent of these extrinsic broadening effects then they should be reported or removed before stating absolute line widths. However when only relative measures of quality are required, e.g. between different parts of a crystal, then only the relative changes in line width need to be considered.

The lifetime of the phonons will be shortened by changes in the anharmonic elastic constants and broadening of the diamond Raman line with increasing temperature has been observed [7,8]. Interaction with crystalline defects, including mass defects in isotopically mixed samples and point, line and planar defects, as well as crystalline boundaries, will also cause phonon decay. The reduction in lifetime will depend on the propagation velocity of the phonons, the cross-section associated with each particular defect and the mean free path of the phonons between defects. This last factor implies that the lifetime is inversely proportional to the defect density raised to a power that reflects the dimension of the defect, cubed for point defects, squared for line defects and linear for planar defects. In small crystals the phonon lifetime will be reduced by a factor proportional to the crystal diameter (assuming spherical crystals), although other effects on the lineshape may also be present. A proportional relationship between the line width and the concentration of substitutional nitrogen impurities in synthetic type Ib stones has been observed [22].

Broadenings may also result from stress gradients crossing the sampled area, in which case the form of the broadening will reflect both the shape of the optical interaction volume and stress distribution. Alternatively, unresolved stress splitting of the diamond line, as described above, may be perceived as broadening, although this effect will be resolved if spectra are obtained under different polarization conditions.

The line shape as mentioned above is naturally Lorentzian, and the lifetime-induced broadening effects will preserve this shape. Extrinsic, systematic, spectrometer and possible stress (or inhomogeneity) related broadening effects will, in general, be a mixture of Gaussian and Lorentzian in form and the measured line shape will be a non-trivial combination of these best described by a Voigt profile. Any observed intrinsic broadening of the diamond Raman line can be safely interpreted as implying reduction in crystalline quality, although additional measurements or careful line fitting processes may be necessary to identify or attempt to quantify the exact cause.

More complex lineshapes can occur in crystallites with sizes in the nanometre range and these have been qualitatively described by a theory of quantum confinement [23,24]. This theory predicts a spectral shift, broadening and asymmetry towards lower relative wavenumbers with decreasing crystallite size. Fano-resonance effects on the Raman lineshape have been observed in crystals that have been doped with large concentrations of boron and careful analysis of this lineshape may reveal the doping levels in the crystal.

## D NON-DIAMOND CARBON

The Raman signal of non-diamond carbon may consist of a variety of different line-shapes. In single crystal graphite a sharp peak at 1580 $cm^{-1}$ is observed and in microcrystalline graphite this is accompanied by a band at around 1350 $cm^{-1}$ whose relative intensity depends on the microcrystallite size [25] and whose exact position depends, unusually for a Raman line, on the wavelength of the incident radiation [26]. The main peak is known as the G band and the lower peak as the D band, whose appearance is attributed to symmetry-breaking effects. Another D' line at 1620 $cm^{-1}$ has also been observed in microcrystalline graphite, although its origin is unclear and its appearance has been interpreted as either a broadening or an upshift in the main Raman line. These lines will be broadened with decreasing crystallite size in a manner analogous to that described above.

In amorphous carbon the Raman spectrum consists of a very broad asymmetric band with a peak at around 1400–1450 $cm^{-1}$. Some authors have observed differences in the spectrum from films with different proportions of $sp^2$ and $sp^3$ bonds and also when using different wavelengths (from UV to IR) for incident radiation. Bands at or around 1210, 1150 and 1080 $cm^{-1}$ have been observed and interpreted variously as nanocrystalline diamond, a tetrahedrally bonded amorphous carbon (T-band) or the signature of a diamond precursor. Measurements from the predicted hexagonal crystalline allotrope of carbon, Lonsdaleite, would be expected to show a Raman band at ~1315–1325 $cm^{-1}$. However, the existence of this crystalline form is not widely established and its formation under CVD conditions should be considered highly unlikely.

It has been observed that Raman spectra from a single diamond film may show widely varying intensities of non-diamond components when observed using lasers from the UV to the IR [27]. In all cases the non-diamond signal from the film at blue/UV wavelengths is greatly reduced, which may tempt authors to assign a greater 'quality factor' to these films. The measured relative efficiency for Raman scattering from graphite is some 55 times that from diamond [28]. Some authors have quoted much higher efficiencies for the Raman scattering from disordered carbon species, although there is some doubt if these measurements took absorption effects into account. The various non-diamond forms of carbon have smaller optical bandgaps than diamond reducing to none at all in the case of 'black', semi-metallic, graphite. Therefore the non-diamond carbons may absorb the incident and scattered light and the resulting Raman signal will be attenuated. Alternatively some authors have ascribed this enhancement effect to resonant Raman scattering in the non-diamond components of the films [29] and although this may indeed play some part, the absorption of the incident and scattered light by the non-diamond species is certainly sufficient to mostly explain the effect.

Other researchers have defined quality factors in a variety of ways, including the ratios of spot intensities and the ratios of areas under the total spectrum at different relative wavelengths. Because the intensity of Raman scattering is directly proportional to the amount of material sampled the relative proportion of components might be measured by curve fitting the spectra, measuring the area under each component and using the relative scattering efficiencies. However, the proportion of graphitic-like carbon is strictly only proportional to the area under the G band, and even then must be corrected for absorption effects. Because of the variability in line width described above, spot intensities do not make such reliable indicators and if possible the spectrum should be line fitted. Raman spectra obtained from diamond films may often also contain a significant background photoluminescence contribution. Although the luminescence spectrum may have implications for diamond quality, it should really be considered separately from analysis of the Raman signal and this may require some fitting procedure to remove the background. The presence of spectral signatures from any non-diamond carbon will always indicate lower quality in diamond films. However, the effects of excitation wavelength on the relative visibility of non-diamond spectral components should be noted and care should be taken accordingly in comparing spectra.

## E CONCLUSION

The Raman spectrum of good quality, stress free diamond will consist of a single, sharp (1–5 $cm^{-1}$) peak at approximately 1332.4 relative wavenumbers. Stress in the diamond may be observed from shifts and possibly splitting in the diamond line and will indicate inhomogeneities and poor quality. Broadening and asymmetry of the diamond line will, in all cases, imply reduction in the quality of diamond, although systematic sources of broadening should be accounted for. When defects, or micro-crystallite size, alone are responsible for the broadening then the increase in line width will be approximately proportional to the defect density, or inversely proportional to the crystallite size.

The appearance of non-diamond lines in the Raman spectrum will imply reduction in crystalline quality. The non-diamond signature may differ widely depending on whether it originates from inclusions of microcrystalline graphite or amorphous carbon. However, the presence of photoluminescence and the effects of absorption at different laser wavelengths should be taken into account. Attempts to quantify the quality of the film from the relative strength of the non-diamond signal must make estimates for the types of inclusions and the optical absorption expected.

## REFERENCES


[1] A.K. Ramdas [ Datareview in this book: *A1.1 Raman, Brillouin and IR spectroscopy of phonons in conventional diamond* ]

[2] C.D.O. Pickard, T.J. Davis, W.N. Wang, J.W. Steeds [ *Diam. Relat. Mater. (Switzerland)* vol.7 (1998) p.238-42 ]

[3] W. Hayes, R. Loudon [ *Scattering of Light by Crystals* (John Wiley and Sons, 1978) ]

[4] D.A. Long [ *Raman Spectroscopy* (McGraw-Hill, 1977) ]

[5] J.L. Birman [ *Encyclopedia of Physics* vol.XXV/2b (Springer-Verlag, Berlin, 1974) ]

[6] R.S. Krishnan [ *Proc. Indian Acad. Sci. (India)* vol.24 (1946) p.45 ]

[7] E. Anastassakis, H.C. Hwang, C.H. Perry [ *Phys. Rev. B (USA)* vol.4 (1971) p.2493-7 ]

[8] W.J. Borer, S.S. Mitra, K.V. Namjoshi [ *Solid State Commun. (USA)* vol.9 (1971) p.1377-81 ]

[9] M.A. Washington, H.Z. Cummins [ *Phys. Rev. B (USA)* vol.15 (1977) p.5840-2 ]

[10] R.M Chrenko [ *J. Appl. Phys. (USA)* vol.63 (1988) p.5873-5 ]

[11] K.C. Hass, M.A. Tamor, T.R. Anthony, W.F. Banholzer [ *Phys. Rev. B (USA)* vol.44 (1991) p.12046 ]

[12] J. Spitzer [ *Solid State Commun. (USA)* vol.88 (1993) p.509 ]

[13] D. Kirillov, G.J. Reynolds [ *Appl. Phys. Lett. (USA)* vol.65 (1994) p.1641-3 ]

[14] K.C. Hass, M.A. Tamor, T.R. Anthony, W.F. Banholzer [ *Phys. Rev. B (USA)* vol.45 (1992) p.7171 ]

[15] R. Vogelgesang, A.K. Ramdas, S. Rodriguez, M. Grimsditch, T.R. Anthony [ *Phys. Rev. B (USA)* vol.54 (1996) p.3989-99 ]

[16] E.S. Zouboulis, M. Grimsditch [ *Phys. Rev. B (USA)* vol.43 (1991) p.12490 ]

[17] H. Herchen, M.A. Capelli [ *Phys. Rev. B (USA)* vol.43 (1991) p.11740 ]

[18] S. Ganesan, A.A. Maradudin, J. Oitmaa [ *Ann. Phys. (USA)* vol.56 (1970) p.556-94 ]

[19] C.D.O. Pickard, T.J. Davis, A. Gilmore, W.N. Wang, J.W. Steeds [ *Diam. Relat. Mater. (Switzerland)* vol.6 (1997) p.1062-6 ]

[20] A.K. McQuillan, W.R.L. Clements, B.P. Stoicheff [ *Phys. Rev. A (USA)* vol.1 (1970) p.628 ]

[21] A. Debernardi, S. Baroni, E. Molinari [ *Phys. Rev. Lett. (USA)* vol.75 (1995) p.1819-22 ]

[22] H. Hanzawa, N. Umemura, Y. Nisida, H. Kanda, M. Okada, M. Kobayashi [ *Phys. Rev. B (USA)* vol.54 (1996) p.3793-9 ]

[23] P.M. Fauchet, I.H. Campbell [ *Crit. Rev. Solid State Mater. Sci. (USA)* vol.14 (1988) p.S79 ]

[24] H. Richter, Z.P. Wang, L. Ley [ *Solid State Commun. (USA)* vol.39 (1981) p.625-9 ]

[25] F. Tuinstra, J.L. Koenig [ *J. Chem. Phys. (USA)* vol.53 (1970) p.1126-30 ]

[26] R.P. Vidano, D.B. Fischbach, L.J. Willis, T.M. Loehr [ *Solid State Commun. (USA)* vol.39 (1981) p.341-4 ]

[27] S.M. Leeds, T.J. Davis, P.W. May, C.D.O. Pickard, M.N.R. Ashfold [ *Diam. Relat. Mater. (Switzerland)* vol.7 (1998) p.233-7 ]

[28] N. Wada, S.A. Solin [ *Physica B (Netherlands)* vol.105 (1981) p.353-6 ]

[29] M. Yoshikawa et al [ *Phys. Rev. B (USA)* vol.46 (1992) p.7169-74 ]

CHAPTER A3

# ELECTRON EMISSION FROM DIAMOND

# A3.1 Electron affinities of diamond surfaces

J. Ristein

March 2000

## A INTRODUCTION

The electron affinity $\chi$ of a solid is defined as the energy gained when an additional electron is brought out of the vacuum into the material. Within the single electron approximation on which the band structure concept of solids is based this is identical to the energy difference between the vacuum level and the lowest unoccupied state inside the solid. For a semiconductor this state is the conduction band minimum (CBM). $\chi$ must strictly be distinguished from the work function $\Phi$ which is defined as the energy difference between the Fermi level $E_F$ and the vacuum level (VL). In a non-degenerate semiconductor the Fermi level is usually situated within the bandgap and thus the work function is larger than the electron affinity. For metals $\chi$ and $\Phi$ are identical.

The surface Fermi level position relative to the electronic states of a semiconductor (say relative to the CBM) usually differs between the volume and the surface. This difference is identical to the so-called surface potential $eV_s$ which is set up by a charge exchange between surface states or defects and a depletion (or accumulation) layer of majority carriers in the bulk of the material. This phenomenon, usually called surface band bending, modifies the work function, but not the electron affinity. On the other hand charge polarization within heteropolar bonds on the surface can modify the electron affinity but has no influence on the surface potential. It nevertheless changes the work function indirectly via $\chi$.

The electron affinity is a crucial quantity for the efficiency of electron emission from the conduction band into vacuum. For a number of conventional p-type semiconductors the combination of a downward surface band bending and a reduction of the (still positive!) electron affinity can lead to so-called effective negative electron affinity when the condition $\chi < |eV_s|$ is fulfilled (FIGURE 1). In this case, the CBM in the semiconductor bulk beyond the space charge layer lies above the VL and thus ballistic electrons excited to the CBM beyond a certain depth can exit into the vacuum. Only when the electron affinity is further reduced to meet $\chi < 0$ can fully thermalized electrons also exit into the vacuum (true negative electron affinity). In this case the threshold for photoelectron emission as a function of photon energy, i.e. the threshold of the photoelectron yield spectrum, coincides exactly with the bandgap energy. In the kinetic energy distribution of secondary electrons, which can be excited by UV light or an external electron beam, the thermalized electrons form a characteristic peak in the secondary electron spectra (NEA peak) which is often taken as a qualitative criterion for true NEA [1]. The relevant definitions as described above are illustrated in FIGURE 1.

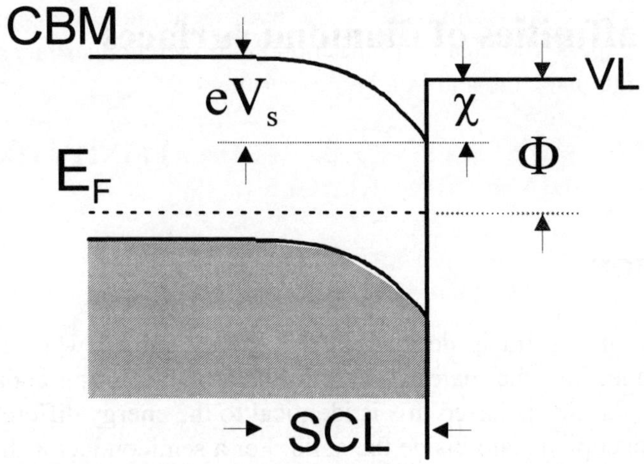


FIGURE 1 Definition of characteristic energies at the surface of a semiconductor. A p-type semiconductor with effective negative but true positive electron affinity has been illustrated to show the combined effect of surface band bending and electron affinity on the work function. SCL denotes a space charge layer, which is in this case made up by hole depletion in the semiconductor.

## B ELECTRON AFFINITY OF THE CLEAN AND THE HYDROGENATED DIAMOND (100) AND (111) SURFACES

The most important crystallographic surfaces of diamond are the (100) and (111) surfaces. After chemical vapour deposition (CVD) growth of diamond the dangling bonds of both surfaces are naturally terminated by hydrogen. This hydrogen termination can also be achieved on polished or cleaved diamond surfaces by exposure to a hydrogen plasma or even by polishing in olive oil. For (111) the dangling bonds of the bulk terminated structure are simply saturated by hydrogen forming an unreconstructed (111)1×1:H surface [2]. On (100) surfaces single bonded symmetrical dimers with hydrogen decoration form a (100)2×1:2H structure [3]. (The notation indicates the number of hydrogen atoms in the surface unit cell, although each surface atom carries only one hydrogen atom.) Both crystallographic surfaces show a true negative electron affinity in the hydrogen-terminated state. This is a unique feature of diamond not achievable on any other semiconductor. Upon vacuum annealing at sufficiently high temperatures both surfaces release their hydrogen termination and reconstruct as clean surfaces with a 2×1 symmetry [2,3]. Concomitant with the loss of the hydrogen termination is a change to positive electron affinity. For the (111) surface experimental data are available which show $\chi = -1.27 \pm 0.07$ eV for the hydrogenated and $\chi = +0.38 \pm 0.07$ eV for the clean surface [4]. The reduction of $\chi$ by hydrogen as compared to the clean surface can be quantitatively attributed to the surface dipole density which is induced by the heteropolar C–H bonds. Band structure calculations yield theoretical values of $\chi = -2.0$ eV for hydrogenated and $\chi = +0.35$ eV for clean diamond (111) [5]. For the diamond (100) surface NEA has been found qualitatively by photoelectron and photoelectron yield spectroscopy. Quantitative values for this surface have only been obtained by band structure calculations giving $\chi = -2.07$ eV for (100)2×1:2H and $\chi = +0.69$ eV for the clean (100)2×1 surface [6]. It should be mentioned that oxygen termination is expected to raise the electron affinity of the clean surfaces. This effect, opposite to that of hydrogen, is again due to the dipole-density induced on the surface, this time, however, with opposite sign due to the larger electronegativity of oxygen as compared to carbon [6].

Although a number of different terminations by hydrogen and hydrocarbon groups have been discussed in theoretical papers, the clean and the monohydrogenated surfaces are so far the only ones which have been unambiguously substantiated by experiment.

## C THE INFLUENCE OF NEGATIVE ELECTRON AFFINITY ON THE FIELD EMISSION FROM DIAMOND SURFACES

The NEA of hydrogen-saturated diamond surfaces is often suspected to be responsible for efficient field emission observed from some diamond-based carbon materials, since NEA allows conduction band electrons from diamond to be released into the vacuum without a remaining energy barrier at the surface [7]. This argument, however, only circumvents the essential problem of field emission, which is the extraction of electrons from the Fermi level. Transport through the conduction band is thus only an intermediate step, which transfers the problem to the back contact. Recent systematic studies of efficient field emitters based on nanocrystalline diamond have shown that the transport of electrons to the surface is not within the conduction band but rather at the Fermi level, probably within highly defective or graphitic grain boundaries [8]. Electrons are then emitted across the surface barrier via the Fowler–Nordheim mechanism [9]. Geometrical field enhancement is mainly responsible for the high efficiency of the material. It can be supported, however, by the comparatively low work function of NEA diamond surfaces as was recently suggested in an inhomogeneous electron emission model [10].

## D CONCLUSION

This Datareview has summarised the information on the electron affinity of diamond surfaces. Hydrogenation produces negative electron affinities on (100) and (111) surfaces. Oxygen is expected to produce positive electron affinities on (100) surfaces, but this is not yet proven. The effect of negative electron affinity on field emission properties is also discussed.

## REFERENCES


[1] R.L. Bell [ *Negative Electron Affinity Devices* (Clarendon, Oxford, UK, 1973) ]

[2] G. Kern, J. Hafner, G. Kresse [ *Surf. Sci. (Netherlands)* vol.366 (1996) p.445; note that figures 5 and 7 have been erroneously exchanged in this publication! ]

[3] G. Kern, J. Hafner, G. Kresse [ *Surf. Sci. (Netherlands)* vol.352-354 (1996) p.745 ]

[4] J.B. Cui, J. Ristein, L. Ley [ *Phys. Rev. Lett. (USA)* vol.81 (1998) p.429-32 ]

[5] J. Robertson, M.J. Rutter [ *Diam. Relat. Mater. (Switzerland)* vol.7 (1998) p.620 ]

[6] M.J. Rutter, J. Robertson [ *Phys. Rev. B. (USA)* vol.57 (1998) p.9241-5 ]

[7] M.W. Geis, J.C. Twichell, T.M. Lyszczarz [ *J. Vac. Sci. Technol. B (USA)* vol.14 (1996) p.2060 ]

[8] O. Gröning, O.M. Küttel, P. Gröning, L. Schlapbach [ *J. Vac Sci. Technol. B (USA)* vol.17 (1999) p.1970 ]

[9] E.L. Murphey, R.H. Good [ *Phys. Rev. (USA)* vol.102 (1956) p.1464 ]

[10] J.B. Cui, J. Ristein, L. Ley [ *Phys. Rev. B. (USA)* vol.60 (1999) p.16135-42 ]

# A3.2 Origin of field emission from diamond: theory and experiment

R. Schlesser and Z. Sitar

December 1999

## A INTRODUCTION

In the past few years, there has been a considerable interest in field emission from diamond, as its mechanical and chemical stability, in conjunction with its negative electron affinity (NEA), made it appear a very promising candidate for commercial, vacuum-microelectronic applications, and in particular for flat-panel displays.

While a substantial technological effort has been made to optimise diamond-based field emitters, the fundamental understanding of underlying mechanisms has been lagging behind. Recent studies employing a combination of classical I–V characterisation and field emission energy distribution (FEED) measurements revealed the origin of field emitted electrons from natural diamond, diamond thin films, and coatings. Theoretical models have been developed to explain experimental data. Electron injection at the back contact/diamond interface, as well as charge transport in diamond have been identified as crucial, current-limiting factors, while the importance of local field enhancement at the cathode back contact and at the emission surface has been stressed in recent work. The anticipated benefits of NEA properties have been re-evaluated in the framework of these findings.

## B EXPERIMENTAL TECHNIQUES

I–V characterisation is the most commonly used analytical technique to study field emitters; data are interpreted in terms of Fowler–Nordheim plots, where $\ln(I/V^2)$ is plotted versus 1/V. The Fowler–Nordheim equation [1], which was originally developed to model the field emission current density from flat, metallic surfaces as a function of the extracting electric field, becomes linear in these coordinates [2,3]:

$$\ln(I/V^2) = c_1 \cdot \frac{1}{V} + c_2 \tag{1}$$

The coefficients $c_1$ and $c_2$ can be approximated as

$$c_1 = -\left(6.44\times 10^7\,\frac{\Phi^{3/2}}{\beta}\right)\ [\mathrm{V}]$$

$$c_2 = \frac{10.4}{\Phi^{1/2}} + \ln\left(\frac{A\beta^2}{\Phi}\right) - 13.4 \tag{2}$$

where I [A] is the emission current, Φ [eV] the work function of the emitter, A [$cm^2$] the emission area, and β [$cm^{-1}$] the local field enhancement factor, which links the applied voltage V [V] to the actual, local electric field strength F [V/cm] at the emission surface:

$$F = \beta V \tag{3}$$

Thus, in a linear Fowler–Nordheim plot, slope ($c_1$) and ordinate intercept ($c_2$) depend on three parameters: $\Phi$, $\beta$, and A. Consequently, linear regression of Fowler–Nordheim plots leads to an underdetermined system of equations. As has been commented recently [4], unrealistically small work functions $\Phi$ have been reported in a number of publications as a result of underestimated field enhancement factors $\beta$.

Recently, FEED measurements have been performed on diamond field emitters. The FEED technique, which was originally used to study metallic cold cathodes (see, for example, [4]), consists of measuring the kinetic energies of electrons emitted under given biasing conditions. FEED spectra are typically displayed on a shifted, kinetic energy scale ($E - E_F$), whose origin coincides with the kinetic energy of electrons directly emitted from the Fermi level of the cathode's back contact:

$$(E - E_F) = E_{kin} - eV + \Phi_A \quad (4)$$

This energy scale implicitly takes into account any changes in the measured electron energy $E_{kin}$ due to changes in the extraction voltage V. The analyzer work function, $\Phi_A$, is a constant that can be determined by independent calibration procedures [5].

In the case of metallic emitters, FEED spectra typically feature a well-defined peak located at the Fermi level, i.e. at $(E - E_F) = 0$. For semiconductors, the location of the FEED peak relative to the back contact's Fermi level identifies the energy levels (conduction band, valence band, surface states) from which electrons are field emitted at the surface: see, for example, [5] for details.

Combining I–V and FEED data yields all relevant emitter parameters, in particular the work function $\Phi$ and the field enhancement factor $\beta$ [6]. Measuring sequences of bias-voltage-dependent FEED spectra (V-FEED) helps identify field-induced band bending and charge transport limitations in wide bandgap cathodes [5].

## C EXPERIMENTAL RESULTS: I–V DATA

A multitude of I–V data on diamond field emitters has been published in recent years. Most of these publications focused on the technological optimisation of threshold fields and maximisation of field emission current densities. Several systematic studies revealed fundamentally relevant facts that contributed to an improved understanding of the field emission mechanism.

One of the most important discoveries in diamond field emission consists of a strong, positive correlation between the performance of field emitters (turn-on field, achievable current density) and the material defect density. Zhu et al [7] showed in a parametric study that increasing doses of post-growth implanted B, C, or Na ions enhanced field emission, while the quality of the samples decreased, as evidenced by Raman spectroscopy. It was concluded that larger defect densities increased the conductivity and/or altered the work function of the samples. A similar behaviour was observed by Park et al [8] in defective MPCVD-grown diamond films. Show et al [9] measured the density of paramagnetic defects in CVD diamond by electron spin resonance (ESR), and confirmed the same trend. At present, insufficient electron transport through the cathode material is considered to be one of the most crucial emission-current-limiting factors.

Givargizov et al [10,11] studied field emission from single, crystalline diamond particles grown by CVD on sharp Si tips. They discussed the importance of local surface morphology inducing strong local fields at the emission surface, as well as the injection process of electrons from a metallic back-contact into a dielectric layer (intrinsic diamond). Geis et al [12] demonstrated that roughening of the back contact interface of natural, type Ib diamond drastically enhanced conduction through the metal/diamond interface, an observation that emphasised the importance of the electron injection mechanism into diamond.

McClure et al [13] compared field emission from Mo tips electrophoretically coated with diamond or graphite. Both types of coatings featured very similar I–V characteristics, and yielded lower turn-on fields than bare Mo tips. The improved emission characteristics were mainly attributed to local field enhancement due to increased surface roughness of the coatings. Since both graphite and diamond coatings yielded similar improvements, it was concluded that diamond's NEA properties did not play a major role in the emission process.

Gröning et al [14] observed that nominally undoped MPCVD diamond films deposited on p-type Si showed stronger field emission after an initial vacuum-arc discharge. AFM images revealed micrometre-sized craters at the film surface; conductivity maps showed enhanced conductivity in the crater region. It was concluded that arc-induced morphology changes enhanced the field emission current. Sowers et al [15] made identical observations in the case of N-doped MPCVD diamond films, and commented that micro-arcing during measurements could lead to non-representative results. These observations confirm that, even in the case of flat cathodes, most of the emission current originates from localised emission areas that feature increased, local defect conduction, and/or substantial field enhancement due to microscopic asperities.

Although most diamond field emitters investigated to date were either intrinsic or p-type, Sugino et al [16] recently reported on field emission from P-doped, n-type diamond. Field emission characteristics were seen to improve with increasing temperature due to an increase in ionised donor concentration. Analysis of I–V data led to the conclusion that the field emission process was primarily limited by internal electron emission at the metal/diamond Schottky barrier.

## D EXPERIMENTAL RESULTS: FEED DATA

Several research groups investigated field emission from a variety of differently prepared diamond samples by FEED. In several cases, the origin of field emitted electrons could be determined experimentally.

Xu et al [17] reported on CVD diamond films deposited on Mo substrates. FEED spectra showed a relatively broad single-peak (FWHM ~ 1 eV) structure, with its maximum at 2–3 eV below the Fermi level of the substrate. Hot electron emission was considered as a peak-broadening mechanism, while the peak shift away from the Fermi level was interpreted in terms of strong electron scattering in the film. Significant changes in the spectra were seen as a function of the sampling region; the emission mechanism was suggested to be based on the creation of local, electroformed conducting channels in the presence of embedded carbon particles.

Yu et al [18] investigated diamond-like carbon coated tungsten tips. Their spectra featured FEED peaks located at the Fermi level, with less peak broadening (FWHM ~ 0.4 eV) than was reported by Xu et al. A slight bias-voltage-dependent peak shift was interpreted as an ohmic voltage drop across the coating.

Bandis and Pate [19] employed a combination of FEED and photoemission spectroscopy to investigate field emission from the (111) 1×1:H surface of a natural, type IIb, single-crystal diamond. Electrons were found to originate from the valence band maximum, at 0.6 eV below the Fermi level (which was located at the B acceptor level); the position of the FEED peaks was independent of the applied bias voltage, indicating that the sample was sufficiently conducting to prevent substantial, field-induced band bending. These results were reproduced subsequently by several research groups.

Schlesser et al [20] investigated field emission from nominally intrinsic, electrophoretically deposited diamond coatings on single Mo tips. V-FEED analysis showed significant peak shifts toward lower kinetic energies, increasing linearly with bias voltage; extrapolation of V-FEED data to the flat-band condition identified the conduction band minimum as the origin of field emitted electrons. The observed peak shifts were interpreted in terms of field-induced band bending, as the undoped coatings were highly resistive and only a few 100 nm thick.

Surface conduction effects were studied using FEED as well. Küttel et al [21] observed field emission from a μm-sized metallic particle placed on a hydrogen-saturated, insulating diamond (100) surface. Large energy shifts in FEED spectra were discussed in terms of charge transport by surface conduction. These peak shifts were proportional to the sample current and were attributed to ohmic charge transport limitations, in contrast to the voltage-proportional shifts observed by Schlesser et al [20], which originated from field-induced band bending.

Gröning et al [22] reported on nanocrystalline CVD diamond thin films. Field emission originated from the Fermi level located at 1 ±0.2 eV above the valence band maximum, indicating that nanocrystalline diamond's Fermi level is pinned 1 eV above the valence band edge by partially filled levels. No significant peak shifts were observed with changing bias voltages, indicating that the applied electric field was not penetrating into these films. A combination of I–V and FEED data led to the conclusion that the films had a large work function of 5.7 eV, and that the observed, low-field electron emission was primarily due to field enhancement yielding local field strengths $10^3$ times larger than the nominal applied field.

## E THEORETICAL MODELS

Several, in part contradictory, theoretical models have been published in recent years. Due to a large variety of diamond-based emitters, and the often undocumented diamond quality used in different experiments, it remains a challenge to cover all observations by one comprehensive theory.

Huang et al [23] modelled diamond field emission in various crystallographic directions, considering the effect of NEA, band bending, image interaction, and surface charges. Their study showed that the field emission current levels observed experimentally could only be explained by postulating the presence of two subbands in the intrinsic bandgap. Defects or impurities may generate such subbands, and thus, their theoretical study emphasised the importance of defect conduction in diamond. A Monte Carlo study published by Cutler et al [24] showed that hot electron transport is achievable for internal electric fields larger than 1 V/μm and quasi-ballistic transport should occur at field strengths greater than 10 V/μm. Lerner et al [25] modelled the performance of metal/diamond cold cathode emitters, employing a Spindt-type structure and substitutionally nitrogen-doped diamond films, a design that had been suggested by Geis et al [26]. They identified the rough surface of the metal injector, as well as the NEA properties of diamond, as essential features for low-threshold field operation.

There appears to be a growing consensus regarding the relevance of the electron injection mechanism into the conduction band of diamond [11,15,20]. Earlier calculations [27] that modelled field emission from a metal covered with a semiconducting thin film showed that the I–V characteristics of such emitters are dominated by the injection process at the metal/diamond interface, if

$$\frac{\chi}{\Phi_b} < \varepsilon^{2/3} \tag{5}$$

where $\chi$ is the electron affinity of the semiconductor, $\Phi_b$ the Schottky barrier height at the metal/semiconductor interface, and $\varepsilon$ the dielectric constant of the semiconductor. In the case of diamond, with $\chi < 1.5$ eV and $\varepsilon = 5.7$, EQN (5) yields a threshold Schottky barrier of 0.5 eV. Hence, for barriers larger than 0.5 eV, even under the assumption of a positive electron affinity of 1.5 eV, the Schottky barrier at the metal/diamond interface limits the emission current, rather than the emission barrier at the diamond/vacuum interface.

Recently, Silva et al [28] presented a 'space charge interlayer' model, which is based on space charge band bending at the back contact due to the presence of a thin, depleted DLC interlayer. Hot electron transport across the depleted DLC layer and subsequent injection into diamond was modelled for both n- and p-type polycrystalline diamond films. The model was found to explain various experimental observations, and was in agreement with V-FEED measurements that evidenced the presence of band bending in thin, polycrystalline coatings [20].

As an alternative field emission mechanism Geis et al [29] proposed the use of metal-diamond-vacuum triple junctions to enhance field emission currents. In such devices, electrons tunnel from a metal into diamond surface states, where they are accelerated prior to emission into vacuum.

## F CONCLUSION

FEED experiments have shown that, depending on the sample quality and type of dopant, diamond field emission may originate from the valence band maximum (type IIb, B-doped samples), from defect levels (e.g. nanocrystalline diamond), from surface states, or from the conduction band minimum (intrinsic, and most likely n-type diamond).

Valence band emission: Due to the large bandgap of diamond ($E_g = 5.5$ eV), and even in the presence of an NEA in the order of $\chi = -1$ eV, field emission originating from the valence band maximum is limited by a work function $\Phi = E_g + \chi \sim 4.5$ eV, which is as large as typical metal work functions.

Conduction band emission: The field emission process may be subdivided into three steps: (1) injection of electrons from the back contact into diamond, (2) electron transport toward the emission area, and (3) field emission at the diamond surface. The emission barrier at the surface is small, but injection of electrons into the conduction band typically limits the extractable field emission current. As a consequence, it appears that the potential benefits of NEA cannot be exploited to date, a situation that could possibly change with the advent of shallow level n-type diamond.

A number of studies have identified local field enhancement as the main current-enhancing factor, improving electron injection at the back contact/diamond interface, as well as electron emission at the diamond/vacuum interface. While the importance of surface conductivity as a transport mechanism remains unclear at present, defect conduction in diamond has been identified as a major beneficial factor to the emission process, thus explaining the reported superiority of lower quality diamond for field emission applications.

## REFERENCES

[1] R.H. Fowler, L.W. Nordheim [ *Proc. R. Soc. Lond. A (UK)* vol.119 (1928) p.173-81 ]
[2] I. Brodie, C.A. Spindt [ *Adv. Electron. Electron Phys. (UK)* vol.83 (1992) p.1-106 ]

[3] I. Brodie [ *Proc. IEEE (USA)* vol.82 (1994) p.1006-34 ]

[4] J.W. Gadzuk, E.W. Plummer [ *Rev. Mod. Phys. (USA)* vol.45 (1973) p.487-548 ]

[5] R. Schlesser, M.T. McClure, B.L. McCarson, Z. Sitar [ *J. Appl. Phys. (USA)* vol.82 (1997) p.5763-72 ]

[6] O. Gröning, O.M. Küttel, P. Gröning, L. Schlepbach [ *J. Vac. Sci. Technol. B (USA)* vol.17 (1999) p.1064-71 ]

[7] W. Zhu et al [ *Appl. Phys. Lett. (USA)* vol.67 (1995) p.1157-9 ]

[8] K.H. Park et al [ *J. Vac. Sci. Technol. B (USA)* vol.16 (1998) p.724-8 ]

[9] Y. Show, F. Matsuoka, M. Hayashi, H. Ito, M. Iwase, T. Izumi [ *J. Appl. Phys. (USA)* vol.84 (1998) p.6351-4 ]

[10] E.I. Givargizov, V.V. Zhirnov, A.N. Stepanova, E.V. Rakova, A.N. Kiselev, P.S. Plekhanov [ *Appl. Surf. Sci. (Netherlands)* vol.87 (1995) p.24-30 ]

[11] E.I. Givargizov, V.V. Zhirnov, A.V. Kuznetsov, P.S. Plekhanov [ *J. Vac. Sci. Technol. B (USA)* vol.14 (1996) p.2030 ]

[12] M.W. Geis, J.C. Twichell, N.N. Efremov, K. Krohn, T.M. Lyszczarz [ *Appl. Phys. Lett. (USA)* vol.68 (1996) p.2294-6 ]

[13] M.T. McClure, R. Schlesser, B.L. McCarson, Z. Sitar [ *J. Vac. Sci. Technol. B (USA)* vol.15 (1997) p.2067-71 ]

[14] O. Gröning, O.M. Küttel, E. Schaller, P. Groning, L. Schlapbach [ *Appl. Phys. Lett. (USA)* vol.69 (1996) p.476-8 ]

[15] A.T. Sowers, B.L. Ward, S.L. English, R.J. Nemanich [ *J. Appl. Phys. (USA)* vol.86 (1999) p.3973-82 ]

[16] T. Sugino, K. Kuriyama, C. Kimura, Y. Yokota, S. Kawasaki, J. Shirafuji [ *J. Appl. Phys. (USA)* vol.86 (1999) p.4635-42 ]

[17] N.S. Xu, Y. Tzeng, R.V. Latham [ *J. Phys. D (UK)* vol.27 (1994) p.1988-91 ]

[18] M.L. Yu, H.-S. Kim, B.W. Hussey, T.H.P. Chang, W.A. Mackie [ *J. Vac. Sci. Technol. B (USA)* vol.14 (1996) p.3797-801 ]

[19] C. Bandis, B.B. Pate [ *Appl. Phys. Lett. (USA)* vol.69 (1996) p.366-8 ]

[20] R. Schlesser, M.T. McClure, W.B. Choi, J.J. Hren, Z. Sitar [ *Appl. Phys. Lett. (USA)* vol.70 (1997) p.1596-8 ]

[21] O.M. Küttel, O. Gröning, L. Schlapbach [ *J. Vac. Sci. Technol. A (USA)* vol.16 (1998) p.3464-70 ]

[22] O. Gröning, O.M. Küttel, P. Gröning, L. Schlapbach [ *J. Vac. Sci. Technol. B (USA)* vol.17 (1999) p.1970 86 ]

[23] Z.H. Huang, P.H. Cutler, N.M. Miskovsky, E. Sullivan [ *Appl. Phys. Lett. (USA)* vol.65 (1994) p.2562-4 ]

[24] P.H. Cutler, Z.-H. Huang, N.M. Miskovsky, P. D'Ambrosio, M. Chung [ *J. Vac. Sci. Technol. B (USA)* vol.14 (1996) p.2020 ]

[25] P. Lerner, P.H. Cutler, N.M. Miskovsky [ *J. Vac. Sci. Technol. B (USA)* vol.15 (1997) p.337-42 ]

[26] M.W. Geis, J.C. Twitchell, T.M. Lyszczarz [ *J. Vac. Sci. Technol. B (USA)* vol.14 (1996) p.2060 ]

[27] A. Kiejna, P. Niedermann, O. Fischer [ *Appl. Phys. A (Germany)* vol.50 (1990) p.331-8 ]

[28] S.R.P. Silva, G.A.J. Amaratunga, K. Okano [ *J. Vac. Sci. Technol. B (USA)* vol.17 (1999) p.557-61 ]

[29] M.W. Geis et al [ *Nature (UK)* vol.393 (1998) p.431-5 ]

# A3.3 Properties of secondary electron emission from diamond: influence of dopants

A. Shih and J.E. Yater

December 1999

## A INTRODUCTION

We confine our discussion of secondary-electron emission to the emission that is stimulated by energetic electrons impinging on a material. The secondary-electron-emission property of a material is often characterised by its secondary-electron yield, δ, which is defined as

$$\delta = I_s / I_o$$

where $I_s$ is the secondary-electron intensity and $I_o$ is the primary-electron intensity. δ is not a single-valued property of a material but instead varies with the primary-electron energy. Beginning at low primary energy, δ increases with increasing energy, passes through a maximum $\delta_{max}$, and then falls off monotonically. The yields from metals and semiconductors are, in general, small, and $\delta_{max}$ varies from 0.5 to 1.8 [1]. The yields from conducting wide-bandgap materials such as diamond are high, and $\delta_{max}$ is often greater than 5 [1]. The precise value of δ depends strongly on the bulk and surface properties of the sample. In this Datareview, we will discuss the influence of dopant type and dopant concentration on δ, and in the following Datareview [2], we will discuss the influence of surface adsorbates on δ.

## B SECONDARY-ELECTRON-EMISSION PROCESS

FIGURE 1 illustrates the secondary-electron-emission process using a three-step model [3,4]: (1) generation of internal secondary electrons by kinetic impact of the primary electrons, (2) transmission of the internal secondary electrons through the material, and (3) escape of the secondary electrons over the vacuum barrier. The presence of dopants influences δ by affecting the transmission process (step 2), while the influence of the surface affects the escape process (step 3).

In a wide-bandgap material, an energetic primary electron slows down mainly through the excitation of valence electrons into the conduction band. Since most of the kinetic energy of the primary electron is converted in the generation of electron–hole pairs, the number of secondary electrons is large. In diamond, a majority of the internal secondary electrons have energies about 0–2 eV above the conduction-band minimum $E_c$. Because of the large bandgap of diamond ($E_{gap}$ = 5.47 eV), collisions between such low-energy secondary electrons and valence electrons are forbidden. As a result, the low-energy secondary electrons have to lose energy through electron–impurity and electron–phonon scattering, which produce small energy losses on the order of 0.1 eV. Hence, the rate of energy loss is small and the resulting electron escape depth is very long. The adsorption of H or Cs on the surface lowers $E_{vac}$ to a position near or below $E_c$, and the large number of low-energy secondary electrons can escape, resulting in a large δ. The properties that make diamond a prolific secondary-electron emitter are the unusually long escape depth of the internal secondary electrons and the low vacuum-barrier height.

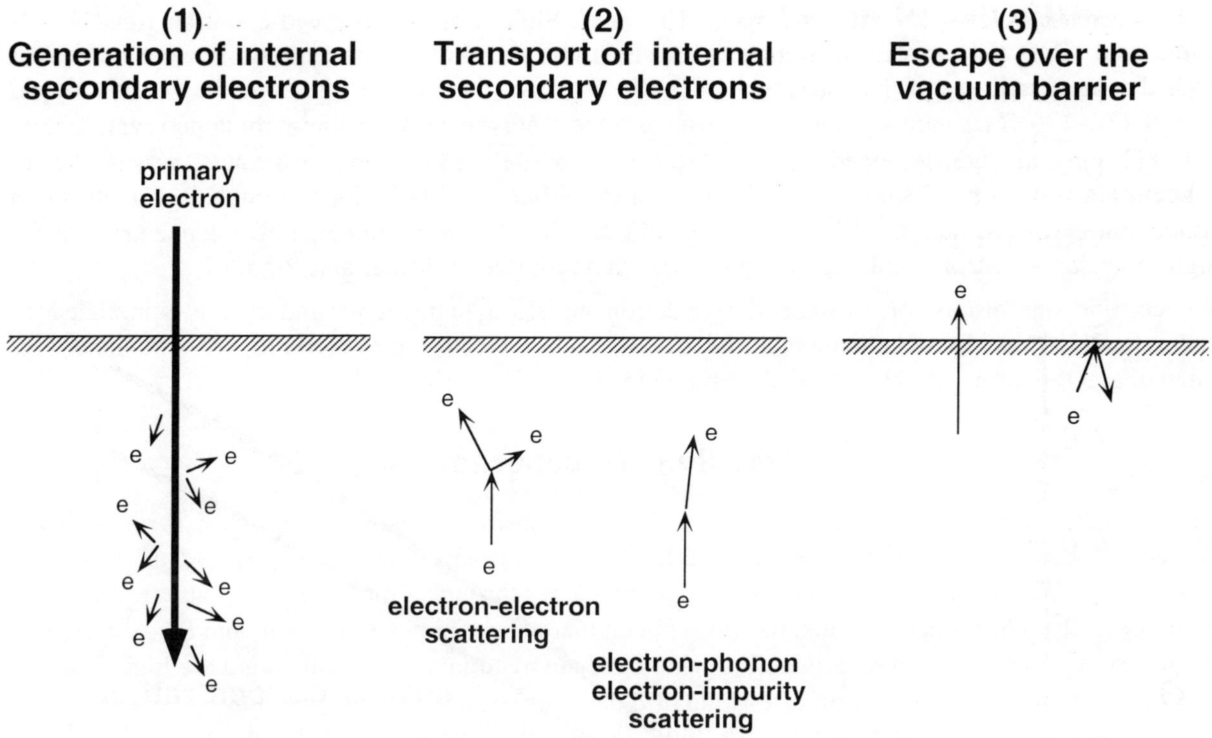


FIGURE 1 A three-step model of the secondary-electron-emission process.

## C N-DOPING

N-doped diamond samples are insulators. However, electrical conductivity is needed to sustain the secondary-electron-emission process. Otherwise, charging of the sample results. If $\delta > 1$, the sample becomes positively charged, resulting in an increased vacuum-barrier height and a decreased yield. If $\delta < 1$, the sample becomes negatively charged, resulting in a slow down or even repulsion of the impinging electrons. With a perfect insulator, $I_s = I_o$ and $\delta = 1$. In fact, $\delta_{max}$ was measured to be 1.07 ±0.05 from an N-doped CVD diamond sample [5].

## D B-DOPING

B-doped diamond is electrically conductive, and it has a high δ when it is terminated with H or Cs. For H-terminated diamond samples, both Miller et al [6] and Shih et al [7] found that the secondary-electron yields are sensitive to the B-doping level. In fact, there is an optimum range of B concentrations for high secondary-electron yields. For example, δ decreases at low B-doping levels due to the insufficient electrical conductivity required to sustain high secondary-electron emission levels. On the other hand, B dopants are sources for electron collisions, which can suppress the yield if the doping concentration becomes high. Specifically, most of the internal secondary electrons in diamond have kinetic energies of about 1 eV. These low-energy electrons are unable to excite valence electrons over the 5.47 eV bandgap, and thus collisions with phonons and impurities become the major source of energy loss. As the doping concentration increases, the collision rate increases and the electron escape depth decreases, resulting in lower electron emission.

Miller and Brandes [6] studied H-terminated homo-epitaxially grown C(100), C(110) and C(111) and found that the secondary-electron yields were greatest for B concentrations of $10^{19}$–$10^{20}$ $cm^{-3}$. The yields decrease below $10^{15}$ $cm^{-3}$ and above $10^{21}$ $cm^{-3}$. Shih et al [7] observed a similar trend for H-terminated CVD polycrystalline samples. In this study, the doping concentrations were correlated with the measured sample resistivities. FIGURE 2 shows that the yields from the heavily doped sample (50–170 Ω cm) are significantly lower than the yields from the moderately doped sample (50–170 kΩ cm) or lightly doped sample (155 kΩ cm). Based on the resistance versus dopant concentration study by Nishimura et al [8], we estimate that the heavily doped sample has an atomic B concentration of less than $6 \times 10^{19}$ $cm^{-3}$. This indicates that the optimum range of B concentrations for high secondary-electron yield is much lower than that reported by Miller and Brandes.

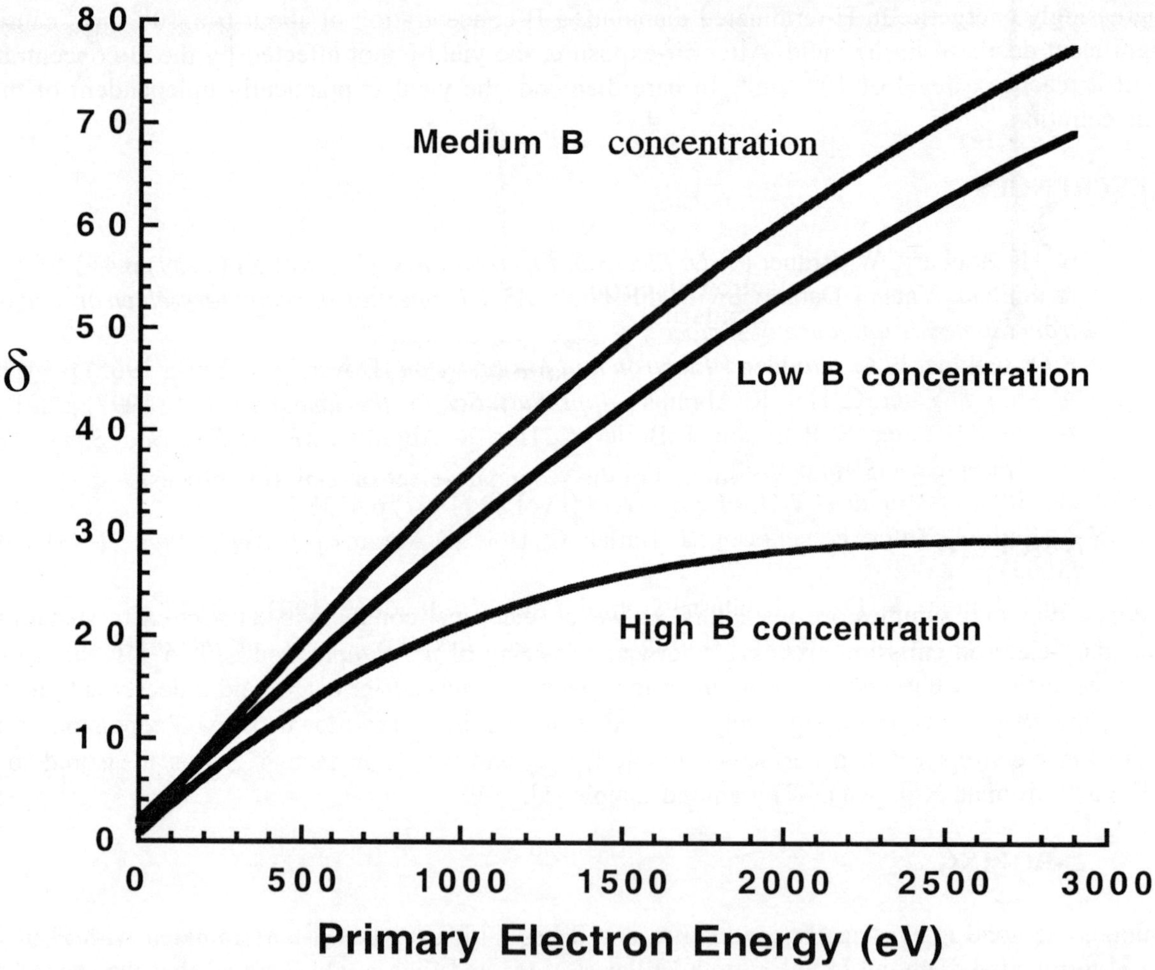


FIGURE 2 Secondary-electron-yield spectra from H-terminated CVD diamond samples with various B concentrations.

The disagreement between the two studies arises from the difference in the sample treatment. Thoms et al [9] showed that hydrocarbon molecules are present on air-exposed diamonds and desorb upon heating. The samples of Miller and Brandes were air-exposed and unheated, and $\delta_{max}$ was about 10. Shih et al [7] found that sample heating reduces the vacuum-barrier height and increases $\delta_{max}$ from about 10 to about 30 or 70, depending on the B-doping level. Because of the higher vacuum-barrier height in the samples of Miller and Brandes, a higher fraction of the secondary electrons was of high energy. However, the escape depths of secondary electrons with high kinetic energy ($E > E_{gap}$) are not sensitive to the B-doping level since, in this case, the energy-loss process is dominated by electron–

electron scattering rather than electron–impurity scattering. This explains why the optimum range for the yields from these samples is higher than the range found for the samples of Shih et al.

## E CONCLUSION

Although the electrical conductivity does not directly affect the secondary-electron generation, transmission or emission processes, it is needed to prevent sample charging by balancing the electron flow. In diamond, B atoms provide the required electrical conductivity but N atoms do not. However, B atoms act also as collision centres to diminish the secondary-electron yield by impeding the transmission of secondary electrons. However, as the vacuum-barrier height increases, the sensitivity of the yields to the B-dopant concentration decreases as the emitted secondary electrons become increasingly energetic. In H-terminated diamond, a B concentration of about $6 \times 10^{19}$ $cm^{-3}$ causes a significant decrease in the yield. After air-exposure, the yield is not affected by the B concentration until it reaches a level of $10^{21}$ $cm^{-3}$. In bare diamond, the yield is practically independent of the B concentration.

## REFERENCES


[1] O. Hachenberg, W. Brauer [ *Adv. Electron. Electron Phys. (UK)* vol.XI (1959) p.413-98 ]

[2] A. Shih, J. Yater [ Datareview in this book: *A3.4 Properties of secondary electron emission from diamond: influence of surface* ]

[3] R.O. Jenkins, W.G. Trodden [ *Electron and Ion Emission* (Dover, New York, 1965) p.54-70 ]

[4] A. Shih, J. Yater, C. Hor, R. Abrams [ *Appl. Surf. Sci. (Netherlands)* vol.111 (1997) p.251 ]

[5] A. Shih, J. Yater, P. Pehrsson, J. Butler, C. Hor, R. Abrams [ *Mater. Res. Soc. Symp. Proc. (USA)* vol.416 (1996) p.461 ]

[6] J. Miller, G. Brandes [ *J. Appl. Phys. (USA)* vol.82 (1997) p.4538 ]

[7] A. Shih, J. Yater, P. Pehrsson, J. Butler, C. Hor, R. Abrams [ *J. Appl. Phys. (USA)* vol.82 (1997) p.1860 ]

[8] K. Nishimura, K. Das, J.T. Glass [ *J. Appl. Phys. (USA)* vol.69 (1991) p.3142 ]

[9] B.D. Thoms, P.E. Pehrsson, J.E. Butler [ *J. Appl. Phys. (USA)* vol.75 (1994) p.1804 ]

# A3.4 Properties of secondary electron emission from diamond: influence of surface

A. Shih and J.E. Yater

December 1999

## A INTRODUCTION

A wide range of secondary-electron yields δ from diamond is reported in the literature [1–8]. Some earlier diamond samples were not of high quality, and δ was correlated with the size of the non-diamond background in the Raman spectra [1]. More recently, high quality diamond samples have been routinely fabricated (followed by an acid cleaning to remove non-diamond surface carbon), but the reported yields still show a large variation. This variation is due largely to the fact that δ is extremely sensitive to the vacuum-barrier height at the diamond surface.

## B SECONDARY-ELECTRON ENERGY DISTRIBUTION

In the previous Datareview [9], we described secondary-electron emission using a three-step process: (1) generation of secondary electrons, (2) transport through the material, and (3) escape over the vacuum barrier. FIGURE 1 presents a model of the secondary-electron escape across the vacuum interface of diamond, where the left and right sides of FIGURE 1 portray the secondary electrons before and after the escape, respectively. The model uses an energy-distribution curve (EDC) to represent the secondary electrons since the energy-distribution characteristics of the electrons underlie the surface sensitivity of the secondary-electron emission. In the EDC, the abscissa is N(E), the number of internal secondary electrons at energy E, where E is measured relative to the conduction-band minimum $E_c$. FIGURE 1 uses two possible positions of $E_{vac}$ to illustrate two entirely different situations for the secondary-electron emission. In the case of H-terminated diamond, where the vacuum level $E_{vac}$ is near to, or below, $E_c$, all the internal secondary electrons can escape and the entire EDC transmits across the vacuum interface. However, in the case of bare diamond where $E_{vac}$ is about 1.5 eV higher than $E_c$, only high-energy secondary electrons can escape. Consequently, the emitted EDC is the high-energy portion of the total EDC. A dominant feature in the total EDC is a large and narrow peak which is typically about 0.5 ±0.2 eV above $E_c$ with a full-width at half-maximum of about 0.65 ±0.1 eV [6,8]. A large fraction of the secondary electrons is contained within the peak. In EDCs taken from H-terminated CVD diamond, the peak contains ~40% and ~90% of the total emitted secondary electrons generated by 500 eV and 3000 eV primary electrons, respectively. Because of the importance of the low-energy secondary-electron contribution to δ, it follows that δ is sensitive to changes in the electron affinity $\chi = E_{vac} - E_c$. At a bare diamond surface, χ is positive and large, resulting in a barrier height that is too high for the electrons contained in the peak to overcome. Consequently, the yields are low (to be discussed in Section C). Adsorbates such as H and Cs lower χ and significantly increase the yield. The key factors that determine the value of χ are the presence of co-adsorbed molecules (Section D), the degree of the adsorbate coverage (Section E), and the surface orientation (Section F).

## C 'BARE' DIAMOND

Hydrogen is present on most of the as-received diamond samples, CVD or natural, and its desorption requires a 950 to 1000°C heating [10]. FIGURE 2 shows the secondary-electron-yield spectrum from a CVD diamond after 15 min of heating at 1000°C. $E_{vac}$ is measured to be about 1.5 eV above $E_c$ [11], and the energy barrier is too high for most of the secondary electrons to overcome (see FIGURE 1). As a result, the yields are low.

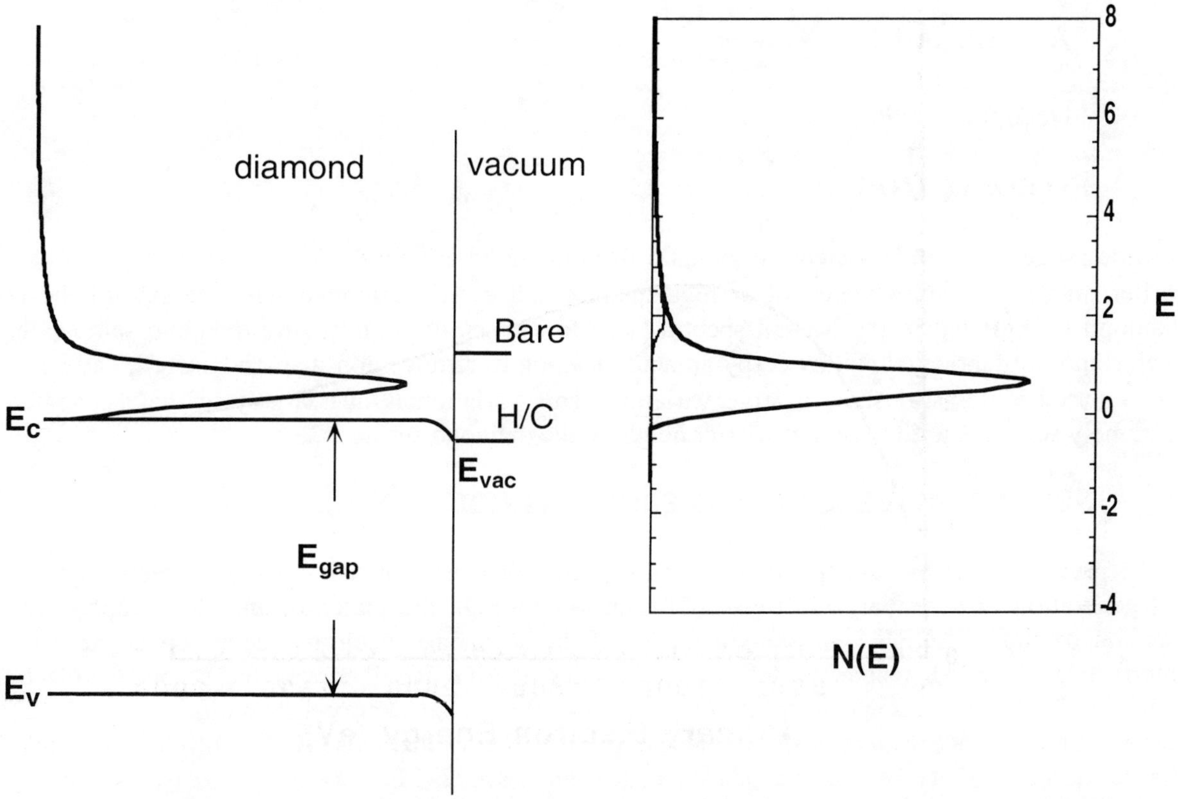


FIGURE 1 Model of the escape of secondary electrons over the vacuum barrier of diamond. Energy-distribution curves are used to show the secondary-electron density at various energies. The left side shows the presence of the secondary electrons in the conduction band. The right side shows the emitted secondary electrons for the two cases, i.e. H-terminated diamond (H/C) and bare diamond (Bare).

Although FIGURE 2 is among the lowest secondary-electron-yield curves obtained from our diamond samples after repeated heating at 1000°C, the desorption of hydrogen may not be complete. Malta et al [2] reported a yield of 1.3 at 3 keV after heating a (100) type IIb diamond at 1000°C for 15 min. In FIGURE 2, we measure a yield of 1.9 at 3 keV. Hence, we will use the term 'bare' diamond to acknowledge the fact that the diamond surface may not be truly bare.

## D CO-ADSORBED MOLECULES

Hydrocarbon molecules are known to be present on air-exposed H-terminated diamond surfaces [10]. The maximum secondary-electron yields from these samples are typically about 10 [6,7]. After thermal desorption of hydrocarbon molecules, the yields increase greatly. Although the desorption of hydrocarbon molecules is complete at 400°C [10], the yields continue to rise until 800°C [6], indicating that other molecules are co-adsorbed on the surface in addition to hydrocarbon molecules.

$E_{vac}$ is often determined by the emission onset in EDC. The EDCs, which are taken after successive heatings, showed no change in the emission onset but only an increase in the height of the low-energy peak. This can be explained by the fact that the emission onset reveals only the lowest value of $E_{vac}$, and the co-adsorbed molecules only raise $E_{vac}$ at parts of the surface area.

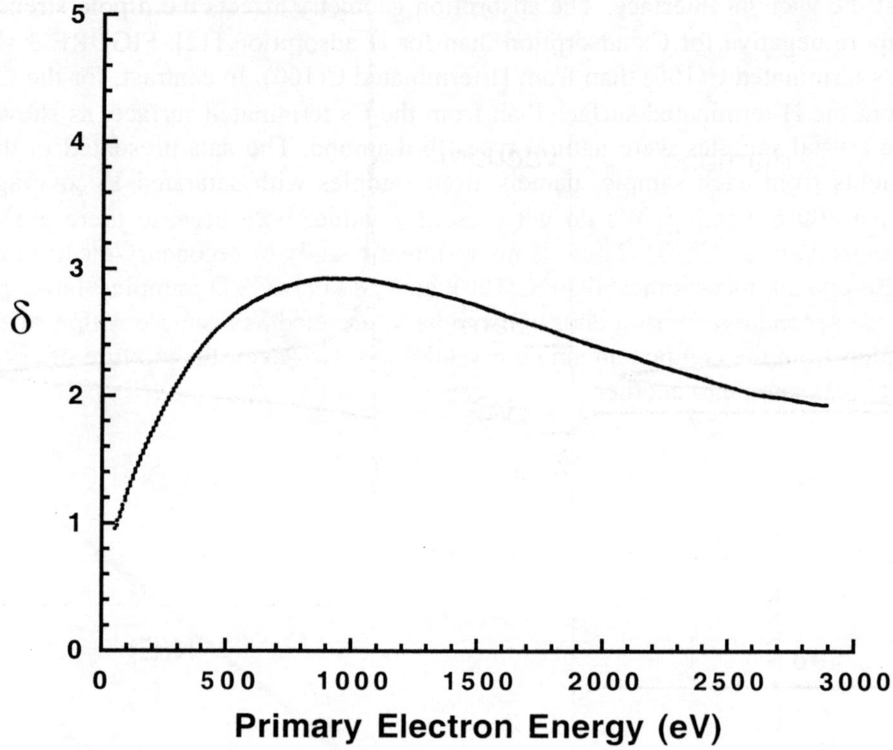


FIGURE 2 A secondary-electron-yield spectrum from a CVD diamond after a 15 min heating at 1000°C; hydrogen desorption from the surface may not be complete.

## E ADSORBATE COVERAGE

Secondary-electron emission has been studied systematically as a function of H or Cs coverage on diamond samples, and the yields showed strong coverage dependence. For hydrogen adsorption, saturated hydrogen coverage results in the highest secondary-electron yield. For Cs adsorption, the yield increases initially with Cs exposure, but in some cases excessive Cs exposure leads to a decrease in the yield, e.g. on C(111) surfaces. The decrease is understood by Cs cluster formation [12].

Many of the CVD diamond samples were H-terminated using a hydrogenation process as the final preparation step. However, air exposure allows hydrocarbons and other molecules to co-adsorb on the surface, partially suppressing the secondary electron yields. Thermal treatments, which desorb the co-adsorbed molecules, may desorb some of the hydrogen as well. Subsequent exposure of the samples to atomic hydrogen significantly increases the secondary-electron yields. The yields reach a plateau at the hydrogen-saturation coverage [6,10].

In contrast to H adsorption, excess Cs atoms may form clusters, which scatter the secondary electrons and decrease the yields. Heating at 200°C desorbs the excessive Cs and increases the yields.

## F SURFACE ORIENTATION

Electrically positive atoms such as H and Cs induce dipole-layer formations on the diamond surface and lower $\chi$ at the vacuum interface. The adsorption geometry affects the dipole strength and $\chi$. For C(100), $\chi$ is more negative for Cs adsorption than for H adsorption [12]. FIGURE 3 shows that $\delta$ is higher from Cs-terminated C(100) than from H-terminated C(100). In contrast, for the C(111) sample, $\delta$ is higher from the H-terminated surface than from the Cs-terminated surface, as shown in FIGURE 4. Both single-crystal samples were natural type-IIb diamond. The data presented in this section are the highest yields from each sample, namely from samples with saturated H coverage or with Cs coverage after a 200°C heating. We do not present $\chi$ values here because there are disagreements about their precise values [12,13]. There is no systematic study of secondary-electron emission from diamond single-crystal faces other than C(100) and C(111). CVD samples have polycrystalline surfaces, and the secondary-electron emission can be interpreted as a superposition of the secondary-electron emission from the component single-crystal faces. However, the mixture of crystal faces may vary from one CVD sample to another.

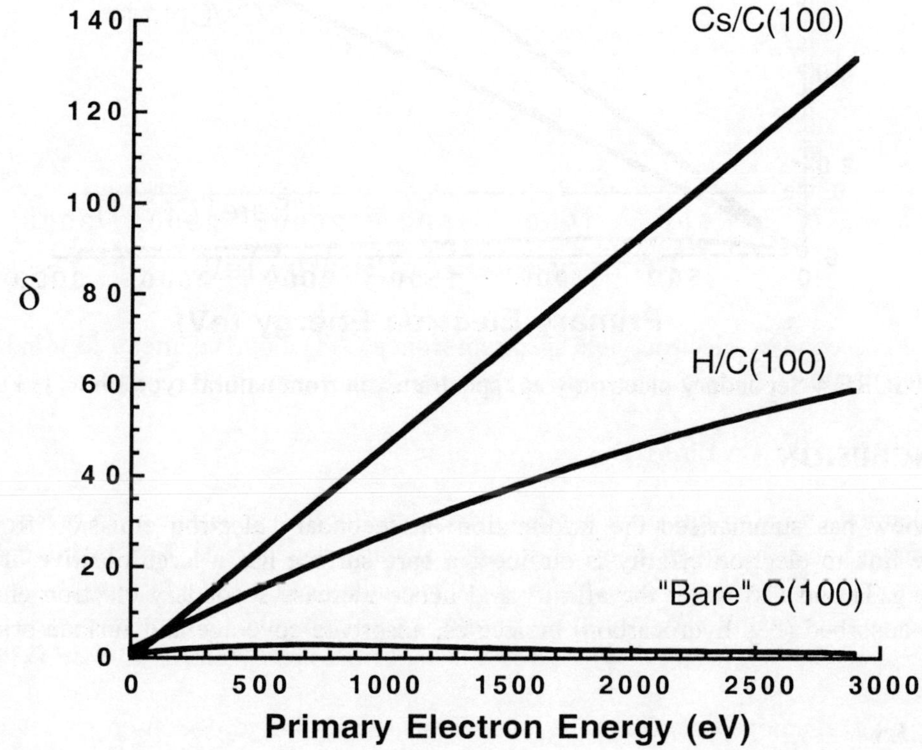


FIGURE 3 Secondary-electron-yield spectra taken from natural type IIb C(100).

In contrast with the yield spectra of 'bare' diamond, the yield spectra from H- or Cs-terminated diamond show a continuous rise over the energy range of the measurements. An energetic primary electron slows down via collisions with valence electrons to create electron–hole pairs. In diamond, most of the created electrons have an energy near $E_c$ which gives rise to the low-energy peak. As the primary-electron energy increases, the number of generated electron–hole pairs increases and the secondary-electron yields increase. However, as the primary electron energy increases, the primary electrons penetrate deeper into the material and the secondary electrons need to travel further to reach the surface. When the penetration depth of the primary electron exceeds the escape depth of the secondary electron, the yield begins to decrease. In the case of 'bare' diamond (with $\chi \sim 1.5$ eV), this decrease in δ occurs around 1000 eV (FIGURE 2). When $\chi < 0$, the low-energy secondary electrons contained in the peak can now be emitted. These electrons have long escape depths because their energies are too low to create electron–hole pairs in diamond. Consequently, the yield shows a continuous increase from 0 to 3 keV.

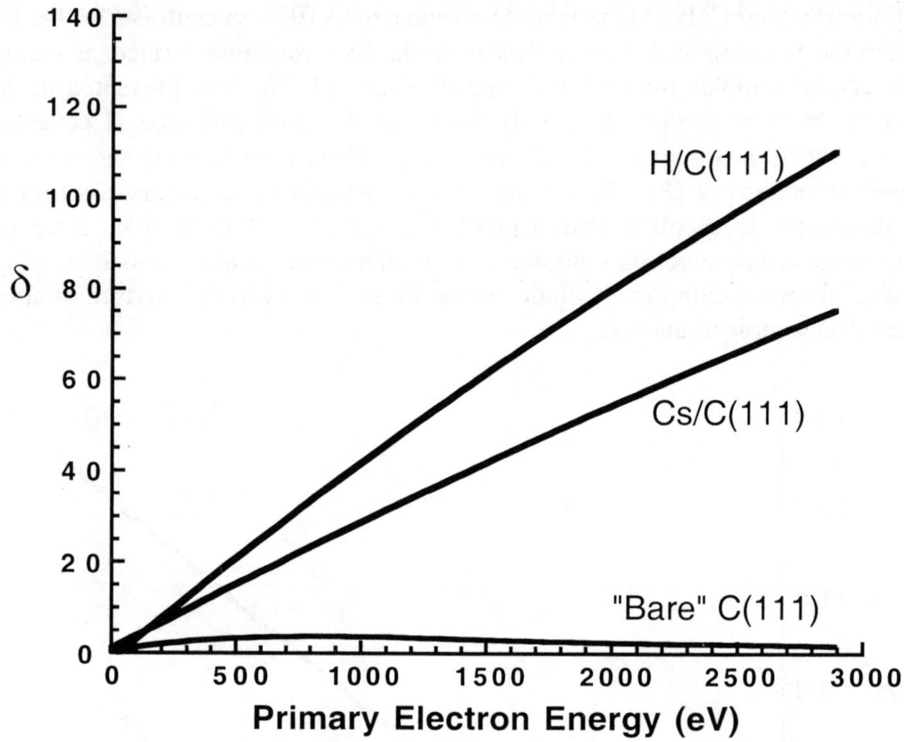


FIGURE 4 Secondary-electron-yield spectra taken from natural type IIb C(111).

## G CONCLUSION

This Datareview has summarised the information on secondary electron emission from diamond surfaces. The link to electron affinity is outlined; a bare surface has a large positive affinity while adsorbates (e.g. H and Cs) lower the affinity and hence increase secondary electron emission. The affects of co-adsorbed (e.g. hydrocarbon) molecules, adsorbate coverage and surface orientation are discussed.

## REFERENCES

[1] G.T. Mearini, I.L. Krainsky, Y.X. Wang J.A. Dayton, R. Ramessham, M.F. Rose [ *Thin Solid Films (Switzerland)* vol.253 (1994) p.151 ]

[2] D.P. Malta et al [ *Appl. Phys. Lett. (USA)* vol.64 (1994) p.1929 ]

[3] G.T. Mearini, I.L. Krainsky, J.A. Dayton [ *Appl. Phys. Lett. (USA)* vol.66 (1995) p.242 ]

[4] S.P. Bozeman et al [ *Diam. Relat. Mater. (Switzerland)* vol.5 (1996) p.802 ]

[5] D.P. Malta, J.B. Posthill, T.P. Humphreys, M.J. Mantini, R.J. Markunas [ *Mater. Res. Soc. Symp. Proc. (USA)* vol.416 (1996) p.311 ]

[6] A. Shih, J. Yater, P. Pehsson, J. Butler, C. Hor, R. Abrams [ *J. Appl. Phys. (USA)* vol.82 (1997) p.1860 ]

[7] J. Miller, G. Brandes [ *J. Appl. Phys. (USA)* vol.82 (1997) p.4538 ]

[8] J.E. Yater, A. Shih, R. Abrams [ *J. Vac. Sci. Technol. A (USA)* vol.16 (1998) p.913 ]

[9] A. Shih, J. Yater [ Datareview in this book: *A3.3 Properties of secondary electron emission from diamond: influence of dopants* ]

[10] B.D. Thoms, P.E. Pehrsson, J.E. Butler [ *J. Appl. Phys. (USA)* vol.75 (1994) p.1804 ]

[11] J.E. Yater, A. Shih, R. Abrams [ *Phys. Rev. B (USA)* vol.56 (1997) p.R4410 ]

[12] J.E. Yater, A. Shih [ submitted for publication ]

[13] L. Diederich, O.M. Kuttel, P. Aebi, L. Schlapbach [ *Surf. Sci. (Netherlands)* vol.418 (1998) p.219 ]

CHAPTER A4

# PROPERTIES AND TECHNOLOGY OF DIAMOND SURFACES

# A4.1 Morphology of diamond surfaces

W.J.P. van Enckevort

March 2000

## A INTRODUCTION

The relation between the structure and the (theoretical) morphology of a crystal is best described by the periodic bond chain theory [1,2]. In this model a crystal is considered as an interwoven network of chains of strong bonds, the so-called periodic bond chains (PBCs). The surfaces of crystals can be classified into three main types: i) F faces consisting of two or more non-parallel, interconnected PBCs within a slice $d_{hkl}$; ii) S faces with one PBC in slice $d_{hkl}$ and iii) K faces with no PBC in slice $d_{hkl}$. The slice thickness $d_{hkl}$ is the interplanar distance between two adjacent net planes (hkl), corrected for the extinction conditions of the space group in question. Normally the F-faces are flat and grow layer-wise, involving steps, two-dimensional nuclei and growth spirals. The S and K faces are rough on an atomic scale and grow by random addition of growth units. In general, the F-faces are the slowest growing ones and therefore determine the shape of a crystal.

For diamond there is only one type of periodic bond chain, which is the zigzag C / C \ C / C \ chain parallel to <110> [3]. Slice $d_{111}$ is the only one in diamond that contains more than one interwoven <110> PBC. This means that {111} is the only F-face. One PBC is present in the slices $\{hhk\}_{h>k}$, therefore the faces in the zone between {111} and {110} are S type. All remaining faces are K type. This implies that the theoretical habit of diamond is the octahedron, which is bounded by the six F-type {111} faces. Steps on these faces should be parallel to <110>, running towards the $<11\bar{2}>$ directions.

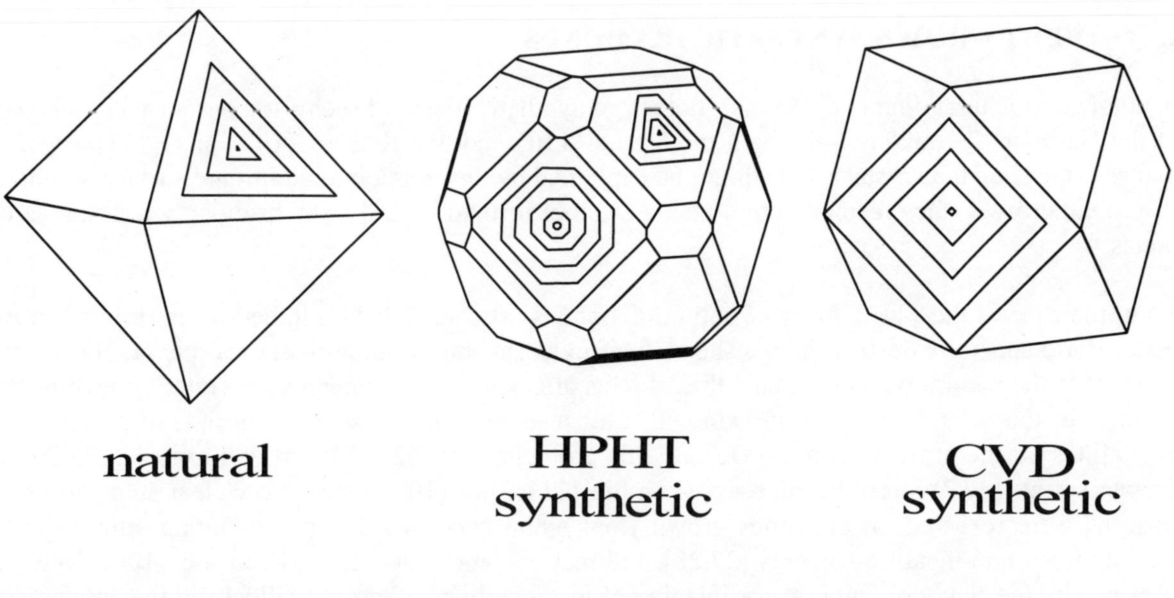


FIGURE 1 Idealised morphologies of natural, HPHT-grown and CVD-grown diamonds. For the {111} and {100} faces characteristic orientations of growth steps are also indicated.

FIGURE 1 shows the observed shapes of as-grown natural diamonds, high temperature high pressure (HPHT) grown synthetics and chemical vapour deposition (CVD) grown diamonds, including the step patterns on the different faces. Only the natural diamonds agree with the theoretical habit.

## B NATURAL DIAMOND

It is generally assumed that almost all natural diamonds were formed in the upper mantle at a depth of 150 to 200 km [4]. Probably the crystals grew from a silicate magma solution [5] at temperatures between 900°C and 1300°C and pressures of about 50 kbar [6]. The original growth form of most natural diamonds is octahedral, as predicted by the PBC theory. However, during the process of ascent to the Earth's surface by volcanic eruption almost all the octahedrally shaped crystals suffered dissolution. In the case of slight dissolution the surfaces became covered by trigonal etch pits [7], which are shown to be related to the surface outcrops of dislocations in the crystals [8–10]. In cases of severe etching the diamonds turned dodecahedral [11–13]. Only a minor fraction ($\approx 10^{-3}$) of the natural diamonds shows the original surface patterns, consisting of shallow, triangular growth hillocks with steps parallel to <110> as shown in FIGURE 1 [8–10,14]. It was shown that the centres of the hillocks are related to dislocations, which indicates spiral growth. In other cases growth steps were generated by enhanced two-dimensional nucleation from foreign fluid or solid 'particles' contacting the crystal surface [14]. From the anisotropy of the step patterns it was concluded that the rate-limiting process in natural diamond growth is the addition of growth units at the kink positions of the growth steps. Further it was deduced that the surface energy of the diamond/magma interface is about 15% of the broken bond energy (5.3 $J/m^2$) of an ideal diamond/vacuum interface [14].

From examination of growth zoning by using X-ray diffraction [15–17] and chemical etching [12], it was concluded that a small number of natural diamonds also showed non-faceted, cuboid faces during their growth [18]. Diamonds whose cuboid morphology has been preserved were found in the Jwangeng mine in Botswana [19]. The rounded shape cubic faces agrees with the K-character of the form {100}. The slow growth rate of the otherwise fast K-form is probably due to the presence of an adsorbed layer, which retards crystallisation. This is confirmed by the different kinds of point defects in the growth sectors related to these faces as compared to those of the {111} faces [20].

## C HPHT-GROWN SYNTHETIC DIAMONDS

In contrast to natural diamond, the morphology of synthetic diamond seems to conflict with the results of the Hartman–Perdok analysis: both the S face {110} and two K-faces {100} and {113} appear as planar F-faces on the crystals. This might be explained by the creation of additional surface stabilizing PBCs, either by surface reconstruction via C–C dimer formation [21] or by bridging via metal–carbon bonds.

As in the case of the {111} faces of natural diamonds, also for HPHT synthetics the original growth patterns are generally destroyed by a shut-off effect at the end of the growth experiment. However, in contrast to the natural diamonds, now the artifacts are caused by a sudden accelerated growth prior to removal of the crystals from the growth cell. This leads to dendrite-shaped imprints of solvent metal crystallites on the growth surfaces. Despite this problem, steps [22–24], growth hillocks [25,26] and growth spirals [27,28] have been observed on the {111} and {100} faces. Very clear step and hillock patterns were revealed on diamonds grown from metal catalyst-solvents containing silicon [26] as well as from non-metallic solvents [27,28]. Indirect evidence for step flow on the {100} faces was obtained by the cathodoluminescence imaging of fossil patterns of growth hillocks in the cubic sectors of synthetic diamonds [29]. For the octahedral growth sectors a direct, one-to-one correspondence between cathodoluminescence fossil patterns and growth hillocks on the {111} faces could be established [26,28]. The observed step directions on the {111} faces are parallel to <110>; on the {100} faces they are parallel to <110> as well as <100>.

## D CVD-GROWN DIAMONDS

### D1 3D Morphology

The habit of CVD-grown polycrystalline diamonds is generally determined by two flat faces, namely {111} and {100}. Therefore the growth form of these crystals ranges from octahedral, via cubo-octahedral to cubic, depending on the relative growth rates of both faces, which is expressed as $\alpha = v_{100}\sqrt{3}/v_{111}$ [30,31]. This anisotropy factor can be controlled by changing the growth conditions, such as the deposition temperature and the gas phase composition [31,32]. Increasing the substrate temperature [31,32] as well as adding oxygen to the gas phase [33] tends to decrease α, favouring cubic crystal shapes. The diamond particles often show multiple twinning, especially on the {111} faces [31], which is largely determined by the α factor [34]. Addition of impurities, such as boron, nitrogen or phosphorus in larger amounts to the growth system [35], as well as large $CH_3/H$ ratios [36] in the gas phase, lead to non-faceted morphologies. In such conditions, the {111} facets roughen most easily, while {100} faces often persist. This leads to dendritic diamond shapes topped by well faceted {100} facets [37]. If the situation worsens, spherical diamond(-like) particles are formed due to increased secondary nucleation [35,36].

### D2 General Aspects

Because of the low pressure of the gas phase during CVD diamond growth, almost no surface artifacts develop due to a shut-off effect at the end of the deposition experiments. Therefore, the original growth patterns are well preserved, which enables their close examination with the help of ex situ optical and scanning probe microscopy. Nearly all of the observations summarised below concern homoepitaxially grown diamond layers. Recently also in situ scanning tunnelling microscopy (STM) of diamond surfaces, i.e. during growth and etching, has successfully been carried out [38]; in this way shut-off effects are completely avoided and time-dependent information is obtained.

The most important features encountered on both the {111} and the {100} faces of CVD-grown diamond are growth hillocks, monoatomic and higher steps and foreign, randomly deposited or twinned diamond nuclei [39–43]. The steps originate from screw dislocations leading to spiral growth [44,45] or are generated by 2D contact nucleation induced by the presence of the foreign diamond crystallites on the surface [38,40–44]. It has also been suggested that the steps on the {001} faces result from the occurrence of repeated two-dimensional nucleation on the top of growth hillocks when the step density is low [46].

The presence of growth hillocks and foreign diamond crystallites on the {100} and {111} surfaces is a nuisance in producing high quality single crystalline diamond layers, e.g. for electronic applications. It was found that using a substrate a few degrees misoriented from the exact crystallographic plane largely improved the crystallographic quality of the grown layers. Such layers grow by misorientation step flow and almost no hillocks or foreign diamond crystallites were found on their surfaces [41,44,46,47].

### D3 {100} Surfaces

Homoepitaxial growth on single crystalline {100} diamond substrates produces diamond layers of the highest perfection. For the {100} surface the occurrence of steps seems to disagree with the results of the periodic bond chain analysis. However, both modelling [48–51] and observation by scanning tunnelling microscopy [48,51-54] have shown that the hydrogen terminated surface is $2\times1$ reconstructed by dimer formation. This leads to a non-zero step energy, which makes the {100} surface behave as an F-face. Both atomic height, $d_{400}$ [52–55], and double height $d_{200}$ [56] steps more or less parallel to the PBC directions <110> have been reported. The single steps where the dimer rows on the upper terrace are parallel to the step edge, are straight, and those where the dimer rows are perpendicular are ragged, showing a high density of kinks [53]. Apart from the lowest steps imaged by STM, optical microscopy also revealed bunched steps and macrosteps, again parallel to <110> [44], especially on thicker diamond layers grown by the flame method [57]. It was shown that nitrogen impurities play a crucial role in the occurrence of step bunching [58].

The rate-limiting step in the {100} diamond growth process is still under debate: surface diffusion towards steps or the integration of the growth units at the step positions? From observed patterns of growth hillocks emerging in misorientation step trains and using kinematic wave theory for step propagation, it was inferred that step integration rather than surface diffusion is rate-limiting in diamond growth [42,59]. This seem to agree with STM observations, which suggest that the surface diffusion length of growth units is very short [48]. On the other hand, however, the occurrence of regular multi-atomic steps as imaged by STM [55] and the need for surface diffusion to explain hillock formation by repeated 2D nucleation [46] point to an important role of surface diffusion.

## D4 {111} Surfaces

Homoepitaxial growth on {111} substrates generally produces poor quality diamond layers, characterised by numerous growth hillocks, twins and cracks [40–42]. Therefore, compared to {100} these layers received less attention in the literature. The {111} layers are very prone to twinning [31,33,34]: the growth of thicker {111} films often leads to multiple-twinned growth layers, finally ending in polycrystalline layers [42,60]. This prohibits the growth of large diamond crystals. Moreover, the thin {111} layers show a network of cracks, which in fact are cleavages parallel to {111} [40–42]. Probably fracture is introduced by stresses resulting from a difference in lattice parameter between the substrate and the grown layers. Films grown on substrates with slight misorientation from {111} did not reveal growth hillocks, but still suffer fracture [42]. By the flame technique, using high substrate temperatures thick {111} layers of better quality can be produced [61–63].

From the shape of the growth hillocks on diamond layers grown by the hot filament method it was deduced that the steps are roughly parallel to $\langle 1\bar{1}0\rangle$, running towards the $\langle 11\bar{2}\rangle$ directions [42] as predicted by the PBC method. A similar step direction has been observed by STM [48,64]. However, for flame-grown diamonds [60,61] the steps were found to propagate in the opposite directions $\langle \bar{1}\bar{1}2\rangle$, which indicates the occurrence of step reconstruction [21]. The steps imaged by STM are of unit height, $d_{111}$ [45,48,64]; kinks on these steps have been reported as well [56]. On the terraces between adjacent steps STM also revealed adsorbates, presumably adsorbed $CH_3$ molecules, which may act as precursors for diamond growth [48,64]. Theoretical and experimental work [48,65] indicates that during CVD growth the {111} surfaces generally are $(1\times1)$ monohydride-terminated, i.e. are close to the bulk structure. However evidence also exists for the occurrence of a (partly) trihydride, i.e. $CH_3$, terminated surface [48,65].

From the development of cleavage steps into a bunched step pattern with a discontinuity in slope at the top and using the kinematic wave theory of step propagation it was suggested that {111} diamond growth is limited by surface diffusion of growth species towards the steps [38,42]. The mean displacement, $X_s$, of adsorbed growth units is estimated to be 4 nm.

## D5 {110}, {113} and Other Surfaces

The {110} surfaces, which are S-type according to the broken bond model, are rough on an atomic [52] and a microscopic scale [40–43]. The surface roughness increases with layer thickness [42]. For very thick {110} layers grown by the flame method a pattern of microfacets develops, the orientations of which are close to the form $\{hhk\}_{h<k}$ [57,62]. Compared to the F-like {100} and {111} surfaces, the {110} faces grow fast [41,42]. Because of this high growth rate, randomly nucleated and twinned diamond crystallites are not formed on these surfaces, because they are buried at an early stage of growth [41,43]. Unfortunately, compared to the {100} faces, the crystalline perfection of the {110} layers is considerably less, as, for instance, was shown by fractography [60]. Recently, the occurrence of planar, stepped {110} faces on arrow-shaped diamond crystallites, grown from ethanol-hydrogen mixtures at atmospheric pressure and very high substrate temperature, have been reported [66]. Possibly the F-behaviour of the {110} face in this case results from surface reconstruction or oxygen adsorption.

Flame growth experiments using cylindrical diamond substrates at temperatures exceeding 1225°C revealed small {113} facets [61–63,67,68]. The striated patterns on these diamond surfaces consist of numerous macro-steps parallel to the PBC direction <110>. The F-like character of the {113} facets is explained by C-C dimer formation perpendicular to the <110> PBCs as a result of surface reconstruction [62,63]. These interactions between the growth units perpendicular to <110> are weak, which makes the {113} surfaces slightly curved. At lower temperatures the surface reconstruction diminishes due to an increased surface coverage by hydrogen atoms. Then the F-character of the {113} faces disappears and a curved band $\{hhk\}_{h<k}$ develops, which is a set of S-faces in the $<1\bar{1}0>$ zone [62,63]. The (chemical) roughening temperature of {113} is estimated to be 1200 ±100°C [67]. Recently {113} facets have also been observed on diamond particles grown by microwave plasma CVD at the high deposition temperature of 1200°C on a Co substrate [69].

Surface orientations other than {111}, {100}, {110} and {113} have not been scrutinised in detail. However, flame growth experiments using cylindrical and spherical diamond substrates showed that all faces in the zone $\{hhk\}_{h<k}$ are S-type and all other possible faces {hkl} behave as rough, i.e. K-type [62,63]. The S-type faces in the zone $\{hhk\}_{h<k}$ (which is different from $\{hhk\}_{h>k}$ as predicted by the PBC method!) appear as curved, striated growth bands, whereas the K-faces are characterised by microfaceted structures composed of $\{hhk\}_{h<k}$ facets [62].

## E CONCLUSION

The examination of as-grown crystal surfaces using optical microscopy, scanning electron microscopy and scanning probe microscopy is a most powerful way to unravel the mechanisms of diamond growth. Assuming only nearest-neighbour interaction and using the PBC method, diamond should grow as an octahedron with steps parallel to <110> on the {111} faces. This was only found for natural diamonds. Man-made diamonds, either by the HPHT method or by CVD, do not obey simple theory. Sometimes they grow as octahedrons, but in most cases as cubo-octahedrons, cubes or other shapes. This is due to the complicated surface processes during crystal growth, involving surface reconstruction, chemical kinetics and adsorption by metal, hydrogen and various kinds of impurity atoms. Therefore, much work remains to be done before diamond growth is fully understood.

## REFERENCES

[1] P. Hartman, W.G. Perdok [ *Acta Crystallogr. (Denmark)* vol.8 (1955) p.49, p.521, p.525 ]

[2] P. Bennema [ in *Handbook of Crystal Growth* vol.1, Ed. D.T.J. Hurle (North Holland, Amsterdam, 1993) ch.7 ]

[3] P. Hartman [ *Z. Kristallogr. (Germany)* vol.121 (1965) p.78 ]
[4] H.O.A. Meyer [ *Am. Mineral. (USA)* vol.70 (1985) p.344 ]
[5] I. Sunagawa [ *J. Cryst. Growth (Netherlands)* vol.99 (1990) p.1156 ]
[6] H.O.A. Meyer [ in *Mantle Xenoliths* Ed. P.H. Nixon (Wiley, Chichester, 1987) p.501-24 ]
[7] S. Tolansky [ *The Microstructures of Diamond Surfaces* (N.A.G. London, 1955) ]
[8] F.C. Frank, A.R. Lang [ in *Physical Properties of Diamond* Ed. R. Berman (Clarendon, Oxford, 1965) p.69-115 ]
[9] I. Sunagawa, K. Tsukamoto, T. Yashuda [ in *Materials Science of the Earth's Interior* Ed. I. Sunagawa (TERRAPUB, Tokyo, 1984) p.331-49 ]
[10] W.J.P. van Enckevort, M. Seal [ *Philos. Mag. A (UK)* vol.57 (1988) p.939-54 ]
[11] M. Seal [ *Phys. Status Solidi (Germany)* vol.3 (1963) p.658 ]
[12] M. Seal [ *Am. Mineral. (USA)* vol.50 (1965) p.105 ]
[13] M. Moore, A.R. Lang [ *J. Cryst. Growth (Netherlands)* vol.26 (1974) p.133 ]
[14] W.J.P. van Enckevort [ *J. Cryst. Growth (Netherlands)* vol.119 (1992) p.177 ]
[15] A.R. Lang [ *J. Cryst. Growth (Netherlands)* vol.24/25 (1974) p.108 ]
[16] A.R. Lang [ in *The Properties of Diamond* Ed. J.E. Field (Academic, London, 1979) p.425-69 ]
[17] S. Suzuki, A.R. Lang [ *J. Cryst. Growth (Netherlands)* vol.34 (1976) p.29 ]
[18] F.C. Frank [ in *Proc. Int. Ind. Diamond Conf. Oxford, Vol I, Science* Ed. J. Burls (Industrial Diamond Information Bureau, London, 1967) p.119-35 ]
[19] C.M. Welbourn, M.T. Rooney, D.J.F. Evans [ *J. Cryst. Growth (Netherlands)* vol.94 (1989) p.229 ]
[20] W.J.P. van Enckevort, E.P. Visser [ *Philos. Mag. (UK)* vol.62 (1990) p.597-614 ]
[21] L.J. Giling, W.J.P. van Enckevort [ *Surf. Sci. (Netherlands)* vol.161 (1985) p.567 ]
[22] S. Tolansky, I. Sunagawa [ *Nature (UK)* vol.184 (1959) p.1526 ]
[23] S. Tolansky, I. Sunagawa [ *Nature (UK)* vol.185 (1959) p.203 ]
[24] H. Kanda, T. Ohsawa, O. Fukunaga, I. Sunagawa [ *J. Cryst. Growth (Netherlands)* vol.94 (1989) p.115 ]
[25] W.J.P. van Enckevort, G.Z. Liu [ PhD Thesis, University of Nijmegen, Netherlands, 1982 ]
[26] H. Kanda, T. Ohsawa [ *Diam. Relat. Mater. (Switzerland)* vol.5 (1996) p.8-12 ]
[27] H. Kanda, M. Akaishi, S. Yamaoka [ *J. Cryst. Growth (Netherlands)* vol.106 (1990) p.471 ]
[28] H. Kanda, M. Akaishi, S. Yamaoka [ *J. Cryst. Growth (Netherlands)* vol.108 (1991) p.421 ]
[29] F.C. Frank, A.R. Lang, D.J.F. Evans, M.L.T. Rooney, P.M. Spear, C.M. Welbourn [ *J. Cryst. Growth (Netherlands)* vol.100 (1990) p.119-35 ]
[30] C. Wild, N. Herres, P. Koidl [ *J. Appl. Phys. (USA)* vol.68 (1990) p.973 ]
[31] C. Wild, R. Kohl, N. Herres, W. Müller-Sebert, P. Koidl [ *Diam. Relat. Mater. (Switzerland)* vol.3 (1994) p.373-81 ]
[32] J.W. Lindsey, J.M. Larson, S.L. Girshick [ *Diam. Relat. Mater. (Switzerland)* vol.6 (1997) p.481-5 ]
[33] R.E. Rawles, W.G. Morris, M.P. D'Evelin [ *Appl. Phys. Lett. (USA)* vol.69 (1996) p.4032-4 ]
[34] G. Knuyt, M. Nesladek, T. Vandevelde, K. Meykens, C. Quaeyhaegens, L.M. Stals [ *Diam. Relat. Mater. (Switzerland)* vol.6 (1997) p.435-9 ]
[35] R. Haubner, S. Bohr, B. Lux [ *Diam. Relat. Mater. (Switzerland)* vol.8 (1999) p.171-8 ]
[36] E. Kondoh, T. Ohta, T. Mitomo, K. Ohtsuka [ *Diam. Relat. Mater. (Switzerland)* vol.3 (1994) p.270-6 ]
[37] J.J. Schermer, F.K. de Theije [ *Diam. Relat. Mater. (Switzerland)* vol.8 (1999) p.2127-39 ]
[38] W.J.P. van Enckevort, G. Janssen, W. Vollenberg, J.J. Schermer, L.J. Giling [ *Diam. Relat. Mater. (Switzerland)* vol.2 (1993) p.997-1003 ]
[39] L. Ackermann, W. Kulisch [ *J. Cryst. Growth (Netherlands)* vol.8 (1999) p.1256-60 ]
[40] M. Kamo, H. Yurimato, Y. Sato [ *Appl. Surf. Sci. (Netherlands)* vol.33/34 (1988) p.553 ]

[41] W.J.P. van Enckevort, G. Janssen, W. Vollenberg, M. Chermin, L.J. Giling, M. Seal [ *Surf. Coat. Technol. (Switzerland)* vol.47 (1991) p.39-50 ]

[42] W.J.P. van Enckevort, G. Janssen, W. Vollenberg, L.J. Giling [ *J. Cryst. Growth (Netherlands)* vol.148 (1995) p.365-82 ]

[43] C. Findeling-Dufour, A. Vignes, A. Chiquel [ *Diam. Relat. Mater. (Switzerland)* vol.4 (1995) p.429-34 ]

[44] W.J.P. van Enckevort, G. Janssen, J.J. Schermer, L.J. Giling [ *Diam. Relat. Mater. (Switzerland)* vol.4 (1995) p.250-5 ]

[45] T.J. Kreutz et al [ *Phys. Rev. B (USA)* vol.51 (1995) p.14554-8 ]

[46] N. Lee, A. Badzian [ *Diam. Relat. Mater. (Switzerland)* vol.6 (1997) p.130-45 ]

[47] N. Lee, A. Badzian [ *Appl. Phys. Lett. (USA)* vol.66 (1995) p.2203-5 ]

[48] H. Kawarada [ *Surf. Sci. Rep. (Netherlands)* vol.26 (1966) p.205-59 ]

[49] J. Furthmüller, J. Hafner, G. Kresse [ *Phys. Rev. B (USA)* vol.53 (1996) p.7334-51 ]

[50] M.D. Winn, M. Rassinger, J. Hafner [ *Phys. Rev. B (USA)* vol.55 (1997) p.5364-75 ]

[51] Th. Frauenhofer, U. Stephan, P. Blandeck, D. Porezag, H.-G. Busmann, W. Zimmermann-Edling [ *Diam. Relat. Mater. (Switzerland)* vol.3 (1994) p.966-74 ]

[52] H.G. Maguire, M. Kamo, H.P. Lang, H.-J. Güntherodt [ *Appl. Surf. Sci. (Netherlands)* vol.75 (1994) p.144-50 ]

[53] Y. Kuang, N. Lee, A. Badzian, T.T. Tsong, T. Badzian, C. Chen [ *Diam. Relat. Mater. (Switzerland)* vol.4 (1995) p.1371-5 ]

[54] Y. Kuang, Y. Wang, N. Lee, A. Badzian, T. Badzian, T.T. Tsong [ *Appl. Phys. Lett. (USA)* vol.67 (1995) p.3721-3 ]

[55] K. Hayashi, S. Yamanaka, H. Okushi, K. Kajimura [ *Diam. Relat. Mater. (Switzerland)* vol.5 (1996) p.1002-5 ]

[56] T. Tsuno, T. Tomikawa, S. Shikata, T. Imai, M. Fujimori [ *Appl. Phys. Lett. (USA)* vol.64 (1994) p.572-4 ]

[57] J.J. Schermer, W.J.P. van Enckevort, L.J. Giling [ *Diam. Relat. Mater. (Switzerland)* vol.3 (1994) p.408-16 ]

[58] F.K. de Theije, J.J. Schermer, W.J.P. van Enckevort [ *Diam. Relat. Mater. (Switzerland)* accepted (2000) ]

[59] W.J.P. van Enckevort, G. Janssen, L.J. Giling [ *J. Cryst. Growth (Netherlands)* vol.113 (1991) p.295-304 ]

[60] G. Janssen, W.J.P. van Enckevort, W. Vollenberg, L.J. Giling [ *Diam. Relat. Mater. (Netherlands)* vol.1 (1992) p.789-800 ]

[61] J.J. Schermer, L.J. Giling, P. Alers [ *J. Appl. Phys. (USA)* vol.78 (1995) p.2376-84 ]

[62] J.J. Schermer, W.J.P. van Enckevort, L.J. Giling [ *J. Cryst. Growth (Netherlands)* vol.148 (1995) p.248-60 ]

[63] J.J. Schermer, W.J.P. van Enckevort, L.J. Giling [ *J. Cryst. Growth (Netherlands)* vol.166 (1995) p.622-7 ]

[64] H. Sasaki, H. Kawarada [ *Jpn. J. Appl. Phys. (Japan)* vol.32 (1993) p.L1771 ]

[65] X.M. Zheng [ *Surf. Sci. (Netherlands)* vol.364 (1996) p.141-50 ]

[66] Y. Aoki, Y. Mitsuoka, Y. Nakamuta [ *Diam. Relat. Mater. (Switzerland)* vol.8 (1999) p.1310-4 ]

[67] G. Janssen, J.J. Schermer, W.J.P. van Enckevort, L.J. Giling [ *J. Cryst. Growth (Netherlands)* vol.125 (1992) p.42-50 ]

[68] K.A. Snail, Z.P. Lu, R. Weimer, J. Heberlein, E. Pfender, L.M. Hanssen [ *J. Cryst. Growth (Netherlands)* vol.137 (1994) p.676-9 ]

[69] K.W. Chae, Y.-J. Baik [ *Diam. Relat. Mater. (Switzerland)* vol.8 (1999) p.1261-6 ]

## A4.2 Friction of diamond surfaces

J.E. Field

November 1999

### A INTRODUCTION

Diamond has a low coefficient of friction, high wear and erosion resistances and is used extensively in industry to 'work' other materials in processes such as polishing, grinding, cutting, machining, milling, sawing and wire drawing. The very high thermal conductivity of diamond allows many of these processes to be performed at high speeds since the heat is quickly dissipated from the interface without loss of hardness or graphitisation of the diamond. Polycrystalline diamond composites (PCDs) can be fabricated which, though they do not have quite the high hardness and thermal conductivity of diamond, have the advantages of greater toughness, more isotropic properties and larger physical dimensions. The low friction, high hardness combination also makes diamond attractive as a bearing material and as a surgeon's knife.

### B FRICTION

Tabor [1], Wilks and Wilks [2], and Tabor and Field [3] have reviewed research on the friction of diamond. The main results, in air, can be summarised as follows.

(a) The friction depends on crystal face and orientation. On the octahedral face, the friction is low and there is practically no frictional anisotropy. On the cube face, polished in the <100> direction, there is a four-fold symmetry in the friction. The lowest friction is along the <011> or cube diagonal direction, the highest along the <100> or cube edge direction.

(b) If the cube face is polished in the <110> direction, the four-fold symmetry is replaced by two-fold symmetry.

(c) In some situations repeated traversal over the same track on the cube face of diamond reverses the order of frictional anisotropy. After a few hundred traversals the friction is higher in the <011> direction and low in the <100> direction.

(d) The frictional anisotropy on the cube face becomes less marked if a conical stylus of large apical angle is used.

(e) The coefficient of friction may or may not depend on load depending on the geometry of the sliding system.

(f) The friction is scarcely changed by the addition of a conventional lubricant, though work by Feng and Field [4,5] has shown that water can reduce the friction coefficient, $\mu$, for diamond sliding on natural diamond or CVD diamond to as low as 0.02. Recent studies by Grillo and Field [6] show that for a low load of 0.5 N, a low sliding velocity of 0.2 mm/s and a pH for the water of $\approx 7$, $\mu$ can be reduced to 0.003.

(g) Over a wide range of experimental conditions the coefficient of friction lies between $\mu = 0.05$ and 0.15; it rarely exceeds $\mu = 0.2$, except under vacuum conditions.

(h) Finally, we note that although the wear rate on the (100) and (110) faces of diamond may depend on the sense of sliding, the friction in such situations does not depend on the sense of sliding.

For diamond sliding on diamond under vacuum conditions, the friction can reach a high value ($\mu \approx 1$), for example if the samples are in vacuum and are cleaned by heating or bombardment or alternatively if there are repeated traversals over the same track. The high friction is accompanied by a very high wear rate [7]. Bowden and Hanwell [7] also showed that if a mono-molecular oxygen film is allowed to reform the friction decreases dramatically. Feng et al [8] have studied the role of various gases in reducing the friction of clean surfaces. Atomic hydrogen had the greatest effect but both atomic and molecular oxygen reduced the friction coefficient significantly. Buckley and co-workers at NASA [9,10] have made a detailed study of the adhesion and friction of diamond in contact with various metals. The results showed that $\mu$ was related to the relative chemical activity of the metals in high vacuum. All the metals studied transferred to the surface of diamond during sliding under a vacuum of $10^{-8}$ N $m^{-2}$. Several authors have considered the dissipative processes when diamond slides on diamond. Samuels and Wilks [11] argue that the vibration of asperities can cause energy losses but consider that an adhesive term is unlikely. The most recent work [4–6] has shown that there must be adhesion and that shear transformation to graphite can occur at high loads and for some sliding directions [12,13].

The repeated sliding of diamond on diamond produces a waxy debris which consists primarily of hydrocarbons [4,14,15]. It suggests that in air the debris is formed by the interaction of the atmosphere and the wearing surface. This idea is reinforced by experiments in vacuum which show a more powdery deposit; it contains little hydrocarbon and graphite and consists mainly of amorphous carbon with small diamond particles.

### B1 Friction of PCDs

In general, these have higher coefficients of friction than diamond for a particular slider substrate configuration. The binder material (for example, cobalt) controls the friction coefficient [16]. A great practical advantage is that the friction is now essentially isotropic.

## C CONCLUSION

A great deal is known about the friction of diamond surfaces. The detailed mechanisms of energy dissipation still remain a subject of debate, but adhesion and shear transformation to graphite are important dissipative mechanisms.

## REFERENCES


[1] D. Tabor [ *in The Properties of Diamond* Ed. J.E. Field (Academic Press, London, UK, 1979) p.325-50 ]

[2] E.M. Wilks, J. Wilks [ *The Properties and Applications of Diamond* (Butterworth-Heinemann, Oxford, UK, 1991) ]

[3] D. Tabor, J.E. Field [ *The Properties of Natural and Synthetic Diamond* Ed. J.E. Field (Academic Press, London, UK, 1992) p.547-71 ]

[4] Z. Feng, J.E. Field [ *Surf. Coat. Technol. (Switzerland)* vol.47 (1991) p.631-45 ]

[5] Z. Feng, J.E. Field [ *J. Phys. D, Appl. Phys. (UK)* vol.25 (1992) p. A33-9 ]

[6] S.E. Grillo, J.E. Field [ *J. Phys. D, Appl. Phys. (UK)* submitted ]

[7] F.P. Bowden, A.E. Hanwell [ *Proc. R. Soc. Lond. A (UK)* vol.295 (1966) p.233-42 ]

[8] Z. Feng, Y. Tzeng, J.E. Field [ *J. Phys. D, Appl. Phys. (UK)* vol.25 (1992) p.1418-24 ]

[9] K. Miyoshi, D.H. Buckley [ *Appl. Surf. Sci. (Netherlands)* vol.6 (1980) p.161-72 ]
[10] S.V. Pepper [ NASA Tech. Memo. 82741, 1-13, 1981 ]
[11] B. Samuels, J. Wilks [ *J. Mater. Sci. (UK)* vol.23 (1988) p.2846-64 ]
[12] F.M. van Bouwelen, J.E. Field, L.M. Brown [ *Ind. Diam. Rev. (UK)* vol.57 (1997) p.21-6 ]
[13] Y.G. Gogotsi, A. Kailer, K.G. Nickel [ *Nature (UK)* vol.401 (1999) p.663-4 ]
[14] M. Seal [ *Ind. Diam. Rev. (UK)* vol.35 (1965) p.111-6 ]
[15] S.E. Grillo, J.E. Field [ *J. Phys. D, Appl. Phys. (UK)* vol.30 (1997) p.2466-72 ]
[16] C.J. Freeman, J.E. Field [ *J. Mater. Sci. (UK)* vol.24 (1989) p.1069-72 ]

# A4.3 Wear and abrasion of diamond surfaces

J.E. Field

November 1999

## A INTRODUCTION

There is a considerable practical interest in the abrasion and wear properties of diamond (for reviews, see Wilks and Wilks [1–5]). Firstly there are the situations where diamond is ground and polished to make jewellery or industrial tools. Secondly, there are the many industrial applications where diamond is used for polishing, grinding, cutting and machining of other materials. Wear is closely related to polishing and erosion. These specialist topics are dealt with in more detail in Datareviews A4.4 and A4.5.

## B WEAR AND ABRASION

The work of Wilks and Wilks has demonstrated three main results about the abrasion resistance of diamond.

1. Its great dependence on the orientation of the facet and the direction of abrasion.
2. The very sharp changes in magnitude that occur near certain planes and directions: a change of a few degrees can more than double the material removal rate.
3. The impossibility of representing the variation of abrasion resistance by a simple vector quantity. In other words, if the direction of traversal is reversed a different wear rate can result.

The technique adapted by Wilks and Wilks for abrasive wear studies involves rotating a cast iron or bonded wheel of 1 cm radius and allowing the V-shaped edge to be loaded against the diamond under test. The cuts produced are then measured using multiple-beam interferometry at spacings of $\lambda/2$, where $\lambda$ is the wavelength of a monochromatic light source (546.1 nm). An important result is that each revolution of the wheel removes approximately the same amount of material over a wide speed range. This suggests that thermal effects at 'hot spots' are not important. An example from their extensive data is given in TABLE 1. Note that there is a factor of 10 difference in the depth of cuts, which implies a difference of over 100 in the rate of removal of material!

TABLE 1 Depths of cuts made by bonded wheels, normalised to a depth of 10 ($\lambda/2$) in the softest direction [3].

| Plane and Direction | | | | | | |
|---|---|---|---|---|---|---|
| Plane | Dodecahedron (011) | | Cube (001) | | Octahedron (111) | |
| Direction | [100] | [100] | [100] | [110] | $[11\bar{2}]$ | $[\bar{1}\bar{1}2]$ |
| Depth of cut ($\lambda/2$) | 10 | 1.8 | 8.5 | 1.1 | 3.9 | 2.1 |

The first attempt at explaining the mechanism of material removal was that of Tolkowsky [6]. His model, subsequently refined and developed by Wentorf [7] and Wilks and Wilks [2,3] is based on micro-cleavage with cleavage on {111} planes dominating. The model proved successful in explaining the main features of the results but this now appears fortuitous. Modern, high resolution studies of diamond surfaces do not support the thesis that micro-cleavage is the dominant mechanism (see Datareview A4.4).

It is unfortunate that some authors have introduced the phrase 'abrasion hardness'. The 'hard' and 'soft' planes and directions for abrasion are not the same as with indentation hardness, which is not surprising since abrasion of diamond can involve brittle processes. For this reason the phrase 'abrasion resistance' is to be preferred.

### B1 Wear at Light Loads

Crompton et al [8] describe experiments on the wear of diamonds rubbing a range of materials at a speed of 88 mm $s^{-1}$. They showed that the rate of wear was largely determined by the hardness of the material. The rates of wear produced on the diamonds by a diamond grinding wheel were comparable with removal rates obtained when polishing diamond on a bonded wheel, as might be expected. Wear was only observed after rubbing had proceeded for an incubation period, the length of which decreased steadily with increasing hardness of the rubbing material. It appears that during this initial period the abrasion is producing fatigue damage in the form of subsurface cracks, and that material only begins to be removed when the damage reaches some critical value, as in observations of the fatigue of diamond by repeated indentations (see EMIS Datareviews Series no.9, Datareview 1.8, Section F6).

### B2 Chemical Wear

In addition to mechanical wear, a diamond can also suffer wear by thermal and chemical processes. The most common of these are graphitisation and oxidation processes. If oxygen, or other oxidising agents, for example sodium nitrate, are present, a black surface coating can form at ~900 K. This is not true graphitisation, which involves the transition of diamond to graphite without the aid of external agents. If diamonds are heated in an inert atmosphere, the onset of graphitisation can be detected at ~1800 K and the rate of graphitisation increases rapidly until at ~2400 K a 0.1 carat octahedron is converted totally in less than three minutes.

The other possible form of chemical attack is by two groups of metals. The members of the first group are avid carbide formers, and include tungsten, tantalum, titanium and zirconium. At high temperatures these will react chemically with diamond to form their respective carbides. The second group includes iron, cobalt, manganese, nickel and chromium, and also the platinum group of metals. In the molten state these metals are true solvents for carbon.

This has important consequences for the machining of ferrous alloys with diamond tools. Hard ferrous metals are usually ground with composites based on cubic boron nitride.

## C CONCLUSION

Diamond exhibits a low wear rate. Natural diamond is very anisotropic in its wear properties. Wear can be by mechanical, thermal and chemical processes. Some metals are solvents for carbon or form carbides and this has implications for the machining of such materials.

## REFERENCES

[1] E.M. Wilks, J. Wilks [ *The Properties and Applications of Diamond* (Butterworth-Heinemann, Oxford, UK, 1991) ]

[2] E.M. Wilks, J. Wilks [ *in Physical Properties of Diamond* Ed. R. Berman (Clarendon Press, Oxford, UK, 1965) p.221-56 ]
[3] E.M. Wilks, J. Wilks [ *J. Phys. D, Appl. Phys. (UK)* vol.16 (1972) p.1772-81 ]
[4] J. Wilks, E.M. Wilks [ *The Properties of Diamond* (Academic Press, London, UK, 1979) p.351-82 ]
[5] J. Wilks, E.M. Wilks [ in *The Properties of Synthetic and Natural Diamond* Ed. J.E. Field (Academic Press, London, UK, 1992) ]
[6] M. Tolkowsky [ DSc Thesis, University of London, 1920 ]
[7] R.H. Wentorf Jr. [ *J. Appl. Phys. (USA)* vol.30 (1959) p.1765-8 ]
[8] D. Crompton, W. Hirst, M.G.S. Howes [ *Proc. R. Soc. Lond. A (UK)* vol.333 (1973) p.435-54 ]

# A4.4 Polishing diamond surfaces

J.E. Field

November 1999

## A INTRODUCTION

The practical details of the mechanical polishing of diamond can be found in texts by Grodzinski [1] and Bruton [2]. Reviews have been given by Wilks and Wilks [3–5]. In general, diamonds are shaped to form jewellery or tools by grinding or polishing on a rotating wheel or 'scaife' consisting either of cast iron replenished with a mixture of diamond powder and oil or of diamond powder bonded into a metal or resin matrix. The rate of removal of material depends to a surprising extent on the plane of the diamond being polished, and the direction of polish. There are so-called 'soft' and 'hard' directions. Removal rates may differ by as much as 100:1 using a bonded scaife, and even more with a cast iron wheel where the harder directions of the diamond tend to dislodge the diamond dust from the scaife. The skilled polisher will know the hard and soft directions from the sound and ease of polishing. As is well known, a gemstone can have an intricate pattern of facets added to optimise the quantity of diamond and the stone's 'brilliance' and 'sparkliness' or 'fire'. The abrasion resistance is clearly not a vector quantity, and information on the abrasion resistance is only available for planes lying in the principal zones, and for positions quite close to them. For an account of these variations, with references to earlier work, see Wilks and Wilks [3].

There is a growing interest in polishing chemical vapour deposition (CVD) diamond. The two main barriers for the commercial exploitation of CVD diamond are the growth times and the difficulty and time involved in achieving a satisfactory polish. The problems are compounded with CVD diamond since the material is polycrystalline. This has stimulated research into a whole range of techniques; see [6] for a review.

## B MECHANICAL POLISHING

A very characteristic feature of the polishing of diamond is that the rate of removal of material is proportional to the speed of the abrading particles. This result was first established by Tolkowsky [7] who measured the loss of weight of diamonds polished under typical commercial conditions, and has since been confirmed by Wilks and Wilks [3,8] using a micro-abrasion tester which yields similar abrasion rates with rather more precision. This relationship means that the amount of material removed is proportional to the number of revolutions of the scaife, irrespective of the speed of the scaife. It follows that the abrasion process is not thermally activated.

The micro-cleavage model as outlined in Datareview A4.3, Section B, is superficially very good at explaining the so-called 'hard' and 'soft' directions. However, it is now possible to study surfaces down to atomic resolution and these do not provide evidence of small octahedra being plucked out. In the 'hard' directions there is evidence of cleavage with the {111} cleavage planes dominating but in the 'soft' directions the surfaces show smooth shallow grooves [9,10]. There is now evidence that in the 'soft' directions shear transformation of the surface layers to graphite takes place [11,12].

## C OTHER POLISHING MECHANISMS

In addition to mechanical polishing, Malshe et al [6] also consider the conversion of diamond to graphite, followed by the removal of the non-diamond forms by micro-chipping or atomic diffusion and the removal of carbon atoms by evaporation, ablation, sputtering or chemical reaction.

Graphite is the stable room temperature form of carbon. Some metals such as iron, cobalt and nickel decrease the activation energy and aid the transformation to graphite. These metals, plus manganese, chromium and the platinum group of metals, are all solvents of carbon when they are in the molten state. Other metals such as tungsten, tantalum, titanium and zirconium are avid carbide formers. Jin and co-workers [13,14] have studied a range of high carbon solubility metals as potential replacements for the conventional iron polishing scaife. Increases of carbon solubility of 25% over that for iron can be achieved but usually at higher than ambient temperatures. The benefits in polishing rates have to be balanced against costs.

Conventional thermo-chemical polishing involves adding an oxidising agent such as potassium nitrate to the polishing scaife [1]. More recently Yoshikawa et al [15–17] have tried polishing with heated metal plates of iron, magnesium, molybdenum and nickel. The polishing rate is controlled by the diffusion of carbon atoms from the diamond into the metal. This rate decreases as the carbon concentration in the metal increases. The rate is increased by operating at higher temperatures and is affected by the gaseous environment. Under vacuum conditions the polishing rate is at its highest and this is recommended at the start of a process. Hydrogen is added at a later stage since, though the polishing rate drops, the surface finish of the diamond is better.

The extensive literature on laser polishing is reviewed by Malshe et al [6]. Sputtering [18,19], plasma etching [20], and the use of ion-beams [21] for polishing have all been attempted.

### C1 Polishing PCD

As expected the polishing of PCDs is more isotropic. TABLE 1 summarises data obtained by Wilks and Wilks [22,23]. The removal rates for polishing on a cast iron scaife are high, but it is essential to maintain the supply of diamond powder [24]. The removal rates with the bonded wheel are much lower when polishing PCDs. Wilks and Wilks suggest that this is because the grits are held in the plane of the wheel and are unable to sink into the metal phase of the PCD. Despite the lower removal rates, PCDs are generally polished by bonded wheels to avoid the difficulties arising from the loss of powder from cast iron scaifes. Acceptable removal rates are generally obtained by using high speeds and applying large loads to the wheel, which in turn requires the use of sufficiently massive polishing machines. Further details are given by Wilks and Wilks [3].

TABLE 1 Removal rates of diamond when polished on a cast iron scaife and on a bonded wheel for three PCDs and single crystal diamond. Wheel speed, 2932 rev min$^{-1}$; load 2015 g wt.

| Material | Direction | Removal rates ($10^{-4}$ mm$^3$ s$^{-1}$) | |
|---|---|---|---|
| | | Cast iron wheel | Bonded wheel |
| PCD (A75); 75 μm grain size | - | - | 0.04 |
| PCD (AX4); 25 μm grain size | - | 12.9 | 0.04 |
| PCD (DXI); 25 μm grain size | - | 19.7 | - |
| Single crystal diamond; | <100> | 5.9 | 3.3 |
| {011} face | <011> | 0.66 | 0.14 |

## D CONCLUSION

There have been major advances in understanding the mechanisms of natural diamond polishing in recent years. A topic of great practical interest is the polishing of CVD diamond. As noted above, a wide range of techniques are being attempted at present.

## REFERENCES


[1] P. Grodzinski [ *Diamond Technology* (NAG Press, London, UK, 1953) ]

[2] E. Bruton [ *Diamonds* (NAG Press, London, UK, 1970) ]

[3] E.M. Wilks, J. Wilks [ *Physical Properties of Diamond* Ed. R. Berman (Clarendon Press, Oxford, UK, 1965) p.221-56 ]

[4] E.M. Wilks, J. Wilks [ *J. Phys. D, Appl. Phys. (UK)* vol.6 (1972) p.1772-81 ]

[5] J. Wilks, E.M. Wilks [ *The Properties of Diamond* (Academic Press, London, UK, 1979) p.351-82 ]

[6] A.P. Malshe, B.S. Park, W.D. Brown, H.A. Naseem [ *Diam. Relat. Mater.* (*Switzerland*) vol.8 (1998) p.1198-213 ]

[7] M. Tolkowsky [ DSc Thesis, University of London, 1920 ]

[8] J. Wilks, E.M. Wilks [ *J. Phys. D, Appl. Phys. (UK)* vol.5 (1972) p.1902-19 ]

[9] M. Couto, W.J.M. van Enckevort, B. Wichman, M. Seal [ *Appl. Surf. Sci.* (*Netherlands*) vol. 62 (1992) p.263-8 ]

[10] M. Couto, W.J.M. van Enckevort, M. Seal [ *J. Hard Mater. (UK)* vol.5 (1994) p.31-48 ]

[11] F.M. van Bouwelen, L.M. Brown, J.E. Field [ *Ind. Diam. Rev. (UK)* vol.57 (1997) p.21-6 ]

[12] S.E. Grillo, J.E. Field [ *J. Phys. D, Appl. Phys. (UK*) vol.30 (1997) p.2466-72 ]

[13] S. Jin, J.E. Graebner, T.H. Tiefel, G.W. Kammlott, G.J. Zydzik [ *Diam. Relat. Mater. (Netherlands)* vol.1 (1992) p.949-53 ]

[14] S. Jin et al [ *Appl. Phys. Lett. (USA)* vol.60 (1992) p.1948-50 ]

[15] M. Yoshikawa [ *Diamond Optics III, SPIE (USA)* vol.1325 (1990) p.210-5 ]

[16] H. Tokura, M. Yoshikawa [ in *Appl. of Diamond Films and Relat. Mater.* (*Netherlands*) Eds. A. Feldman et al (1991) p.227-32 ]

[17] H. Tokura, A. Hirata, D. Liang, M. Yoshikawa [ in *Appl. of Diamond Films and Relat. Mater.* (*Netherlands*) Eds. A. Feldman et al, 3rd International Conference NIST-SP-885 (1995) p.253-8 ]

[18] P.K. Backmann, D. Leers, D.U. Wiechert [ *Diam. Relat. Mater.* (*Switzerland)* vol.2 (1993) p.683-8 ]

[19] H. Funamoto, O. Koseki [ *Surf. Coat. Technol. (Switzerland)* vol.47 (1991) p.474-8 ]

[20] C. Vivensany, L. Ferlazzo-Manim, M.F. Ravet, G. Turban, F. Rousseaux, A. Gieguel [ *Diam. Relat. Mater. (Switzerland)* vol.5 (1996) p.840-5 ]

[21] A. Hirata, H. Tokura, M. Yoshigawa [ *Thin Solid Films (Switzerland)* vol.212 (1992) p.43-7 ]

[22] J. Wilks, E.M. Wilks [ *The Properties and Applications of Diamond* (Butterworth-Heinemann, Oxford, UK, 1992) ]

[23] J. Wilks, E.M. Wilks [ in *The Properties of Synthetic and Natural Diamond* Ed. J.E. Field (Academic Press, London, UK, 1992) ]

[24] M.P. Hitchiner, E.M. Wilks, J. Wilks [ *Wear (Switzerland)* vol.94 (1984) p.103-20 ]

# A4.5 Erosion of diamond surfaces

J.E. Field

November 1999

## A INTRODUCTION

All forms of diamond have practical use in erosive environments. The erosion resistances of bulk natural diamond and poly-crystalline diamond PCD composites are the highest yet recorded [1–3]. The erosion resistance of chemically vapour deposited (CVD) diamond continues to improve.

## B SOLID PARTICLE EROSION

With sand as the erodent, there is negligible erosion of natural diamond until velocities in excess of ~100 m $s^{-1}$ are attained. Then the steady state erosion proceeds at a rate of ~ 0.05 mg per kg of erodent. Bulk diamond first forms a network of cracks and material loss only starts when cracks intersect. PCDs erode primarily by loss of the binder material (usually cobalt). Long time sand erosion studies of a De Beers PCD (Syndite) at 250 m $s^{-1}$ show that the erosion rate is about 8000 times less than for stainless steel and about 120 times less than that of tungsten carbide [4].

The solid particle erosion of single crystal diamond has been the subject of a number of studies [1–3,5]. The mechanisms for material removal, and the steady-state gravimetric erosion rate (defined as mass loss in mg per kg of erodent used) and its variation with impact velocity have been determined; see TABLE 1. The effect of crystal orientation on impact damage has also been considered [6]. Recently, there have been a number of studies on the erosion of CVD diamond. There is growing industrial interest in the erosion performance of CVD diamond, both as a free-standing material and in coating form, particularly for aerospace, biomedical and mechanical wear applications. Research on free-standing CVD diamond has highlighted some important differences in the behaviour of this bulk material as compared to single crystal diamond, particularly the effect of grain size and orientation on the mechanism and rate of removal [6–10]; see TABLE 1. Used as a protective coating, its performance on many different metal and ceramic substrates has been studied [11–19], focusing on coating integrity, the effect of coating thickness and growth conditions on the erosion rate and also the problems associated with poor interfacial adhesion.

TABLE 1 Erosion by quartz particles 300–600 μm in diameter [20].

| | Single crystal diamond | Free-standing CVD diamond |
|---|---|---|
| Removal mechanism | Multiple intersections of ring cracks | Intersection of ring cracks plus grain pull-out and chipping |
| Example erosion rate at 114 m $s^{-1}$/mg $kg^{-1}$ | 0.04 ±0.01 for {100} face | 2.3 ±0.1* |
| Velocity exponent, n ($E=kV^n$) | 1.0 ±0.05 | 3.3 ±0.3* |

*Note that the erosion of CVD diamond varies with sample quality, grain size and growth orientation.

## C RAIN EROSION

Diamond is currently being considered as an IR 'window' material in aircraft and missile applications. The velocity at which impacts of rain cause strength loss is known as a 'threshold' velocity, and values have been measured for a wide range of IR transparent materials; see TABLE 2. The data are standardised to the threshold velocity for impacts with 2 mm water drops [21–25]. FIGURE 1 shows a correlation between damage threshold velocity and fracture toughness. The threshold is for the onset of short circumferential cracks around the impact area.

TABLE 2 Rain erosion threshold data [22,24].

| Materials | Multiple impact threshold velocity (0.8 mm jet) (m $s^{-1}$) | Multiple impact threshold velocity (2 mm drop) (m $s^{-1}$) |
|---|---|---|
| Zinc sulphide | 130 | 190 |
| Magnesium fluoride | 205 | 290 |
| Spinel | 350 | 460 |
| Sapphire (25 mm disc) | 420–430 | ~535 |
| Sapphire (6 mm disc) | 420–440 | ~540 |
| Natural diamond | ~530 | ~600 |
| High temperature high pressure (HTHP) diamond | ~525 | ~600 |
| Amorphic diamond on ZnS (1 μm) | ~160 | ~235 |
| ZnS/D 20%, 0 – 1 μm (a) | 190 | 270 |
| ZnS/D 10%, 1 – 3 μm (a) | 140 | 205 |
| ZnS/D 10%, 0 – 1 μm (a) | 140 | 205 |
| CVD diamond (b) | 200–350 | 285-450 |

(a) This series of composites was produced by Raj and co-workers [27,28]. 20%, 0 – 1 μm expresses the amount of diamond (D) in the composite and the diamond particle size.
(b) See text.

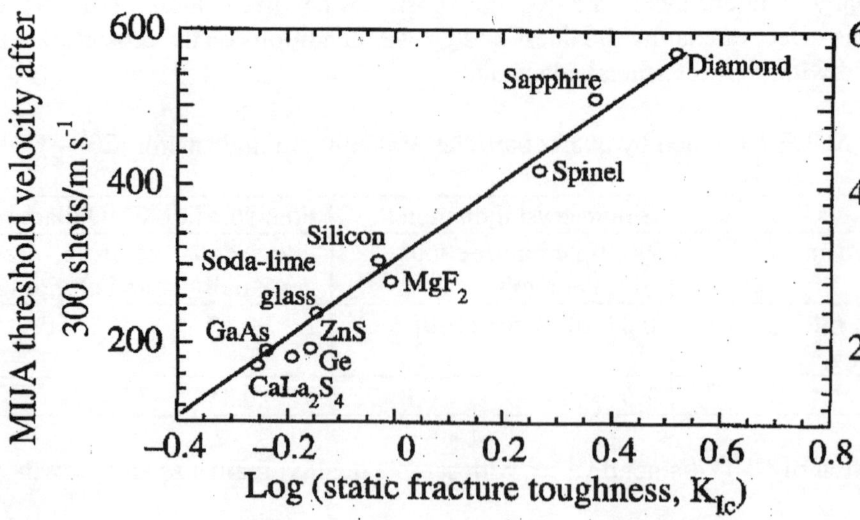


FIGURE 1 The absolute damage threshold velocities for 2 mm diameter water drops as a function of fracture toughness.

The value for natural diamond is based on only one specimen, a 6 mm diameter, 1 mm thick type II diamond. Such a small specimen would produce strong stress wave reflections from the boundaries which would keep the 'threshold' below that for a larger specimen. The HTHP synthetic diamond gave similar values to natural diamond [25] but was 9 mm × 9 mm × 3.5 mm thick. Due to its size and shape fewer damaging wave reinforcements occurred and the threshold was thought to be close to a true bulk value for this type of diamond. The amorphous diamond layer gives some protection to the zinc sulphide substrate. The zinc sulphide/diamond composites also performed better than zinc sulphide.

CVD diamond has improved in quality during the last decade from a threshold value of ~ 200 m $s^{-1}$ to a value within about 200 m $s^{-1}$ of that for natural diamond. However, various forms of damage can take place with thin CVD free-standing diamonds. In samples less than ~ 1.1 mm thick, the first form of damage is at the rear surface and the threshold depends on grain size and sample thickness [25,26]. Since the grain size is smallest near a nucleation surface, the threshold values are usually higher if the nucleation (rather than a growth) surface is at the rear. In thin samples, the first form of damage of the impacted face is an enlarged ring crack caused by reinforcements between the front surface Rayleigh wave and reflected bulk waves. For samples greater than about 1.2 mm thick, short circumferential cracks form around the impact area (as with other bulk brittle materials) giving the values listed in TABLE 2.

## D CONCLUSION

Diamond in its various forms is very erosion resistant. If transparency is not a requirement, PCDs are of interest in very erosive environments. For IR windows and domes, CVD diamond has great potential, and it is now possible to fabricate CVD diamond domes of several centimetres diameter with very good erosion resistances to both solid particle and liquid drop impact.

## REFERENCES

[1] I.P. Hayward, J.E. Field [ *J. Hard Mater. (UK)* vol.1 (1990) p.53-64 ]
[2] Z. Feng, J.E. Field [ *J. Hard Mater. (UK)* vol.1 (1990) p.273-87 ]
[3] R.A. Vaughan, A. Ball [ *J. Hard Mater. (UK)* vol.2 (1991) p.257-69 ]
[4] P.L. Kaye, J.E. Field, [ *J. Hard Mater. (UK)* vol.4 (1993) p.167-75 ]
[5] E.J. Coad, C.S.J. Pickles, G.H. Jilbert, J.E. Field [ *Diam. Relat. Mater. (Switzerland)* vol.5 (1996) p.640-3 ]
[6] R.H. Telling, J.E. Field [ *Int. J. Refract. Hard Met. (UK)* vol.16 (1998) p.269-76 ]
[7] Z. Feng, Y. Tzeng, J.E. Field [ *Thin Solid Films (Switzerland)* vol.212 (1992) p.35-42 ]
[8] R.H. Telling, J.E. Field [ *Diam. Relat. Mater. (Switzerland)* vol.8 (1998) p.850-4 ]
[9] R.H. Telling, J.E. Field [ *Wear (Switzerland)* vol.233-235 (1999) p.666-73 ]
[10] J.H. Kim, D.S. Lim [ *Diam. Relat. Mater. (Switzerland)* vol.8 (1999) p.865-70 ]
[11] M.V. Kral, J.L. Davidson, J.J. Wert [ *Wear (Switzerland)* vol.166 no.1 (1993) p.7-16 ]
[12] E.J. Coad, C.S.J. Pickles, G.H. Jilbert, J.E. Field [ *Diam. Relat. Mater. (Switzerland)* vol.5 no.6-8 (1996) p.640-3 ]
[13] D.W. Wheeler, R.J.K. Wood [ *Wear (Switzerland)* vol.233-235 (1999) ]
[14] D.W. Wheeler, R.J.K. Wood [ *Wear (Switzerland)* vol.229 no.1 (1999) p.523-36 ]
[15] A. Alahelisten, P. Hollman, S. Hogmark [ *Wear (Switzerland)* vol.177 (1994) p.159-65 ]
[16] S. Amirhaghi, H.S. Reehal, E. Plappert, Z. Bajic, R.J.K. Wood, D.W. Wheeler [ *Diam. Relat. Mater. (Switzerland)* vol.8 (1999) p.845-9 ]
[17] T. Grogler, E. Zeiler, O. Plewa, S.M. Rosiwal, R.F. Singer [ *Surf. Coat. Technol. (Switzerland)* vol.112 no.1-3 (1999) p.129-32 ]

[18] G. Heinrich, T. Grogler, S.M. Rosiwal, R.F. Singer [ *Surf. Coat. Technol. (Switzerland)* vol.94-95 (1997) p.514-20 ]

[19] A.J. Miller, D.M. Reece, M.D. Hudson, C.J. Brierley, J.A. Savage [ *Diam. Relat. Mater. (Switzerland)* vol.6 no.2-4 (1997) p.386-9 ]

[20] A.R. Davies, R.H. Telling [ Internal Report, Cavendish Lab., Univ. of Cambridge, UK, 1999 ]

[21] C.R. Seward, C.S.J. Pickles, R. Marrah, J.E. Field [ *Proc. SPIE (USA)* vol.1760 (1992) p.280-90 ]

[22] J.E. Field, E. Nicholson, C.R. Seward, Z. Feng [ *Philos. Trans. R. Soc. Lond. A (UK)* vol.342 (1993) p.261-75 ]

[23] C.R. Seward, J.E. Field, E.J. Coad [ *J. Hard Mater. (UK)* vol.5 (1994) p.49-62 ]

[24] E.J. Coad, C.S.J. Pickles, C.R. Seward, G.H. Jilbert, J.E. Field [ *Proc. R. Soc. Lond. A (UK)* vol.454 (1998) p.213-38 ]

[25] E.J. Coad, J.E. Field [ *Proc. SPIE (USA)* vol.3060 (1998) p.169-80 ]

[26] J.E. Field [ *Wear (Switzerland)* vol.233-235 (1999) in press ]

[27] D.S. Farquhar, R. Raj, S.L. Phoenix [ *J. Am. Ceram. Soc. (USA)* vol.73 (1990) p.3074-80 ]

[28] L.A. Xue, D.S. Farquhar, T.W. Noh, A.J. Sievers, R. Raj [ *Acta Metall. Mater. (USA)* vol.38 (1990) p.1743-52 ]

# A4.6 Etching of diamond

W.J.P. van Enckevort and F.K. de Theije

March 2000

## A INTRODUCTION

In the first instance the research on diamond etching was aimed at understanding the genesis of natural diamonds. In these studies attention was given to the dodecahedral shape of many rough stones as well as to the 'trigon controversy', which lasted more than two decades. The rounded {110} form of the dodecahedrons was explained as being the result of a severe post-growth dissolution or graphitisation of octahedrally shaped diamonds when they were brought up from the upper mantle region towards the Earth's surface [1–3]. On the {111} surfaces of diamonds which suffered less dissolution, often numerous pits, the so-called 'trigons', were encountered. Originally they were thought to result from growth processes [4,5], but in 1958 it was suggested that the features are etch pits [6]. The etch pit theory was confirmed by Lang and Frank [7,8] who related the depressions to dislocations by using X-ray diffraction topography in combination with optical microscopy. Their observations were confirmed in later dislocation studies of natural diamond [9,10], by laboratory etching experiments [11,12] and theory [13]. Now it is well established that the original growth form of most natural diamonds is octahedral, showing only shallow growth hillocks on the {111} surfaces [7,9,14].

Diamond can be etched by reduction of the carbon matrix to hydrocarbons using hydrogen and related compounds as well as by oxidation to CO or $CO_2$ utilising oxygen, water or molten $KNO_3$. Although substantial efforts have been made in understanding both ways of diamond etching, the precise mechanisms are still to be uncovered. Section B1 reviews the reduction etching, Section B2 the oxidative etching.

Since the 1960s, etching has also been used as a tool for revealing the occurrence and distribution of point, linear and planar defects in natural and high temperature/high pressure grown (HPHT) synthetic diamond crystals. Methods, results and backgrounds for characterizing diamonds by etching are summarised in Section C.

Soon after the use of the chemical vapour deposition (CVD) technique for diamond growth became widespread, efforts were made in processing the grown material by gas phase etching. In this way the rough, as-grown diamond surfaces are smoothed and graphitic or amorphous carbon (a-C) in the grain boundaries between the diamond particles is removed. Pattern etching of diamond surfaces using lithographic methods and shaping of diamonds applying various etch techniques is important for the development of diamond electronics and diamond based micro-mechanics. Applications of diamond etching are treated in Section D.

## B MECHANISMS OF ETCHING

Without activation, no chemical etchant attacks diamond at room temperature. At elevated temperatures diamond is no longer chemically inert and it is eroded by oxygen and oxygen compounds, molten [15] and solid metals [16] and by hydrogen. If the temperature exceeds 2000 K the metastable diamond spontaneously turns into graphite at ambient pressures. At lower temperatures, however, graphitisation can be catalysed by contacting diamond surfaces with solid or molten metals [15–17]. The fundamental research of diamond etching was mostly focused on understanding the chemical attack of the crystals in reducing or oxidizing ambients, which is essential for understanding CVD diamond growth [18] and for processing of diamond.

### B1 Hydrogen Etching

Diamond is etched by the attack of monoatomic hydrogen, either neutral or ionised, i.e. H or $H^+$. Since molecular hydrogen splits up in relevant amounts only at temperatures beyond ≈ 2000 K, i.e. when diamond spontaneously converts to graphite, no thermochemical etching studies of diamond using $H_2$ have been reported. All investigations reported concentrated on etching utilizing hydrogen radicals or ions produced by hot filament [19], plasma [20–22] or another method [23,24] of $H_2$ activation.

Surface topography showed that plasma-activated hydrogen etching generates flat {100} surfaces [21] covered by monoatomic steps parallel to [110] [20], which indicates dimer formation during etching [26]. This was confirmed by low energy electron diffraction (LEED), showing (2×1) reconstruction of the plasma polished {100} faces [19,27]. From indirect evidence it was suggested that apart from surface reconstruction also diffusion of hydrocarbon species on the crystal surface plays a major role in plasma-activated diamond etching [20,21]. In contrast to hydrogen activation by plasma methods, atomic hydrogen produced by passing molecular hydrogen along a hot filament generated {111} oriented nanofacets on {100} oriented diamond surfaces. This was established by infrared absorption spectroscopy and LEED [19]. The same was found for the {110} faces, whereas {111} remained intact. Roughening of {100} diamond by atomic hydrogen etching has also been reported by Komatsu et al [23] using a different method of activation. The actual reason for this disparity is not clear. Possibly fast $H^+$ ions [28], which occur in hydrogen plasmas, but are not produced in the hot filament experiments, induce the observed smoothing. A stabilisation of {100} diamond by $H^+$ ion bombardment perpendicular to these cubic faces has indeed been reported [24].

Little theoretical work has been carried out on diamond etching using activated hydrogen. A recent quantum mechanical and kinetic Monte Carlo study of the occurrence of hydrogen etching during CVD diamond growth showed that, in contrast to the case for {110} and {111} faces, etching plays a key role during the growth of the {100} surfaces [18]. It was shown that the probability of $CH_2$ removal from the {100} surface decreases for isolated, step-site and surface $CH_2$ groups, respectively. Although not concluded by the authors, this implies that, like growth, also etching of {100} diamond must proceed layer-wise involving steps.

In short, despite it being the 'simplest' system possible, little is known about hydrogen etching of diamond.

### B2 Oxygen Etching

Diamond is quite susceptible to oxidation: compared to hydrogen, oxygen etching sets in at lower temperatures and proceeds at much higher rates, and activation to atomic or ionic oxygen is not needed.

Thermochemical etching of diamond can be carried out by using molecular oxygen, water vapour or molten $KNO_3$. Etching studies of diamond by gases at 400°C and above showed that molecular oxygen etches faster than water vapour [29,30], whereas molecular hydrogen did not show any observable etching activity [31]. Diamond removal using molecular oxygen depends on temperature and oxygen pressure [32–36], the crystallographic orientation of the diamond surface [12,32–36], the presence of traces of water vapour [12] or fluorine [37] and the occurrence of twinning and graphitic or a-C inclusions in the crystals [29,33]. The oxidation rate of single crystal diamond is shown to be fastest for the {111} surface, followed by {110}, with {100} exhibiting the slowest etch rate [36].

First order and second order diamond oxidation kinetics have been reported [36–39], with activation energies ranging from 172 to 260 kJ/mol [32–41]. Since the bond strength of one C-C bond is 349 kJ/mol, this suggests that the rate limiting step of diamond oxidation involves breakage of maximally one C–C bond.

In many cases the morphology of oxidised single crystal diamond surfaces is determined by etch pits. If etched at lower temperatures, these patterns conflict with the results of the first order broken bond periodic bond chain (PBC) analysis of diamond [42]. This theory predicts the {111} surfaces to be covered by the so-called 'negative' etch pits with steps parallel to <110> and pointing towards $[11\bar{2}]$ and neither steps nor pits should appear on the {100} faces. However, dry oxygen etching experiments show the reverse, namely well developed etch pits on {100} and a rough {111} surface [40]. Adding water vapour to the oxygen leads to a dramatic smoothing of the {111} surfaces, resulting in shallow 'positive' pits with monoatomic <110> steps pointing towards $[\bar{1}\bar{1}2]$, which is opposite to the prediction of the PBC theory [41]. On {100} again etch pits bounded by <110> steps were found. Etching in molten $KNO_3$ produced the same morphology as etching in an oxygen-water mixture [41]. The controversy between the PBC results and observations is explained by the formation of bridging ether (C–O–C) and carbonyl (>C=O) surface groups, stabilizing the {100} surfaces and the $[\bar{1}\bar{1}2]$ steps on the {111} faces [40,41,43]. Etching in oxygen-water mixtures at temperatures exceeding 900°C resulted in a 180° rotation of the pits on {111} [12], which agrees with the PBC analysis. A 'right', i.e. 'negative', orientation of trigons was also observed for diamonds etched in high pressure water at T = 1100–1500°C [44] and on natural diamonds that were etched during their ascent to the Earth's surface by volcanic eruption [7,8].

Thermochemical oxygen etching at temperatures above 900°C produces a protecting, nano-crystalline graphite layer on the diamond surfaces, which lowers the rate of material removal [36]. The graphite crystallites exhibit a preferential orientation which is determined by the symmetry of the underlying diamond surface. This leads to an optical anisotropy of the graphite film [45]. Diamond etching by molecular oxygen seems to proceed in two steps [12]: first, formation of a graphitic layer by $C_{dia} + 1/2\ O_2 \rightarrow CO_{ads}$ (= adsorbed carbon monoxide) followed by $CO + CO_{ads} \rightarrow C_{gra} + CO_2\uparrow$; second, subsequent oxygen etching of the graphite layer. At T < 900°C the second step proceeds faster and a clean diamond surface is formed. A relative rate increase of the first step leads to graphite formation at T > 900°C.

To unravel the mechanism of oxidative diamond etching a considerable amount of effort has been made. It was quickly realised that oxygen chemisorbed on a diamond surface desorbs as CO and to a lesser extent as $CO_2$ [38,46] which is different from hydrogen, which generally leaves the diamond surface as $H_2$ [38]. Oxygen etching of diamond powders at atmospheric pressure and subsequent analysis using infrared spectroscopy revealed bridging keton (C=O), ether (C–O–C), lactone [C(O)–O–C] and carboxylic anhydride [C(O)–O–C(O)] structures on the crystallite surfaces [46]. At temperatures from 300 to 380°C the surface is oxidised by the formation of keton and ether bridges, which develop into lactone and carboxylic anhydride structures at higher temperatures ranging from 380 to 480°C. These latter complexes decompose at temperatures exceeding 480°C leading to keton groups and $CO_{(2)}$. The continuous adsorption of oxygen and desorption of $CO_{(2)}$ species in fact comprises the 'burning' of diamond. Unfortunately this investigation involving powders is not surface specific. Much research has been devoted to adsorbed oxygen complexes on {001} single crystal diamond faces, both by theory (see for instance [43,47,48]) and by experiment [38,49,50]. The most important surface species encountered during these studies are the bridging keton and ether groups on a (1×1) or (2×1) reconstructed surface. Probably during etching there was half or one chemisorbed oxygen atom per surface carbon. From a surface topographic study of etched {001} diamond plates it was concluded that the abstraction of $CO_{(2)}$ from the kink-site of a <110> growth step is the rate limiting step of the thermal oxidation process [40].

Addition of water or HF was shown to slow down the oxidation of {110} diamond; $F_2$ blocks the process completely at temperatures below 900°C [37]. This was explained by the formation of C–H, C–OH and C–F surface compounds, which hinder or prevent $O_2$ attack and thus retard diamond etching.

Diamond etching using activated, reactive ions and radicals is, depending on the kinetic energy of the incident particles, in fact a combination of physical sputtering and chemical etching. The process is predominantly physical if the crystal surface is bombarded with high energy (1.5 keV) $H_2^+$ ions [25] or $Ar^+$ ions [51]. Then etching proceeds via surface amorphisation and subsequent atom removal by further collisions. However, all the work done on diamond etching using activated oxygen, i.e. atomic or ionic oxygen, points to chemical etching. Like thermal etching this proceeds by a continued chemisorption of oxygen at the diamond surface followed by desorption as $CO_{(2)}$. The surface chemistry is activated by the chemical and kinetic energy of the impinging oxygen radicals or ions generated by RF [52,53], microwave [54] and ion beams [51] or by foreign ions such as $Xe^+$ [55] and electrons [56]. An important consequence is that the temperature of the surfaces being etched can be kept very low, down to 100°C [53], 0°C [55], or even –140°C [57]. Most research on activated oxygen etching of diamond is aimed at technology, and therefore a detailed understanding of the surface processes is lacking. Probably these are comparable to those occurring during thermal oxidation of diamond.

## C DEFECT REVEALING

### C1 Dislocations

In almost all cases diamond etching is irreversible, i.e. the driving force for dissolution approaches infinity. However, despite this extremely large undersaturation, etching of the flat diamond faces {111} and {100} mostly proceeds layer by layer and no kinetic roughening occurs. This is caused by 'kinetic smoothing', a statistical process that decreases surface roughness in the case of dissolution [58]. In the presence of stresses induced by crystal defects, such as dislocations, stacking faults and grain boundaries, a local, preferential 2D nucleation of monolayer etch pits occurs, which is followed by lateral expansion of growth steps. This leads to the development of etch pits or etch grooves visible by optical microscopy. Examination of these features formed after etching provides information on the defect structure in the crystals [59].

Virtually all dislocation etchants for diamond are based on oxidation. Most used is molten $KNO_3$ at 525°C and upwards, which produces well formed triangular pits on the {111} faces [5,41,60] and square pits on {001} [41]. By etching of matched pairs of {111} cleavage faces Patel and Tolansky [61] showed that there was a correlation between the configuration of point-bottomed pits on opposing faces of each pair. This means that the pits are related to dislocations. Recent $KNO_3$ etching experiments uncovered two types of etch pits on {100}, which are related to two different types of dislocations ending on this face [41]. Pits on {111} and/or {001} diamond were also found after etching in water [44], oxygen [40,41,62], molten metal at ultrahigh pressures [63], in an $NaNO_2$ melt [64], in wet air or oxygen at low pressures [12,40,41], in an oxidising propane [65] or acetylene flame [66] and in molten rock [7,11]. In a number of these cases [7,44,61] the pits are shown to be formed at the surface outcrops of dislocations. However, if diamonds are etched in oxygen at elevated temperatures [40,41] the number of shallow, point-bottomed etch pits exceeds the number of dislocations by two orders of magnitude. Possibly the formation of these pits is not sustained by defects, but is caused by a repeated two-dimensional nucleation at the bottom of the pits when the step density is low [67]. If so, this implies that surface diffusion is important under these etch conditions [68].

Some work has been done on dislocation etching using a reducing hydrogen plasma [69]. Since here the etching rate is very low, this technique is quite suitable for the defect characterisation of the individual crystallites in polycrystalline CVD grown films.

### C2 Extended Defects

Due to differences in diamond removal rate caused by a non-uniform distribution of impurities, growth sectors and growth bands can be imaged as surface areas of different height after etching or polishing [70]. In addition, the outcrops of two-dimensional defects such as sector boundaries, growth bands, twins and stacking faults can be revealed as etch grooves [71]. In contrast to dislocation etching, the height differences and grooves are best produced on 'rough' faces. For diamond this means faces other than {111} [42]. It has been shown that etching of polished {100} [72] and {110} [73,74] faces readily produces patterns associated with extended defects.

Molten $KNO_3$ during 15 minutes at approximately 700°C is a very suitable etchant. Etching of polished {100} faces of natural diamonds revealed a puzzling pattern of straight and curved growth bands and sector boundaries, which indicates a very complicated growth history of the crystals [72]. The curved growth bands are due to periods of dissolution during growth, and the curved sector boundaries are explained by the occurrence of non-faceted cuboid growth [75]. $KNO_3$ etching of {110} faces polished on HPHT grown diamonds showed the occurrence of {111}, {100}, {110} and {113} growth sectors of complex shape [73]. It was shown that the mono-substitutional nitrogen impurity in the synthetic crystals enhances the dissolution rate. For the boron impurity the situation is more complex. Slight diamond etching using an oxidising propane flame was shown to uncover the surface outcrops of stacking faults and twin lamellae [65].

## D APPLICATIONS

Although the mechanisms of diamond etching are not fully understood, etching has found widespread use in the 'new diamond technology'. The applications of etching can be divided into two main categories: first, diamond material preparation by cleaning and surface smoothing; second, processing of diamond by pattern etching and three-dimensional shaping. In this, dry etching methods are employed, mostly using an activated gas phase.

### D1 Cleaning and Smoothing

It is well known that non-diamond carbon phases such as amorphous carbon and graphite etch faster than diamond [53]. This holds for activated hydrogen [31] and oxygen [76] etching as well as for thermochemical oxidation [29]. Such preferential etching can be used to 'wash out' the non-diamond carbon phases from CVD grown diamond films and to characterise how these phases are distributed in the layers. Cleaning of diamond films has been carried out by etching in an RF- [53] or microwave- [54,76] activated oxygen plasma, by heating in molecular oxygen [29] or by exposure to hydrogen atoms [31]. Plasma oxygen gives the best selectivity in combination with a fast removal rate [53]. Another use of oxidative surface cleaning is the brief in situ etching of the single crystal diamond substrate using the oxygen-rich part of the acetylene-oxygen flame prior to diamond deposition by the combustion flame technique [66]. This pre-growth etching results in homoepitaxial diamond layers of improved quality. The preferential oxidation of graphite as compared to diamond has also been utilised for the production of thin, free-standing homoepitaxial diamond films using a lift-off technique [77]. Here, first a {100} diamond substrate surface is damaged by $O^+$ implantation, then this surface layer is completely graphitised by annealing at 1200°C. After growth of a single crystalline CVD diamond on top of this layer, this film is separated from its substrate by thermal oxidation of the graphite layer in between.

The use of single and polycrystalline diamond as a substrate material for diamond electronics, heat spreaders, optical windows and X-ray masks requires smooth surfaces of submicrometre roughness. For single crystal diamond these surfaces are mostly produced by conventional polishing techniques using diamond powder [79]. Because of the more or less random crystallite orientation in polycrystalline diamond films, these layers are not easily planed by abrasive methods. Here, chemical etching methods are preferred. For instance, mechanical polishing can be used in conjunction with chemicals to enhance the diamond removal rate. Examples are hot metal lapping [79] in which the diamond layers are thinned by metal carbide formation involving solid state diffusion [80], and chemically assisted polishing and planing (CAMPP) where mechanical polishing using an alumina plate is used in combination with oxidising compounds, such as $KNO_3$ and KOH [81]. Dry etching techniques have also been proved to be successful. For example, smoothing of polycrystalline CVD layers has been realised by ion beam polishing and reactive ion etching through the use of a planarising layer of photoresist, Ti and Si oxides [82–84] or gold [85]. Smoothing of thin CVD grown diamond films can also be carried out without a covering layer, using a beam of $Ar^+/O^+$ or $Ar^+/N_2^+$ ions [86]. An alternative for smoothing polycrystalline diamond layers is by making use of the 'polishing' action of a laser beam [81]. An extended review of methods for polishing and planing diamond films is given by Malshe et al [81].

### D2 Pattern Etching and Shaping

The chemical inertness and hardness of diamond makes it difficult to remove surface layers in a smooth, damage-free manner without affecting any masking material. Since wet chemical etchants, such as molten $KNO_3$ and molten metals, attack nearly all kinds of masking layers, one has to resort to dry etching techniques for producing predefined etch patterns in diamond. This can be realised in two different ways: first, one can etch the whole surface after application of an etch-resistant mask; second, one can use a narrow, scanning beam that activates etching of a small surface area. In the second case no mask is needed.

Most patterning work is carried out by chemical etching using oxygen plasmas generated by microwave excitation [54,87,88] or RF excitation [89–91], thermally activated oxygen [91] or oxygen activated by synchrotron radiation [57]. For masking, oxidation resistant materials, such as $SiO_2$ [88–90] and $Si_3N_4$ [89–92] or metals like Al [55], Ni [57,54], Ti [91] and Au [87], are used, which are applied onto the diamond surfaces by photoresist techniques. During etching on top of the base metals, Ti, Al, a thin oxide layer is formed which protects the underlying metal from further corrosion. At the last stage of the lithographic process the masks are removed by wet or dry chemical etching. Pattern etching was also achieved by oxidation of Al/Si/Ni masked single crystal diamond, using $NO_2$ gas activated by a beam of $Xe^+$ ions [55]. Finally, bias-assisted etching of diamond in a hydrogen plasma using Au, Pt or Ti/Au masking layers yields well defined etch patterns [28]. It has been suggested that electron emission from the uncovered diamond film leads to degradation (graphitisation) of the surface, which is then quickly etched away.

The use of masks is avoided by scanning the diamond surface using a focused electron beam in a low pressure oxygen [56] or hydrogen [93] atmosphere. In this way, chemical etching is locally activated and well defined etch patterns can be produced. An alternative for the electron beam is applying a $Ga^+$ ion beam, which activates $XeF_2$ as an etching agent [94]. Laser-assisted (thermal) diamond oxidation has been shown to be only suitable for patterning nanocrystalline diamond films, which are easily heated up because of their relatively low thermal conductivity [95].

Alternative methods for selective etching are excimer laser ablation in combination with a metal mask [96] and catalytic etching in an $H_2$ atmosphere using thin film patterns of Pt, Ni and Fe [97]. Structures of polycrystalline diamond films on top of a silicon substrate can also be produced by selected area CVD deposition. In this case the substrate is pretreated in several manners to obtain predefined, local areas of enhanced crystallite nucleation [92,98,99].

In general, diamonds are shaped by using polishing [78] or laser cutting techniques. However, for obtaining sharp cutting edges or ultra-sharp diamond tips, etching can be helpful. Ion sputter techniques [100,101] and ion beam etching [102,103] using different configurations can produce diamond edges and tips with sub-micrometre radii. Ion sputtering is used to generate tips and edges with apex angles exceeding 90°, whereas ion milling is best suited to shape radii with apex angles less than 60° [104], which is more challenging. Reactive ion etching of CVD-grown films in an oxygen plasma can produce a high density of columnar structures, which make the diamond film 'porous' [105,106]. The ensemble of sharp tips obtained in this way enhances the field emission characteristics of a diamond film to a great extent [105].

## E CONCLUSION

Understanding diamond etching is a highly relevant issue in diamond science and technology. It provides insight into the complex surface chemistry of diamond, it helps to unravel its defect structure and it makes it possible to clean and smooth diamond films and to process them into more or less complicated structures.

## REFERENCES


[1] M. Moore, A.R. Lang [ *J. Cryst. Growth (Netherlands)* vol.26 (1974) p.133-9 ]

[2] M. Seal [ *Phys. Status Solidi (Germany)* vol.3 (1963) p.658-64 ]

[3] J. Wilks, E. Wilks [ *Properties and Applications of Diamond* (Butterworth-Heinemann Ltd, Oxford, 1991) p.120-6 ]

[4] S. Tolansky [ *The Microstructures of Diamond Surfaces* (N.A.G. Press Ltd, London, 1955) ]

[5] S. Tolansky [ in *Physical Properties of Diamond* Ed. R. Berman (Clarendon, Oxford, 1965) ch.3 ]

[6] F.C. Frank, K.E. Puttick, E.M. Wilks [ *Philos. Mag. (UK)* vol.3 (1958) p.1262-72 ]

[7] F.C. Frank, A.R. Lang [ in *Physical Properties of Diamond* Ed. R. Berman (Clarendon, Oxford, 1965) p.140-54 ]

[8] A.R. Lang [ *Proc. R. Soc. Lond. A (UK)* vol.278 (1964) p.234-42 ]

[9] I. Sunagawa, K. Tsukamoto, T. Yashuda [ in *Materials Science of the Earth's Interior* Ed. I. Sunagawa (TERRAPUB, Tokyo, 1984) p.331-49 ]

[10] W.J.P. van Enckevort, M. Seal [ *Philos. Mag. A (UK)* vol.57 (1988) p.939-54 ]

[11] F.C. Frank, K.E. Puttick [ *Philos. Mag. (UK)* vol.3 (1958) p.1273-9 ]

[12] T. Evans, D.H. Sauter [ *Philos. Mag. (UK)* vol.6 (1961) p.429-40 ]

[13] J.C. Angus, Th.J. Dyble [ *Surf. Sci. (Netherlands)* vol.50 (1975) p.157-159 ]

[14] W.J.P. van Enckevort [ *J. Cryst. Growth (Netherlands)* vol.119 (1992) p.177-94 ]

[15] S. Jin et al [ *Appl. Phys. Lett. (USA)* vol.65 (1994) p.2675-7 ]

[16] W.-S. Lee, Y.-J. Baik, K.Y. Eun, D.-Y. Yoon [ *Diam. Relat. Mater. (Switzerland)* vol.4 (1995) p.989-95 ]

[17] S. Evans, M.R. Ney [ *J. Hard Mater. (UK)* vol.1 (1990) p.169-81) ]

[18] See for instance C.C. Battaile et al [ *J. Chem. Phys. (USA)* vol.111 (1999) p.4291-9 ]

[19] C.-L. Cheng, H.-C. Chang, J.-C. Lin, K.-J. Song, J.-K. Wang [ *Phys. Rev. Lett. (USA)* vol.78 (1997) p.3713-6 ]

[20] K. Hayashi, S. Yamanaka, H. Okushi, K. Kajimura [ *Diam. Relat. Mater. (Switzerland)* vol.5 (1996) p.1002-5 ]

[21] R.E. Rawles, S.F. Komarov, R. Gat, W.G. Morris, J.B. Hudson, M.P. D'Evelyn [ *Diam. Relat. Mater. (Switzerland)* vol.6 (1997) p.791-5 ]

[22] N. Lee, A. Badzian [ *Appl. Phys. Lett. (USA)* vol.66 (1995) p.2203-5 ]

[23] S. Komatsu et al [ *J. Vac. Sci. Technol. A (USA)* vol.16 (1998) p.749-53 ]

[24] W.J. Zhang, X. Jiang, Y.B. Xia [ *J. Appl. Phys. (USA)* vol.82 (1997) p.1896-9 ]

[25] A. Hoffman, P.J.K. Paterson, S. Prawer [ *Nucl. Instrum. Methods Phys. Res. B (Netherlands)* vol.52 (1990) p.63-7 ]

[26] L.J. Giling, W.J.P. van Enckevort [ *Surf. Sci. (Netherlands)* vol.161 (1985) p.567 ]

[27] B.D. Thoms, M.S. Owens, J.E. Butler, C. Spiro [ *Appl. Phys. Lett. (USA)* vol.65 (1994) p.2957-9 ]

[28] B.R. Stoner, G.J. Tessmer, D.L. Dreifus [ *Appl. Phys. Lett. (USA)* vol.62 (1993) p.1803-5 ]

[29] N. Uchida, T. Kurita, K. Uematsu, K. Saito [ *J. Mater. Sci. Lett. (UK)* vol.9 (1990) p.249-50 ]

[30] N. Uchida, T. Kurita, H. Ohkoshi, K. Uematsu, K. Saito [ *J. Cryst. Growth (Netherlands)* vol.114 (1991) p.565-8 ]

[31] N. Uchida, T. Kurita, K. Uematsu, K. Saito [ *J. Mater. Sci. Lett. (UK)* vol.9 (1990) p.251-2 ]

[32] Q. Sun, M. Alam [ *J. Electrochem. Soc. (USA)* vol.139 (1992) p.933-6 ]

[33] A. Joshi, R. Nimmagadda, J. Herrington [ *J. Vac. Sci. Technol. A (USA)* vol. 8 (1990) p.2137-42 ]

[34] R.R. Nimmagadda, A. Joshi, W.L. Hsu [ *J. Mater. Res. (USA)* vol.5 (1990) p.2445-50 ]

[35] S. Dallek, L. Kabacoff, M. Norr [ *Thermochim. Acta (Netherlands)* vol.192 (1991) p.321-6 ]

[36] T. Evans [ in *The Properties of Diamond* Ed. J.E. Field (Academic Press, London, 1979) p.403-24 ]

[37] C.J. Chu, C. Pan, J.L. Margrave, R.H. Hauge [ *Diam. Relat. Mater. (Switzerland)* vol.4 (1995) p.1317-24 ]

[38] R.E. Thomas, R.A. Rudder, R.J. Markunas [ *J. Vac. Sci. Technol. A (USA)* vol.10 (1992) p.2451-7 ]

[39] T. Evans, C. Phaal [ *Proc. 5th Biennial Conf. on Carbon* (Pergamon Press, New York, 1962) vol.1 p.147-53 ]

[40] F.K. de Theije, N.J. van der Laag, M. Plomp, W.J.P. van Enckevort [ *Philos. Mag. A (UK)* vol.80 (2000) p.725-45 ]

[41] F.K. de Theije, O. Roy, N.J. van der Laag, W.J.P. van Enckevort [ *Diam. Relat. Mater. (Switzerland)* accepted for publication (2000) ]

[42] W.J.P. van Enckevort [ Datareview in this book: *A4.1 Morphology of diamond surfaces* ]

[43] S. Skokov, B. Weiner, M. Frenklach [ *Phys. Rev. B (USA)* vol.55 (1997) p.1895-901 ]

[44] H. Kanda, S. Yamaoka, N. Setaka, H. Komatsu [ *J. Cryst. Growth (Netherlands)* vol.38 (1977) p.1-7 ]

[45] W.J.P. van Enckevort [ *J. Appl. Crystallogr. (Denmark)* vol.20 (1987) p.11-5 ]

[46] T. Ando, K. Yamamoto, M. Ishii, M. Kamo, Y. Sato [ *J. Chem. Soc. Faraday Trans. (UK)* vol.89 (1993) p.3635-40 ]

[47] X.M. Zheng, P.V. Smith [ *Surf. Sci. (Netherlands)* vol.262 (1992) p.219-34 ]

[48] J.L. Whitten, P. Cremaschi, R.E. Thomas, R.A. Rudder, R.J. Markunas [ *Appl. Surf. Sci. (Netherlands)* vol.75 (1994) p.45-50 ]

[49] L.M. Struck, M.P. D'Evelyn [ *J. Vac. Sci. Technol. A (USA)* vol.11 (1993) p.1992-7 ]

[50] M.Z. Hossain et al [ *Surf. Sci. (Netherlands)* vol.436 (1999) p.63-71 ]

[51] T.J. Whetten, A.A. Armstead, T.A. Grzybowski, A.L. Ruoff [ *J. Vac. Sci. Technol. A (USA)* vol.2 (1984) p.477-80 ]

[52] G.S. Sandhu, W.K. Chu [ *Appl. Phys. Lett. (USA)* vol.55 (1989) p.437-8 ]

[53] A. Joshi, R. Nimmagadda [ *J. Mater. Res. (USA)* vol.6 (1991) p.1484-90 ]

[54] R. Ramesham, B.H. Loo [ *J. Electrochem. Soc. (USA)* vol.139 (1992) p.1988-93 ]

[55] N.N. Efremow, M.W. Geis, D.C. Flanders, G.A. Lincoln, N.P. Economou [ *J. Vac. Sci. Technol. (USA)* vol.3 (1985) p.416-8 ]

[56] J. Taniguchi, I. Miyamoto, N. Ohno, S. Honda [ *Jpn. J. Appl. Phys. (Japan)* vol.35 (1996) p.6574-8 ]

[57] H. Ohashi, E. Ishiguro, T. Sasano, K. Shobatake [ *Appl. Phys. Lett. (USA)* vol.68 (1996) p.3713-5 ]

[58] E. van Veenendaal, P. van Beurden, W.J.P. van Enckevort, E. Vlieg, J. van Suchtelen, M. Elwenspoek [ *J. Appl. Phys. (USA)* submitted (2000) ]

[59] K. Sangwal [ *Etching of Crystals: Theory, Experiment and Application, Defects in Solids* Ed. S. Amelinckx (North Holland, Amsterdam, 1987) vol.15 ]

[60] A.R. Patel [ *Physica (Netherlands)* vol.27 (1961) p.1097 and vol.28 (1962) p.44-51 ]

[61] A.R. Patel, S. Tolansky [ *Proc. R. Soc. Lond. A (UK)* vol.243 (1957) p.41-7 ]

[62] Y. Yamaoka, H. Kanda, N. Setaka [ *J. Mater. Sci. (UK)* vol.15 (1980) p.332-6 ]

[63] S. Tolansky [ *Proc. R. Soc. Lond. A (UK)* vol.270 (1962) p.443-51 ]

[64] A.R. Patel, T.C. Patel [ *J. Appl. Crystallogr. (Denmark)* vol.4 (1971) p.207 ]

[65] D.P. Malta, J.B. Posthill, R.A. Rudder, G.C. Hudson, R.J. Markunas [ *J. Mater. Res. (USA)* vol.8 (1993) p.1217-9 ]

[66] J.J. Schermer, F.K. de Theije, L.J. Giling [ *J. Cryst. Growth (Netherlands)* vol.165 (1996) p.387-401 ]

[67] F.K. de Theije, W.J.P. van Enckevort [ work in progress ]

[68] T. Surek, J.P. Hirth, G.M. Pound [ *J. Cryst. Growth (Netherlands)* vol.18 (1973) p.20-8 ]

[69] X. Jiang, C. Rickers [ *Appl. Phys. Lett. (USA)* vol.75 (1999) p.3935-7 ]

[70] E.M. Wilks, J. Wilks [ *Wear (Switzerland)* vol.81 (1982) p.329-46 ]

[71] B. Dam, W.J.P. van Enckevort [ *J. Cryst. Growth (Netherlands)* vol.51 (1981) p.607-23 ]

[72] M. Seal [ *Am. Mineral. (USA)* vol.50 (1965) p.105-23 ]

[73] Z. Zheng, H. Kanda, T. Ohsawa, S. Yamaoka [ *J. Mater. Sci. Lett. (UK)* vol.9 (1990) p.331-3 ]

[74] N.S. Pandaya, S. Tolanski [ *Proc. R. Soc. Lond. A (UK)* vol.225 (1954) p.40-8 ]

[75] F.C. Frank [ in *Proc. Int. Ind. Diamond Conf. Oxford, vol.1: Science* Ed. J. Burls (Industrial Diamond Information Bureau, London, 1967) p.119-35 ]

[76] I. Watanabe, K. Haruta, Y. Shimamura [ *Jpn. J. Appl. Phys. (Japan)* vol.33 (1994) p.2035-40 ]

[77] R. Locher et al [ *Diam. Relat. Mater. (Switzerland)* vol.6 (1997) p.654-7 ]

[78] J.E. Field [ Datareview in this book: *A4.4 Polishing diamond surfaces* ]

[79] H. Tokura, M. Yoshikawa [ in *Applications of Diamond Films and Related Materials* Eds. Y. Tzeng, M. Yoshikawa, M. Murakawa, A. Feldman (Elsevier, Amsterdam, 1991) p.241 ]

[80] S. Jin et al [ *Appl. Phys. Lett. (USA)* vol.60 (1992) p.1948-50 ]

[81] A.P. Malshe, B.S. Park, W.D. Brown, H.A. Naseem [ *Diam. Relat. Mater. (Switzerland)* vol.8 (1999) p.1198-213 ]

[82] D.F. Grogan, T. Zhao, B.G. Bovard, H.A. Macleod [ *Appl. Opt. (USA)* vol.31 (1992) p.1483-7 ]

[83] B.G. Bovard, T. Zhao, H.A. Macleod [ *Appl. Opt. (USA)* vol.31 (1992) p.2366 ]

[84] C. Vivensang, L. Ferlazzo-Manin, M.F. Ravet, G. Turban, F. Rousseaux, A. Gicquel [ *Diam. Relat. Mater. (Switzerland)* vol.5 (1996) p.840-4 ]

[85] G.M.R. Sirineni, H.A. Naseem, A.P. Malshe, W.D. Brown [ *Diam. Relat. Mater. (Switzerland)* vol.6 (1997) p.952-8 ]

[86] S. Ilias et al [ *Diam. Relat. Mater. (Switzerland)* vol.5 (1996) p.835-9 ]

[87] S.J. Pearton, A. Katz, F. Ren, J.R. Lothian [ *Electron. Lett. (UK)* vol.28 (1992) p.822-4 ]

[88] S.A. Grot, R.A. Ditzio, G.Sh. Gildenblat, A.R. Badzian, S.J. Fonash [ *Appl. Phys. Lett. (USA)* vol.61 (1992) p.2326-8 ]

[89] O. Dorsch, M. Werner, E. Obermeier, R.E. Harper, C. Johnston, I.M. Buckley-Golder [ *Diam. Relat. Mater. (Netherlands)* vol.1 (1992) p.277-80 ]

[90] L. Schäfer, A. Bluhm, C.-P. Klages, B. Löchel, L.-M. Buchmann, H.-L. Huber [ *Diam. Relat. Mater. (Switzerland)* vol.2 (1993) p.1191-6 ]

[91] K.K. Chan, G.A.J. Amartunga, T.K.S. Wong [ *Diam. Relat. Mater. (Netherlands)* vol.1 (1992) p.281 ]

[92] A. Masood, M. Aslam, M.A. Tamor, T.J. Potter [ *J. Electrochem. Soc. (USA)* vol.138 (1991) p.L67-8 ]

[93] J. Taniguchi, I. Miyamoto, N. Ohno, K. Kantani, M. Komuro, H. Hiroshima [ *Jpn. J. Appl. Phys. (Japan)* vol.36 (1997) p.7691-5 ]

[94] J. Taniguchi, N. Ohno, S. Takeda, I. Miyamoto, M. Komuro [ *J. Vac. Sci. Technol. B (USA)* vol.16 (1998) p.2506-10 ]

[95] V.G. Ral'chenko, K.G. Korotushenko, A.A. Smolin, E.N. Loubnin [ *Diam. Relat. Mater. (Switzerland)* vol.4 (1995) p.893-6 ]

[96] C. Johnston, P.R. Chalker, I.M. Buckley-Golder, P.J. Marsden, S.W. Williams [ *Diam. Relat. Mater. (Switzerland)* vol.2 (1993) p.829-34 ]

[97] V.G. Ralchenko et al [ *Diam. Relat. Mater. (Switzerland)* vol.2 (1993) p.904-9 ]

[98] J. Narayan, X. Chen [ *J. Appl. Phys. (USA)* vol.71 (1992) p.3795-801 ]

[99] T. Inoue et al [ *J. Appl. Phys. (USA)* vol.67 (1990) p.7329-36 ]

[100] I. Miyamoto, N. Taniguchi [ *Bull. Jpn. Soc. Precis. Eng. (Japan)* vol.16 (1982) p.181-2 ]

[101] I. Miyamoto, N. Taniguchi [ *Bull. Jpn. Soc. Precis. Eng. (Japan)* vol.17 (1983) p.195-6 ]

[102] I. Miyamoto, K. Kawata, M. Kimura, N. Kanekama [ *J. Mater. Sci. Lett. (UK)* vol.7 (1988) p.1175-7 ]

[103] A.N. Stepanova, E.I. Givargizov, L.V. Bormatova, V.V. Zhirnov, E.S. Mashkova, A.V. Molchanov [ *J. Vac. Sci. Technol. B (USA)* vol.16 (1998) p.678-80 ]

[104] I. Miyamoto, S.T. Davies, K. Kawata [ *Nucl. Instrum. Methods Phys. Res. B (Netherlands)* vol.39 (1989) p.696-9 ]

[105] H. Shiomi [ *Jpn. J. Appl. Phys. (Japan)* vol.36 (1997) p.7745-8 ]

[106] O. Dorsch, M. Werner, E. Obermeier [ *Diam. Relat. Mater. (Switzerland)* vol.4 (1995) p.456-9 ]

CHAPTER A5

# PROPERTIES OF DEFECTS IN DIAMOND: NITROGEN AND HYDROGEN

# A5.1 Theory of aggregation of nitrogen in diamond

R. Jones and J.P. Goss

March 2000

## A INTRODUCTION

Modelling defects using atomistic and quantum-mechanical techniques provides us with a powerful method for identifying the microscopic structure of centres observed in experiment. Properties including the energies and lifetimes of optical transitions [1], local vibrational modes and vibrational resonances [2] have proved particularly powerful for nitrogen in diamond. However, processes involving the motion of nitrogen through the lattice are still not fully understood, and the literature is noticeably sparse concerning the modelling of these problems. Several levels of theory can be applied to the simulation of defects in crystalline material, ranging from potentials gained from parametrisation using experimental observations, to so-called ab initio theories which use no experimental input, the details of which are in the literature.

## B NITROGEN IN DIAMOND

Nitrogen is the most common impurity found as point defects, and is found in all types of diamond. The form in which most nitrogen is present is often used to classify the material. In natural type Ib and CVD diamonds, N is present as a substitutional defect, present as either a neutral or a positively charged species. The neutral defect does not possess tetrahedral symmetry but undergoes a trigonal distortion with N and one of the C atoms moving away from each other along ⟨111⟩ [3–6] (FIGURE [1]). This is responsible for the yellow/brown colouration of synthetic diamonds grown without nitrogen getters.

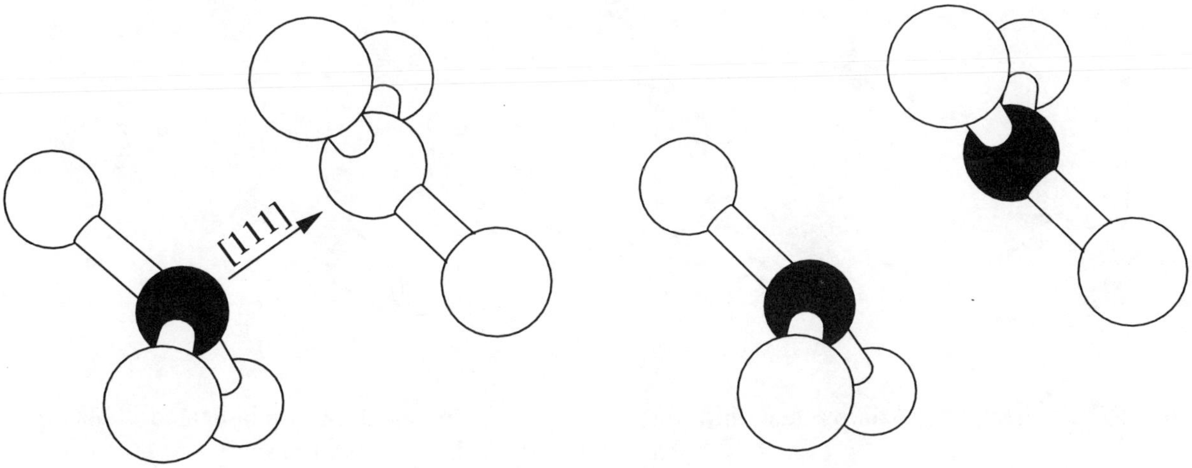


FIGURE 1 Schematics of the single substitutional nitrogen (left) and A-centre (right) showing the dilation of the bond adjacent to nitrogen. Nitrogen atoms are indicated in black.

Modelling [2] using ab initio local density functional cluster theory [7] successfully accounted for the vibrational modes of the defect, as well as its reorientation barrier [8]. The migration barrier by exchange with a carbon neighbour has been calculated to be 6.3 eV [6].

The aggregation of N into an A-centre, which is a complex made up from two neighbouring N atoms at adjacent lattice sites (FIGURE 1), is well established experimentally and occurs during extended heat treatments [9]. The defect was identified by Davies [10] from the symmetry of the 3.8 eV absorption line and its behaviour in diamonds containing $^{15}N$. Theory [2] confirmed that the A-centre gives vibrational resonances at 1282, 1203, 1093 and 480 $cm^{-1}$.

Further aggregation of nitrogen leads to the B-centre which is made up from four nitrogen atoms surrounding a lattice vacancy (FIGURE 2). These $VN_4$ defects, also formed during extended anneals, gave vibrational bands around the observed ones at 1332, 1171, 1093, 1003 and 780 $cm^{-1}$. It is thought that the aggregation of four nitrogen atoms makes it energetically favourable to form a vacancy-interstitial pair resulting in the B-centre and a self-interstitial, which in turn aggregate to form the well-known (001)-orientated platelets [11]. Material that contains mainly aggregated nitrogen is termed type Ia, and may possess predominantly A-centres (type IaA), B-centres (type IaB) or a mixture (type IaAB). The first-principles modelling suggested that the B-centre, like the A-centre, should be optically active. However, no absorption band has been so far identified. It is not clear whether the A-centres migrate as a unit to form the B-centres, or whether they dissociate during the annealing stage and form $VN_4$ from individual N atoms.

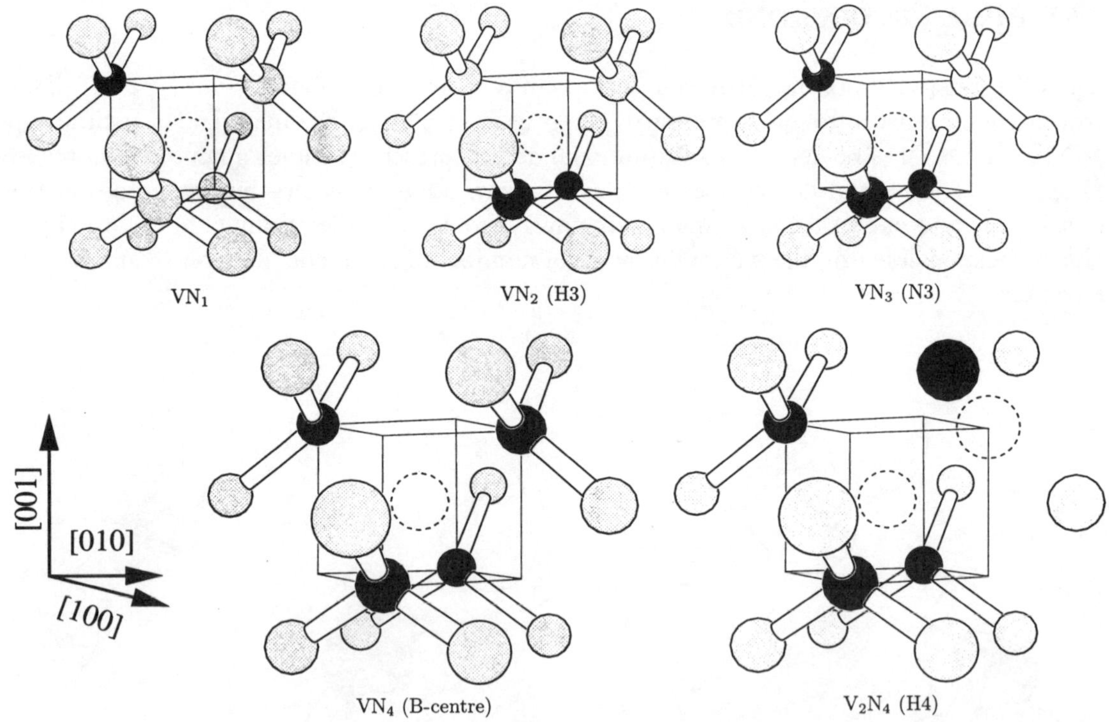


FIGURE 2 Schematics of the vacancy-nitrogen complexes. Nitrogen atoms are indicated in black, carbon in grey and the vacant sites by a dashed empty circle.

Other complexes made up from nitrogen and vacancies are also present in the material, especially when the samples have been irradiated (see FIGURE 2). Many vacancy-nitrogen defects have been modelled. The $VN_2$ or H3 centre and $V_2N_2$ or H4 centres are considered in [2,6,12], while the optical properties of VN and $VN_3$ have been discussed in [1,13]. It has been suggested that vacancies may enhance the motion of nitrogen through the lattice [14], and the activation barrier for this process has been estimated at 4.5 eV [6], lower than that of concerted exchange. A peculiarity surrounds the optical lines of the $VN_3$ defect. This centre gives rise to a set of lines N2–N4. However, more than the manifold states arising from the $sp^3$ orbitals on the atoms bordering the vacancy are required to explain the optical properties of this complex. Additional levels are found from the first principles modelling [13].

Nitrogen interstitials have been described in a combined theoretical and experimental study [15]. There are some peculiarities associated with the centre, not least of which is the very high thermal stability of the experimental observables associated with this centre. The properties of the Hla peak are highly dependent on the material type (Ia or Ib), but in all cases an anneal to 1400°C is required to reduce the intensity of this local mode around 1450 $cm^{-1}$. However, this is still likely to be the most mobile form of nitrogen with an estimated migration barrier of 1 eV [6], and it is possible that self-interstitials catalyse the aggregation of nitrogen via migration of interstitial nitrogen.

Synthetic material grown via the high-temperature, high pressure method often contains relatively large concentrations of Ni. Recent work has focused on nickel–nitrogen–vacancy centres where it appears that substitutional Ni can kick out an interstitial and spontaneously form a nickel-vacancy centre which is stabilised by nitrogen. These complexes have also been modelled [16], although processes behind the aggregation of nitrogen to the Ni-vacancy complex remain to be explored.

## C CONCLUSION

In conclusion, modelling of point defects involving N is fairly advanced and reliable structures, energetics, and vibrational modes can be found although the calculation of optical transition energies continues to cause problems. The current understanding of the processes governing the motion of nitrogen through diamond on the other hand is less well developed.

## REFERENCES


[1] J.P. Goss, R. Jones, S.J. Breuer, P.R. Briddon, S. Öberg [ *Phys. Rev. Lett. (USA)* vol.77 (1996) p.3041-4 ]

[2] R. Jones, P.R. Briddon, S. Öberg [ *Philos. Mag. Lett. (UK)* vol.66 (1992) p.67-74 ]

[3] P.R. Briddon, R. Jones, M.I. Heggie [ *Proc. Int. Conf. on New Diamond Science and Technology* Washington, DC, 23-27 Sept. 1990 (Mater. Res. Soc., USA, 1991) p.63 ]

[4] P.R. Briddon, R. Jones, M.I. Heggie [ *Mater. Sci. Forum (Switzerland)* vol.83-87 (1991) p.457-62 ]

[51 S.A. Kajihara, A. Antonelli, J. Bernholc, R. Car [ *Phys. Rev. Lett. (USA)* vol.66 (1991) p.2010-3 ]

[6] A. Mainwood [ *Phys. Rev. B (USA)* vol.49 (1994) p.7934-40 ]

[7] R. Jones, P.R. Briddon [ *Semicond. Semimet. (USA)* vol.51A (1998) ch.6 ]

[8] S.J. Breuer, P.R. Briddon [ *Phys. Rev. B (USA)* vol.53 (1996) p.7819-22 ]

[9] T. Evans, A. Qi, J. Maguire [ *J. Phys. C, Solid State Phys. (UK)* vol.14 (1981) p.379-84 ]

[10] C. Davies, M.H. Nazaré, M.F. Hamer [ *Proc. R. Soc. Lond. A (UK)* vol.351 (1976) p.245 ]

[11] P.J. Fallon, L.M. Brown, J.C. Barry, J. Bruley [ *Philos. Mag. A (UK)* vol.72 (1995) p.21-37 ]

[12] R. Jones, V.J.B. Torres, P.R. Briddon, S. Öberg [ *Mater. Sci. Forum (Switzerland)* vol.143-7 (1994) p.45-50 ]

[13] R. Jones, J.P. Goss, P.R. Briddon, S. Öberg [ *Phys. Rev. B (USA)* vol.56 (1997) p.1654-6 ]

[14] A.T. Colins [ *J. Phys. C (UK)* vol.13 (1980) p.2641 ]

[15] I. Kiflawi, A. Mainwood, H. Kanda, D. Fisher [ *Phys. Rev. B (USA)* vol.54 (1996) p.16719-26 ]

[16] J.P. Goss, B.J. Coomer, R. Jones, P.R. Briddon, S. Öberg [ *Proc. De Beers Diamond Conf.* Oxford, 1999, p.10.1 ]

# A5.2 Aggregates of nitrogen in diamond

I. Kiflawi and S.C. Lawson

December 1999

## A INTRODUCTION

Nitrogen is the major impurity in both natural and synthetic diamond. It is found in dispersed single-substitutional form and in aggregated form: namely A centres, B centres and voidites. Apart from voidites, which can only be observed with the aid of transmission electron microscopy (TEM), all the other aggregates have signatures in optical spectroscopy. A schematic presentation of the different types of diamond classified by their nitrogen aggregation states is shown in FIGURE 1. A full presentation is found in the reviews by Nazare [1] and Woods [2].

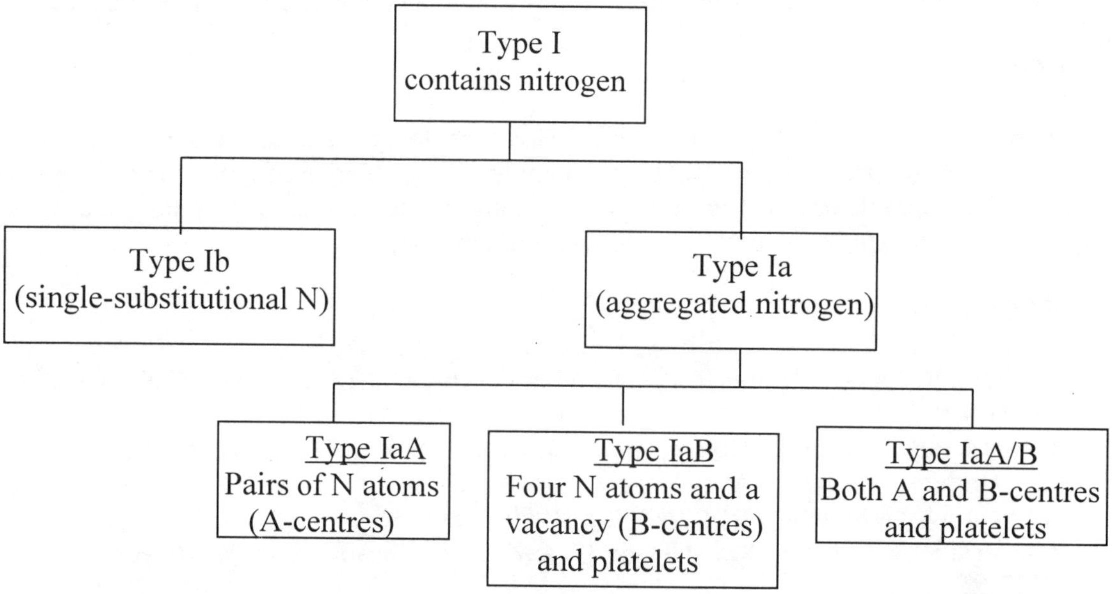


FIGURE 1 Schematic presentation of the different types of diamond.

During the nitrogen aggregation process other nitrogen containing centres may also be produced. These centres are usually associated with vacancies and give rise to defects such as the 575 and 637 nm systems (associated with single nitrogen atoms) and the H2, H3, H4 and N3 centres (associated with higher aggregation states of nitrogen). The concentration of nitrogen in these centres constitutes a very small fraction of the total nitrogen concentration and they are not discussed further here.

## B THE NITROGEN AGGREGATION SEQUENCE

It is currently accepted that nitrogen is incorporated into the diamond lattice as single-substitutional atoms. In nature, nitrogen aggregates are formed as a result of an aggregation process that takes place over geological time-scales at the temperatures which prevail within the earth's upper mantle. This view was supported by laboratory experiments in which diamonds were annealed at high temperatures and the various aggregates were produced. FIGURE 2 shows the nitrogen aggregation sequence and the laboratory conditions under which these processes take place. The first aggregation sequences, i.e.

conversion of C to A-centres and A to B-centres and platelets, are reviewed by Woods [3]. At the time of that review there was no laboratory evidence for the last stage of the aggregation process, namely the conversion of platelets into dislocation loops and voidites. There was also uncertainty regarding the chemical composition of the platelets and the source of the nitrogen in the voidites.

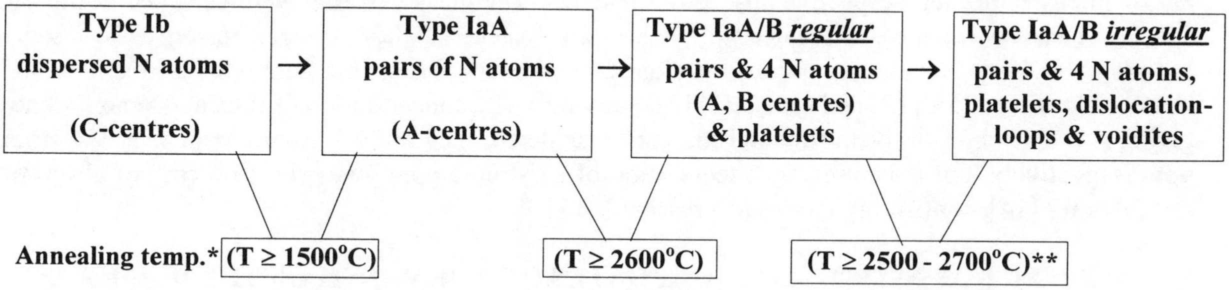


* The temperature at which aggregation occurs depends on many factors such as the presence of impurities other than nitrogen and the applied pressure. The presence of vacancies or Ni and Co enhances the aggregation process [16–18]. Also, a higher aggregation rate is achieved if lower pressures are used during annealing [19].

** The temperature at which this process takes place is dependent on whether the temperature and the applied pressure maintained during the annealing of the diamond specimen is in the graphite or the diamond stable region of the carbon phase diagram; the low end of the temperature range corresponds to the graphite region [4].

FIGURE 2 Nitrogen aggregation sequence.

Evans et al [4] annealed diamonds containing platelets, and observed using TEM that the platelets were converted into dislocation loops. This conversion was accompanied by the disappearance of the X-ray extra reflections and the IR absorption features which were associated with the presence of the platelets. This demonstrated that these features are indeed caused by the platelets. Also, octahedral voidite-like defects were produced. Kiflawi and Bruley [5] examined the chemical composition of these defects and found them to contain molecular nitrogen, as in the case of natural voidites. This demonstrated that the last stage of the aggregation process can be achieved under laboratory conditions. Kiflawi and Bruley also found that the concentration of nitrogen in the laboratory-formed voidites was similar to that which was in the platelets before they were transformed [5].

## C PLATELETS

Although platelets are not a form of nitrogen aggregate they are very closely connected with the aggregation of nitrogen in diamond and at one time they were thought to be composed of nitrogen. Recent studies [6–8] have shown that nitrogen is only incorporated in the platelets as an impurity, with quoted concentrations lying between 0 and 50%. Evans et al [9] suggested that the nitrogen may be decorating the platelets. However, IR, TEM and EELS measurements conducted on platelet-containing diamond after high temperature annealing, up to 2650°C, showed no change in the nitrogen concentration or in the shape of the platelets [8]. This would suggest that the nitrogen forms part of the structure of the platelets, rather than simply decorating them. However, changes were observed in their characteristic IR absorption, namely a broadening of the B' peak and the disappearance of the associated peak at 328 $cm^{-1}$. These changes were attributed to possible modification of the platelet structure caused by the high temperature annealing [8].

## D ASSAYING OF NITROGEN IN DIAMOND

Since the last Datareview on this subject [10], further measurements have been carried out in order to determine the conversion factor between the concentration of nitrogen in the various aggregates and the IR intensity of their respective absorption features. The measurements were confined to the C, A and B-centres which have characteristic IR spectra. Instead of the inert gas fusion assay used by Woods et al [11], mass spectroscopy was used to measure the concentration of nitrogen in well-characterised type IaA [12] and type IaB [13] diamonds. The concentration of A-centres and B-centres giving rise to 1 $cm^{-1}$ of absorption at 1282 $cm^{-1}$ was determined at 16.1 atomic ppm and 79.4 atomic ppm respectively. For C-centres, a concentration of 25 atomic ppm gives rise to 1 $cm^{-1}$ of absorption at 1130 $cm^{-1}$ [14], confirming the result presented in [10].

## E EFFECT OF NITROGEN AGGREGATES ON THE WIDTH OF THE DIAMOND RAMAN LINE

Dispersed single-substitutional nitrogen and nitrogen aggregates affect the line width of the diamond Raman line [14]. The relationship between the nitrogen concentration and the Raman line width was found to follow a linear regression represented by $1.60 + 1.52 \times 10^{-3}$ N $cm^{-1}$ for C-centres and $1.57 + 0.97 \times 10^{-3}$ N $cm^{-1}$ for A-centres [15].

## F CONCLUSION

The different types of diamond classified by their nitrogen aggregation states have been shown. The nitrogen aggregation sequence, platelets and assaying of nitrogen in diamond have been discussed. The relationship between the nitrogen concentration and the Raman line width in diamond was found to follow a linear regression.

## REFERENCES


[1] M.H. Nazare [ in *Properties and Growth of Diamond* Ed. G. Davies (INSPEC, IEE, London, UK, 1994) ch.3 p.85-7 ]

[2] G.S. Woods [ in *Properties and Growth of Diamond* Ed. G. Davies (INSPEC, IEE, London, UK, 1994) ch.3 p.88-97 ]

[3] G.S. Woods [ in *Properties and Growth of Diamond* Ed. G. Davies (INSPEC, IEE, London, UK, 1994) ch.3 p.98 ]

[4] T. Evans, I. Kiflawi, W. Luyten, G. Van Tendeloo, G.S. Woods [ *Proc. R. Soc. Lond. A (UK)* vol.449 (1995) p.295 ]

[5] I. Kiflawi, J. Bruley [ *Diam. Relat. Mater. (Switzerland)* in press (January 2000) ]

[6] J. Bruley [ *Philos. Mag. Lett. (UK)* vol.66 (1992) p.47 ]

[7] P.J. Fallon, L.M. Brown, J.C. Barry, J. Bruley [ *Philos. Mag. A (UK)* vol.72 (1995) p.21 ]

[8] I. Kiflawi, J. Bruley, W. Luyten, G. Van Tendeloo [ *Philos. Mag. B (UK)* vol.78 (1998) p.299 ]

[9] T. Evans, G.S. Woods, C.D. Clark, H. Kanda [ *Abstracts De Beers Diamond Conf.* Reading, 1990 (unpublished) p.20 ]

[10] G.S. Woods [ in *Properties and Growth of Diamond* Ed. G. Davies (INSPEC, IEE, London, UK, 1994) ch.3 p.101 ]

[11] G.S. Woods, G.C. Purser, A.S.S. Mtimkulu, A.T. Collins [ *J. Phys. Chem. Solids (UK)* vol.51 (1990) p.1191-7 ]

[12] S.R. Boyd, I. Kiflawi, G.S. Woods [ *Philos. Mag. B (UK)* vol.69 (1994) p.1149 ]

[13] S.R. Boyd, I. Kiflawi, G.S. Woods [ *Philos. Mag. B (UK)* vol.72 (1995) p.351-61 ]

[14] I. Kiflawi, A.E. Mayer, P.M. Spear, J.A. van Wyk, G.S. Woods [ *Philos. Mag. B (UK)* vol.69 no.6 (1994) p.1141-7 ]

[15] N.V. Surovtsev, I.N. Kupriyanov, V.K. Malinivsky, V.A. Gusev, Yu.N. Pal'yanov [ *J. Phys., Condens. Matter (UK)* vol.11 (1999) p.4767-74 ]

[16] A.T. Collins [ *J. Phys. C (UK)* vol.13 (1980) p.2641 ]

[17] I. Kiflawi, H. Kanda, A. Mainwood [ *Diam. Relat. Mater. (Switzerland)* vol.7 (1998) p.327-32 ]

[18] D. Fisher, S.C. Lawson [ *Diam. Relat. Mater. (Switzerland)* vol.7 (1998) p.299 ]

[19] I. Kiflawi, H. Kanda, D. Fisher, S.C. Lawson [ *Diam. Relat. Mater. (Switzerland)* vol.6 (1997) p.1643-9 ]

# A5.3 The nitrogen interstitial in diamond

I. Kiflawi and S.C. Lawson

December 1999

## A INTRODUCTION

Orbital cluster calculations suggest that the nitrogen interstitial in diamond prefers a <001> split or bond-centred site [1]. However, until very recently there was no experimental evidence for the existence of such a point defect.

## B OPTICAL PROPERTIES

Collins et al [2] studied the effect of changing the host isotope to $^{13}C$ and found that a local-mode peak at 1450 $cm^{-1}$ shifts to 1423.4 $cm^{-1}$. Woods and Collins [3] studied the effect on this peak of doping diamonds with mixed nitrogen isotopes. They found that the peak is split into two and only two components and suggested that the transition occurs at a C–N split interstitial. However, Kiflawi et al [4] studied this defect (called the H1a centre) in diamonds grown with different ratios of $^{12}C$ and $^{13}C$ and identified it with one nitrogen atom and two equivalent carbon atoms. Two possible configurations were considered; namely, substitutional nitrogen with two carbon atoms in <110> split interstitial positions and a bond-centred interstitial nitrogen. Approximate molecular orbital calculations identified the centre as a bond-centred nitrogen interstitial slightly displaced off the centre of the axis between the two carbon atoms but still equidistant from each of them [4].

This nitrogen interstitial can be produced by annealing electron or neutron irradiated nitrogen containing diamonds [4,5], or as a consequence of the nitrogen aggregation process in diamond in the presence of transition metal impurities (see Datareveiw A6.2). The nitrogen interstitials are very mobile in the diamond lattice and contribute to the nitrogen aggregation process [4].

Isoya et al [6] identified the electron paramagnetic resonance centre, NIRIM-4, as a boron-nitrogen, $[B\text{-}N]^+$, <100> split interstitial. This centre is also formed by electron irradiation of nitrogen-containing boron doped diamonds [7].

## C CONCLUSION

In conclusion, it is of note that many unassigned photoluminescence and cathodoluminescence features are present in irradiated diamond, some of which may be associated with nitrogen interstitials [8]. Clearly further research is required in this area.

## REFERENCES


[1] C. Weigel, D. Peak, J.W. Corbett, G.D. Watkins, R.P. Messmer [ *Phys. Status Solidi B (Germany)* vol.63 (1974) p.131 ]

[2] A.T. Collins, G. Davies, H. Kanda, G.S. Woods [ *J. Phys. C (UK)* vol.21 (1988) p.1363 ]

[3] G.S. Woods, A.T. Collins [ *J. Phys. C (UK)* vol.15 (1982) p.L949 ]

[4] I Kiflawi, A. Mainwood, H. Kanda, D. Fisher [ *Phys. Rev. B (USA)* vol.54 (1996) p.16719 ]

[5] G.S. Woods [ *Philos. Mag. B (UK)* vol.50 (1984) p.673 ]

[6] J. Isoya, H. Kanda, Y. Morita [ *Phys. Rev. B (USA)* vol.56 (1997) p.6392 ]

[7] J. Isoya, H. Kanda, Y. Morita [ *Advances in New Diamond and Technology, 4^th^ Int. Conf. on New Diamond and Technology* Kobe, Japan, 18-22 July 1994 (My, Tokyo, 1994) p.351 ]
[8] J.W. Steeds et al [ *Diam. Relat. Mater. (Switzerland)* vol.8 (1999) p.94-100 ]

# A5.4 Neutral ($[N_S]^0$) and ionised ($[N_S]^+$) single substitutional nitrogen in diamond

M.E. Newton

February 2000

## A INTRODUCTION

Nitrogen is the most prevalent impurity in diamond. The neutral single substitutional nitrogen centre (P1 or $[N_S]^0$) has been extensively studied by EPR (electron paramagnetic resonance) and ENDOR (electron nuclear double resonance) [1–6], and it has long been accepted that the nitrogen forms four bonding orbitals with neighbouring carbon atoms, whilst the remaining unpaired electron (S = ½) goes into a highly localised anti-bonding orbital directed along one of the N–C directions. $[N_S]^0$ is a deep donor, and is not useful as an n-type dopant, but often donates an electron to other defects in its vicinity. The infrared absorption spectrum from $[N_S]^+$ has recently been identified, and is described in Section F.

## B EPR AND ENDOR MEASUREMENTS ON THE $[N_S]^0$ CENTRE – ELECTRONIC g-VALUE, $^{14}N$, $^{15}N$ AND $^{13}C$ HYPERFINE PARAMETERS AND $^{14}N$ QUADRUPOLE PARAMETERS

Smith et al [1] first observed the $[N_S]^0$ (P1) centre, and identified nitrogen because it is the only naturally occurring atom with an almost 100% abundant I = 1 isotope. Smith et al [1] also observed $^{13}C$ satellites from a unique carbon neighbour; the atomic model of a single substitutional nitrogen atom, with a unique N–C bond elongated to accommodate the unpaired electron localised in an N–C anti-bonding orbital was established. EPR measurements have established that the electronic g-value is axially symmetric about the unique N–C bond, but the departure from an isotropic g-value is very small, $(g_{\parallel}+2g_{\perp})/3 = 2.0024$ and $(g_{\perp}-g_{\parallel}) = +0.00003(1)$. The data of Zhang et al [7] were used to determine these parameters. The analysis and the values for $g_{\parallel}$ and $g_{\perp}$ in their original paper [7] are incorrect. The finding that $g_{\parallel} < g_{\perp}$ is in accordance with theory [8]. $[N_S]^0$ is often observed in polycrystalline diamond films grown by chemical vapour deposition, in low concentrations, such that the EPR lines are narrow and at microwave frequencies of around 9.5 GHz the central transition shows a small asymmetry. This asymmetry is predominately due to the off-diagonal hyperfine terms in the spin Hamiltonian; there is only a very small contribution from the anisotropic g-matrix.

$^{14}N$ and $^{15}N$ hyperfine parameters for the $[N_S]^0$ centre have been determined accurately by ENDOR [5,6] and are given in TABLE 1. Measurement of the nuclear g-values by ENDOR confirmed unambiguously that nitrogen is responsible for the observed hyperfine splitting. A small nitrogen hyperfine anomaly was identified as originating from the different zero point amplitudes of $^{14}N$ and $^{15}N$ [6]. $^{13}C$ hyperfine coupling of the unpaired electron with neighbouring carbon atoms in seven different positions has been identified [5,6]. The measured $^{13}C$ hyperfine parameters are given in TABLE 2 [5,6]. The labelling of the sites associated with each of the coupling parameters in TABLE 2 refers to FIGURE 1. The assignment of a hyperfine coupling to a specific site is determined by the intensity of the $^{13}C$ EPR satellites (number of equivalent neighbours) and the magnitude and symmetry of the interaction. When there is significant uncertainty in the assignment possible alternatives are given in TABLE 2. Analysis of the hyperfine coupling with simple theory indicates a localisation of ≈ 25% of the unpaired electron on the nitrogen, and ≈ 75% on the unique carbon ($C_a$); the localisation on other carbon neighbours is very small. The $^{14}N$ quadrupole interaction is axially symmetric about the <111> axis of the defect and $P_{||} = -3.973(1)$ MHz [6]. This value has been shown to be in good accord with theory [9]. Simple analysis of the quadrupole and hyperfine interactions suggests that the unique N–C bond is about 25–30% longer than the normal C–C bond length.

TABLE 1 $^{14}N$ and $^{15}N$ hyperfine parameters for the $[N_S]^0$ centre determined by Cox et al [6]. The directions of principal values of hyperfine matrices are given as [θ,ϕ], where θ is measured from the [001] crystal axis, and ϕ from the [100] axis while rotating about [001].

| $^{14}N$ | $^{15}N$ |
|---|---|
| $A_{\|\|}$ = 114.032(3) MHz \|\| [54.74,45] | $A_1$ = 159.730(7) MHz \|\| [54.74,45] |
| $A_{\perp}$=81.318(2) MHz | $A_2$ = 113.838(6) MHz |

TABLE 2 $^{13}C$ hyperfine parameters for the carbon atoms surrounding the nitrogen atom of the $[N_S]^0$ centre. The hyperfine interaction is given in units of MHz. The directions of principal values of hyperfine matrices are given as [θ,ϕ], where θ is measured from the [001] crystal axis, and ϕ from the [100] axis while rotating about [001]. The labelling of the atoms refers to FIGURE 1 and the text.

| $C_1$ (a) [6] | $C_2$ (d) [6] | $C_3$ (c) [6] | $C_4$ (b) [6] | $C_5$ (g or e) [6] | $C_6$ (f) [5] | $C_7$ (e or g) [5] |
|---|---|---|---|---|---|---|
| $A_{\|\|}$ = 338.171(5) \|\| [54.74,45] | $A_1$ = 30.921(5) \|\| [90,315] | $A_1$ = 26.488(5) \|\| [90,315] | $A_1$ = 10.638(4) \|\| [90,315] | $A_1$ = 11.757(3) \|\| [71.5,33.2] | A = 4.1 | A = 2.7 |
| $A_{\perp}$ = 139.531(6) | $A_2$ = 40.292(4) \|\| [58.66,45] | $A_2$ = 22.771(4) \|\| [52.36,45] | $A_2$ = 14.153(4) \|\| [59.19,45] | $A_2$ = 8.579(3) \|\| [138.6,101.0] | | |
| | $A_3$ = 31.662(4) \|\| [31.34,225] | $A_3$ = 25.319(4) \|\| [37.64,225] | $A_3$ = 10.618(5) \|\| [30.81,225] | $A_3$ = 8.122(3) \|\| [55.0,137.0] | | |

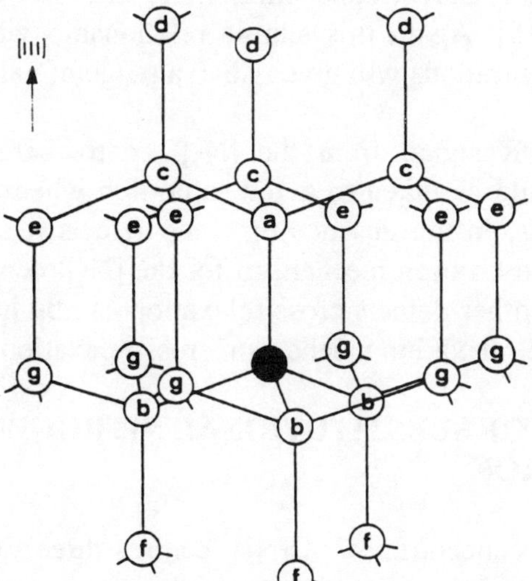


FIGURE 1 Carbon atom sites around a single substitutional nitrogen atom (filled circle). The labelling of the carbon atoms refers to the notation used in the text and tables. (Figure adapted from Barklie and Guven [5].)

## C MOTIONAL AVERAGING OF THE $[N_S]^0$ CENTRE AT HIGH TEMPERATURES

Above 600 K motional averaging starts to occur for the $[N_S]^0$ centre as the unpaired electron hops between the four N–C bonds. At 1200 K the EPR spectrum is truly isotropic, since the unpaired electron is hopping so quickly that it sees an average of all possible sites. Measurements in the temperature range 600–1200 K determine that the reorientation activation energy is 0.7(1) eV [10,11]. The rate of anneal of stress-induced ordering of the $[N_S]^0$ centre has been measured in the temperature range 78–200 K [12]. The results show a marked deviation from the high temperature Arrhenius-type behaviour. It is concluded that at low temperature reorientation of the centres occurs by tunnelling between thermally populated excited vibrational states [12].

## D EPR LINE WIDTH, SPIN-LATTICE RELAXATION AND SPIN-SPIN RELAXATION MEASUREMENTS ON THE $[N_S]^0$ CENTRE

In synthetic type Ib diamond it has been shown that for concentrations of the $[N_S]^0$ centre higher than about 10 parts per million carbon atoms (ppm) the EPR linewidth is a linear function of concentration [13]. For an $[N_S]^0$ concentration C (ppm) the peak to peak EPR linewidth (in Gauss) is

$$\sqrt{(1.2\times10^{-4}C^2+0.014)}$$

This relation holds for homogeneous samples; if different paramagnetic species are present then the C should be the total paramagnetic defect concentration. Isoya showed that for inhomogeneous synthetic diamonds no simple relationship exists [14]. Van Wyk et al showed that the spin-spin relaxation rate ($1/T_2$) for the $[N_S]^0$ centre is a linear function of total paramagnetic impurity concentration over a wide variation of concentration (0.05–400 ppm) [13]. $T_2$ (in seconds) follows the simple concentration dependence:

$$1/T_2 = 1.4\times10^4 C$$

C (ppm) is the total paramagnetic defect concentration. $T_2$ does not change significantly within the temperature range 4–300 K [15]. Again this simple relationship will not hold in inhomogeneous samples where a range of concentrations will give rise to a range of values for $T_2$.

At low concentrations the EPR signal from the $[N_S]^0$ centre saturates easily (slow spin-lattice relaxation), and great care should be taken to avoid saturation when using continuous wave EPR to determine defect concentrations. In the absence of other defects it appears that spin-orbit phonon-induced tunnelling is the main relaxation mechanism for the $[N_S]^0$ centre in the temperature range 4–300 K [15]. In the presence of other defects cross relaxation can be important and if the other centre has a more effective spin-lattice relaxation mechanism cross relaxation can be dominant [15].

## E SEPARATED PAIRS OF SUBSTITUTIONAL NITROGEN ATOMS AND TWO PHOTON ABSORPTION

In diamonds containing a high concentration of $[N_S]^0$ centres three EPR spectra have been observed that originate from distinct coupled (separation > third nearest neighbour) pairs of substitutional nitrogen atoms [16]. In these three cases the interaction between the two unpaired electrons is sufficiently large to give rise to resolved weak satellites centred on g = 2.0024. When the external Zeeman field is oriented parallel to a <100> crystallographic axis the hyperfine interaction with the two nitrogen atoms gives rise to a characteristic 1:2:3:2:1 pattern for each of these centres.

Another group of lines was observed at approximately half the magnetic field of the $[N_S]^0$ centre. This spectrum is a superposition of several spectra originating from weakly coupled distant pairs of substitutional nitrogen atoms (separation greater than about 0.7 nm). The coupling is so weak that the satellites from the component spectra cannot be observed, but the forbidden transitions from all these pairs contribute to the half field spectrum. The hyperfine splitting from two substitutional nitrogen atoms produces an easily identifiable spectrum [16].

In diamonds containing a high concentration of $[N_S]^0$ centres a replica of the $[N_S]^0$ EPR spectrum has been observed at approximately twice the magnetic field of the usual $[N_S]^0$ EPR spectrum. This spectrum was attributed to a two-photon transition at isolated $[N_S]^0$ centres [16].

## F OPTICAL ABSORPTION OF THE $[N_S]^0$ AND $[N_S]^+$ CENTRES

The $[N_S]^0$ centre gives rise to absorption in the infrared, visible and ultraviolet regions. In diamond containing $[N_S]^0$ grown from elements in natural isotopic abundance the infrared absorption has a maximum at 1130 $cm^{-1}$ and a sharp peak at 1344 $cm^{-1}$ [17]. The ratio of heights of the 1130 and 1344 $cm^{-1}$ peaks depends markedly on the resolution used to record the spectrum. The peak at 1330 $cm^{-1}$ shifts to lower energy by about 15 $cm^{-1}$ on replacing $^{14}N$ with $^{15}N$ [18]. The frequency ratio 1130/1115 = 1.013 is close to the value expected for a C–N vibration. In contrast there is no detectable effect on the sharp mode at 1344 $cm^{-1}$ on changing the isotope of nitrogen. This mode does not involve motion of the nitrogen atom; in diamonds made of $^{13}C$ this mode shifts to 1292 $cm^{-1}$ confirming that it is a pure carbon vibration associated with the strengthened $C_c$–$C_a$ bonds (see FIGURE 1) [19]. It was shown that 1 $cm^{-1}$ of absorption at 1130 $cm^{-1}$ is produced by 25 ±2 ppm of $[N_S]^0$ centres [20]. The infrared absorption spectrum from the $[N_S]^0$ centre is shown in FIGURE 2(a).

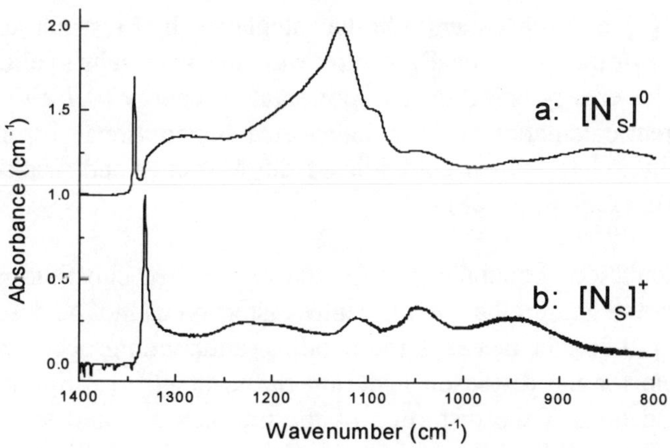


FIGURE 2 Infrared absorption spectra in the defect-induced one phonon region of diamond due to (a) $[N_S]^0$ and (b) $[N_S]^+$ centres. The concentration of $[N_S]^0$ is 25 ppm (spectrum offset vertically by 1.0 $cm^{-1}$ for clarity) and the concentration of $[N_S]^+$ is 5.5 ppm. Spectra recorded at a resolution of 1 $cm^{-1}$.

In diamonds that contain a suitable acceptor (e.g. boron, vacancy, substitutional nickel) the $[N_S]^0$ can donate an electron to a nearby acceptor. The diamagnetic $[N_S]^+$ centre is thus formed. This centre forms four identical bonds with neighbouring carbon atoms and sits on the lattice site. The infrared absorption spectrum from the $[N_S]^+$ has recently been identified, and is shown in FIGURE 2(b) [21]. In this work it was shown that 1 $cm^{-1}$ of absorption at 1332 $cm^{-1}$ is produced by 5.5 ±1 ppm of $[N_S]^+$ centres. Other defects may also give rise to absorption at 1332 $cm^{-1}$, but the $[N_S]^+$ component is uniquely identified by further peaks at 1046 and 950 $cm^{-1}$.

Absorption from the $[N_S]^0$ centre in the visible starts at about 1.7 eV and rises continuously to the band edge. It has been known for many years [22] that the integrated area of broad band centred on 270 nm (4.6 eV) and superimposed on this absorption is correlated with the $[N_S]^0$ concentration. However, care should be taken when using this technique as it is difficult to accurately subtract the background absorption. The absorption continuum that starts at about 1.7 eV has been decomposed into two broad bands with peaks at 3.3 and 3.9 eV, at room temperature [23]. These are polarised perpendicular (3.3 eV) and parallel (3.9 eV) with respect to a trigonal <111> axis and the 4.6 eV band is polarised parallel to this axis [23]. At 77 K sharp UV optical absorption lines can be observed in diamonds containing the $[N_S]^0$ centre. A sharp ZPL at 4.059 eV (with phonon replicas at 4.120 and 4.179 eV) has been shown to occur between totally symmetric states of a trigonal defect [24]. The 4.6 eV band is mainly due to vibronic transitions at this defect [24]. Weaker features have also been observed, a ZPL at 4.567 eV (with phonon replicas at 4.626 and 4.673 eV) that originates from a centre with trigonal symmetry, and three very sharp ZPLs at 4.184, 4.191 and 4.197 eV each with phonon replicas at about 36 meV higher energy [24].

## G PHOTOCONDUCTIVITY

In a single crystal type Ib diamond photoconductivity measurements indicated the existence of a deep donor level at about 1.7 eV below the conduction band [25]. This measurement has been used (along with the onset of optical absorption) to justify placing the nitrogen donor level at 1.7 eV below the conduction band. More recent measurements on diamond grown by CVD suggest that the photoionisation energy of $[N_S]^0$ is 2.2 eV [26].

## H THEORY

In the original EPR paper [1], and subsequently in the calculation by Messmer and Watkins [27], it was asserted that the trigonal symmetry of the P1 centre was due to a Jahn–Teller distortion. However, estimates of the Jahn–Teller energy based on the reorientation energy (0.7 eV) are suspiciously large. Subsequently, self-consistent calculations on the undistorted substitutional nitrogen defect were shown to predict an $A_1$ antibonding state, which contradicted the Messmer and Watkins calculation of a $T_2$ state, unstable to a Jahn–Teller splitting [28].

More recent ab initio calculations conclude that for the undistorted substitutional nitrogen there is a singly–occupied $A_1$ state in the gap, so that the off centre distortion cannot be described by manifestation of the Jahn–Teller effect [29,30]. In fact it is the bonding–antibonding occupancy in one of the N–C bonds which is responsible for the distortion lowering the symmetry to trigonal. Although the recent calculations differ on the details of the distortion of the nitrogen and surrounding carbon atoms, two points are clear: (i) the extension of the unique N–C bond is large, both calculations predicting extensions of about 25% over the normal C–C bond length; and (ii) the relaxation of the neighbouring carbon atoms is significant. The calculations of Briddon and Jones [31] have given insight into the origins and structure of the one phonon IR absorption associated with the $[N_S]^0$.

## I CONCLUSION

The $[N_S]^0$ centre has been extensively studied and many of its properties are well understood. The precise position of the $[N_S]^0$ donor level is still being debated in the literature, and the interaction of $[N_S]^0$ with other defects via charge transfer is attracting considerable attention. The identification of the infrared absorption spectrum of $[N_S]^+$ is important because is has been shown that substantial amounts of nitrogen can be incorporated in this form.

**REFERENCES**


[1] W.V. Smith, P.P. Sorokin, I.L. Gelles, G.L. Lasher [ *Phys. Rev. (USA)* vol.115 (1959) p.1546 ]

[2] J.H.N. Loubser, L. du Preez [ *Br. J. Appl. Phys. (UK)* vol.16 (1965) p. 457 ]

[3] R.J. Cook, G.H. Whiffen [ *Proc. R. Soc. Lond. A (UK)* vol.295 (1966) p.99 ]

[4] R.J. Cook [ *J. Sci. Instrum. (UK)* vol.43 (1966) p.548 ]

[5] R.C. Barklie, J. Guven [ *J. Phys. C, Solid State Phys. (UK)* vol.14 (1981) p.3621 ]

[6] A. Cox, M.E. Newton, J.M. Baker [ *J. Phys., Condens. Matter (UK)* vol.6 (1994) p.551 ]

[7] S. Zhang, S.C. Ke, M.E. Zvanut, H.T. Tohver, Y.K. Vohra [ *Phys. Rev. B (USA)* vol.49 (1994) p.15392]

[8] G.D. Watkins, J.W. Corbett [ *Phys. Rev. A (USA)* vol.134 (1964) p.1359 ]

[9] O.D. Tucker, M.E. Newton, J.M. Baker [ *Phys. Rev. B (USA)* vol.50 (1994) p.15586 ]

[10] L.A. Shul'man, I. M. Zaritskii, G.A. Podzyarei [ *Sov. Phys.-Solid State (USA)* vol.8 (1967) p.1029 ]

[11] J.H.N. Loubser, W.P. Van Ryneveld [ *Br. J. Appl. Phys. (UK)* vol.18 (1967) p.1029 ]

[12] C.A. Ammerlaan, E.A. Burgemeister [ *Phys. Rev. Lett. (USA)* vol.47 (1981) p.954 ]

[13] J.A. Van Wyk, E.C. Reynhardt, G.L. High, I. Kiflawi [ *J. Phys. D, Appl. Phys. (UK)* vol.30 (1997) p.1790 ]

[14] J. Isoya, C.P. Lin, M.K. Bowman, J.R. Norris, S. Yazu, S. Sato [ *Science and Technology of New Diamond – Proc. 1st Int. Conf. on New Diamond Science and Technology* Ed S. Saito (KTK-Terra, Tokyo, 1998) ]

[15] E.C. Reynhardt, G.L. High, J.A. Van Wyk [ *J. Chem. Phys. (USA)* vol.109 (1998) p.8471 ]

[16] V.A. Nadolinny et al [ *Phys. Rev. B (USA)* vol.60 (1999) p.5392 ]

[17] R.M. Chrenko, H.M. Strong, R.E. Tuft [ *Philos. Mag. B (UK)* vol.11 (1965) p.313 ]

[18] A.T. Collins, G.S. Woods [ *Philos. Mag. (UK)* vol.46 (1982) p.77 ]

[19] C.D. Clark, A.T. Collins, G.S. Woods [ in *Properties of Synthetic and Natural Diamond* Ed. J.E. Field (Academic Press, London, UK, 1992) p.35 ]

[20] G.S. Woods, J.A. Van Wyk, A.T. Collins [ *Philos. Mag. B (UK)* vol.62 (1990) p.589 ]

[21] S.C. Lawson, D. Fisher, D.C. Hunt, M.E. Newton [ *J. Phys., Condens. Matter (UK)* vol.10 (1998) p.6171 ]

[22] H.B. Dyer, F.A. Raal, L. du Preez, J.H.N. Loubser [ *Philos. Mag. (UK)* vol.11 (1965) p.763 ]

[23] J. Koppitz, O.F. Schirmer, M. Seal [ *J. Phys. C, Solid State Phys. (UK)* vol.19 (1986) p.1123 ]

[24] M.H. Nazaré, A.J. Neves [ *J. Phys. C, Solid State Phys. (UK)* vol.20 (1987) p.2713 ]

[25] R.G. Farrer [ *Solid State Commun. (USA)* vol.7 (1969) p.685 ]

[26] J. Rosa, M. Vanecek, M. Nesladek, L.M. Stals [ *Diam. Relat. Mater. (Switzerland)* vol.8 (1999) p.721 ]

[27] R.P. Messmer, G. Watkins [ *Phys. Rev. B (USA)* vol.7 (1973) p.2568 ]

[28] G.B. Bachelet, B.A. Baraff, M. Shluter [ *Phys. Rev. B (USA)* vol.24 (1981) p.4736 ]

[29] S.A. Kajihara, A. Antonelli, J. Bernholc, R. Car [ *Phys. Rev. Lett. (USA)* vol.66 (1991) p.2010 ]

[30] P.R. Briddon, M.I. Heggie, R. Jones [ *Proc. 2nd Int. Conf. on New Diamond Science and Technology* Eds. R. Messier, T.J. Glass, J.E. Butler, R. Roy (Materials Research Society, Pittsburg, 1991) p.63 ]

[31] P.R. Briddon, R. Jones [ *Physica B (Netherlands)* vol.185 (1993) p.179 ]

# A5.5 Influence of nitrogen content on the mechanical properties of diamond

E.J. Brookes and R.D. Daniel

January 2000

## A INTRODUCTION

The mechanical strength of diamond is influenced by a number of significant factors, including the applied stress system, the ambient temperature and the degree of both internal (impurities) and external (surface finish) perfection. The criterion by which materials are considered to be brittle is that they do not exhibit bulk plastic deformation when stressed to failure at normal ambient temperatures and pressures. Fracture occurs when a certain level of stress is applied and the mode of failure will be that which requires the smallest stress. Materials, where the bonding is predominantly covalent or where there is a substantial degree of covalent bonding, have a large inherent lattice resistance to dislocation motion and failure occurs at low stresses, below the theoretical fracture stress. Diamond, as with any other crystalline material, can fail by either brittle fracture, cleavage, or in a ductile mode, flow by a shear process, and recent work has demonstrated that significant plasticity and creep occurs in diamond at temperatures developed in routine applications such as metal cutting and forming. It is this aspect of deformation together with the effect of the variation in nitrogen content that forms the basis of recent research and will be discussed here.

Despite the availability of a reasonable amount of information, the correlation between the mechanical strength of diamonds and their nitrogen content has yet to be unravelled. Clearly, however, the behaviour of a given diamond depends critically on the conditions of measurement, which must be controlled for the purposes of accurate comparison. Measurements of the compressive strength of small synthetic diamonds indicate that doping with nitrogen increases the strength, whilst other work reports that the strength of small synthetic diamond crystals decreases with nitrogen content [1,2]. Chrenko and Strong [3], measuring the abrasion resistance of synthetic type Ib diamond, found a decrease of a factor of approximately 10 as the nitrogen content increased from 0.1 to 200 ppm. On the other hand Wilks and Wilks [4] reflect that there is no discernible difference between good quality synthetic and natural diamond when subjected to abrasion tests. For the past three decades, research work on the brittle fracture of diamond has been dominated by the experimental work of Field and his research group at the University of Cambridge. However, in [5] Field reviews recent work, which suggests some evidence of a decrease in strength with increasing platelet concentration in type Ia diamond, but offers no definite correlation between nitrogen defects and strength.

The apparent contradiction in results between the various groups may be due to the inhomogeneous nature of the samples that were tested. It is likely that it is not just the simple case of nitrogen concentration but the fact that nitrogen type, position and experimental temperature could affect results, together with additional effects due to the presence of dislocation networks and stress/strain conditions. However, with large, good quality synthetic diamond now routinely produced, a better understanding of the effect of impurity elements in terms of their modification of mechanical properties is now possible.

## B INDENTATION HARDNESS OF DIAMOND

Indentation techniques are the main methods employed for measuring the strength of many materials. The property measured is the plastic strength and the measurement is in terms of a unit of pressure, which can be compared directly with other mechanical properties. Definitive measurements of the room temperature indentation hardness of diamond were originally reported by Knoop [6] and repeated by Brookes [7]. The measurements were based on indentations made using the Knoop inverted pyramidal indenter, producing good quality, rhombohedral-shaped, indentations, without any apparent cracking. Subsequently Brookes [8] described the anisotropy in the micro-hardness of different types of diamond and showed that nitrogen containing diamonds were significantly softer than diamonds containing dislocation networks and that the hardness of diamond is significantly anisotropic at room temperature. Novikov and Dub [9] used Vickers indenters to determine the fracture toughness of single crystal diamond and demonstrated similar room temperature $K_{Ic}$ values for both type Ia and type Ib synthetic diamonds. Sumiya et al and Miyoshi [10,11] have more recently reported an assessment of the room temperature hardness anisotropy of both natural and synthetic diamond and are in agreement with Chrenko and Strong [3] in that, for {001}<100>, the hardness increases as the nitrogen content decreases for type Ia and type Ib diamond. However, for synthetic type IIa diamond, Sumiya [10] reports that the anisotropy is reversed, with the <110> direction on the {001} plane found to be extremely hard, in fact almost impossible to indent.

Hardness measurements at increasing temperatures indicate that hardness decreases as a function of temperature, with a corresponding increase in plastic deformation above a specific temperature, depending on the type of diamond [12]. At elevated temperatures slip lines are visible, some cracking is developed and indentations, examined after etching, substantiated the conclusion that dislocation motion was not possible except at high temperature. In general, temperatures approaching one half to two thirds the melting point are required to induce macroscopic plastic deformation by slip, the specific temperature being dependent on the applied stress system [13]. Indentation techniques have been used to identify the temperature at which the stress required for plasticity in diamond becomes less than for fracture, i.e. a brittle ductile transition temperature or BDT. Whilst there is some variation in actual temperatures, macroscopic plastic deformation is only observed at temperatures in excess of 750°C, depending on the type of diamond and the method of measurement [14,15]. The exception to this is yield in a single crystal diamond at room temperature during high pressure (1.7 Mbar) experiments, where one low nitrogen diamond containing no platelets deformed plastically by a significant amount [16].

There are a number of drawbacks to the hardness test in evaluating the mechanical properties of ultra-hard materials. In making hardness indentations on diamond with diamond, the apexes of the indenters either fracture or plastically deform with slip lines observed on the indenter facets. Strength measurements of covalent materials in the temperature range below the brittle ductile transition, via indentation techniques, are far less reproducible than measurements in the ductile:quasi-brittle regime, where there is extensive plastic deformation preceding fracture. Essentially, the presence, characteristics, density and orientation of flaws influence the decrease in brittle strength below the theoretical strength. Above the BDT temperature, the deformation mechanisms are similar to those in any strong solid and it is expedient to use an indentation method; however, the representative strain beneath a rigid indenter is high ($\approx$ 8%) and it is likely that the fracture stress will be exceeded.

The hardness of single crystals is affected by many factors, including the constraints imposed by the limited number of slip planes and by the level of impurities, in particular point defects. Impurities can affect dislocation motion either by pinning the existing dislocations or by acting as lattice resistance to the movement of fresh dislocations. To more fully understand the mechanics of plastic deformation under conditions of point loading, a flow stress or critical resolved shear stress criterion, namely the stress required to move freshly introduced dislocations through the crystal lattice on the most favourably oriented slip plane, is required. However, the hardness test in its standard form does not permit the evaluation of a critical resolved shear stress, as the representative strain developed beneath the indenter almost always exceeds both the fracture stress and the critical resolved shear stress. It must also be borne in mind that the value of critical resolved shear stress ($\tau_C$), in particular for covalently bonded materials, depends on the experimental conditions such as strain rate, temperature and the initial dislocation density and purity of the material. However, in general, a decrease in both experimental temperature and crystal purity and an increase in strain rate and initial dislocation density result in an increase in the value of critical resolved shear stress of a material. Recently, a soft impressor technique, i.e. a modification of the hardness test, has been developed, where the tensile stresses which cause brittle failure are not attained but the resolved shear stresses exceed those required for dislocation initiation and multiplication [14]. The evaluation of the applied mean contact pressure necessary for the initiation and multiplication of dislocations is possible which permits an estimation of the critical resolved shear stress. Indirectly, the measurement of the rosette wing length associated with a plastic indentation or impression will reflect the level of lattice resistance and indicate the hardness or flow stress of the crystal.

## C THE SOFT IMPRESSOR TECHNIQUE

First described by Brookes and Green [17] the application of a suitable applied pressure, which is a function of the impressor material, is provided through softer, blunted, cones. When impressing any single crystal above the BDT, primary slip systems operate, dislocations are observed and the resultant level of deformation of the specimen is dependent on the nature of the solid. The deformation or change in surface topography of a polished surface can be detected with good accuracy by optical reflection or transmission optical interferometry or by classical dislocation etch pitting techniques. Thus, the minimum mean pressure to initiate permanent deformation can be determined. Soft impressors ensure that stress concentrations are not developed at the tip of sharp indenters or in the peripheral contact of rigid spherical indenters, such as to produce cracking, and data can be obtained from small specimens and volumes, reducing the probability of a few random but effective stress concentrators dominating the behaviour.

Using the technique of applying reduced contact pressures, plastic deformation without fracture can be induced in ultra-hard, covalently bonded materials, such as diamond. Under these experimental conditions, significant dislocation mobility and multiple intersecting slip were observed above a specific BDT temperature in a range of natural and synthetic diamonds. This approach was then employed to provide an estimate of the critical resolved shear stress of different types of diamond above 770°C [14,18]. In addition, an investigation of impression creep in diamond at temperatures up to 1500°C was made. It was shown that the application of reduced levels of strain could distinguish quite clearly between different types of diamond. The technique was shown to be particularly sensitive to diamonds with different levels of nitrogen and in general it was observed that nitrogen-containing diamond deforms plastically and fractures more easily, for a given set of experimental conditions, than diamonds with very low levels of nitrogen but high dislocation densities [18].

Point defects in metallic materials, whether vacancies or interstitials, will tend to influence the resistance to dislocation movement, thus having a drastic effect on the mechanical properties. The presence of certain dopants in semiconducting materials can be shown to increase the dislocation mobility in n-type silicon, or to decrease the dislocation mobility in p-type silicon [19]. Accompanying these effects, the critical resolved shear stress is either decreased or increased respectively. Roberts et al [20] have shown that doping semiconductor materials with small concentrations of electrically active solutes results in significant changes in dislocation velocity, yield stress, hardness and fracture behaviour in the temperature range in which plastic flow becomes important. Similarly, the presence of nitrogen point defects in diamond must hinder the movement of dislocations through the lattice, effectively inducing a solution hardening effect.

In order to make comparisons when evaluating the strength of diamond critical attention must be paid to the growth conditions, together with considerations of the normal load, the stress axes, crystallographic orientation, growth sectors and the homologous temperature. To further our understanding of plastic deformation and creep, the mechanical processes involved in the formation of an impression by a softer indenter, and to make meaningful comparisons between diamond types, the classification of each individual diamond is very important. The aim of the recent work has been to isolate the effect of varying the nitrogen concentration from the effect of grown-in dislocations by studying well-characterised type Ib diamonds containing different concentrations of the single substitutional nitrogen and applying the soft impressor technique [21,22]. Impressions were made on {001} planes in carefully selected {001} growth sectors, in diamonds grown under the same conditions but with varying concentrations of nitrogen. These results are summarised in FIGURE 1 in the form of a schematic diagram of temperature versus threshold mean contact pressure for type Ib diamond with 100 ppm nitrogen and 850 ppm nitrogen. At room temperature and in the elastic regime (I), there is no discernible dislocation movement and Hertzian fracture occurs on {111} cleavage planes, just outside the contact area at a sufficiently high contact pressure. In the plastic regime (II), dislocation movement is observed on all {111} <110> slip systems and the minimum mean flow pressure is less than that required to produce {111} ring cracks. If cracks are formed in this regime, they tend to lie on {110} planes and are considered to be initiated by dislocation reactions. The flow stress falls rapidly over this temperature regime but, ultimately, becomes relatively temperature insensitive in the third regime (III). The form of the schematic is similar for all types of diamond; however, the I:II transition or B-D-T temperature decreases with increasing nitrogen content while the onset of regime III is at a lower temperature for lower nitrogen content. The inset photographs show interferograms of impressions made at 1400°C, with a normal load of 125 N, a contact pressure of 4.5 GPa and a 300 s dwell time. Clearly, significantly more plastic deformation is observed in a synthetic type Ib diamond containing 100 ppm single substitutional nitrogen than in one containing 850 ppm nitrogen, for the same experimental conditions.

## D NITROGEN DEFECTS AND IMPRESSIONS

The basis of the distinction between types I and type II diamond depends on the presence or absence of nitrogen in the lattice. Each natural diamond is unique and a rigid distinction between the various types of diamond is not always possible. Natural type Ia diamonds generally contain more than one type of impurity or defect, are of mixed character and are often very complex. In synthetic type Ib diamond the amount of nitrogen can vary between wide limits according to the growth conditions, but is predominantly in single substitutional sites. Indeed, the distribution can vary within a diamond, as nitrogen substitution in the growing diamond occurs along specific crystallographic directions. Satoh et al [23] and Burns et al [24] have shown the existence of sectional growth in synthetic diamond plates and measured the nitrogen concentrations for the different growth sectors. They have shown that the nitrogen concentration within a diamond crystal is dependent on the growth sector, the synthesis temperature and the metal catalyst used.

In synthetic type Ib diamond, the isolated single substitutional and simple nitrogen aggregate are the most abundant defects, although the number of optical centres in diamond involving nitrogen is extensive. Studies of the photoluminescence (PL) and cathodoluminescence (CL) topography of impressions made in synthetic type Ib diamonds have identified the optically active defects. The dominant defect centres created by plastic deformation are the 575 nm defect (N–V) and the 1.945 eV defect, which is the N–V defect with an associated charge. The optical centres are created when vacancies, produced by the non-conservative motion of dislocations moving on intersecting slip systems, diffuse away from the dislocation line and interact with the nitrogen defects. Under cathodoluminescence, the 575 nm defect can clearly be seen decorating slip lines. The 1.945 eV defect does not cathodoluminesce, but recent photoluminescence work has shown that it also decorates areas of plastic deformation [21,25]. In that work it was shown that the grown-in H3 (N–V–N) defect responsible for the bright green luminescence in (001) growth sectors is modified by dislocation activity.

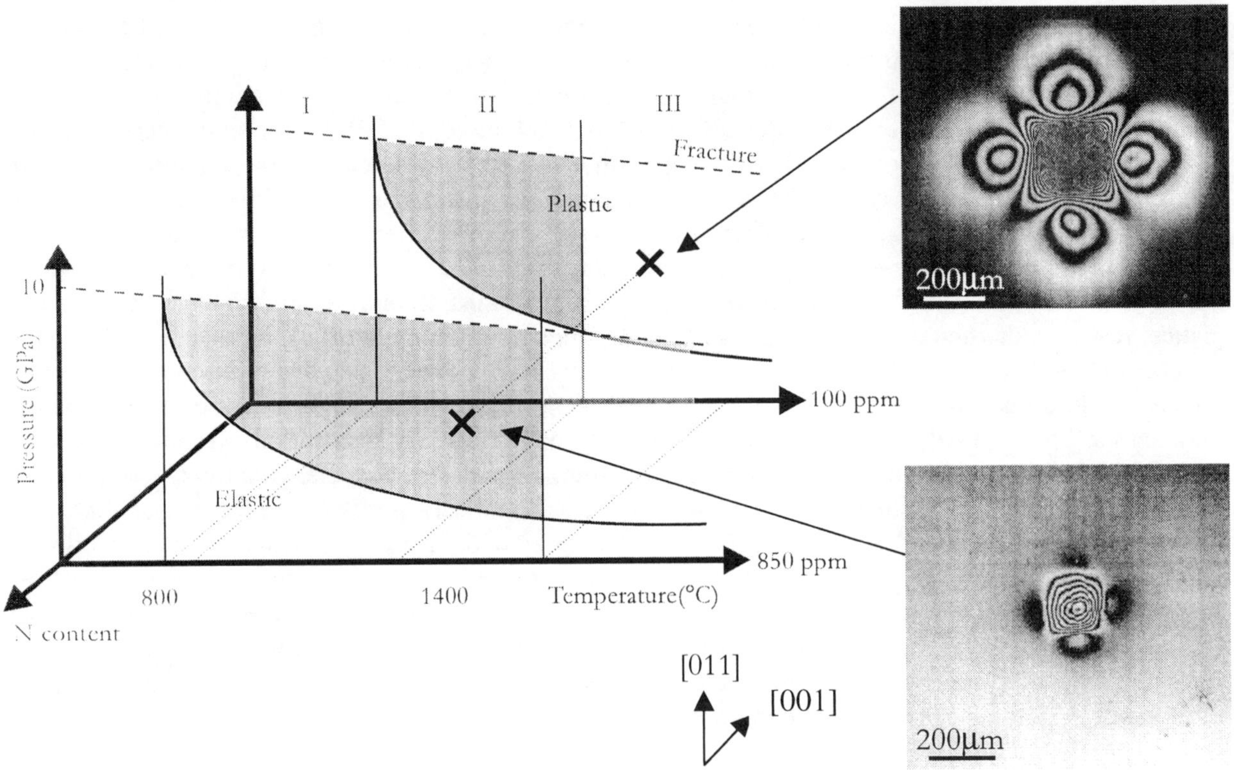


FIGURE 1 A schematic of the minimum mean contact pressure to produce plastic deformation in synthetic type Ib diamond with nitrogen concentrations 100 ppm and 850 ppm. The inset interferograms are of impressions made at 1400°C, contact pressure 4.5 GPa and dwell time 300 s.

## E DISCUSSION

It is clear from previous and current work that nitrogen concentration plays a significant role in the mechanical properties of diamond. At room temperature, a reduction in the concentration of single substitutional diamond increases the indentation hardness of the diamond. There is also some evidence to suggest that the state of the nitrogen in the lattice influences the deformation mechanism in type Ia diamonds in that they have higher indentation hardness than type Ib diamond. Indentation hardness is anisotropic regardless of the nitrogen concentration or type but the anisotropy is reversed in synthetic type IIa diamond where the <110> direction on the {001} plane is shown to be extremely hard. It has been suggested that in high purity synthetic diamond, containing few nitrogen impurities, the heterogeneous formation of dislocations is almost impossible whereas in natural type Ia and synthetic type Ib diamond dislocations can be generated under specific conditions of Knoop indentation, thus lowering the hardness [26].

At temperatures just sufficient to allow dislocation initiation and movement, above regime II in nitrogen-containing diamond, diamonds with very low nitrogen content require a higher contact pressure to initiate multiple intersecting slip than diamonds with higher levels of nitrogen. In other words, increased concentrations decrease the critical resolved shear stress. When the experimental temperature is high enough (1400°C), however, and dislocation multiplication with both cross-slip and climb becomes possible, the effect of nitrogen is reversed and an increase in nitrogen content is reflected in a decrease in the volume of plastic deformation associated with impressions. Increased levels of nitrogen increase the critical resolved shear stress of synthetic type Ib diamond at 1400°C, i.e. diamonds with a low nitrogen content require a lower contact pressure to exceed the critical resolved shear stress than diamonds with significantly higher levels of nitrogen. Therefore it is reasonable to assume that nitrogen plays an active role in the mechanism of plastic deformation by dislocation motion. Recent spectroscopic studies have identified the optically active defects associated with the impressions and that nitrogen-containing defects are both created and annihilated by dislocation activity. It is not clear, however, exactly what role the nitrogen impurity takes and why its presence aids dislocation motion at lower temperatures but restricts it at higher temperatures, when cross slip and climb processes are activated.

## F CONCLUSION

Essentially, type Ia diamond containing the A-defect is harder than a type Ib diamond containing single substitutional diamond. The more nitrogen present, the lower the hardness at room temperature, whereas at elevated temperatures the effect of nitrogen concentration is dependent on the plastic deformation mechanisms.

In summary, it is clear that the concentration of single substitutional nitrogen defects does affect the mechanical properties of diamond, most significantly at temperatures above the brittle:ductile transition temperature. Therefore it is essential that systematic investigations must include a study of the crystallographic defects and related optical colour centres, together with the evaluation of the critical resolved shear stress. The control and reproducibility of growth conditions, sample morphology and diamond crystal growth sectors should open the way to a better understanding of the influence of defects related to impurity atoms such as nitrogen in diamond.

## REFERENCES


[1] G.B. Bokii, N.F. Kirova [ *So. Phys.-Crystallogr. (USA)* vol.20 (1975) p.386-8 ]

[2] G.B. Bokii, N.F. Kirova, V.I. Nepsha [ *Sov. Phys.-Dokl. (USA)* vol.24 (1979) p.83-4 ]

[3] R.M. Chrenko, H.M. Strong [ *Report No. 75CRD089* (General Electric Company, Schenectady, New York, 1975) ]

[4] E.M. Wilks, J. Wilks [ *Properties and Applications of Diamond (UK)* (1982) ]

[5] J.E. Field [ *The Properties of Natural and Synthetic Diamond* (Academic Press, 1992) p.473-515 ]

[6] F. Knoop, C.G. Peters, W.B. Emerson [ *J. Res. Natl. Bur. Stand. (USA)* vol.23 (1939) p.33-61 ]

[7] C.A. Brookes [ *Nature (UK)* vol.228 (1970) p.660-1 ]

[8] C.A. Brookes [ *The Properties of Diamond* (Academic Press, 1979) p.383-402 ]

[9] N.V. Novikov, S.N. Dub [ *Sverkhtverdye Materialy (USSR)* vol.14 (1992) p.5-11 ]

[10] H. Sumiya, N. Toda, S. Satoh [ *Diam. Relat. Mater. (Switzerland)* vol.6 (1997) p.1841-6 ]

[11] K. Miyoshi [ *Diam. Films Technol. (Japan)* vol.8 (1998) p.153-72 ]

[12] V.I. Trefolov, Yu.V. Mil'man, O.N. Grigoriev [ *Prog Cryst. Growth Charact. (UK)* vol.6 (1988) p.226-77 ]

[13] T. Evans, J. Sykes [ *Philos. Mag. (UK)* vol.29 (1972) p.135-47 ]

[14] C.A. Brookes, E.J. Brookes, V.R. Howes, S.G. Roberts, C.P. Waddington [ *J. Hard Mater. (UK)* vol.1 (1991) p.3-24 ]

[15] N.V. Novikov, S.N. Dub, V.I. Mal'nev [ *J. Hard Mater. (UK)* vol.4 (1993) p.19-27 ]

[16] H.K. Mao, P.M. Bell, K.J. Dunn, R.M. Chrenko, R.C. DeVries [ *Rev. Sci. Instrum. (USA)* vol.50 (1979) p.1002-9 ]

[17] C.A. Brookes, P. Green [ *Nature (UK)* vol.246 (1973) p.119-22 ]

[18] E.J. Brookes [ PhD Dissertation, University of Hull, UK, 1979 ]

[19] P.B. Hirsch, P. Pirouz, S.G. Roberts, P.D. Warren [ *Philos. Mag. B (UK)* vol.52 (1985) p.761-86 ]

[20] S.G. Roberts, P. Pirouz, P.B. Hirsch [ *J. Mater. Sci. (UK)* vol.20 (1985) p.1739-47 ]

[21] E.J. Brookes, J.D. Comins, R.D. Daniel, R.M. Erasmus [ *Diam. Relat. Mater. (Switzerland)* in press (2000) ]

[22] R.D. Daniel [ PhD Dissertation, University of Hull, UK, in preparation (2000) ]

[23] S. Satoh, H. Sumiya, K. Tsuji, Y. Shuji [ *Science and Technology of New Diamond (USA)* (1990) p.351-5 ]

[24] R.C. Burns et al [ *J. Cryst. Growth (Netherlands)* vol.104 (1990) p.257-79 ]

[25] E.J. Brookes, A.T. Collins, G.S. Woods [ *J Hard Mater. (UK)* vol.4 (1993) p.97-105 ]

[26] P. Humble, R.H.J. Hannick [ *Nature (UK)* vol.273 (1978) p.37-9 ]

# A5.6 Influence of nitrogen on the surface morphology of CVD grown diamond

F.K. de Theije and W.J.P. van Enckevort

February 2000

## A INTRODUCTION

Nitrogen is an impurity that plays an essential role in the growth and defect structure of natural and synthetic diamonds. Unfortunately, natural diamonds exhibiting the original, as-grown surface patterns are extremely rare (less than $1:10^3$ [1]). Moreover, synthetic diamonds grown by the high-pressure high-temperature method exhibit, apart from a few exceptions, surface patterns that are deteriorated by imprints of metal dendrites. Therefore, no correlation between the nitrogen content of the mother liquid and the surface morphology of these diamonds grown at high pressures could be established. In the case of diamonds grown from the gas phase however, the original, as-grown surfaces are unaffected and the surface morphology could be related to the nitrogen impurity content in the gas phase.

The addition of nitrogen to the gas phase has a large influence on the morphological and defect properties of chemical vapour deposited (CVD) diamond films. Small amounts of nitrogen can change the film texture to a pattern of crystallites with rectangular coplanar {001} facets, thereby reducing the surface roughness and improving the crystalline quality [2,3]. Furthermore, it generally induces an increase in the deposition rate by a factor of 2 to 5 [3–8]. Large amounts of nitrogen cause deterioration of the films [3,6,9–11] or even inhibit diamond growth [12]. From many experiments, it was concluded that the influence of nitrogen on the growth and morphology of the {001} and {111} faces is completely different. Therefore, first the effects of nitrogen on homoepitaxially grown {001} and {111} layers will be discussed (Section B), before its influence on polycrystalline films, containing both {111} and {001} facets, is elaborated in Section C.

## B HOMOEPITAXIAL DIAMOND DEPOSITION

### B1 The {001} Face

The influence of nitrogen on the morphology of the {001} surface is studied for homoepitaxial diamond films grown by hot filament chemical vapour deposition (HFCVD) [4,13]. Furthermore, Schermer et al [14] have investigated homoepitaxial films grown by flame deposition, in which nitrogen is present in the gas phase in a radially changing concentration due to in-diffusion from the ambient atmosphere. Films grown on {001} diamond surfaces in the presence of nitrogen by HFCVD have a completely different morphology from undoped films. Whereas the latter are quite smooth with many square, shallow growth hillocks and grown-in polycrystallites, the morphology of the doped films is determined by the presence of step bunches [4,13] (see FIGURE 1). In this case, nitrogen acts as an impurity impinging on a growing surface, thereby destabilising uniform step trains by hampering the propagation of individual steps, leading to a step bunching instability [15]. To study this effect, a mesoscopic Monte Carlo model has been used [13], in which a parameter set, consistent with the observations from the diamond growth experiments, including line tension, anisotropy in step propagation, roughness of steps and impurity strength, is presented. Macrosteps due to step bunching are also observed on homoepitaxial {001} diamond films grown by flame deposition [14].

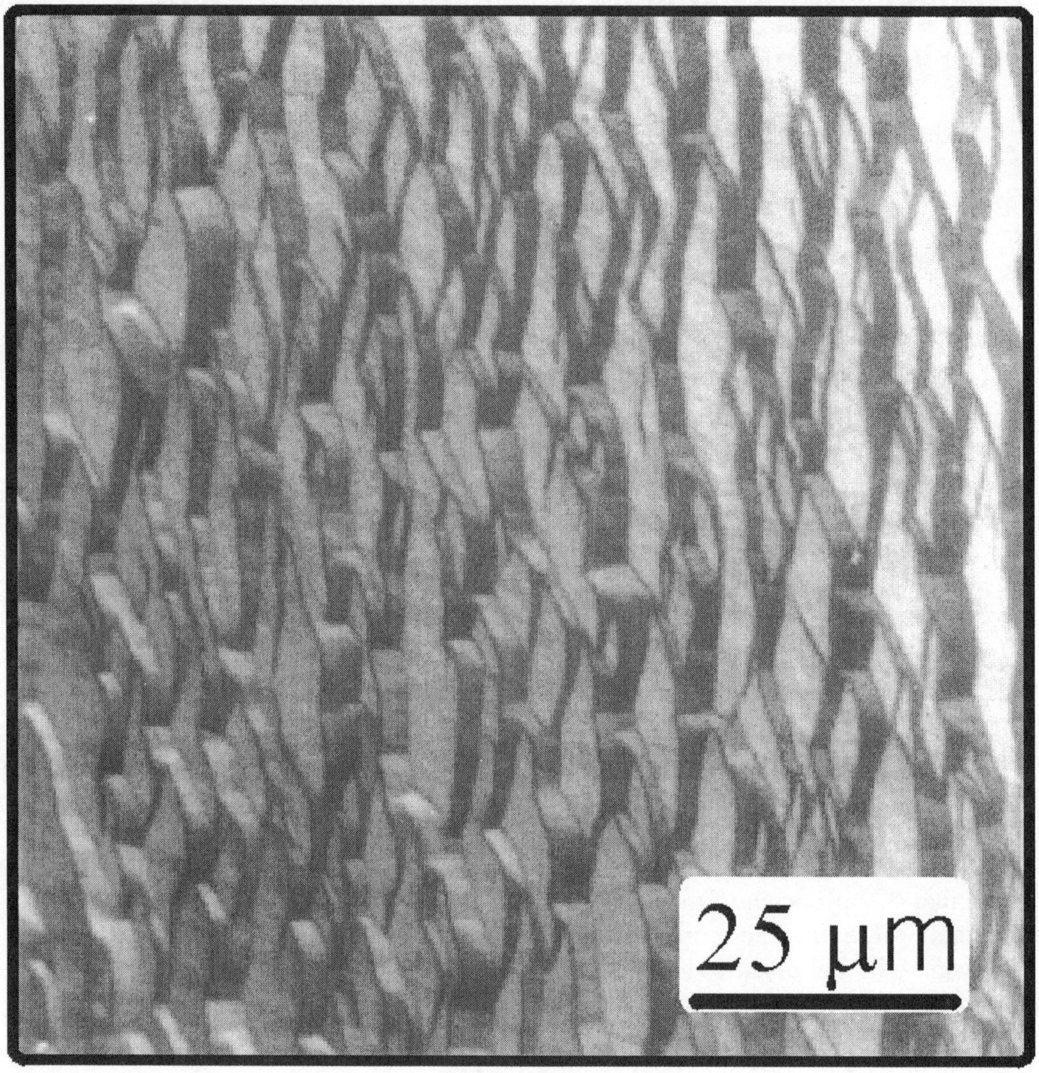


FIGURE 1 Step bunches on a homoepitaxial diamond film grown by HFCVD with nitrogen in the gas phase.

The addition of small amounts of nitrogen has been found to increase the growth rate of the {001} faces by a factor of 2, whereas on {111} only a minor effect is observed [4,8]. On the other hand, Samlenski et al [16] found an increase of only 30% for {001} and 20% for {111} faces.

## B2 The {111} Face

Homoepitaxial {111} diamond films grown by HFCVD also show a different morphology upon nitrogen addition. Without nitrogen addition, these surfaces are smooth with many shallow growth hillocks and a few microtwins. However, upon nitrogen addition a significant increase in twin formation is observed [4]. This was confirmed by observations on homoepitaxial {111} flame grown films, where during growth the flame was 'contaminated' by some nitrogen in diffusing from the ambient. These surfaces also show many twins [17] or even seem to be polycrystalline in nature [18].

# C POLYCRYSTALLINE DIAMOND

## C1 {001} Texturing

The addition of nitrogen during growth can result in the formation of {001} textured diamond films with a decreased surface roughness [2–4,19–21], a feature that is desirable for industrial applications. An example of a {001} textured film developed upon nitrogen addition during flame deposition is shown in FIGURE 2 [3]. Nitrogen in the gas phase causes the growth of multiple twinned crystallites and favours the formation of planar {001} faces at the expense of rough {111} faces. The {111} faces, which incorporate 3 to 4 times more nitrogen as compared to {001} [22], deteriorate by multiple twinning [4]. In addition, the growth rate in the <100> direction is increased with respect to <111> (see Section B). According to the van der Drift competitive growth model of polycrystalline layers [23], the film texture is determined by the ratio of growth rates along <100> and <111>. Wild et al [19] introduced a parameter $\alpha$, which is defined as this ratio multiplied by a factor $\sqrt{3}$, to describe crystallite morphologies and experiments concerning oriented polycrystalline films. It was shown that an $\alpha$ value between 2.5 and 3 is necessary to obtain {001} textured films [8].

FIGURE 2 {001} textured film grown by flame deposition in the presence of nitrogen.

The range of experimental conditions (N/C ratio, carbon concentration, growth temperature, gas pressure) for which {001} texturing develops is quite narrow; a small change in the process conditions may lead to a total loss of texturing [3,4]. Some general rules are the following. The effect of nitrogen addition on diamond growth is enhanced at larger carbon concentrations [2,3]. Therefore, the amount of nitrogen necessary to produce {001} texture is less for higher carbon concentrations in the gas phase [2]. For low carbon concentrations (≤1%) no {001} texture is found [2,6,7,11,24,25]. If the carbon concentration exceeds this lower limit of 1%, the ratios for which {001} texture is found are N/C ~ 0.1–10%, depending on the deposition conditions, the source of nitrogen (e.g. $N_2$ or $NH_3$) and the gas phase activation.

The presence of oxygen in the gas mixture also influences the effect of nitrogen. Yu et al [26] found that for small oxygen concentrations the diamond films tend to form {001} texture as the nitrogen concentration increases, whereas for higher oxygen levels the {111} texture is dominant and is not affected by the addition of nitrogen. Baranauskas et al [27] found that, when growing diamond from ethanol/hydrogen gas using HFCVD and very high nitrogen amounts (up to 20 vol%), this process also results in {001} textured films.

## C2 Deterioration of the Diamond Films

Very high nitrogen concentrations were found to deteriorate the diamond film and decrease the growth rate [3,6,7,9–11]. Afzal et al [7] and Eccles et al [11] attributed this to the increased amount of CN, which would decrease the concentration of growth promoting $CH_3$ radicals in the gas phase. The CN may competitively adsorb onto vacant diamond growth sites [11] and then induce increased twinning and stacking faults [7]. Bohr et al [6] suggested that the large supply of CN and HCN enhances the abstraction of adsorbed H, leading to reconstruction of the diamond surface. This deteriorates the diamond phase purity and influences the growth rate. Schermer et al [3] attributed the deterioration of the crystallites to the obstruction of the step growth mechanism of diamond, resulting in highly twinned crystallites, amorphous features and a lower growth rate.

## C3 Influence on the Growth Rate

Many groups found an increase in diamond growth rate upon addition of small amounts of nitrogen by a factor 2 to 5, for HFCVD [4,6,7,27], flame deposition [3] and microwave CVD [5,12]. Others reported a decrease (Bohr et al [6], for lower filament temperatures) or no effect at all [21]. This diversity in growth rate results again demonstrates the small process window upon nitrogen addition to a CVD system. This is also reflected by the results of Haubner et al [24], who studied the influence of substrate temperature, filament temperature and gas pressure on HFCVD diamond growth rates. Without nitrogen, the variations in growth rate are much smaller than upon addition of nitrogen.

It is suggested that surface processes are responsible for the increased growth rate. An increased generation of growth steps especially on {001} is proposed [5,8]. Cao et al [4] argued that different adsorption of $NH_x$ and possibly $CNH_x$ radicals on the {111} and {001} surfaces is responsible for the different effects of nitrogen addition on the growth rates of both faces. In particular these species may break part of the dimers on (2×1) reconstructed {001} surfaces. Another surface process is proposed by Stiegler et al [28] who suggested that nitrogen causes enhanced 2d-nucleation during growth. However, the fact that nitrogen leads to step bunching [13] inevitably implies that surface nitrogen retards the motion of steps and thus decreases the growth rate. The increased growth rate has been explained by the model of defect induced stabilisation [29]. This model predicts that the incorporation of nitrogen donors charges the vacancies in the growing diamond surface, thereby reversing the thermodynamic stability of diamond over graphite. However, the fact that {111} incorporates more nitrogen compared to {001} [22] would lead to a larger enhancement of the growth rate of {111} faces compared to {001}. Another mechanism, concerning sub-surface nitrogen, has been proposed by Frauenheim et al [30]. They explain the increased growth rate of {001} by charging of the (2×1) reconstructed {001} surfaces by electrons from substitutional sub-surface nitrogen, which acts as a donor. Hereby, the surface reconstruction bond lengthens and weakens, whereby an ideal adhesion point for a $CH_2$-species is created. After addition of the $CH_2$ growth unit, the electron is transferred to another reconstruction bond and the process is repeated again. In this way, one nitrogen electron can catalyse the incorporation of numerous growth units. This sub-surface process can explain the selective increased growth rate on {001} and the repeated catalytic action of nitrogen disproves the objections [5,6,11] that sub-surface processes are not possible because of the very low incorporation probability of nitrogen [22].

### C4 Gas Phase Effects

Several authors discuss the effect of nitrogen on the composition of the activated gas phase [6,7,24,31–33]. Klein-Douwel et al [32] visualised the distribution of CN in an oxy-acetylene flame by using laser-induced fluorescence and found a maximum CN signal in the annulus of enhanced growth, a cylindrical region where the growth rate is highest and the morphology consists of columnar crystallites with {001} top facets. From thermodynamic calculations, HCN and CN are found to be the most important nitrogen species [6,7]. Furthermore, it was found that CN and HCN can abstract adsorbed H atoms, thus creating vacant growth sites at the diamond surface. Bohr et al [6] also proposed that HCN and CN could decrease the carbon supersaturation and thus improve diamond quality [24]. Tsang et al [33] proposed that $N_2$ in a $CH_4/H_2$ gas mixture acts as a catalyst for the destruction of H during HFCVD, thereby reducing the etching of the non-diamond phase. For diamond deposition by using a flat combustion flame, which uses an $O_2/C_2H_2$ gas mixture, thermodynamic equilibrium calculations indicate that NO is the dominant nitrogen species [21].

## D CONCLUSION

The role of nitrogen on the growth and the surface morphology of CVD diamond continues to be of interest to the diamond researchers. Small amounts of nitrogen can change the film texture towards {001}, thereby reducing the surface roughness and improving the crystalline quality, whereas high nitrogen concentrations deteriorate the diamond film. The process window for {001} texturing is quite narrow; in particular the carbon amount must exceed a minimum concentration of 1%. Small amounts of nitrogen generally induce an increase in the deposition rate by a factor of 2 to 5 by increasing the rate in the <001> directions, whereas the growth rate in the <111> directions is hardly altered. Charging of the dimer bond on the (2×1) reconstructed {001} surfaces by substitutional sub-surface nitrogen is found to be the most probable explanation at this time.

## REFERENCES


[1] W.J.P. van Enckevort [ *J. Cryst. Growth (Netherlands)* vol.119 (1992) p.177-94 ]

[2] R. Locher, C. Wild, N. Herres, D. Behr, P. Koidl [ *Appl. Phys. Lett. (USA)* vol.65 (1994) p.34-6 ]

[3] J.J. Schermer, F.K. de Theije [ *Diam. Relat. Mater. (Switzerland)* vol.8 (1999) p.2127-39 ]

[4] G.Z. Cao, J.J. Schermer, W.J.P. van Enckevort, W.A.L.M. Elst, L.J. Giling [ *J. Appl. Phys. (USA)* vol.79 (1996) p.1357-64 ]

[5] W. Müller-Sebert, E. Wörner, F. Fuchs, C. Wild, P. Koidl [ *Appl. Phys. Lett. (USA)* vol.68 (1996) p.759-60 ]

[6] S. Bohr, R. Haubner, B. Lux [ *Appl. Phys. Lett. (USA)* vol.68 (1996) p.1075-7 ]

[7] A. Afzal, C.A. Rego, W. Ahmed, R.I. Cherry [ *Diam. Relat. Mater. (Switzerland)* vol.7 (1998) p.1033-8 ]

[8] C. Wild, R. Locher, P. Koidl [ *Mater. Res. Soc. Symp. Proc. (USA)* vol.416 (1996) p.75-80 ]

[9] L. Bergman, M.T. McClure, J.T. Glass, R.J. Nemanich [ *J. Appl. Phys. (USA)* vol.76 (1994) p.3020-7 ]

[10] E. Boettger, A. Bluhm, X. Jiang, L. Schäfer, C.-P. Klages [ *J. Appl. Phys. (USA)* vol.77 (1995) p.6332-7 ]

[11] A.J. Eccles et al [ *Thin Solid Films (Switzerland)* vol.343-344 (1999) p.627-31 ]

[12] S. Jin, T.D. Moustakas [ *Appl. Phys. Lett. (USA)* vol.65 (1994) p.403-5 ]

[13] F.K. de Theije, J.J. Schermer, W.J.P. van Enckevort [ *Diam. Relat. Mater. (Switzerland)* vol.9 (2000) accepted for publication ]

[14] J.J. Schermer, W.J.P. van Enckevort, L.J. Giling [ *Diam. Relat. Mater. (Switzerland)* vol.3 (1994) p.408-16 ]

[15] F.C. Frank [ in *Growth and Perfection of Crystals* Eds. R.H. Doremus, B.W. Roberts, D. Turnbull (Wiley, New York, 1958) p.411 ]

[16] R. Samlenski, J. Schmälzlin R. Brenn, C. Wild, W. Müller-Sebert, P. Koidl [ *Diam. Relat. Mater. (Switzerland)* vol.4 (1995) p.503-7 ]

[17] J.J. Schermer, L.J. Giling, P. Alers [ *J. Appl. Phys. (USA)* vol.78 (1995) p.2376-84 ]

[18] G. Janssen, W.J.P. van Enckevort, W. Vollenberg, L.J. Giling [ *Diam. Relat. Mater. (Netherlands)* vol.1 (1992) p.789-800 ]

[19] C. Wild et al [ *Diam. Relat. Mater. (Switzerland)* vol.2 (1993) p.158-68 ]

[20] K. Janischowsky, M. Stammler, L. Ley [ *Diam. Relat. Mater. (Switzerland)* vol.8 (1999) p.179-84 ]

[21] C.A. Wolden, C.E. Draper, Z. Sitar, J.T. Prater [*Diam. Relat. Mater. (Switzerland)* vol.7 (1998) p.1178-83 ]

[22] R. Samlenski, C. Haug, R. Brenn, C. Wild, R. Locher, P. Koidl [ *Diam. Relat. Mater. (Switzerland)* vol.5 (1996) p.947-51 ]

[23] A. van der Drift [ *Philips Res. Rep. (Netherlands)* vol.22 (1967) p.267-88 ]

[24] R. Haubner, S. Bohr, B. Lux [ *Diam. Relat. Mater. (Switzerland)* vol.8 (1999) p.171-8 ]

[25] P. Kania, G. Francz, P. Oelhafen [ *Diam. Relat. Mater. (Switzerland )* vol.3 (1994) p.696-701 ]

[26] Z. Yu, U. Karlsson, A. Flodström [ *Thin Solid Films (Switzerland)* vol.342 (1999) p.74-82 ]

[27] V. Baranauskas, B.B. Li, A. Peterlevitz, M.C. Tosin, S.F. Durrant [ *J. Appl. Phys. (USA)* vol.85 (1999) p.7455-8 ]

[28] J. Stiegler, A. Bergmaier, J. Michler, S. Laufer, G. Dollinger, E. Blank [ *Thin Solid Films (Switzerland)* vol.352 (1999) p.29-40 ]

[29] Y. Bar-Yam, D.T. Moustakas [ *Nature (UK)* vol.342 (1989) p.786-7 ]

[30] Th. Frauenheim et al [ *Diam. Relat. Mater. (Switzerland)* vol.7 (1998) p.348-55 ]

[31] A. Badzian, T. Badzian, S. Tong Lee [ *Appl. Phys. Lett. (USA)* vol.62 (1993) p.3432-4 ]

[32] R.J.H. Klein-Douwel, J.J. Schermer, J.J. ter Meulen [ *Diam. Relat. Mater. (Switzerland)* vol.7 (1998) p.1118-32 ]

[33] R.S. Tsang, C.A. Rego, P.W. May, M.N.R. Ashfold, K.N. Rosser [ *Diam. Relat. Mater. (Switzerland)* vol.6 (1997) p.247-54 ]

# A5.7 Hydrogen related vibrational and electronic transitions in diamond

A.M. Zaitsev

March 2000

## A INTRODUCTION

Hydrogen is a very common impurity in diamond. In natural diamonds its concentration may attain a value of $10^{21}$ $cm^{-3}$. Especially high content of hydrogen can be found in CVD diamond films [1]. Most of the hydrogen in diamond is believed to be in inclusions and, consequently, in an optically inactive state [2]. However, some hydrogen does form covalently bonded defects with carbon and impurity atoms giving rise to vibrational and electronic transitions detected with optical methods.

## B ABSORPTION VIBRATIONAL TRANSITIONS

Local vibrations of hydrogen-related defects normally have relatively high energy influencing predominantly optical modes of the vibrational spectrum of diamond. However, there are low energy defect-induced modes assigned to hydrogen-related defects.

The most characteristic vibrational feature of hydrogen-rich natural diamonds is a narrow Lorentzian-shaped line at 3107 $cm^{-1}$. Its strength in type Ia diamonds may be as high as 13 $cm^{-1}$ [3,4]. The line is very intense in diamonds of a light grey colour; its intensity correlates with the content of cloud-like inclusions [5,6]. This feature is also readily observed in hydrogen-rich CVD diamond films (preferentially in high-quality homoepitaxial films [1]). In $^{13}C$:$^{14}N$ diamonds the line shifts to 3098 $cm^{-1}$. A weak Lorentzian-shaped line at 3098 $cm^{-1}$ (about 1% intensity of the 3107 $cm^{-1}$ line intensity) always accompanies the 3107 $cm^{-1}$ line, and it is attributed to vibrations localized on naturally present $^{13}C$ isotopes [3,7,8]. In CVD films the vibration is probably localized at interfacial surfaces forming C-H bonds [9]. The 3107 $cm^{-1}$ line is attributed to carbon–hydrogen vibrations in the cis-form of the di-substituted ethylene group –CH=CH– [5,10], or in the vinilidene group >C=$CH_2$ (vibration of the $sp^2$ bonds) [3,5]. This vibration is interpreted as a stretching one (s). The corresponding bend vibration (b) gives a line at 1405 $cm^{-1}$. The s and b vibrations produce the combinations 2b (first overtone at 2786 $cm^{-1}$), 3b (second overtone at 4168 $cm^{-1}$) and s+b at 4496 $cm^{-1}$ [3,4,9,11–13]. The line persists on annealing at temperatures above 2350°C [14]. It is believed that the 3107 $cm^{-1}$ vibration involves only one carbon atom. A tentative atomic model of the centre implies hydrogen atoms locating in vacancies or dislocation lines [1] (FIGURE 1).

In some type Ia natural diamonds sharp peaks at 687, 1170, 1547, 2920, 3236, 4703, 4157 and 4165 $cm^{-1}$ are observed. All of them are attributed to hydrogen-related vibrations. The 3235 $cm^{-1}$ vibration is also active in Raman scattering. Besides these narrow lines, natural diamonds may also reveal several relatively broad hydrogen-related features at 1240, 8615, 8255, 7850 and 7500 $cm^{-1}$ [4,15,16]. The 1240 $cm^{-1}$ band is readily observed in brown high-hydrogen diamonds showing type Ib character. This feature appears to correlate with the 2.6 eV absorption band [3,17] (FIGURES 1 and 2).

A very complex band spreading in a spectral range from 500 to 1550 $cm^{-1}$ is observed in natural high-hydrogen diamonds of light grey colour [4]. The band consists of many narrow lines: the most intense lines located at the wings of the band are at 1545, 1495, 1430, 1010, 870, 770 and 710 $cm^{-1}$. The band is tentatively ascribed to interstitial hydrogen atoms in different configurations. Interestingly, vibrations of energies 1380 and 1250 $cm^{-1}$ localizing at substitutional hydrogen atoms in the diamond lattice have been predicted in [18,19] (FIGURE 2).

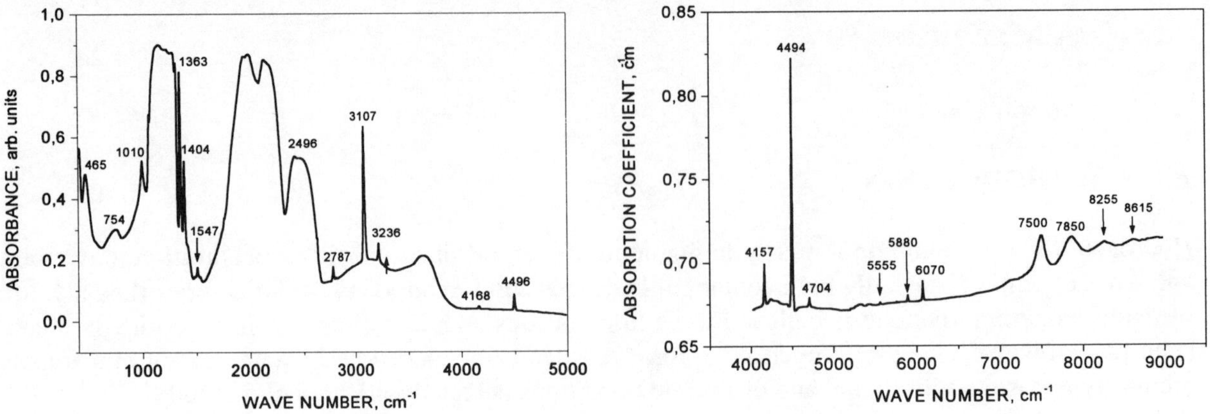


FIGURE 1 IR absorption spectra of hydrogen-rich natural diamonds [4,15].

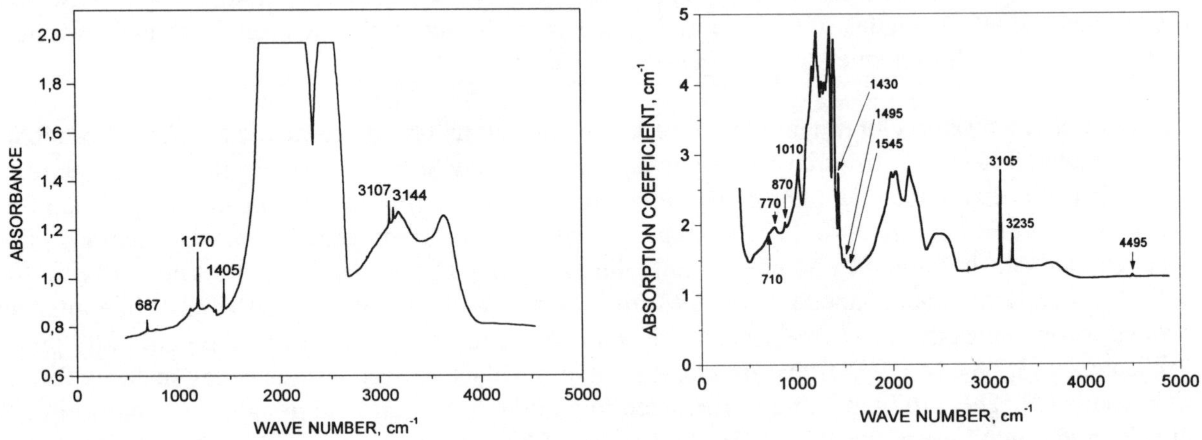


FIGURE 2 IR absorption spectra of a green-to-yellow type IaAB+Ib natural diamond (left) exhibiting sharp lines of hydrogen related vibrations, and of a light grey hydrogen-rich natural diamond (right) [4,16].

The most characteristic hydrogen-related feature of CVD diamond films is the doublet at 2923 $cm^{-1}$ and 2854 $cm^{-1}$. The spectral range of the doublet ranges from 2800 to 3000 $cm^{-1}$; the FWHM of each band is 30 to 50 $cm^{-1}$ (FIGURE 3). The spectral position of the doublet may differ somewhat in films of different structural perfection. This feature is more intense in <111>- and less intense in <110>-textured films [22]. The doublet is especially strong in brownish polycrystalline CVD films [23–26]. The 2854 $cm^{-1}$ band is readily observed in single-crystal films deposited onto {110}- and {111}-oriented substrates (it is absent from the {100}-oriented films). Addition of oxygen considerably suppresses the hydrogen-related absorption in CVD diamond films [22] (FIGURE 3). It is also reduced with increasing film thickness due to an increase in average grain size resulting in reduction of the grain boundary contribution (the centres responsible for this doublet are supposed to be localized at grain boundaries) [22,27]. Fine structure of the doublet may reveal narrow lines at 2820, 2831, 2845, 2884 and 2923 $cm^{-1}$ in $^{12}C$:H CVD films; in $^{13}C$:H films the corresponding lines are at 2810, 2820, 2834, 2845, 2870 and 2908 $cm^{-1}$. In $^{12}C$:D deuterated CVD films three additional relatively broad bands peaking at 2075, 2120 and 2170 $cm^{-1}$ can be distinguished [1]. All of the lines coincide well with those present in α-C:H DLC films (this implies that hydrogen in diamond films is mostly incorporated in defective regions, grain boundaries or interfaces [28]). The doublet is attributed to asymmetrical and symmetrical C–H stretching modes of $sp^3$-bonded methylene groups $-CH_2-$ [5,20,24,28–36]. The concentration of $CH_2$ groups in nominally undoped CVD films is 1000 ppm [35]. The total hydrogen concentration $N_H$ can be assessed as [20,24,26,37,38]:

$$N_H \sim (1 \div 10) \times 10^{17} \quad [\mathrm{cm}^{-3}] \quad \int_{2800\mathrm{cm}^{-1}}^{3000\mathrm{cm}^{-1}} \mu(\nu)d\nu \text{, or } N_H \sim 10^{20} [\mathrm{cm}^{-3}] \mu_{2840\mathrm{cm}^{-1}} \quad [\mathrm{cm}^{-1}]$$

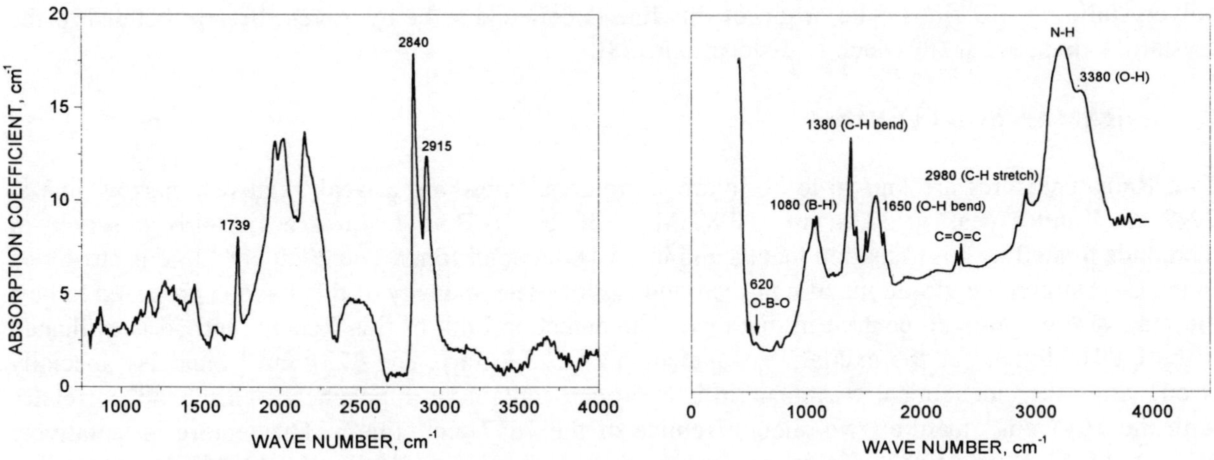


FIGURE 3 Absorption spectra of a free-standing flame-grown CVD diamond film exhibiting strong hydrogen-related absorption [20] (left), and a CVD diamond film from a gas mixture with addition of $CO_2$ [21] (right).

A sharp peak at 1252 $cm^{-1}$ is observed in some high-hydrogen homoepitaxial CVD films deposited onto {110}- and {111}-oriented diamond substrates. This peak strongly correlates with the 2949 $cm^{-1}$ band. The feature is possibly induced by substitutional hydrogen atoms (tentatively an H–C–H bending mode) [18,19,33] (FIGURE 4).

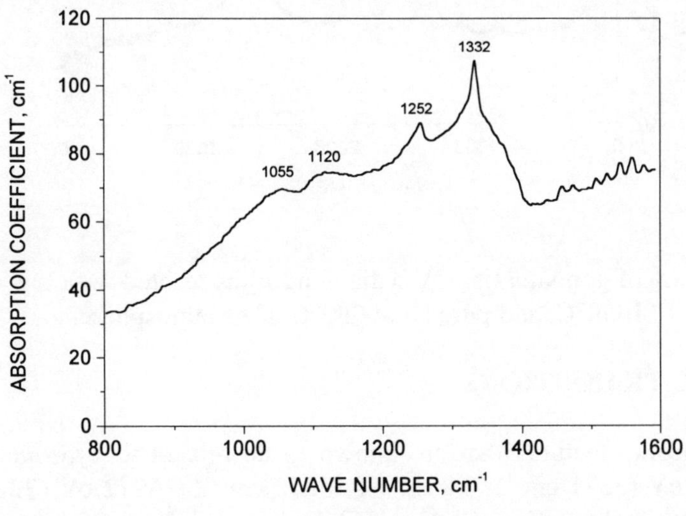


FIGURE 4 Absorption spectrum of a flame-grown epitaxial CVD diamond film grown on a {110} oriented diamond substrate [33].

Hydrogenated diamond surfaces reveal the following absorption feature: a peak at 2900 $cm^{-1}$ (FWHM of 40 $cm^{-1}$) observed from (100) surfaces. Deuterated diamond surfaces reveal this feature in a spectral range from 2050 to 2250 $cm^{-1}$. This vibration is attributed to a C–H stretching mode of $sp^3$-hybridized bonds [39,40]. A peak at 2860 $cm^{-1}$ (FWHM ~ 25 $cm^{-1}$) is observed from clean (111) surfaces partially covered with hydrogen [41]. (111)-1×1 reconstructed surfaces may also reveal a peak at 2840 $cm^{-1}$ (FWHM of 40 $cm^{-1}$). This feature is attributed to C–H symmetrical stretching vibrations of $CH_3$ species [39]. Hydrogenated (111) surfaces of natural type IIa diamonds exhibit a 2830 $cm^{-1}$ relatively narrow peak (FWHM of 10 $cm^{-1}$) attributed to a C–H stretch mode on the ideal (111) diamond surface [39,41]. Probably this very feature is observed at 2822 $cm^{-1}$ from polycrystalline CVD films. The origin of this line is believed to be hydrogen directly bonded to the crystalline diamond at sites such as dislocations [28].

## C RAMAN SCATTERING

Two Raman features are known to be hydrogen-related. These are a weak relatively narrow line at 2922 $cm^{-1}$ and a peak at 3234 $cm^{-1}$ (FWHM of 50 $cm^{-1}$). Both features are readily observed in diamonds treated by hot transition metals and in CVD diamond films. The 2922 $cm^{-1}$ line is attributed to the C–H stretching vibrations of non-diamond carbon. The intensity of this band is proposed to be a measure of the hydrogen content in diamond. The detection limit of this Raman method is evaluated to be 0.001 atom% (at the excitation wavelength of 228.9 nm). The 3234 $cm^{-1}$ band is especially strong after thermochemical treatment in a hydrogen-containing atmosphere. This band correlates with the 1617 $cm^{-1}$ feature (two-phonon replica of the 1617 $cm^{-1}$ line?). The feature is tentatively attributed to C–H vibrations of hydrogen atoms bound to graphite inclusions [42] (FIGURE 5).

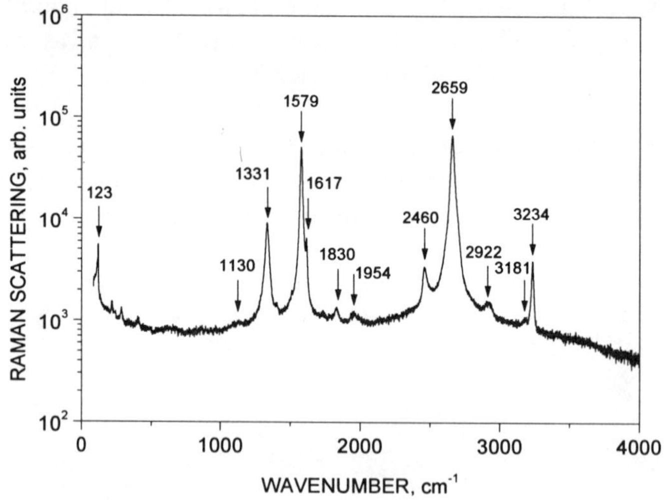


FIGURE 5 Raman spectra of good quality CVD diamond films reacted with low carbon steel at about 1000°C, and pure Ni at 900°C in Ar atmosphere.

## D ELECTRONIC TRANSITIONS

There are several electronic features that are known to be related to hydrogen. In the IR spectral region they are 0.6906 eV (5571 $cm^{-1}$), 0.8515 eV (6870 $cm^{-1}$), 0.8972 eV (7238 $cm^{-1}$), and 0.9130 eV (7366 $cm^{-1}$) observed in absorption spectra of CVD diamond films. The lines are relatively broad (FWHM around 50 $cm^{-1}$). A Gaussian shape of the lines indicates non-homogeneous broadening. The lines shift by about 10 $cm^{-1}$ to lower energies and decrease by about 3 times in intensity as the temperature increases from 80 to 300 K implying a rigid atomic structure of the corresponding defects [1]. No shift is observed for the 0.6906 eV line in $^{13}C$ diamonds, whereas the 0.8515, 0.8972 eV lines and the 0.9130 eV line shift by 0.2 meV and 1.5 meV respectively. The hydrogen nature of these lines has been proven by doping CVD films with deuterium. The isotopic data imply the incorporation of one hydrogen atom in each of the 0.8515, 0.8972 and 0.9130 eV centres and a two hydrogen atom structure of the 0.6906 eV centre [1,7] (FIGURE 6).

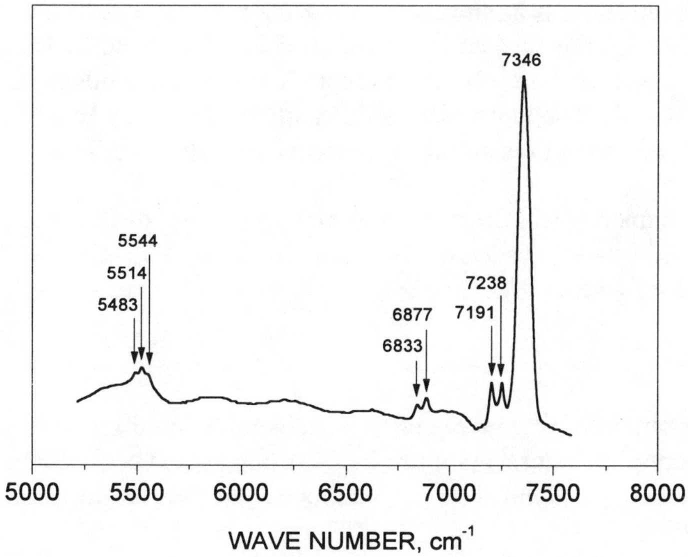


FIGURE 6 Absorption spectrum of a homoepitaxial CVD diamond film deuterated to 50% [1].

The 2.20 eV and 2.271 eV centres are hydrogen-related features in luminescence of natural diamond and CVD diamond films (the 2.20 eV centre is also seen in absorption). The 2.20 eV centre is accompanied by broader lines peaking at 555 and 545 nm, which may possibly be phonon side bands of the centre. The 2.271 eV centre is a characteristic feature created in type IIa diamonds by $H^+$ ion implantation. This centre is readily created by high energy ion irradiation in CVD diamond films. In the samples implanted with $D^+$ ions the line shifts to 2.273 eV. The 2.271 eV centre anneals at temperatures above 1100°C implying an interstitial atomic structure [4,5,9,43,44] (FIGURE 7).

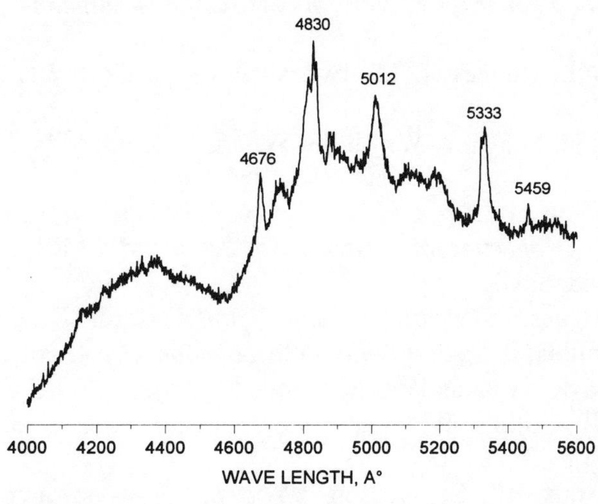


FIGURE 7 CL spectrum (80 K) taken from a nominally undoped CVD diamond film irradiated with 164.4 MeV $^{12}C^{+5}$ ions at a dose of $10^{14}$ $cm^{-2}$.

The hydrogen-related Green Band with a maximum at 2.3 eV (FWHM of 0.5 eV) is a pronounced luminescence feature of hydrogenated diamond films. This band must be distinguished from the boron-related 2.3 eV band having almost the same spectral position. The band is produced in CVD films by H-plasma treatment [45,46]. It disappears after oxidation in acids and subsequent annealing at a temperature of 400°C in $N_2$ atmosphere. The band intensity changes strongly with temperature showing a thermal activation energy of 0.3 eV. The spatial distribution of the defects responsible for the hydrogen-related Green Band is homogeneous over the sample [47]. The Green Band is attributed to the defects responsible for the surface electrical conductivity of the hydrogenated CVD diamond films (possibly D–A pairs with boron as the acceptors and some unidentified donors with energy levels of $E_c$–2.25 eV) [46–48]. The luminescence intensity of the Green Band has been proposed to be used for optical characterization of the surface electrical conductivity of hydrogenated diamond [47].

In some hydrogen-rich diamonds (they may be coloured grey-violet or chameleon) the following sharp hydrogen-related features can be observed in absorption: 2.767 eV; 2.792 eV; 2.817 eV; 2.87 eV; 2.916 eV; 3.07 eV; 3.26 eV [4,5].

## E CONCLUSION

Hydrogen is an important impurity in diamond considerably modifying its optical and electrical properties. The hydrogen-related absorption can be as strong as that caused by nitrogen. Due to this it has been proposed to refer to diamond showing dominant hydrogen-related absorption (in particular at 2900 $cm^{-1}$) as type IIc [20].

## REFERENCES


[1] F. Fuchs, C. Wild, K. Schwarz, P. Koidl [ *Diam. Relat. Mater. (Switzerland)* vol.4 (1995) p.652-6 ]

[2] R.J. Sweeney, V.M. Prozesky, K.S. Viljoen, S. Connell [ *Nucl. Instrum. Methods Phys. Res. B (Netherlands)* vol.158 (1999) p.582-7 ]

[3] G.S. Woods, A.T. Collins [ *J. Phys. Chem. Solids (UK)* vol.44 (1983) p.471-5 ]

[4] E. Fritsch, K. Scarratt, A.T. Collins [ in *New Diamond Science and Technology* (MRS Int. Conf. Proc., 1991) p.671-6 ]

[5] J.E. Field (Ed.) [ *The Properties of Natural and Synthetic Diamond* (Academic Press, London, 1992) ]

[6] C.M. Welbourn, M.L. Rooney, D.J.F. Evans [ *J. Cryst. Growth (Netherlands)* vol.94 (1989) p.229-52 ]

[7] F. Fuchs, C. Wild, K. Schwarz, W. Müller-Sebert, P. Koidl [ *Appl. Phys. Lett. (USA)* vol.66 (1995) p.177-9 ]

[8] E. Fritsch, K.V.G. Scarratt [ *Proc. SPIE (USA)* vol.1146 (1989) p.201 ]

[9] G. Davies (Ed.) [ *Properties and Growth of Diamond* EMIS Datareviews Series No. 9 (INSPEC, IEE, London, UK, 1994) ]

[10] E. Pereira, T. Rodrigues, L. Pereira [ *Diamond Conf.* Oxford, UK, 1987 ]

[11] G. Davies, A.T. Collins, P. Spear [ *Solid State Commun. (USA)* vol.49 (1984) p.433 ]

[12] A.T. Collins, G. Davies, H. Kanda, G.S. Woods [ *J. Phys. C. (UK)* vol.21 (1988) p.1363-76 ]

[13] A. Mainwood, A.T. Collins, P. Woad [ *Mater. Sci. Forum (Switzerland)* vol.143-147 pt.1 (1993) p.29-33 ]

[14] M.R. Brozel, T. Evans, R.F. Stephenson [ *Proc. R. Soc. Lond. A (UK)* vol.361 (1978) p.109-27 ]

[15] N. Ferrer, J.M. Nogues-Carulla [ *Diam. Relat. Mater. (Switzerland)* vol.5 (1996) p.598-602 ]

[16] I.M. Reinitz, E. Fritsch, J.E. Shigley [ *Diam. Relat. Mater. (Switzerland)* vol.7 (1998) p.313-6 ]

[17] A.T. Collins, K. Mohammed [ *J. Phys. C (UK)* vol.15 (1982) p.147-58 ]

[18] A.V. Kurdumov, V.G. Malogolovets, N.V. Novikov, A.H. Piljankevich, L.A. Shulman [ *Polymorphous Modification of Carbon and Boron Nitride* (Metallurgija, Moscow, 1994) p.318 (in Russian) ]

[19] V.G. Malogolovets, N.I. Nikityuk [ *Superhard Materials* no.5 (1981) p.28-31 (in Russian) ]

[20] G. Janssen, W. Vollenberg, J. Giling, W.J.P. van Enckevort, J.J.D. Schaminee, M. Seal [ *Surf. Coat. Technol. (Switzerland)* vol.47 (1991) p.113-26 ]

[21] Chia-Fu Chen, Sheng Hsiung Chen [ *Advances in New Diamond Science and Technology* Eds. S. Saito, N. Fujimori, O. Fukunaga, M. Kamo, K. Kobashi, M. Yoshikawa (MYU, Tokyo, 1994) p.709-12 ]

[22] S. Haq, D.L. Tunnicliffe, J.A. Savage [ *Diam. Relat. Mater. (Switzerland)* vol.3 (1994) p.593-7 ]

[23] R.S. Sussman et al [ *Diam. Relat. Mater. (Switzerland)* vol.3 (1994) p.303-12 ]

[24] R. Erz, W. Dötter, K. Jung, H. Ehrhardt [ *Diam. Relat. Mater. (Switzerland)* vol.4 (1995) p.469-72 ]

[25] R.S. Sussmann [ *Ind. Diam. Review (UK)* no.2 (1993) p.63-72 ]

[26] Z. Yin, Z. Akkerman, B.X. Yang, F.W. Smith [ *Diam. Relat. Mater. (Switzerland)* vol.6 (1997) p.153-8 ]

[27] H. Sternschulte et al [ *Diam. Relat. Mater. (Switzerland)* vol.4 (1995) p.1189-92 ]

[28] B. Dischler, C. Wild, W. Müller-Sebert, P. Koidl [ *Physica B (Netherlands)* vol.185 (1993) p.217-21 ]

[29] P. John, D.K. Milne, I.C. Drummond, M.G. Jubber, J.I.B. Wilson, J.A. Savage [ *Diam. Relat. Mater. (Switzerland)* vol.3 (1994) p.486-91 ]

[30] G. Janssen, W.J.P. van Enckevort, J.J.D. Schaminee, W. Vollenberg, J. Giling, M. Seal [ *J. Cryst. Growth (Netherlands)* vol.104 (1990) p.752-7 ]

[31] X.X. Bi, P.C. Eklund, J.G. Zhang, A.M. Rao, T.A. Perry, C.P. Beetz [ *J. Mater. Res. (USA)* vol.5 (1990) p.811-7 ]

[32] C. Wild, N. Heres, J. Wagner, P. Koidl, T. Anthony [ *Proc. Optical and Structural Characterization of CVD Diamond, 1st Int. Symp. on Diamond and Diamond-Like Films* (The Electrochem. Soc. Pennington, NJ, 1989) ]

[33] G. Janssen, W.J.P. van Enckevort, W. Vollenberg, L.E. Giling [ *Diam. Relat. Mater. (Netherlands)* vol.1 (1992) p.789-800 ]

[34] T. Ando, M. Ishii, M. Kamo, Y. Sato [ *Diam. Relat. Mater. (Switzerland)* vol.4 (1995) p.607-11 ]

[35] L.D. Kislovskii, B.V. Spitsyn [ *Sov. Phys.-Crystallogr. (USA)* vol.25 (1989) p.239-40 ]

[36] K.M. McNamara, K.M.M. Rutledge, B.E. Scruggs, K.K. Gleason [ *J. Appl. Phys. (USA)* vol.77 (1995) p.1459-62 ]

[37] W.J.P. van Enckevort [ in *Synthetic Diamond: Emerging CVD Science and Technology* Eds. K.E. Spear, J.P. Dismukes (John Wiley & Sons, Inc., 1994) p.307-53 ]

[38] D.K. Basa, F.W. Smith [ *Thin Solid Films (Switzerland)* vol.192 (1990) p.121 ]

[39] T. Ando et al [ in *Advances in New Diamond Science and Technology* Eds. S. Saito, N. Fujimory, O. Fukunaga, M. Kamo, K. Kobashi, M. Yoshikawa (MYU, Tokyo, 1994) p.461-4 ]

[40] K. Ushizawa, M.H. Gamo, Y. Kikuchi, I. Sato, T. Ando [ *Phys. Rev. B (USA)* vol.60 (1999) p.R5165-8 ]

[41] R.P. Chin, J.Y. Huang, Y.R. Shen, T.J. Chuang, H. Seki, M. Buck [ *Phys. Rev. B (USA)* vol.45 (1992) p.1522 ]

[42] R.W. Bormett, S.A. Asher, R.E. Witoski, W.D. Partlow, R. Lizewski, F. Pettitt [ *J. Appl. Phys. (USA)* vol.77 (1995) p.5916-23 ]

[43] A.A. Gippius, V.S. Vavilov, A.M. Zaitsev, B.S. Zhakupbekov [ *Physica B (Netherlands)* vol.116 (1983) p.187-92 ]

[44] A.A. Gippius, A.M. Zaitsev, V.S. Vavilov [ *Sov. Phys.-Semicond. (USA)* vol.16 (1982) p.256-61 ]

[45] K. Hayashi, T. Seguchi, H. Okushi [ *Mater. Sci. Forum (Switzerland)* vol.258-263 (1997) p.745 ]

[46] K. Hayashi, H. Watanabe, S. Yamanaka, T. Sekiguchi, H. Okushi, K. Kajimura [ *J. Appl. Phys. (USA)* vol.81 (1997) p.744 ]

[47] K. Jakoubovskii, G.J. Adriaenssens [ *Characterization of the broad green band luminescence in CVD and synthetic Ib diamond* to be published ]

[48] K. Iakoubovskii, G.J. Adriaenssens [ *Luminescence Excitation Spectra in Diamond* to be published ]

# A5.8 Hydrogen in CVD diamond: EPR

M.E. Newton

February 2000

## A INTRODUCTION

In recent years the incorporation of hydrogen into polycrystalline diamond films grown by chemical vapour deposition (CVD) has attracted considerable attention. To date only one paramagnetic defect, the H1 centre [1–5], has been shown to involve hydrogen and it has been demonstrated that this is not a defect incorporated into the bulk diamond but found on hydrogen decorated grain boundaries or in intergranular material. No paramagnetic defect involving hydrogen has yet been shown to be incorporated into the diamond lattice.

## B EPR STUDIES ON H1

The H1 electron paramagnetic resonance (EPR) centre [1,2] has been observed by many workers in diamond films grown by a variety of different techniques of chemical vapour deposition (CVD) [3]. The H1 centre is sometimes called the 'g = 2.0028' defect, since this is its g-value, but it has been shown that other S = 1/2 defects in polycrystalline CVD diamond also have g = 2.0028 [2,3]. The incorporation of hydrogen in the H1 centre is deduced from satellites that can be observed when EPR measurements are made at microwave frequencies above about 8 GHz (at lower microwave frequencies the EPR lineshape is very different [3]). The position and intensity of the satellites as a function of microwave frequency suggests that they arise from the forbidden nuclear-spin flip transitions of a hydrogen atom weakly coupled to the unpaired electron spin [4]. Analysis of the EPR lineshape between 9.8 and 35 GHz led Zhou et al [1,2] to conclude that H1 was a single well defined defect consisting of an unpaired electron in a carbon dangling orbital coupled to a hydrogen atom about 0.2 nm away. The hydrogen hyperfine coupling parameters are $A_{\parallel} = 27.5(2.5)$ MHz and $A_{\perp} = -5.5(2.5)$ MHz [1–3]. Zhou et al [2] proposed that H1 was formed by a hydrogen atom entering a stretched C–C bond at a grain boundary, allowing the carbons to relax back, one bonding to the hydrogen and the other with an unpaired electron primarily localised in its dangling bond.

Pulsed and continuous wave EPR measurements show that the H1 EPR centre has a much higher spin lattice relaxation rate (>100 times) than the single substitutional nitrogen centre known to be incorporated into the diamond lattice [3].

The concentration of H1 centres can be reduced by annealing above 1700 K [5]. Annealing above 1700 K also produces changes in the CH infrared absorption bands seen in polycrystalline CVD diamond [5].

## C ENDOR STUDIES ON H1

$^{1}$H ENDOR (electron nuclear double resonance) measurements on the H1 centre indicate that within 0.2 – 1.0 nm of the defect there is a significant concentration of hydrogen atoms. Although this strong $^{1}$H matrix ENDOR was observed no ENDOR signal was detected from the near neighbour unique hydrogen identified by EPR [5].

## D CONCLUSION

All the experimental evidence is consistent with H1 being located on hydrogen decorated grain boundaries or in intergranular material rather than in bulk diamond. The proposal that H1 is formed by a hydrogen atom entering a stretched C–C bond at a grain boundary (or in intergranular material), allowing the carbons to relax back, one bonding to the hydrogen and the other with an unpaired electron primarily localised in its dangling bond, is consistent with the experimental data.

## REFERENCES

[1] X. Zhou, G. Watkins, K.M. McNamara-Rutledge [ *Mater. Sci. Forum (Switzerland)* vol.196-201 (1996) p.825 ]

[2] X. Zhou, G. Watkins, K.M. McNamara-Rutledge, R.P. Messmer, S. Chawla [ *Phys. Rev. B (USA)* vol. 54 (1996) p.7881 ]

[3] D.F. Talbot-Ponsonby et al [ *Phys. Rev. B (USA)* vol.57 (1998) p.2264 ]

[4] S.L. Holder, L.G. Rowan, J.J. Krebs [ *Appl. Phys. Lett. (USA)* vol.64 (1994) p.1094 ]

[5] D.F. Talbot-Ponsonby, M.E. Newton, J.M. Baker, G.A. Scarsbrook, R.S. Sussmann, A.J. Whitehead [ *Phys. Rev. B (USA)* vol.57 (1998) p.2302 ]

CHAPTER A6

# PROPERTIES OF DEFECTS IN DIAMOND: TRANSITION METALS AND SILICON

# A6.1 Optical and EPR properties of transition metals in diamond

A.J. Neves and J.C. Lopes

January 2000

## A INTRODUCTION

Transition metals, namely alloys of Ni, Co and Fe, are used as 'solvent-catalysts' in the synthesis of diamond at high-pressure high-temperature conditions, the so-called HPHT diamonds. It is known that Ni and Co become incorporated in the lattice during the growth process and produce optical and EPR centres [1–3].

In as-grown samples, nickel is detected in two forms: interstitial nickel in a positive charge state and substitutional nickel in a negative charge state [4,5]. Though both forms may coexist in a given sample [5], usually only one is observed, depending strongly on the amount of nitrogen [5].

Nitrogen is a common impurity in HPHT diamonds unless special measures are taken to exclude it. This is accomplished by adding elements such as Zr or Ti to the growth environment [6,7]. With these nitrogen getters, the nitrogen content may be varied continuously from 0 to 300 ppm by appropriate choice of the alloy composition [8].

Single substitutional nitrogen, $N^0_S$, acts as a donor in diamond; therefore its concentration controls the Fermi level and the charge state of the active defects. Consequently, the observed Ni-related features depend on the composition of the metal alloy and the concentration of donor and acceptor (boron) impurities.

In samples annealed at high temperatures the nitrogen becomes mobile and aggregates to form substitutional pairs. Besides the adjustment of the Fermi level, nitrogen can be trapped by Ni or Co and form complexes that originate new optical and EPR active centres.

Since the exhibited relative intensities of the different nickel-related absorption bands, as well as the EPR centres observed, depend on the amount of $N^0_S$ the following discussion is organised in terms of the nitrogen concentration magnitude.

## B NICKEL AND LOW NITROGEN-CONTENT DIAMONDS

In diamonds with low nitrogen concentrations, less than 10 ppm, two Ni-interstitial EPR centres are observed. The NIRIM1 EPR centre with effective spin S = 1/2 and an isotropic g-factor of 2.0112 is assumed to be an isolated tetrahedral $Ni^+$ atom [4]. The NIRIM2 centre with an effective spin S = 1/2 has trigonal symmetry, with $g_{\|} = 2.3285$, $g_{\perp} \sim 0$. This defect is identified as interstitial nickel in a positive charge state ($3d^9$) with a nearby disturbance (perhaps a vacancy) responsible for the trigonal symmetry [4].

The optical spectra of these diamonds display several multiplet optical transitions at 1.22, 1.40, 2.56 and 3.1 eV [2,9,10]. Very little is known about the 1.22 eV and the 3.1 eV centres which are both observed in absorption. The 1.22 eV zero-phonon line (ZPL) exhibits several unresolved dichroic transitions. The centres' observed behaviour under illumination — the 1.22 eV line grows at the expense of the 1.40 eV system — led Lawson et al [10] to suggest that the two centres are different charge states of the same defect.

The 1.40 eV system, observed in absorption and luminescence, is a doublet with ZPLs at 1.401 and 1.404 eV. Temperature dependence measurements show that the transitions occur between a split ground state and an excited state of a single defect [2,11]. In high resolution spectra of the system Nazaré et al [11,12] observed isotope splitting on the ZPL, and showed that the measured relative intensities of the split components agreed with the abundance ratios of naturally occurring Ni isotopes, which undoubtedly identified a Ni atom as a constituent of the defect. Stress and Zeeman perturbation studies have established that the defect has trigonal symmetry, that the ground state is an orbital doublet which is split by spin-orbit coupling, and that both the ground doubly degenerate and the excited singlet states have effective spins S = 1/2 [11]. These results strongly suggest that the 1.40 eV optical transitions must arise from the same defect as the NIRIM2 EPR signal [4,11]. This assignment is supported by the angular dependence and g-value of EPR signals optically detected in the magnetic circular dichroism of the absorption (MCDA) at 1.40 eV [13]. Recent MCDA studies on the 1.404 eV line have confirmed the interpretation of the optical results and a model was put forward to explain the unusual features of the g-values observed for the ground ($g_{\perp}$~0) and excited ($g_{\parallel}$~0) states [14].

The 2.56 eV luminescence band has a very complex ZPL structure resulting from transitions from five excited levels, separated by 1.6 meV, to two ground state levels, split by 0.4 meV [15]. ODMR studies on this band found either an isotropic resonance at a g-value identical to the $Ni_S^-$ [16] or this one plus the resonance associated with single substitutional nitrogen [17]. The latter results have been interpreted in terms of the luminescence coming from two processes, one involving nitrogen and the other being related to nickel. However, there is insufficient evidence to decide if the luminescence occurs at the $Ni_S^-$ defect or if the resonance is detected by an indirect mechanism.

When this kind of diamond is heat-treated at high temperatures (T > 1800°C) the 1.40 eV system is annealed out and the visible absorption becomes a featureless continuum that increases with energy [18].

## C NICKEL AND HIGH NITROGEN-CONTENT DIAMONDS

When diamonds are grown from a nickel solvent-catalyst without a nitrogen getter they have a higher nitrogen content (> 50 ppm) and exhibit a yellow colour. An EPR signal (W8) with g = 2.0319 ascribed to substitutional $Ni^-$ (S = 3/2, $3d^7$) [5,19] is typical of these diamonds, and two centres with ZPLs at 1.885 and 2.51 eV [1,9] are dominant in the absorption spectra. A photo-EPR investigation showed that the $Ni_S^-$ EPR signal can decrease with optical illumination; the onset of this phenomenon is observed at 2.47 eV, suggesting that the acceptor level is located ~3 eV above the valence band [20].

The ZPL of the 1.885 eV centre is made up of three transitions from a common ground state to three excited states at a rhombic I defect. Stress experiments have shown the presence of an additional lower-lying ground state level, split by 5.8 meV; transitions from this level to the excited states are symmetry forbidden [21].

The 2.51 eV ZPL is also made up of three lines arising from a common ground state. At 77 K the fine structure can no longer be separated; stress results at that temperature are compatible with a tetrahedral symmetry [22]. A tentative study to correlate the optical bands to the W8 centre was not able to produce a definite answer, maybe due to the inhomogeneous distribution of Ni in the samples [23].

When these kinds of diamond are annealed at temperatures in excess of 1500°C the 1.885 eV and 2.51 eV systems decay in strength and a new series of absorptions appear [18]. At the first stage of the heat treatment, the 1.693 eV vibronic band, which occurs at a rhombic I centre [24], is the outstanding feature. Accompanying this band there is a complex series of lines between 2.2 and 2.6 eV. At higher temperatures the 1.693 eV band and some of these lines anneal out. The photoluminescence spectra present a similar complex behaviour; a great number of lines appear after annealing at 1650°C, some of which decrease in intensity or disappear completely when the temperature is increased. Only the photoluminescence bands with ZPLs at energies of 1.562, 1.648, 2.369 (S2), 2.495 (S3) and 2.535 (S2) eV remain when samples are annealed at temperatures higher than 2200°C [25-28]. This complicated pattern indicates that several separate species are being produced during the annealing. Since the heat-treatment is carried out in a temperature range in which nitrogen is mobile, the different defects were attributed to complex forms of nickel and nitrogen. Studies of distribution of Ni-related luminescence centres in pre- and post-annealed diamond seem to confirm that the S2 and S3 defects need both Ni and nitrogen to exist [29].

The formation of complexes has been confirmed by EPR studies carried out by Nadolinny, Yelisseyev and co-workers [25–27], who discovered eight EPR centres (NE1–NE8). They showed that the NE4 ($D_{3d}$ symmetry) defect, ascribed to a nickel atom at the centre of a divacancy, acts as the nucleus for aggregation of nitrogen atoms. As successive (up to four) nitrogen atoms are captured in different configurations, the other NE centres are formed. The g-tensors have monoclinic I symmetry for all these defects with the exception of NE2, for which the symmetry is $C_1$ [27]. Another EPR signal has also been assigned to a nickel and nitrogen, the RM1 centre (S = 1/2, $C_{1h}$) [30]. Recently new EPR centres have been found in annealed synthetic diamonds. Four centres (AB1–4) with S = 1/2, g ≈2, of which two are trigonal and the other two have rhombic I symmetry, have been tentatively assigned to defects containing nickel, but nitrogen is not part of them [31]. The AB5 centre, with effective spin state S = 1, is also trigonal and a possible model is a substitutional $Ni^{2-}$ with a nitrogen atom as next neighbour [32].

These optical and EPR features have all been observed in synthetic diamonds. It is known that transition metals are also found in some inclusions inside natural diamonds, and some of these nickel-related defects have been reported in natural samples. An EPR centre (S = 3/2, g = 2.02) so far only found in Argyle blue diamonds has been suggested to be $Ni_S^-$ with an $N^+$ on the fourth nearest neighbour site [33].

## D COBALT

Some complexes of Co and nitrogen are optically and magnetically active in synthetic diamond. In HPHT diamonds grown using a cobalt-containing solvent-catalyst only one optical feature is assigned to Co in as-grown samples: the luminescence band with ZPL at 1.989 eV [3]. However, after high-temperature annealing, several luminescence systems with ZPL energies at 2.135, 2.207, 2.277, 2.367 and 2.590 eV, an absorption band with ZPL at 1.852 eV [3], and an EPR centre have been reported. As with nickel, the mobile nitrogen atoms combine with cobalt to produce active centres. No cobalt-related defects have been reported in diamonds grown with nitrogen getters. The emission bands are seen in both photo- and cathodoluminescence and most of them have long radiative decay times (~0.1 ms) [3].

Uniaxial stress studies on the 2.367 eV line showed that this transition occurs at a trigonal centre between a non-degenerate ground state and a doubly degenerate excited state. The defect is suggested to be a Co-N substitutional pair [34].

In these samples Twitchen et al [35] observed the O4 EPR centre, with monoclinic I symmetry and a hyperfine splitting from a nucleus with I = 7/2. They suggested that the observed large width of the lines is due to a hyperfine interaction with a nearby nitrogen atom. Based on this hypothesis, they proposed a model for the Co-N complex where a Co atom is situated in the middle of a divacancy with an N atom at a nearest-neighbour site.

Photo- and cathodoluminescence topography studies of annealed diamonds grown from cobalt showed that the emitting centres are restricted to {111} sectors, similarly to nickel-related defects [36]. This has been confirmed by X-ray fluorescence spectroscopy [37], which also established, from the measured concentrations of Ni and Co in crystals grown from alloys of these metals, that the cobalt impurities are more difficult to incorporate in the lattice.

## E CONCLUSION

Nickel and cobalt are incorporated into the lattice during diamond growth; both metals are trapped in a similar way given that all the optical features associated with Ni and Co are confined exclusively to the octahedral growth sectors. Although other transition metals, such as Fe and Mn, are used as solvent-catalysts, so far there have been no reports that they form active optical or paramagnetic defects in synthetic diamonds.

There is nowadays a wealth of experimental and some theoretical knowledge concerning nickel-related optical and paramagnetic defects [38–40], but we still lack an overall model to account for the diversity of features exhibited and the interplay between nickel and nitrogen.

## REFERENCES


[1] A.T. Collins, M.P. Spear [ *J. Phys. D, Appl. Phys. (UK)* vol.15 (1982) p.L183-7 ]

[2] A.T. Collins, M.P. Spear [ *J. Phys. C, Solid State Phys. (UK)* vol.16 (1983) p.963-73 ]

[3] S.C. Lawson, H. Kanda, K. Watanabe, I. Kiflawi, Y. Sato, A.T. Collins [ *J. Appl. Phys. (USA)* vol.79 (1996) p.4348-57 ]

[4] J. Isoya, H. Kanda, Y. Uchida [ *Phys. Rev. B (USA)* vol.42 (1990) p.9843-52 ]

[5] J. Isoya, H. Kanda, J.R. Norris, J. Tang, M.K. Bowman [ *Phys. Rev. B (USA)* vol.41 (1990) p.3905-13 ]

[6] R.H. Wentorf [ *J. Phys. Chem. (USA)* vol.75 (1971) p.1833-7 ]

[7] H.M. Strong, R.M. Chrenko [ *J. Phys. Chem. (USA)* vol.75 (1971) p.1838-43 ]

[8] H. Kanda, T. Sekine [ in *Properties and Growth of Diamond* Ed. G. Davies (INSPEC, IEE, London, UK, 1994) p.415-26 ]

[9] A.T. Collins, H. Kanda, R.C. Burns [ *Philos. Mag. B (UK)* vol.61 (1990) p.797-810 ]

[10] S.C. Lawson, H. Kanda, M. Sekita [ *Philos. Mag. B (UK)* vol.68 (1993) p.39-46 ]

[11] M.H. Nazaré, A.J. Neves, G. Davies [ *Phys. Rev. B (USA)* vol.43 (1991) p.14196-205 ]

[12] G. Davies, A.J. Neves, M.H. Nazaré [ *Europhys. Lett. (Switzerland)* vol.9 (1989) p.47 ]

[13] Th. Pawlik, C. Noble, J.-M. Spaeth [ *J. Phys., Condens. Matter (UK)* vol.10 (1998) p.9833-40 ]

[14] P.W. Mason, F.S. Ham, G.D.Watkins [ *Phys. Rev. B (USA)* vol.60 (1999) p.5417-28 ]

[15] E. Pereira, L. Pereira, D.M. Hofmann, W. Stadler, B.K. Meyer [ *Radiat. Eff. Defects Solids (Switzerland)* vol.135 (1995) p.143 ]

[16] E. Pereira et al [ *Diam. Relat. Mater. (Switzerland)* vol.4 (1994) p.53-8 ]

[17] M.H. Nazaré, P.W. Mason, G.D. Watkins [ *Phys. Rev. B (USA)* vol.51 (1995) p.16741-5 ]

[18] S.C. Lawson, H. Kanda [ *J. Appl. Phys. (USA)* vol.73 (1993) p.3967-73 ]

[19] J.H.N. Loubser, W.P. Ryneveld [ *Nature (UK)* vol.211 (1966) p.517 ]

[20] D.M. Hofmann et al [ *Phys. Rev. B (USA)* vol.50 (1994) p.17618-20 ]

[21] M.H. Nazaré, L.M. Rino [ *Mater. Sci. Eng. B (Switzerland)* vol.21 (1993) p.329-32 ]

[22] M.H. Nazaré, J.C. Lopes, H. Kanda [ *Mater. Res. Soc. Symp. Proc. (USA)* vol.339 (1994) p.625-30 ]

[23] A.T. Collins, H. Kanda, J. Isoya, C.A.J. Ammerlaan, J.A. van Wyk [ *Diam. Relat. Mater. (Switzerland)* vol.7 (1998) p.333-8 ]

[24] A.J. Neves, M.H. Nazaré, J.C. Lopes, H. Kanda [ *Physica B (Netherlands)* vol.273-274 (1999) p.636-9 ]

[25] V.A. Nadolinny, A.P. Yelisseyev [ *Diam. Relat. Mater. (Switzerland)* vol.3 (1993) p.17-21 ]

[26] V.A. Nadolinny, A.P Yelisseyev, O.P. Yurieva, B.N. Feigelson [ *Appl. Magn. Reson. (Austria)* vol.12 (1997) p.543-54 ]

[27] V.A. Nadolinny et al [ *J. Phys., Condens. Matter (UK)* vol.11 (1999) p.7357-76 ]

[28] I.N. Kupriyanov, V.A. Gusev, Yu.M. Borzdov, A.A. Kalinin, Yu.N. Pal'yanov [ *Diam. Relat. Mater. (Switzerland)* vol. 8 (1999) p.1301-9 ]

[29] H. Kanda, K. Watanabe [ *Diam. Relat. Mater. (Switzerland)* vol.8 (1999) p.1463-9 ]

[30] R.I. Mashkovtsev, Yu.N. Palyanov [ *Solid State Commun. (USA)* vol.111 (1999) p.397-402 ]

[31] A.J. Neves et al [ *Physica B (Netherlands)* vol.273-274 (1999) p.651-4 ]

[32] A.J. Neves et al [ *Diam. Relat. Mater. (Switzerland)* in press ]

[33] C.J. Noble, Th. Pawlik, J.-M. Spaeth [ *J. Phys., Condens. Matter (UK)* vol.10 (1998) p.11781-93 ]

[34] K. Johnston et al [ *Physica B (Netherlands)* vol.273-4 (1999) p.647-50 ]

[35] D.J. Twitchen, J.M. Baker, M.E. Newton, K. Johnston [ *Phys. Rev. B (USA)* vol.61 (2000) p.9-11 ]

[36] H. Kanda, K. Watanabe [ *Diam. Relat. Mater. (Switzerland)* vol.6 (1997) p.708-11 ]

[37] X. Jia, S. Hayakawa, W. Li, Y. Gohshi, M. Wakatsuki [ *Diam. Relat. Mater. (Switzerland)* vol.8 (1999) p.1895-9 ]

[38] J.E. Lowther [ *Phys. Rev. B (USA)* vol.51 (1995) p.91-6 ]

[39] L. Paslovsky, J.E. Lowther [ *J. Phys., Condens. Matter (UK)* vol.4 (1992) p.775-84 ]

[40] Y. Jinlong, Z. Manhong, W. Kelin [ *Phys. Rev. B (USA)* vol.49 (1994) p.15525-30 ]

# A6.2 The effect of transition metals (TM) on the aggregation kinetics of nitrogen in diamond

S.C. Lawson and I. Kiflawi

December 1999

## A INTRODUCTION

Nickel and cobalt point-defects can become incorporated into the {111} growth sectors of synthetic diamond grown by the high-temperature high-pressure (HTHP) method using Co or Ni solvent-catalysts [1–4]. It is also found that the rate of aggregation of nitrogen in such growth sectors is significantly higher than in the Ni and Co free {001} growth sectors [2,5]. Indeed the aggregation of nitrogen takes place at much lower temperatures when TM impurities are present. Chrenko et al [6] were the first to achieve aggregation of single nitrogen atoms to A-centres under laboratory conditions and, based on a small number of measurements, they found that the reaction follows second order kinetics. This result would be expected if no other factors were playing a role in the aggregation process, and it was accepted by Evans and Qi in their subsequent work on N aggregation [7]. However, recent studies on the aggregation kinetics in Ni [8] and Ni and Co [9] containing diamonds reveal significant deviations from second order kinetics, except when the samples contain very low concentrations of Co [5].

## B A MECHANISM FOR TM ENHANCED AGGREGATION OF NITROGEN

Evidence was presented that Ni and Co assist in the aggregation process by promoting the formation of nitrogen interstitials [9,10]. It was proposed that at the temperatures where aggregation occurs in the laboratory ( ≥ 1500°C), the large substitutional Ni or Co atoms move to a lower energy configuration in which they essentially acquire a vacancy and release a carbon interstitial. When these mobile interstitials migrate to substitutional nitrogen atoms, they exchange positions creating nitrogen interstitials which can then migrate relatively rapidly to form A-centres [10]. A strong indication that this process is taking place under laboratory annealing is the appearance of the local mode at 1450 $cm^{-1}$. This mode has been associated with a nitrogen interstitial [8,9]. The somewhat complex aggregation kinetics may therefore be expected since the equation describing the rate of loss of single nitrogen atoms would have three terms. One term would be proportional to the rate of release of carbon interstitials, another would be proportional to the concentration of nitrogen available for the exchange between interstitial carbon and substitutional nitrogen and the third term would represent the normal second order kinetics of aggregation. Numerical solution of the equations used to fit the above model has yielded good agreement with the experimental results [8].

To date the kinetics of vacancy-assisted aggregation [11,12] have not been experimentally determined. Any differences between the vacancy enhanced and TM enhanced aggregation kinetics could yield useful information regarding the fundamental mechanisms involved.

## C CONCLUSION

Recent studies on the aggregation kinetics of nitrogen in diamonds containing transition metals have shown significant deviations from second order kinetics. Ni and Co assist in the aggregation process by promoting the formation of nitrogen interstitials. The kinetics of vacancy-assisted aggregation have not yet been determined experimentally.

**REFERENCES**


[1] A.T. Collins, H. Kanda, R.C. Burns [ *Philos. Mag. B (UK)* vol.61 (1990) p.797 ]

[2] S. Satoh, H. Sumiya, K. Tsuji, S. Yazu [ *Science and Technology of New Diamond* (Terra Scientific Publishing Co., 1990) p.351 ]

[3] S.C. Lawson, H. Kanda [ *J. Appl. Phys. (USA)* vol.73 no.8 (1993) p.3967 ]

[4] S.C. Lawson, H. Kanda, K. Watanabe, I. Kiflawi, Y. Sato, A.T. Collins [ *J. Appl. Phys. (USA)* vol.79 (1996) p.4348 ]

[5] I. Kiflawi, H. Kanda, D. Fisher, S.C. Lawson [ *Diam. Relat. Mater. (Switzerland)* vol.6 (1997) p.1643 ]

[6] R.M. Chrenko, R.E. Tuft, H.M. Strong [ *Nature (UK)* vol.270 (1977) p.141 ]

[7] T. Evans, Z. Qi [ *Proc. R. Soc. Lond. A (UK)* vol.381 (1982) p.159 ]

[8] I. Kiflawi, H. Kanda, A. Mainwood [ *Diam. Relat. Mater. (Switzerland)* vol.7 (1998) p.327 ]

[9] D. Fisher, S.C. Lawson [ *Diam. Relat. Mater. (Switzerland)* vol.7 (1998) p.299 ]

[10] I. Kiflawi, A. Mainwood, H. Kanda, D. Fisher [ *Phys. Rev. B (USA)* vol.54 (1996) p.16719 ]

[11] A.T. Collins [ *J. Phys. C (UK)* vol.13 (1980) p.2641 ]

[12] A. Mainwood [ *Phys. Rev. B (USA)* vol.9 (1994) p.7934 ]

# A6.3 Silicon in HPHT and CVD diamond

I. Kiflawi and S.C. Lawson

December 1999

## A INTRODUCTION

One of the first references to optically-active silicon in diamond concerns the observation of a zero-phonon line (ZPL) at 1.681 eV which appears in the cathodoluminescence spectra of diamond implanted with $Si^+$ ions [1]. This feature was not observed when other implantation species were used. A quadratic dependence of the intensity of this feature on implantation dose suggested that the centre contains two Si atoms [1], a model which has since been superseded. Later, this feature was shown to be a common characteristic of diamond grown using CVD techniques. Here, silicon substrates are routinely used and the walls of the CVD reaction chambers are often manufactured from quartz, giving two possible sources for the silicon. In an effort to grow CVD diamonds free from Si, Sagakuchi et al [2] found that the addition of oxygen reduces its incorporation. Using HPHT methods, Hong et al [3] were able to synthesise silicon-doped diamond directly from SiC.

Since the publication of *EMIS Datareviews Series No. 9* (1994), detailed spectroscopic measurements have been carried out on the 1.681 eV centre and these have yielded clear evidence of its association with silicon. A theoretical model for the defect has also been produced which is in good agreement with experimental results. This Datareview focuses on these latest advances.

## B OPTICAL CHARACTERISATION OF SILICON

The first unambiguous identification of this centre with silicon became possible with the availability of silicon-doped single-crystal HPHT grown diamonds [4]. Using such diamonds, which were grown by adding 10% Si to the graphite in the growth cell [5], it was shown that the 1.681 eV ZPL is actually composed of twelve lines [4]. In the good quality, relatively stress-free crystals which were used these lines could be resolved in PL and absorption measurements carried out at temperatures below 40 K [4]. The twelve-line fine structure is observed close to 1.682 eV, and it can be divided into three groups, each group consisting of four components [4]. The relative intensities of absorption for the three groups of lines were found to be the same as the ratios of the natural abundance of the three known silicon isotopes [4], providing good confirmation for the Si model. The four peaks within each group are due to transitions between a split excited state and a split ground state [4]. A similar energy level diagram was determined from uniaxial stress experiments [6,7]; however, no conclusions could be drawn from these measurements regarding the symmetry of the centre. A similar ZPL splitting was observed in good quality CVD diamond [6,8].

Khong et al [9] measured the luminescence decay time of the 1.682 eV system and observed a two stage decay process. An initial slow decay was observed for the first 3 ns followed by a fast decay with a measured decay time of about 2.4 ns. They attributed this behaviour to the presence of energy reservoirs coupled with the excited state of the 1.682 eV centre.

The dominant phonon energies for this system lie at 65 and 130 meV from the main 1.682 eV ZPL [10]. Comparing the intensity of the ZPL and the intensity of the side band, the Huang-Rhys factor has been calculated and values of 0.26 [8] and 0.1 [11] were obtained, indicating that the centre has a weak coupling to the diamond lattice.

## C A MODEL FOR THE 1.682 eV CENTRE

Collins et al [8] conducted isochronal annealing measurements on CVD diamond films and monitored the changes in the absorption due to neutral vacancies (GR1), and the silicon-centre. They observed that the intensity of the 1.682 eV system increases when the intensity of the GR1 system decreases. This result, coupled with the observation that the intensity of the 1.682 eV system is an order of magnitude higher in electron irradiated and annealed silicon containing samples [4,12], strongly indicates that vacancies are involved in the formation of the silicon centre. However, the PL intensity of the 1.682 eV system in the irradiated samples was found to decrease in the annealing range of 900–1200°C and then to subsequently increase again for annealing temperatures up to 2200°C [4,12]. Initially this result cast doubt over the Si-vacancy interpretation [4], but it was later shown to be caused by absorption of the Si related emission by a non PL active peak with ZPL at 1.679 eV. This peak was shown to increase in intensity and then to totally anneal out in the temperature range where the 1.682 eV system decreases [13]. The 1.679 eV peak, and additional peaks at 1.691 and 1.711 eV, are not related to silicon [13]. The 1.711 eV peak has also been observed in polycrystalline CVD diamond [14].

Bicai and Shangda [15] performed theoretical modelling on silicon centres in diamond and concluded that it is energetically unfavourable for such impurities to become incorporated in either substitutional or interstitial positions. However, Goss et al [16] performed similar calculations adding vacancies to the silicon impurities and obtained very good theoretical agreement with the experimentally observed spectra. Their model consists of a Si atom sitting mid-way between two vacancies [16]. A suggestion that the 1.682 eV system is due to a di-silicon centre [10] has not been substantiated and an earlier suggestion that the centre may involve nitrogen has not been confirmed by subsequent experimental results. In fact contrary to this, a topographic study of the PL spectra from various locations of silicon-doped diamond single crystals established that nitrogen rich regions corresponded to regions where the silicon luminescence was significantly reduced [13]. This was attributed to the fact that nitrogen effectively competes with silicon for vacancies, as evidenced by the detection of luminescence from nitrogen-vacancy related centres in these samples [13]. In consequence, the best Si-doped diamonds were grown using nitrogen getters [5].

## D CONCLUSION

Detailed spectroscopic measurements have been carried out on the 1.681 eV centre in diamond yielding clear evidence of its association with silicon. A model for this centre has been developed.

## REFERENCES


[1] V.S. Vavilov, A.A. Gippius, A.M. Zaitsev, B.V. Deryagin, B.V. Spitsyn, A.E. Aleksenko [ *Sov. Phys.-Semicond. (USA)* vol.14 (1980) p.1078 ]

[2] I. Sagakuchi, M. Nishitani-Gamo, K.P. Loh, H. Haneda, S. Hishita, T. Ando [ *Appl. Phys. Lett. (USA)* vol.71 (1997) p.1 ]

[3] S. Kong, H. Kanda, L. Gou [ *Chin. Sci. Bull. (China)* vol.41 no.3 (1996) ]

[4] C.D. Clark, H. Kanda, I. Kiflawi, G. Sittas [ *Phys. Rev. B (USA)* vol.51 (1995) p.16681 ]

[5] G. Sittas, H. Kanda, I. Kiflawi, P.M. Spear [ *Diam. Relat. Mater. (Switzerland)* vol.5 (1996) p. 866 ]

[6] H. Sternschulte et al [ *Diam. Relat. Mater. (Switzerland)* vol.4 (1995) p.1189 ]

[7] H. Sternschulte, K. Thonke, R. Sauer, P.C. Münzinger, C. Michler [ *Phys. Rev. B (USA)* vol.50 (1994) p.14554 ]

[8] A.T. Collins, L. Allers, C.J.H. Wort, G.A. Scarsbrook [ *Diam. Relat. Mater. (Switzerland)* vol.3 (1994) p.932 ]

[9] Y.L. Khong, A.T. Collins, L. Allers [ *Diam. Relat. Mater. (Switzerland)* vol.3 (1994) p.1023 ]
[10] A.A. Gorokhovsky, A.V. Turukhin, A.R. Alfano, W. Phillips [ *Appl. Phys. Lett. (USA)* vol.66 (1995) p.43 ]
[11] M.C. Rossi, S. Salvatori, F. Galuuzzi, R.M. Montereali, F. Somma [ *Diam. Relat. Mater. (Switzerland)* vol.6 (1997) p.1564 ]
[12] C.D. Clark, C.D. Dickerson [ *Surf. Coat. Technol. (Switzerland)* vol.47 (1991) p.336 ]
[13] I. Kiflawi, G. Sittas, H. Kanda, D. Fisher [ *Diam. Relat. Mater. (Switzerland)* vol.6 (1997) p.146 ]
[14] L. Allers [ PhD Thesis, King's College London, 1995 ]
[15] Pan Bicai, Xia Shangda [ *Phys. Rev. B (USA)* vol.49 (1994) p.11444 ]
[16] J.P. Goss, R. Jones, S.J. Breuer, P.R. Briddon, S. Öberg [ *Phys. Rev. Lett. (USA)* vol.77 (1996) p.3041 ]

CHAPTER A7

# PROPERTIES OF DEFECTS IN DIAMOND: VACANCY AND SELF-INTERSTITIAL

# A7.1 Modelling electronic states of the vacancy in diamond

J.E. Lowther

May 2000

## A INTRODUCTION

The most fundamental defect in diamond is the lattice vacancy. In its simplest form it is thought to be an empty carbon site with surrounding $sp^3$ orbitals of four carbons directed toward it forming $a_1$ and $t_2$ one-electron states. The defect has a very similar form in silicon and a variety of other covalent materials, but unlike in these materials the effects of electron correlation and lattice dynamics in diamond are quite unique. This leads to a defect with molecular-type properties exhibiting complex electron correlation with sophisticated dynamic Jahn–Teller features [1]. The interplay between electron correlation and Jahn–Teller interactions is very important in deciding the ultimate ground state of the defect.

A model of the vacancy in diamond was first derived by Coulson and Kearsley [2] and essentially is group theoretical in foundation. It postulates that electrons occupy the $a_1$ and $t_2$ one-electron states with electron correlation leading to a series of multiplets. Molecular details of electron correlation remain, to date, quantitatively poorly described despite many attempts at their description having been made, usually within the framework of Hartree–Fock where standard quantum chemical techniques were applied to evaluate the various multi-centre integrals involving the $a_1$ and $t_2$ molecular orbitals [2,3]. The approach treats the vacancy as a large molecule and, as such, neglects the fact that the defect resides in a solid state lattice and most likely is a deep defect with a strong localized character. To locate the one-electron levels relative to the valence and conduction band edges semi-empirical techniques, again based on quantum chemistry, were applied. Initially Huckel theory was used [4] within a cluster-model description in which a large number of atoms is used to represent the carbon diamond structure. A variety of other cluster-model approaches have been used at other approximate self-consistent levels of quantum chemistry [5–7], using non-self-consistent tight binding approaches [8] or, more recently, using local density theory [9,10]. Problems involving the termination of a finite number of atoms used to represent an infinite solid will always remain in such an approach although the 'surface saturator' techniques used do seem to circumvent the problem. To avoid the surface problem other methods have been introduced and applied to the vacancy in diamond. First is the Greens function approach [11] which used either pseudopotential techniques or an empirical tight binding scheme. More recently a state-of-the-art local density plane wave approach has been applied [12,13] to study the diamond vacancy.

The results of all the calculations follow a pattern, namely that a defect state of $t_2$ symmetry emerges in the forbidden energy gap of the diamond band structure with another state of $a_1$ symmetry lying below it. This is in accord with the original Coulson and Kearsley model of the vacancy [2], but qualitative differences relating to the position of the $t_2$ level and especially the $a_1$ level remain. The precise location of the $a_1$ level is important because under certain energies of photo-excitation free carriers are produced [14,15].

From the point of view of group theory the $a_1$ and $t_2$ levels provide a one-electron basis for the many-electron description of the electron multiplets associated with the vacancy. The neutral vacancy has a ground state $^1E$ with optical transitions to another state $^1T_2$ giving rise to optical absorption at 1.673 eV and called GR1. Such states arise from two-electron multiplets appropriate to the $t_2 \otimes t_2$ direct product. Likewise in the negatively charged vacancy three electrons give rise to a ground state $^4A_2$

with optical absorption to a state $^4T_1$ being associated with the ND1 optical transition at 3.165 eV. It has also been possible to describe many properties of vacancy-related defects within the context of group theory [16] without recourse to computational modelling.

It is now certain that many other optical properties of diamond are associated with the vacancy. The system of zero-phonon sidebands and related bands labelled GR1-8 (GRn) is associated with a vacancy taking a neutral form, i.e. only four electrons on an adjacent bond of each C atom [17,18]. This optical excitation starts at 1.673 eV and extends to almost 3 eV. After that the vacancy takes a negatively charged state and the ND1 optical system emerges which has a zero-phonon line at 3.149 eV [15]. As shown by the experiments of Vermeulen et al [14] there are subtle changes in photoconductivity associated with the GRn and ND1 system strongly implying that somewhere in this spectrum is photoionization. Collins [19] investigated the GR2-8 transitions concluding that the origin of these transitions is complex—partly associated with purely electronic transitions and possibly with some evidence of vibronic replicas as is the case for the GR1 system [15]. But the uncertainty as to the origin of GR2-8 is compounded by the experimental observation [15] that all lines correspond to a series of apparent excited states with the same tetrahedral symmetry, which was contrary to a previous attempt [20] to describe the origin of the lines by extending the Coulson–Kearsley model of four directed $a_1$ and $t_2$ orbitals to consider the next nearest twelve dangling bond orbitals.

Many defects in diamond are also associated with a lattice vacancy. Nitrogen forms many complexes around that vacancy to the extent that four nitrogens about the vacant site are effectively providing enough electrons for an overall charge neutrality. Silicon [21–23], phosphorus [24] and metals [25] are only some elements that can pair up with the vacancy and it is now well known that the vacancy plays an important part in the diffusion of defects through diamond [26]. The lattice vacancy does indeed play a very important role in the properties of diamond.

In the computer modelling of the lattice vacancy in diamond, or in fact all tetrahedrally coordinated covalent solids, the $a_1$ and $t_2$ states always emerge in the forbidden energy gap as being associated with the vacancy. Such one electron states play an invaluable role in constructing simple models of a variety of semiconductor defects highlighting the importance of the diamond vacancy itself. The $a_1$ and $t_2$ states readily lead to models for vacancies which are essentially group theoretical in basis although qualitative details of the system are not at all understood.

The present work will focus on examining various models that have been applied to understand the interactions present at the vacancy in diamond and will introduce very recent calculations of the vacancy which suggest that some earlier ideas of the lattice vacancy may need to be revised.

## B CLUSTER MODEL CALCULATIONS

Conventional cluster model calculations of the lattice vacancy are based either on quantum chemistry approaches, Huckel theory [4,5], or on higher levels of approximation such as CNDO [6] or MNDO [27] (complete or medium neglect of differential overlap, respectively). Some degree of self consistency can therefore be built into the method even within these semi-empirical methods. The local density approximation has also been used in a variety of defect studies in diamond. This approach is conveniently parametrized for the analytical evaluation of multicentre integrals using gaussian functions as applied in the GAMESS [28] or AIMPRO [9] codes. Early pseudopotentials [29] can also be written in terms of gaussian functions providing very convenient input for AIMPRO. Moreover, AIMPRO has been extended to include spin restriction crudely allowing for the multiplet structure that is observed at the neutral and negatively charged vacancy.

In FIGURE 1 we show a typical modest cluster of atoms that is used to simulate defects in diamond.

Here it is a 65-atom cluster with one atom missing representing the vacancy. The surfaces are saturated with either pseudo-hydrogenic-atoms having properties appropriate to diamond or simply hydrogen atoms. Hence, in effect the cluster is 28 C atoms and 36 surface H atoms.

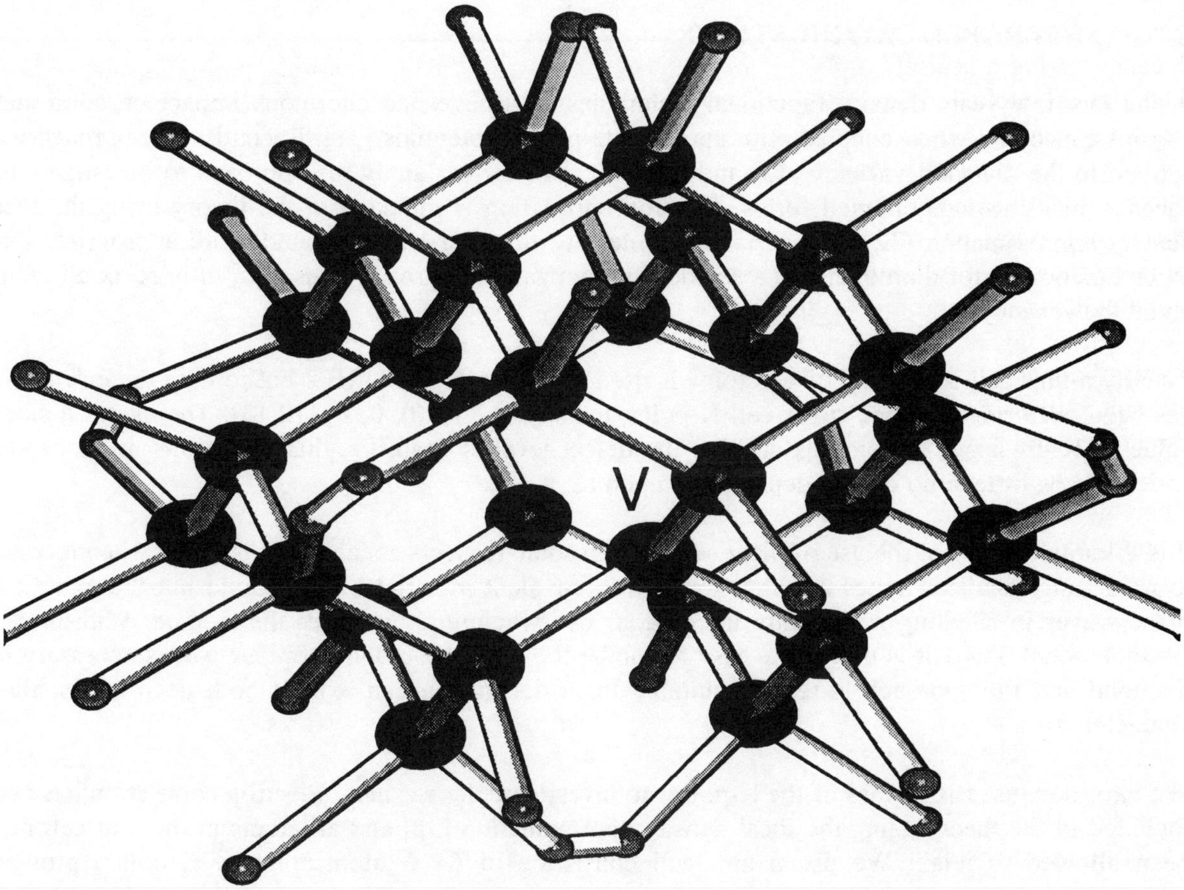


FIGURE 1 A typical 65-atom cluster used to represent diamond. The diamond vacancy is indicated.

Some kind of geometry optimization procedure is used to relax the cluster to a lowest energy configuration and the process repeated with a defect structure located within the cluster. In approaches such as CNDO semi-empirical parameters have been chosen to reproduce the bulk properties [6] and thereafter such parameters are used in defect modelling. As shown by Mainwood [6] the choice of parameters is critical for modelling the relaxation at the diamond vacancy and sometimes an inward relaxation or an outward relaxation can be found. Conventional atomic parameters as used in an MNDO calculation which is a more sophisticated semi-empirical approach than CNDO have predicted an outward relaxation [27].

Breuer and Briddon [10] applied the AIMPRO pseudopotential code to a 304-atom cluster ($C_{190}H_{114}$) using bare Coulomb potentials for the surface saturators and the early pseudopotentials [29] for C. This result for the vacancy also showed an outward relaxation of atoms about the vacant lattice site consistent with earlier quantum chemical calculations. Although these authors considered the position of the electronic ground state of the neutral vacancy they also pointed out that the ground and first excited states of the vacancy are associated with the $t^2_2$ configuration and as such cannot be found within their model. This is unlike the case for the negative vacancy, which corresponds to a change in one-electron configuration from $a_1^2t_2^3$ to $a_1t_2^4$. The AIMPRO code obtained an energy of 3.3 eV for this difference in good agreement with the experimental energy of the ND1 band (3.165 eV). Thus while it is likely that electronic structure calculations will predict the ND1 optical properties quite well because the transition involves a change in occupancy of the one-electron states, the GR1 optical properties remain elusive because of the present difficulty in handling electron correlations.

## C SUPERCELL CALCULATIONS

In the last few years density functional techniques [30] have had enormous impact on solid state physics especially when coupled with appropriate pseudopotentials [31]. Recently the approach was applied to the diamond vacancy [13] and later we shall further apply the approach to investigate the vacancy in its various charged forms. Electron correlation is included in the theory using the local density approximation [32], this form of the density functional being found more appropriate than other functionals for diamond, and a standard minimization approach is used to fully relax all atoms about the vacancy.

Usually a unit cell consisting of 64 atoms is used as shown in FIGURE 2 but to date the calculation has been performed for one super-cell K-point, namely at K = (0, 0, 0) [12,13]. The use of a single value of K presumes to some extent that the defect level is deep, i.e. that it exhibits no dispersion which means little or no energy dependence upon K.

It is acknowledged that the use of plane waves is in some respects conceptually strange when it comes to modelling localized states as often found in a covalent material like diamond but the benefit of plane waves in handling multicentre interactions overwhelmingly justifies the method. With modest computational power it is now quite easy to handle the many thousands of plane waves necessary for diamond and the approach is readily suitable for a defect that can exhibit both deep and shallow characteristics.

We have now used 10 values of the K-points to investigate the vacancy. Electron correlation has been included in the theory using the local density approximation [32] and all atoms in the unit cell have been allowed to relax. We use a unit cell consisting of 64 C atoms; however, unlike previous approaches where the calculation was performed for one super-cell K-point, namely at K = (0, 0, 0) [12,13], we have now used 10 values of the K-points that correspond to a [2 × 2 × 2] Monkhorst–Pack grid [33]: these are given in TABLE 1.

TABLE 1 Supercell K-points (in units of 2π/a where a is the supercell lattice parameter) as used in the present calculation.

| | | | | | | | | |
|---|---|---|---|---|---|---|---|---|
| 1 | 0.00 | 0.00 | 0.00 | | 6 | 0.00 | 0.50 | 0.50 |
| 2 | 0.00 | 0.00 | 0.25 | | 7 | 0.25 | 0.25 | 0.25 |
| 3 | 0.00 | 0.00 | 0.50 | | 8 | 0.25 | 0.25 | 0.50 |
| 4 | 0.00 | 0.25 | 0.25 | | 9 | 0.25 | 0.50 | 0.50 |
| 5 | 0.00 | 0.25 | 0.50 | | 10 | 0.50 | 0.50 | 0.50 |

This grid of K-points does not formally retain an overall tetrahedral symmetry and this has the apparent effect of leading to a lowering in degeneracy of the $t_2$ vacancy-associated deep level. Yet out of these special K-points two have tetrahedral symmetry, namely K = (0, 0, 0) and K = (½, ½, ½), and these points will be referred to when considering dispersion of the $t_2$ level. Carbon pseudopotentials appropriate to neutral C as $2s^22p^2$ were generated using the approach of Troullier and Martins [34] with plane waves to an energy cutoff of 64 Ryd used throughout.

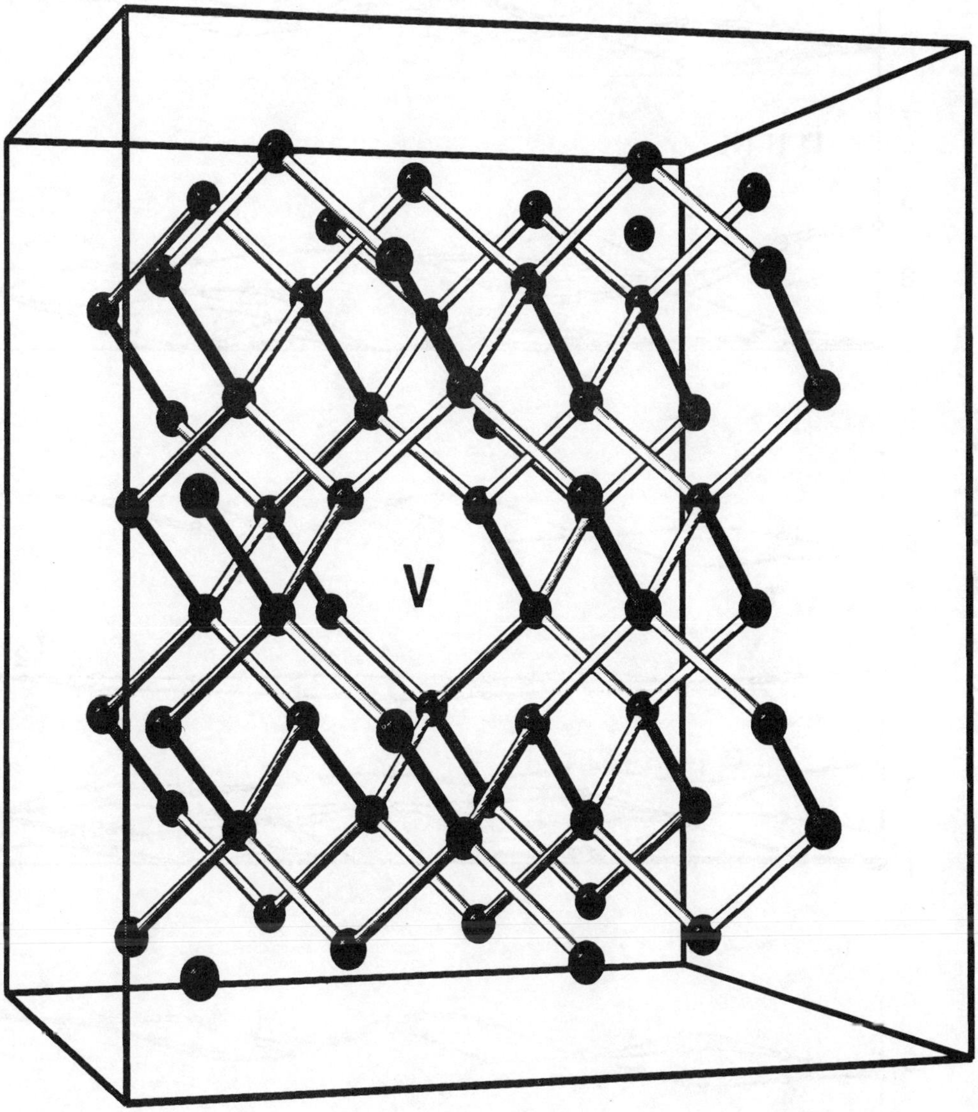


FIGURE 2 A 64-atom super-cell used to represent diamond. The diamond vacancy is indicated.

In FIGURE 3 we show the calculated bulk and defect levels for the neutral ($V^0$) and negatively charged ($V^-$) vacancy. In agreement with cluster model calculations, we find the emergence of the $a_1$ and $t_2$ levels. Both $a_1$ and $t_2$ levels have some dispersion (i.e. dependence on K) compared with the bulk energy bands, although for the $t_2$ level the energy difference between the two extreme K-points with full tetrahedral symmetry is found to be 0.15 eV with the energy at K = (0, 0, 0) lying above that at K = (½, ½, ½) in both cases. The Fermi energy of the system passes through the $t_2$ level and as such an asymmetric distribution of localized charge is distributed over the defect associated Fermi surface. We also point out the slight dispersive nature of the $a_1$ level: this lies at the top of the valence band and therefore is strongly resonant with delocalized states and therefore susceptible to autoionization processes as pointed out recently [35].

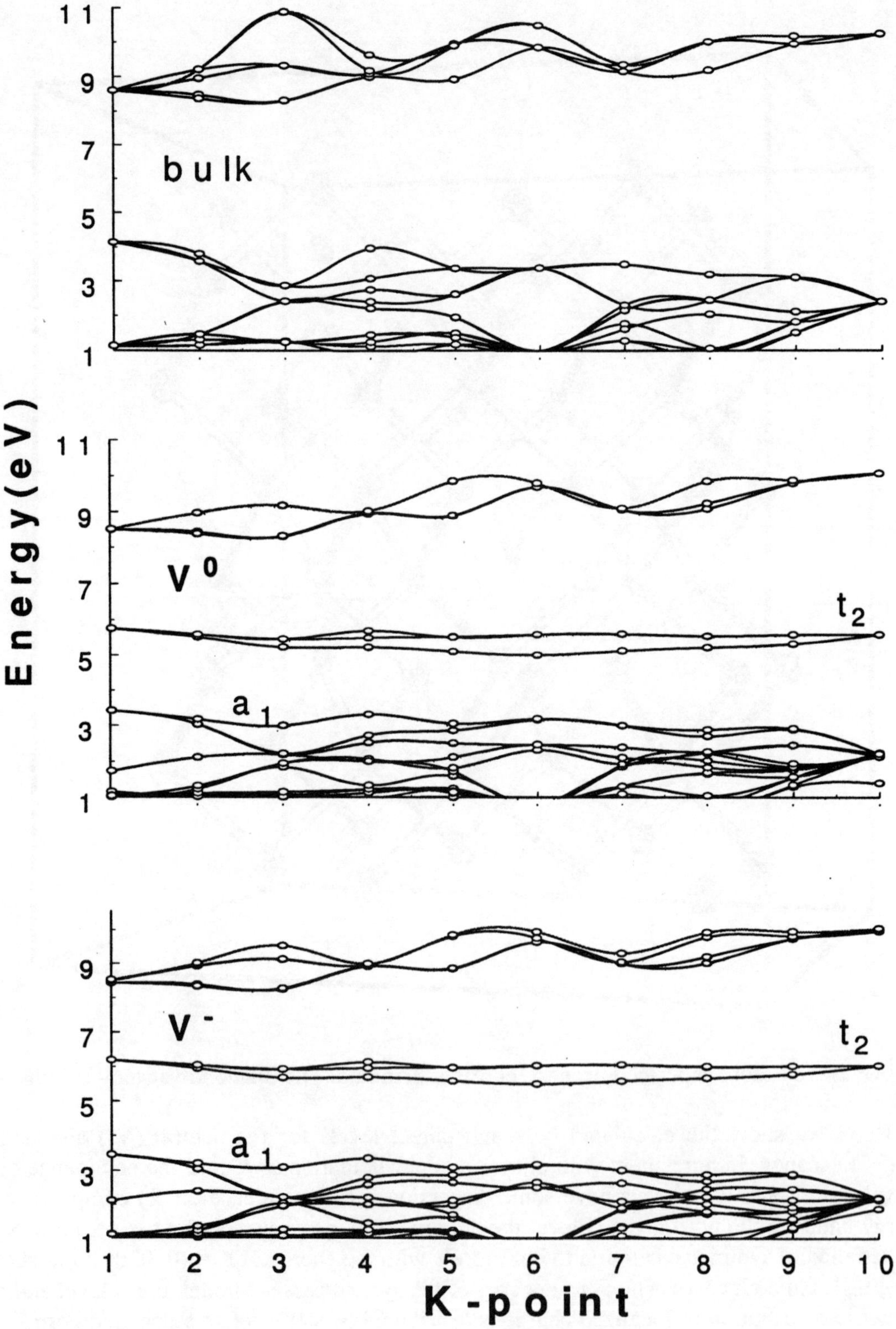


FIGURE 3 Energy band structure of the bulk and the neutral and negatively charged diamond vacancies.

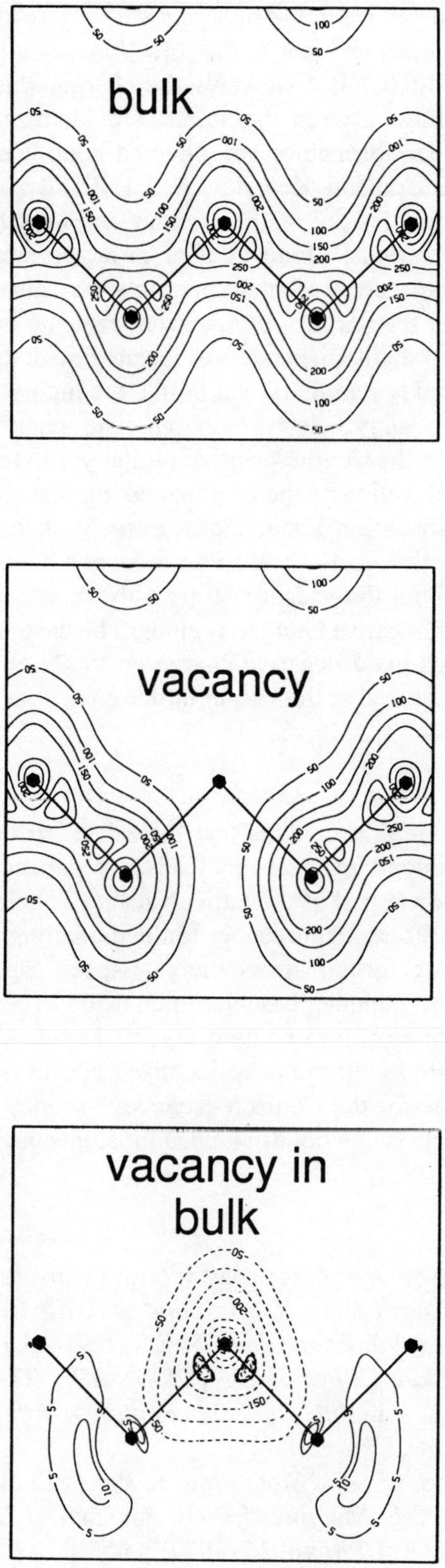


FIGURE 4 Charge density of the bulk and the neutral vacancy, and the difference between the bulk and neutral vacancy. Continuous lines are for electron charge, dashed lines are for hole charge relative to the bulk material. Note that electron contours start from 50 me/(bohr$^3$) whereas hole contours start from 5 me/(bohr$^3$).

We now examine the charge distribution in the bulk and vacancy related supercells. In FIGURE 4 we show the bulk charge density: clearly evident is the directional bonding between atoms as observed many times previously. Next in FIGURE 4 we show the charge density for the neutral vacancy. As with the bulk material the bonding between the atoms is evident but with very little charge at the vacancy atom site. There is a clear absence of any directed bond character toward the vacant lattice site, as basically assumed in the Coulson–Kearsley model. Finally in FIGURE 4 we show what, in essence, corresponds to the charge density of the vacancy as viewed in diamond, i.e. the difference between the two previous calculations. Quite evident now are strong regions of positive charge located at the vacant lattice site. Somewhat surprisingly we also find other regions of charge localization but lying just outside the vacancy tetrahedron, near the back-bonds of each of the four C atoms. This originates from the slight dispersion associated with the defect levels and will form resonant symmetry states of $a_1$ and $t_2$ symmetry, but unlike the highly localized regions at the vacancy lattice site these are regions of negative electron charge. The magnitude of charge in this region is almost an order of magnitude less than in the positive region yet is far larger in spatial extent and no doubt sensitive to the outward relaxation of the C atoms taking place about the vacant lattice site. We also investigated the charge distribution about the negatively charged vacancy and also about the neutral vacancy trapped at an N defect. Both of these defects have been identified in diamond [15]. The positive localized regions about the vacant site are only very slightly changed whereas there are slight increases in the back-bond negative charged regions. The case where N is next to the vacancy is most dramatic with the enormous extra negative charge on the N atom hardly affecting the positive localized distribution of charge situated at the vacant lattice site.

## D CONCLUSION

We have seen that computer modelling predicts that there is a strong localization of charge centred about a vacant lattice site in diamond and that this leads to localized states of $a_1$ and $t_2$ symmetry in accord with the Coulson–Kearsley model of the lattice vacancy. There are however subtle differences now emerging between the two different approaches to the modelling of the lattice vacancy which use cluster-models or supercells. The former stress the real-space aspects of the problem; the latter extends the real-space approach to include phase variation as observed through a band structure. New results extending the supercell approach to contain several K-points in the supercell Brillouin zone show a slight dispersion associated with the deep localized $a_1$ and $t_2$ levels and that this could have important conceptual significance for the Coulson–Kearsley vacancy model which presumed that the localization arises only from electrons on bonds situated on atoms near the vacant site.

## REFERENCES


[1] A.M. Stoneham [ *Theory of Defects in Solids* (Oxford University Press, Oxford, 1975) ]

[2] C.A. Coulson, M.J. Kearsley [ *Proc. R. Soc. Lond. A (UK)* vol.241 (1957) p.433 ]

[3] C.A. Coulson, F.P. Larkins [ *J. Phys. Chem. Solids (UK)* vol.32 (1971) p.2245 ]

[4] R.P. Messmer, G.D. Watkins [ *Phys. Rev. B (USA)* vol.7 (1973) p.2578 ]

[5] M. Astier, J.C. Bourgoin [ in *Int. Conf. on Radiation Effects in Semiconductors* Ed. J.H. Albany (IOP, London, UK 1978) p.180 ]

[6] A.M. Mainwood [ *J. Phys. C, Solid State Phys. (UK)* vol.12 (1979) p.2543 ]

[7] T.L. Estle, S. Estreicher, D.S. Marynick [ *Phys. Rev. Lett. (USA)* vol.58 (1987) p.1547 ]

[8] J.E. Lowther [ *Phys. Rev. B (USA)* vol.36 (1977) p.483 ]

[9] P.R. Briddon, R. Jones [ *Physica B (Netherlands)* vol.185 (1993) p.179 ]

[10] S.J. Breuer, P.R. Briddon [ *Phys. Rev. B (USA)* vol.51 (1995) p.6984 ]

[11] G.B. Bachelet, G.A. Baraff, M. Schluter [ *Phys. Rev. B (USA)* vol.24 (1981) p.4736 ]

[12] J. Bernholc, S.A. Kapihara, C. Wang, A. Antonelli, R.F. Davis [ *Mater. Sci. Eng. B (Switzerland)* vol.11 (1992) p.265 ]

[13] L.H. Li, J.E. Lowther [ *J. Phys. Chem. Solids (UK)* vol.58 (1997) p.1607 ]
[14] L.A. Vermeulen, C.D. Clark, J. Walker [ *Inst. Phys. Conf. Ser. (UK)* vol.23 (1975) p.294 ]
[15] G. Davies [ *Nature (UK)* vol.269 (1977) p.498 ]
[16] J.E. Lowther [ *J. Phys. Chem. Solids (UK)* vol.45 (1984) p.127 ]
[17] C.D. Clark, J. Walker [ *Proc. R. Soc. Lond. (UK)* vol.334 (1973) p.241 ]
[18] G. Davies, M.M. Martin, M.H. Nazare, D. Shaw, M.F. Thomaz [ *J. Phys. C, Solid State Phys. (UK)* vol.20 (1987) p.L13 ]
[19] A.T. Collins [ *J. Phys. C, Solid State Phys. (UK)* vol.11 (1978) p.1 ]
[20] J.E. Lowther [ *Philos. Mag. B (UK)* vol.36 (1977) p.483 ]
[21] C.D. Clark, H. Kanda, I. Kiflawi, G. Sittas [ *Phys. Rev. B (USA)* vol.51 (1995) p.16681 ]
[22] J.P. Goss, R. Jones, S.J. Breuer, P.R. Briddon, S. Oberg [ *Phys. Rev. Lett. (USA)* vol.77 (1996) p.3041 ]
[23] K. Iakoubovskii, G.J. Adriaenssens, M. Nesladek [ *J. Phys., Condens. Matter (UK)* vol.12 (2000) p.189 ]
[24] R. Jones, J.E. Lowther, J. Goss [ *Appl. Phys. Lett. (USA)* vol.69 (1996) p.2489 ]
[25] B.P. Doyle, J.K. Dewhurst, J.E. Lowther, K. Baruth-Ram [ *Phys. Rev. B (USA)* vol.57 (1998) p.4965 ]
[26] G. Davies, S.C. Lawson, A. Mainwood, S.J. Sharp [ *Phys. Rev. B (USA)* vol.46 (1992) p.13157 ]
[27] J.E. Lowther [ 1997, unpublished ]
[28] M.W. Schmidt et al [ *J. Comput. Chem. (USA)* vol.14 (1993) p.1347 ]
[29] G.B. Bachelet, D.R. Hamann, M. Schluter [ *Phys. Rev. B (USA)* vol.26 (1982) p.4199 ]
[30] D.P. Joubert (Ed.) [ *Density Functional Theory* (Springer-Verlag, Heidelberg, Germany, 1998) ]
[31] M. Bockstede, A. Kley, J. Neugebauer, M. Scheffler [ *Comput. Phys. Commun. (Netherlands)* vol.107 (1997) p.187 ]
[32] D.M. Ceperley, B.J. Adler [ *Phys. Rev. B (USA)* vol.45 (1980) p.566 ]
[33] H.J. Monkhorst, J.D. Pack [ *Phys. Rev. B (USA)* vol.13 (1976) p.5188 ]
[34] N. Troullier, J.L. Martins [ *Phys. Rev. B (USA)* vol.43 (1991) p.1993 ]
[35] J.E. Lowther [ *Phys. Rev. B (USA)* vol.48 (1993) p.11592 ]

# A7.2 Theory of the self-interstitial in diamond

A. Mainwood

May 2000

## A INTRODUCTION

Until very recently, none of the self-interstitials had been identified experimentally in any of the semiconductors [1,2], although many of them had been modelled theoretically. For this reason, theoretical modelling was used to determine the properties that might be seen experimentally. This approach contrasts with the more common procedure where the defect is identified experimentally first, and its structure confirmed or supported by theoretical work.

Much of the experimental work on the self-interstitial in diamond has used EPR spectroscopy (the R2 EPR centre) and this is explained in detail in Datareview A7.5. Further experimental work, using optical spectroscopy (the absorption lines at 1.685 eV and 1.859 eV), is also described in the same Datareview. This Datareview will describe the theoretical modelling of the self-interstitial and relate it to the experimental observations.

## B MODELS OF THE SELF-INTERSTITIAL

The first models of the self-interstitial in diamond gave rather ambiguous results [3], but when more systematic modelling was undertaken, it was discovered that the split <001> configuration was the most stable in all charge states [4]. This is in contrast to the situation in silicon, where the <110> split structure is predicted to be energetically favourable [5]. A more detailed model was presented by Breuer and Briddon [6] using the local density approximation code AIMPRO [7]. They showed that there were two degenerate p-like one-electron orbitals orthogonal to the <001> axis and to the line joining the nearest neighbours to each of the split atoms (see FIGURE 1). The many-electron states, which are created when these orbitals are occupied by two electrons, are $^3A_2$, $^1B_1$, $^1B_2$, and $^1A_1$. Breuer and Briddon [6] calculated the splitting to be 0.55 eV between the lowest energy $^1B_1$ and the $^3A_2$ state, while the other two were about 1 eV higher still. A later estimate of the same splittings [8] by the same method, gave 0 and 1.4 eV, respectively. The migration energy of the split interstitial was calculated to be 1.7 eV, in good agreement with experimental data [2].

There are some discrepancies with the experimental data. The spin 1 state (the $^3A_2$ state of Breuer and Briddon) is observed to be 50 meV higher in energy than the ground (spin 0, probably $^1B_1$) state [2]. Mainwood [9] calculated the energies of the many-electron states using the pure p-like orbitals on the two atoms. She found that the lowest two S = 0 states ($^1B_1$ and $^1B_2$) should be separated by energies of ±J from the S = 1 state, where J is the overlap integral between the one-electron p-states, and estimated that energy to be of the order of 100 meV.

In addition, it is difficult to see how transitions between the three S = 0 states could account for the observed optical transitions at 1.685 and 1.859 eV, particularly because the lower state of the former transition is about 6 ± 2 meV above another ground state [10]. An observation that Breuer and Briddon make, that the centre is unstable against a distortion of the structure of the <001> interstitial from $D_{2d}$ to $D_2$ symmetry, begins to indicate the solution to this problem. However, it took a careful comparison of the shifts in energy of the optical lines in $^{13}C$ diamond to disentangle the apparently conflicting data [11].

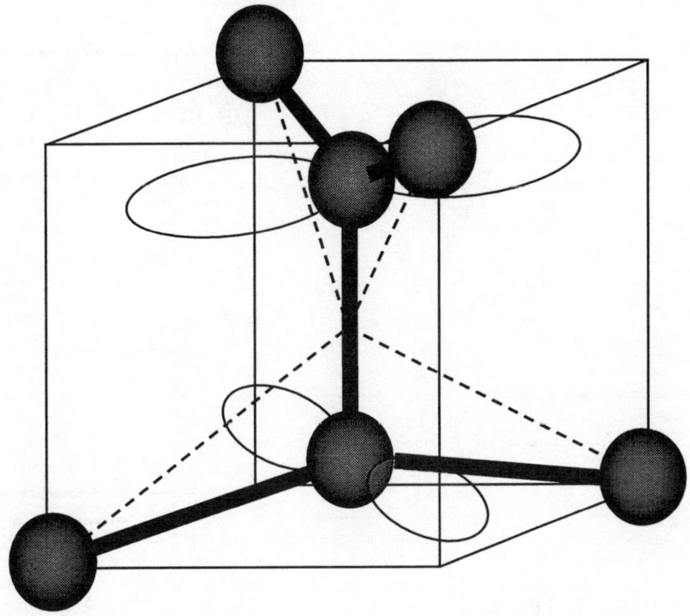

FIGURE 1 The structure of the <001> split self-interstitial. The two p-like orbitals, which give the defect its electronic properties, are shown.

The zero phonon lines of almost all optical transitions are at higher energies of 1–8 meV in $^{13}C$ diamond than their equivalents in $^{12}C$ material. The 1.859 eV line is an exception to this, with a negative shift of 5.2 ± 0.06 meV [11]. Davies et al [11] therefore suggest that this transition is a local mode replica of a forbidden transition, with a local mode energy of 169 meV (1363 $cm^{-1}$). The transitions are shown in FIGURE 2(b). However, using the preliminary uniaxial stress data of [10], Davies et al [12] suggested that it was necessary to have two electronic states of the same symmetry if the observed vibronic coupling and optical transitions were to exist. In the theoretical model of Breuer and Briddon [6] in $D_{2d}$ symmetry, there are no such pairs of states. The small distortion to $D_2$ symmetry transforms the $^1B_1$, $^1B_2$, and $^1A_1$ states to $^1A$, $^1B_1$, and $^1A$. Now the states are of the symmetries shown in FIGURE 2, with $B_1$ coupling between the lowest two reducing the splitting to 6 meV, and the electric dipole transitions shown (also transforming as $B_1$) being allowed.

## C LOCAL MODES OF VIBRATION

Breuer and Briddon [6] find local modes of vibration at the energies 2023 $cm^{-1}$ (A mode, infrared inactive), 1590 $cm^{-1}$ and 1600 $cm^{-1}$ (B modes, infrared active). These have not been positively identified with any experimentally observed infrared modes, but the 1570 $cm^{-1}$ line has been suggested as a plausible candidate [9] for the B modes. The 5RL centre, with a zero phonon line at 4.582 eV, has local modes at 1912, 1630 and 1557 $cm^{-1}$ (among others) [13] and these also have similar properties to the self-interstitial. Both these defects anneal out at 900°C—much higher than would be expected for an interstitial. If one of the split pair of atoms is replaced by $^{13}C$ the modes [6] are predicted to shift to 1984, 1560 and 1595 $cm^{-1}$, respectively, and if both atoms are $^{13}C$, they shift further to 1944, 1557 and 1564 $cm^{-1}$. There is no estimate of the energy of the local mode described in Section B.

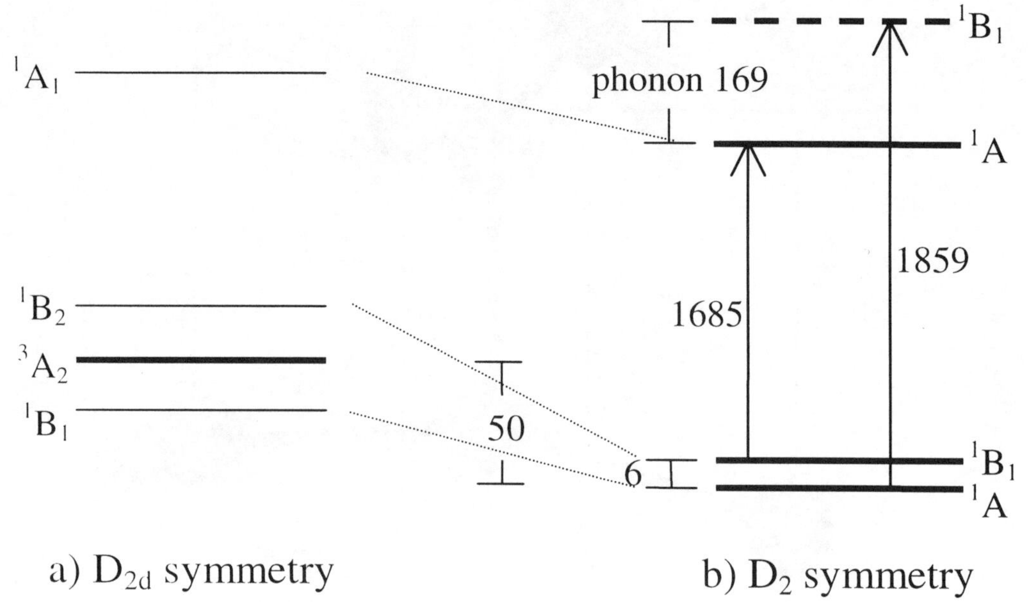


FIGURE 2 Schematic diagram (not to scale) of the energy levels of the split self-interstitial a) in $D_{2d}$ symmetry (which is a stable average configuration for the spin = 1 state), b) in $D_2$ symmetry (stable for the spin = 0 states). The transitions shown by vertical arrows are the only allowed electric dipole transitions and give rise to the observed optical absorption lines. All energies are in meV.

## D AGGREGATES OF SELF-INTERSTITIALS

Before the R2 EPR centre had been associated with the single interstitial, the di-interstitial had already been identified with another spin = 1 EPR centre, R1 [14]. The structure appeared to be that shown in FIGURE 3, with $C_{2h}$ symmetry. Calculations showed that the spin = 1 configuration was lower in energy by 0.2 eV than the spin = 0 equivalent. Also, the migration energy of the di-interstitial was calculated to be almost the same as that of the single interstitial, accounting for the experimental data, which show that R1 and R2 anneal out at similar temperatures (see Datareview A7.5).

The O3 EPR centre [15] originates from a spin = 1 defect with $C_2$ symmetry, which has the structure of three split interstitials at next neighbour positions (see Figure 2 of Datareview A7.5). Coomer et al [8] showed that this structure was stable and had the properties observed. They extended these interstitial aggregates to look at larger aggregates of four or more interstitials. A very stable arrangement of four interstitials, initially with parallel axes, arranged in a square, re-bonds such that all atoms are 4-coordinated with bond angles close to the ideal tetrahedral angle. Similar stable clusters of 4 interstitials are observed in silicon [16]. Further aggregation of these structures would lead to platelets similar to those observed in natural diamond.

## E EPR TENSORS

Since the spin-spin interaction tensor, D, is a function of the wavefunctions of the unpaired electrons, it can be calculated directly from the wavefunctions derived from the modelling of defects. Coomer et al [8] have developed this approach with great success for interstitial defects with spin 1. Comparisons between their predictions and the experimental values are shown in TABLE 1 for the single self-interstitial (R2), the interstitial pair (R1) and the tri-interstitial complex (O3).

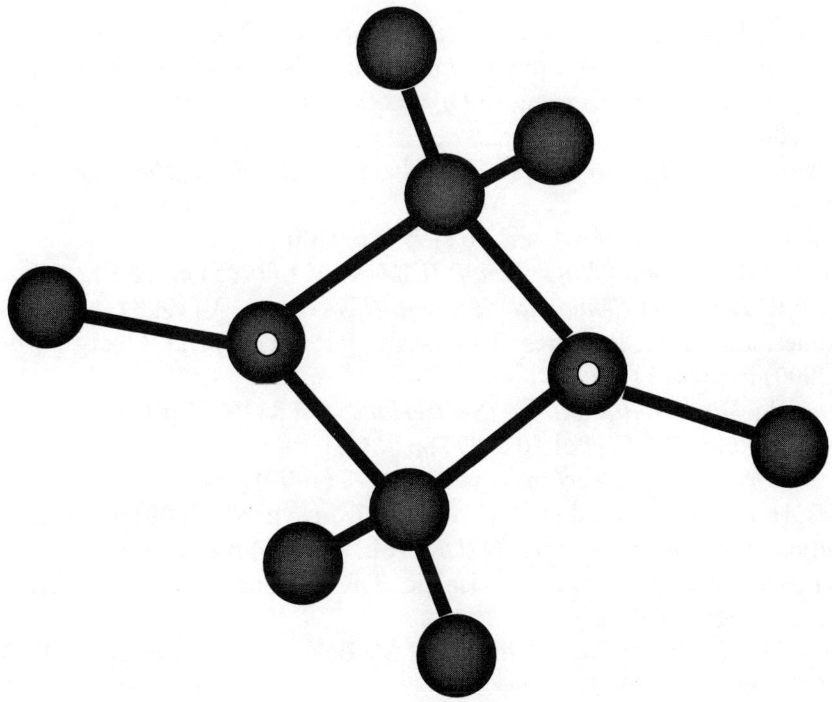

FIGURE 3 The structure of the di-interstitial which gives rise to the R1 EPR signal. The unpaired electron orbitals are on the two atoms marked with white dots, and are oriented perpendicular to the paper.

TABLE 1 The calculated and experimental principal values (GHz), in descending order, and principal directions ($n_1$, $n_2$, $n_3$) for the spin-spin tensor, D, for single, di- and tri-interstitial complexes.

| | | $D_1$ | $D_2$ | $D_3$ | $n_1$ | $n_2$ | $n_3$ |
|---|---|---|---|---|---|---|---|
| Single | calc. [8] | 1.48 | 1.48 | −2.94 | $[1\bar{1}0]$ | [110] | [001] |
| interstitial | R2 [2,17] | ±1.39 | ±1.39 | ∓2.78 | $[1\bar{1}0]$ | [110] | [001] |
| Di-interstitial | calc. [8] | 1.33 | 0.96 | −2.30 | $[1\bar{1}0]$ | [0.21,0.21,0.95] | [0.67,0.67,−0.30] |
| | R1 [14] | 1.41 | 1.40 | −2.81 | $[1\bar{1}0]$ | [0.21,0.21,0.95] | [0.67,0.67,−0.30] |
| Tri-interstitial | calc. [8] | 0.48 | 0.34 | −0.82 | [100] | [0.0,0.30,0.95] | [0.0,0.95,−0.30] |
| | O3 [15] | 0.46 | 0.35 | −0.81 | [100] | [0.0,0.30,0.95] | [0.0,0.95,−0.31] |

## F CONCLUSION

Theoretical modelling has been a major influence on the understanding of self-interstitials in diamond. Their aggregation, to form the observed di- and tri-interstitial defects, can be extrapolated to show the formation of larger platelets similar to those observed in natural diamond.

**REFERENCES**

[1] G.D. Watkins, K. Chow [ *Physica B (Netherlands)* vol.273-274 (1999) p.7-14 ]
[2] D.J. Twitchen et al [ *Physica B (Netherlands)* vol.273-274 (1999) p.644-6 ]
[3] C. Weisel, D. Peak, J.W. Corbett, G.D. Watkins, R.P. Messmer [ *Phys. Rev. B (USA)* vol.8 (1973) p.2906 ]
[4] A. Mainwood, F.P. Larkins, A.M. Stoneham [ *Solid-State Electron. (UK)* vol.21 (1978) p.1431 ]
[5] D.J. Chadi [ *Phys. Rev. B (USA)* vol.46 (1992) p.9400 ]
[6] S.J. Breuer, P.R. Briddon [ *Phys. Rev. B (USA)* vol.51 (1995) p.6984 ]
[7] R. Jones, P.R. Briddon [ *Semicond. Semimet. (USA)* vol.51A (1998) ch.6 ]
[8] B.J. Coomer, J.P. Goss, R. Jones, T.D. Shaw, P.R. Briddon, S. Oberg [ *Phys. Rev. B (USA)* vol.61 (2000) in press ]
[9] A Mainwood [ *Diam. Relat. Mater. (Switzerland)* vol.8 (1999) p.1560 ]
[10] J. Walker [ *J. Phys. C (UK)* vol.10 (1977) p.3867 ]
[11] G. Davies [ *Physica B (Netherlands)* vol.273-274 (1999) p.14-23 ]
[12] G. Davies, H. Smith, H. Kanda [ *Phys. Rev. B (USA)* vol.61 (2000) in press ]
[13] A.T. Collins, P.M. Spear [ *J. Phys. C (UK)* vol.19 (1986) p.6845 ]
[14] D.J. Twitchen, M.E. Newton, J.M. Baker, T.R. Anthony, W.F. Banholzer [ *Phys. Rev. B (USA)* vol.54 (1997) p.6988 ]
[15] D.C. Hunt, D.J. Twitchen, M.E. Newton, J.M. Baker, T.R. Anthony, W.F. Banholzer [ *Phys. Rev. B (USA)* in press (2000) ]
[16] N. Aria, S. Tekeda, M. Kohyama [ *Phys. Rev. Lett. (USA)* vol.78 (1997) p.4265 ]
[17] D.C. Hunt et al [ *Phys. Rev. B (USA)* in press (2000) ]

# A7.3 The vacancy in diamond

G. Davies

April 2000

## A INTRODUCTION

Two charge states, neutral and singly negative, are definitely known for the vacancy in diamond. The singly positive state has not yet been definitively identified. (An early identification [1] is now reassigned to the negative vacancy [2]; a recent possible identification using EPR is discussed in Datareview A7.4). In Sections B and C the optical properties of the two known charge states are reviewed. Much of the recent information on these centres has been derived from or supported by the electron paramagnetic resonance measurements of the neutral vacancy [3], the negative vacancy [2], the divacancy [4], and the self-interstitial [5] and di-self-interstitial [6]. These EPR data are discussed fully in Datareviews A7.4 and A7.5. Conversion of the vacancies between the two charge states is discussed in Section D, and the effects of the electron-phonon coupling on the centres, including the isotope effects, are considered in Sections E and F. The current understanding of the production of vacancies and their destruction through annealing is described in Sections G and H, and recent theoretical work in Section I. One increasing realisation is that vacancies influence attempts to dope or use diamond, and some relevant data are covered in Section J.

## B IDENTIFICATION AND PROPERTIES OF $V^0$

The neutral vacancy can be identified by the presence of the 'GR1' optical band, seen either in absorption or in luminescence. The band is characterised by a zero-phonon line at 1673 meV. It is produced by a transition from a doubly-degenerate orbital (E) state to a triply-degenerate (T) excited state, in a tetrahedral centre [7,8]; the vacancy does not undergo a static distortion to a lower symmetry. We define the strength A of the absorption in a zero-phonon line as the integral over the extent of the zero-phonon line:

$$A = \int d(h\nu)\, \mu(\nu)$$

where $\mu(\nu)$ is the absorption coefficient, in $cm^{-1}$, measured at 77 K and integrated over the zero-phonon absorption line, and $h\nu$ is the photon energy in meV. In these units the concentration in $cm^{-3}$ of neutral vacancies is [9]

$$[V^0] = A_{GR1} / (1.2 \pm 0.3) \times 10^{-16} \qquad (1)$$

The lifetime of the excited T level before de-excitation is

$$\tau = 2.55 \pm 0.1 \text{ ns} \qquad (2)$$

in the limit of low temperature, decreasing to 1 ns at room temperature and 0.4 ns at 470 K [10]. The temperature dependence is interpreted as implying that a non-radiative channel exists, and even in the limit of low temperature the GR1 band is estimated to be only ~1% efficient as a luminescence centre, in terms of photons out to photons absorbed [10]. This inefficiency contrasts with the near 100% efficiency of most vacancy-nitrogen centres. In addition to its zero-phonon line, the GR1 absorption band consists of a structured continuum to higher energy, resulting from the electron–phonon interaction, which is discussed in Section E. Consistent with the short lifetime, non-linearity in the GR1 absorption begins at high powers, of ~ 100 $MW/cm^2$ [11].

The neutral vacancy also produces a set of sharp lines, the GR2, 3... lines, each of which is a transition from the same E ground state of $V^0$ as for the GR1 band, and each terminates in a triply-degenerate excited state [12,13]. Photo-excitation into these excited states produces luminescence from the GR1 band. That is, the excited $V^0$ de-excites internally to the excited $T_2$ of the GR1 band before radiative emission occurs from the GR2, 3... levels. There is also a de-excitation process through charge release from the GR2, 3... levels that is discussed in the next paragraph. The relative probabilities of these de-excitation processes are not known. When the GR1 level is populated via the GR2, 3... levels the luminescence decay time is unchanged from EQN (2) [10].

There is no photoconductivity produced by the GR1 band. The GR2, 3... lines give rise to photoconductivity that has been positively identified as being produced by holes ($h^+$) [14]. The ionisation process is thus

$$V^0 + h\nu \rightarrow V^- + h^+, \qquad h\nu > 2880 \text{ meV}$$

with an undetermined upper limit on the photon energy (but greater than 3200 meV [14]). A possible mechanism for this effect is given in Section I.

The EPR signal of $V^0$ arises from a $^5A_2$ state. This state is observed only in diamonds that contain $V^-$ centres. The $^5A_2$ EPR state of $V^0$ is populated by photo-excitation [3], that is, $V^- + h\nu \rightarrow V^0\,(^5A_2) + e^-$ (see also Section D).

## C OPTICAL PROPERTIES OF $V^-$

Radiation damage of pure diamond results in the vacancies being produced only in the neutral charge state, and hence being observed in the GR1 optical absorption band; this result has recently been confirmed by positron annihilation experiments [15,16]. In nitrogen-containing type I diamond, an additional optical absorption band (the ND1 band) appears with its zero-phonon line at 3149 meV. The ND1 band occurs between a non-degenerate orbital (A) state and a triply degenerate (T) state of the tetrahedral centre [17]. In type Ib diamond (containing isolated nitrogen atoms with a donor energy of about 1.7 eV) the ND1 band dominates over the GR1 band. These results suggest that the ND1 band occurs at the negatively charged vacancy, $V^-$ [18], as confirmed recently by positron annihilation experiments [15,16]. Further confirmation comes from the correlation reported [19] between the strength of the ND1 absorption and the EPR signal from the negative vacancy, giving the relation between the concentration in $cm^{-3}$ of $V^-$ centres and the strength of the ND1 zero-phonon absorption (defined as in Section B) as [19]

$$[V^-] = A_{ND1} / (4.8 \pm 0.2) \times 10^{-16} \tag{3}$$

In diamonds that are recognisably type Ia, with nitrogen concentrations [N] exceeding $2 \times 10^{19}$ $cm^{-3}$, the ratio of GR1 to ND1 absorption appears to be invariant for all [N] at [20,21]

$$A_{ND1} \sim 0.7\ A_{GR1} \tag{4}$$

where the strength of the ND1 zero-phonon line is defined as in Section B. Converting EQN (4) into concentration values, the ratio of $V^-$ to $V^0$ production during 'room temperature' electron irradiation of type Ia diamond is

$[V^-] \sim 0.2\ [V^0]$

Why a fixed ratio is observed is not known.

Photoconductivity is observed from $V^-$ [14] (see Section D) and there is no known luminescence band. Excitation into the ND1 band increases the strength of the GR1 band of $V^0$, implying that the charge emission from $V^-$ is of an electron; there does not appear to have been a direct measurement of the sign of the carrier.

The EPR signal arises from the A orbital ground state of $V^-$, and ENDOR studies show that it is a $^4A_2$ state [2]. Density-functional theory predicts a $^4A_2 \rightarrow {}^4T_1$ transition at 3.3 eV [22].

## D INTERCONVERSION OF $V^0$ AND $V^-$

We have noted that

$$V^0 + h\nu \rightarrow V^- + h^+, \qquad h\nu > 2880 \text{ meV}$$

Conversely, photons absorbed into the ND1 band of $V^-$ reduce the ND1 absorption [23] and increase the GR1 ($V^0$) absorption. When the rate of change of the GR1 band is plotted as a function of the energy of the photons absorbed in the ND1 band, the plot is that of the absorption spectrum of the ND1 band [24], demonstrating that a simple conversion process is occurring:

$$V^- + h\nu \rightarrow V^0 + e^-, \qquad h\nu > 3150 \text{ meV}$$

with an undetermined upper limit on the photon energy, but greater than 3360 meV [24]. Heating in the dark at 500°C, below the onset of appreciable vacancy migration (Section H), reverses the changes by reducing the GR1 intensity and increasing the ND1 intensity [23]. When vacancies are destroyed by annealing, charge transfer also occurs, and will be discussed in Section H.

Fermi-level modifications of the charge states are also possible. Lawson et al [25] have shown that in type Ib diamonds (with isolated N atoms) both negative and neutral vacancies are observed, along with $N^+$ centres. Lawson et al could measure the concentration of $N^+$ centres, and assuming that each $V^-$ is produced by the reaction

$$V^0 + N \rightarrow V^- + N^+$$

the concentration of $V^-$ centres agrees closely with the value in EQN (3). It follows that the vacancies are negatively charged by a 'Fermi-level' effect from the (deep) donors formed by the N atoms. Charge cycling under optical excitation can therefore involve several defects, and the exact dynamics have not yet been thoroughly investigated (but see for example [26] for recent studies of vacancy-nitrogen centres).

## E VIBRONIC (ELECTRON–PHONON) COUPLING

Uniaxial stress measurements on the optical bands of $V^0$ [7,8] and of $V^-$ [17] have shown that the centres are tetrahedral; this result is confirmed by the EPR measurements on $V^0$ [3] and $V^-$ [2]. Consequently, neither $V^0$ nor $V^-$ undergoes a static lowering of its symmetry. However, a dynamic distortion is still possible and definitely occurs at $V^0$ [27]. The doubly degenerate E ground electronic state interacts with E modes of vibration. The T excited electronic state is allowed to couple to both E and T modes of vibration. It was observed by Lowther [28] (see also [29]) that the coupling to E modes makes the 1665 meV optical transition of $V^0$ into an allowed transition, and consequently the relative strengths of the 1673 and 1665 meV lines give the strength of the T × e coupling. Stress-induced dichroism measurements on the GR1 luminescence band show that the E modes have quanta centred on 40 meV and T modes centred on ~100 meV [30]. A good approximation to the Jahn–Teller effects in the E and T electronic states of $V^0$, involving E ('e') and T ('t') modes of vibration, is given by the following values [30]:

$$S_{Exe} = 2.5,\ S_{Txe} = 0.15,\ S_{Txt} = 0.3,\ h\omega_e = 40\ \text{meV},\ h\omega_t = 100\ \text{meV}$$

Here, the S parameters are the energy reductions through the linear electron-phonon terms indicated by the subscripts in units of the effective mode quanta $h\omega_e$ and $h\omega_t$. As at other centres in diamond, significant Jahn–Teller coupling only occurs in the ground electronic state of the centre.

At $V^0$ the electron-phonon coupling to totally symmetric A modes is negligible, consistent with the weak perturbations produced by hydrostatic stresses [7,8].

At $V^-$, the ground state is an orbital singlet [2,17] and so cannot couple to symmetry-lowering phonons. Any Jahn–Teller coupling in the excited T state is small, and the vibronic properties may be fitted using only coupling to A modes, with an energy reduction through the linear electron-phonon coupling of S = 3.1 and an effective phonon quantum of about 80 meV [31].

## F ISOTOPE EFFECTS

Isotope substitution results in the sharp lines of the vacancy changing in energy as in TABLE 1. The reference energies are for natural diamond, which is 98.9% $^{12}C$ and 1.1% $^{13}C$, measured at low temperature (in practice 77 K is low enough).

TABLE 1 Changes in energy of sharp-line structure at the vacancy on isotope substitution from $^{12}C$ to $^{13}C$.

| Label | Energy in $^{12}C$ (meV) | Energy in $^{13}C$ relative to $^{12}C$ (meV) | Ref |
|---|---|---|---|
| $V^0$ : GR1 | 1673 | +2.9 | [32] |
| $V^0$ : GR2 | 2880 | +7.0 | [33,34] |
| $V^0$ : GR3 | 2887 | +7.2 | [33,34] |
| $V^0$ : GR6a | 2958 | +7.2 | [33,34] |
| $V^0$ : GR6b | 2961 | +7.0 | [33,34] |
| $V^0$ : GR7a | 2976 | +7.1 | [33,34] |
| $V^0$ : GR7b | 2982 | +7.4 | [33,34] |
| $V^0$ : GR8a | 2998 | +6.9 | [34] |
| $V^0$ : GR8b | 3002 | +6.9 | [34] |
| $V^0$ : GR8c | 3005 | +6.9 | [34] |
| $V^-$ : ND1 | 3150 | + 4.0 ±0.2 | [33] |

We note that the similar shifts of the GR2, 3... lines confirm that they are electronic in origin and do not involve vibrational quanta (whose energies would be reduced to lower energy on replacing $^{12}C$ by $^{13}C$).

The several contributions to the isotope shifts of zero-phonon lines in diamond have been discussed recently [9]. The simplest to calculate is the effect of the fractional change in lattice volume ($\Delta V/V$ =

$-4.5 \times 10^{-4}$, at low temperature) [35] on changing the isotope from $^{12}C$ (or natural diamond, which is 98.9% $^{12}C$) to $^{13}C$. It produces a shift in the zero-phonon line of [a $(C_{11} + 2 C_{12}) \Delta V/V$] where $a(s_{xx} + s_{yy} + s_{zz})$ is the change in energy of the line as measured by applying stresses with tensor components $(s_{xx} + s_{yy} + s_{zz})$. The elastic constants of diamond are $(C_{11} + 2 C_{12}) = 13.3 \pm 0.15 \times 10^{11}$ Pa [36]. Since a = 3.56 meV / GPa for the ND1 line [17], the volume change shifts that line by 2 meV. The shifts for the other lines are an order of magnitude smaller since a = –0.25 meV / GPa for the 1665 meV line, a = 0.3 meV / GPa for the 1673 meV line [7,8], and a = 0.33 meV / GPa is the mean shift for the other GR lines [13]. This is one reason for the smaller shift of the GR1 line compared to the ND1 line (TABLE 1).

The vacancy, like the self-interstitial (see Datareviews A7.2 and A7.5), is an unusual defect in that it consists entirely of atoms of one species. Consequently, all the modes of vibration of a vacancy are changed, in the harmonic approximation, by $\sqrt{12/13}$ in a $^{13}C$ diamond relative to a $^{12}C$ diamond. At $V^0$, this change gives an opportunity to measure the effects of isotope substitution on a Jahn–Teller active system (Section E). In the $E \times e$ vibronic ground state of $V^0$ (Section E) the Jahn–Teller coupling reduces the zero-point energy in the doubly-degenerate vibrational e modes from $h\omega_e$ to $h\omega_e/2$ as the Jahn–Teller coupling increases. Since there is negligible Jahn–Teller coupling in the excited states (Section E), the optical absorption transition terminates in an excited state with $h\omega$ zero-point energy, and the transition energy includes a contribution + $h\omega_e/2$. Increasing the vibrational mass by changing the isotopes from $^{12}C$ to $^{13}C$ decreases this contribution. For $h\omega_e$ = 40 meV, the effect is to lower the transition energy, on isotope substitution, by a contribution $(1 - \sqrt{12/13})\, h\omega_e/2 = 0.8$ meV, reducing further the isotope effect on the GR1 line. The same type of mechanism may be important in producing the large isotope shift of the GR2, 3... lines. It is suggested that the excited state is formed from a core consisting of $V^+$ with an effective-mass electron orbiting it (Section I). Strong Jahn–Teller coupling in this excited state (in the $V^+$ centre) will enhance the isotope effect.

A further contribution to the shift comes from the electron-phonon coupling that is quadratic in the displacement in the modes; its effect may be estimated from the temperature-dependence of the zero-phonon lines [32] and typically amounts to a few meV.

## G PRODUCTION OF VACANCIES

To be observed a vacancy has to be stable at the time of observation. Typically the observation is made significantly later than the damage was produced, and any reduction in the concentrations of vacancies during that time reduces their apparent rate of production. In particular, 'correlated recombination' in which the displaced atom returns to the site it has vacated is very probable. The production of radiation damage has been surveyed by Palmer [37]. He emphasised that until the mobility of the self-interstitial was understood, the displacement mechanism would not be understood. Recently, confirmation of the identification of the self-interstitial [5] and understanding of the di-interstitial [6] has opened the way for a reconsideration of the displacement. The current situation is now outlined; however it has to be noted that substantial work is currently in progress.

The production of the vacancy in pure diamond has been studied in most detail using electrons of about 2 MeV as the irradiation source. Only neutral vacancies, giving the GR1 absorption band, are produced in pure diamond according to the optical spectra, and to the positron annihilation data [15]. Burgemeister et al [38] have established that the rate of producing the neutral vacancies by 1.5 MeV electrons over the range 0 to $8 \times 10^{19}$ $cm^{-2}$ is slightly sub-linear in dose, and is the same at 250 K and 500 K. The temperature independence has been confirmed by Twitchen et al [39] (using 2 MeV electrons and temperatures of 100 and 350 K). However, at lower beam energies, of 0.6 MeV, the data of Burgemeister et al [38] suggest that the production of $V^0$ is about 2/3 lower at 250 K than at 500 K.

The experimentally determined production rate of $V^0$ in pure diamond when irradiated with 2 MeV electrons at 100 or 350 K is 1.5 ±0.1 $cm^{-1}$ [39]. Again, we emphasise that this is the rate of producing stable end-products, rather than the displacement rate of the carbon atoms.

In type Ia diamonds, the pairs of nitrogen atoms are closely separated: a concentration of $10^3$ atomic ppm can be visualised as a cubic array of pairs of atoms separated by about 15 C-C bond lengths: all points in the crystal are then in close proximity to the nitrogen. Any interaction of the self-interstitial with the nitrogen will result in a modification of the correlated recombination and hence a change in the observed production rate. Vacancies generated near the nitrogen will be in an environment of increased strain, resulting in broader zero-phonon lines. During thermal annealing of the vacancies at 700°C a rapid loss of vacancy concentration occurs as some of the vacancies are unambiguously trapped by the nitrogen pairs (producing an increase in the 'H3' absorption band); this rapid annealing is accompanied by a fast reduction in the width of the GR1 zero-phonon line [21]. The fraction annealing through this fast process was found to be between 0.47 and 0.8 of the total vacancy concentration, depending on the annealing temperature but not on the (high) nitrogen concentration [21]. The implication is that there is a considerably increased rate of production near the nitrogen pairs; from the linewidth the faster annealing vacancies are within about 0.7 nm of a nitrogen pair. This region only contains 3% of the volume of the diamond, but appears to produce over 40% of the stable vacancies. The figures may not be reliable, since they are based partly on linewidth arguments, but the implication is that the pairs of nitrogen have a major influence on the production of stable damage. Other radiation damage products of currently unknown structure have been shown to be generated, by the same radiation dose, in concentrations that are proportional to the local nitrogen concentration [40].

Direct observation of localised production has not yet been achieved. Irradiation in a 2 MeV electron microscope has allowed damage aggregates (of unknown species) to be observed, but under irradiation at high temperature (1300°C) when the vacancies and self-interstitials are mobile, but this technique does not allow the initial damage product to be observed [41]. Confocal techniques have recently been widely used on vacancy-related centres, but the spatial resolution is only of the order of 1 µm (e.g. [42,43]). Radiation-induced diffusion during 5 MeV $He^+$ room-temperature irradiation has been suggested to occur over distances of several micrometres [44].

In contrast to type Ia diamonds, in type Ib diamonds, with single N atoms of donor ionisation energy about 1.7 eV, the production rate of neutral and negative vacancies totals only 0.5 ±0.05 $cm^{-1}$ for electrons of 1.9 MeV [19].

Most recent work on radiation damage has used electrons in the MeV range as the damaging agent. For applications in nuclear-particle detectors heavier particles are more significant. Recent estimates of the damage from 1 MeV neutrons is that 68 ±7 vacancies are initially produced per neutron $cm^{-1}$, and 35 ±7 survive the immediate correlated recombination [45].

## H ANNEALING THE VACANCIES

Some vacancies are destroyed during annealing by recombination with self-interstitials. In this case, the interstitial is the mobile species, with a migration energy determined from measurements of the optical signal from the self-interstitial of 1.68 ±0.15 eV [46], or 1.6 ±0.2 eV measured from the EPR signal of the self-interstitial [47]. Some of the self-interstitials migrate to vacancies, resulting in a reduction of the vacancy concentration by about 20 to 30% of its initial value (but actual values for the self-interstitial loss and the vacancy loss are not available in the published papers) [46,21].

A second annealing stage occurs when the vacancy becomes mobile. In pure diamond, the vacancies will aggregate to form a divacancy, recognised by the 'TH5' optical band [48] and the R4/W6 EPR centre [4]. In these diamonds the vacancies exist only in the neutral charge state, and so the measured activation energy for the loss of vacancies gives the migration energy of $V^0$. The value is 2.4 ±0.3 eV [21].

In type I (nitrogen-containing) diamond the vacancies migrate to trapping centres, which are confirmed to be the nitrogen centres [21], with a migration energy of 2.3 ±0.3 eV [21]. In these diamonds the vacancies are present in both neutral and negative charge states (Section C). The similarity in the migration energies in pure and nitrogen-doped diamond implies that the migration occurs in the neutral charge state. During the annealing, thermal equilibrium is re-established between the depleted neutral vacancies and the negative vacancies [21], and the detailed annealing behaviour may be understood in terms of mobile $V^0$ and static $V^-$ centres. For this process to be possible, the ground states of the $V^0$ and $V^-$ centres must be within a few kT of each other, where T is the annealing temperature (~800°C). This condition has been shown to be valid for the expected values of the Fermi energy [49].

In type Ia diamonds the nitrogen may be present either as substitutional pairs of atoms ('A' aggregates) or as groups of four N atoms ('B' aggregates); these structures compete for capture of the mobile Vs with capture radii R for V at A or at B in the ratio

$$R(V,A) / R(V,B) = 1.7$$

so that the smaller A aggregate has a larger capture radius [9]. For the single nitrogen atoms, the capture radius is estimated at about 20 times larger [50]. There does not appear to be any barrier for capture of the Vs at the nitrogen centres, in agreement with prediction [51].

The annealing stages in pure and type Ib diamonds have recently been observed using positron annihilation [15].

Other impurities are also expected to trap mobile vacancies (see Section J); it has been observed experimentally that annealing CVD irradiated diamond produces an increase in the 1.685 eV Si-V concentration [52].

## I THEORY OF VACANCIES

The energy of the vacancy is strongly determined by the interaction between the electrons at the vacancy, to the extent that the GR1 transition has ground and excited states derived from the same one-electron configuration ($a^2t^2$) [53]. Estimates of the energy levels suggest that the stable charge state is $V^+$ for a Fermi energy up to about 1 eV above the valence band (and so amply covering the Fermi energy of the boron acceptors), $V^0$ for mid-gap Fermi energies, and $V^-$ for Fermi energies within about 2 eV of the conduction band [49]. However, for type I diamond, the difference in the energy of the ground states is small enough that at temperatures used in typical annealing experiments (~800°C) both $V^0$ and $V^-$ may be populated, as observed experimentally (Sections C and H; [49]). The many GR2, 3... states (listed in TABLE 1) can be incorporated in this framework if one electron is excited sufficiently to leave a positively charged core ($V^+$) with the electron in an effective-mass orbital trapped by the Coulomb attraction to the core [53]. However, counter-intuitively, from the energies of the states, the weakly bound electron must be located close to the valence band, rather than the conduction band (the ionisation limit for a weakly-bound electron). An alternative proposal is that the GR2, 3… lines originate in a one electron transition at the neutral vacancy (an $a^2t^2 \rightarrow a^1t^3$ one-electron transition) followed by electron capture from the valence band into the vacant a state [54]; however this process involves states in the continuous valence band, which it can be argued would lead to considerably broader optical transitions than are observed [55].

The electronic structure of the vacancy can be used as the basis for understanding the properties of nitrogen-vacancy centres, using the approach that the additional N atoms lower the symmetry of the centre and also contribute additional electrons [56], although as we have seen it requires extension for the excited states of $V^0$ and for some other centres [22].

The size and sign of the lattice relaxation around the vacancy is still in dispute. Li and Lowther [57] calculate a 10% inward relaxation of the nearest neighbours and a 2% outwards relaxation of the next neighbours, with similar figures given by Joubert et al [58]; Breuer and Briddon [59] calculate a 13% outwards relaxation of the four immediate neighbours, and Zywietz et al [60] calculate a 7% outward relaxation. The sign is expected to be important in any elastic-energy barrier to interaction with another defect, as well as determining the perturbation of any defects that are near a vacancy. Analysis of the EPR signals of $V^0$ and $V^-$ suggests that the nearest neighbours in both cases (in the $^5A_2$ and $^4A_2$ states, respectively, used for the measurements) relax outwards by 15 and 10% (see Datareview A7.4). Note that from the strengths of the vibronic coupling (Section E) it is known that the bond lengths may differ by about 5% in different electronic states.

## J DEFECT ENGINEERING

Given the small interatomic spacing in diamond, impurities tend to cause local lattice expansion. This expansion can be compensated by the presence of a vacancy. Single substitutional nitrogen and pairs of nitrogen atoms create lattice expansion which is tolerated by the diamond (although both act as traps for mobile vacancies), but the 'N3' optical ('P2' EPR) centre which has 3 substitutional N atoms also contains one vacancy, and the 'B' aggregate of 4 nitrogen atoms also contains one vacancy. Other cases have become well-established recently, including the Si-V centre common in CVD diamond [61]. Silicon, along with nitrogen, in CVD diamond acts as a trap for the grown-in vacancies, producing the characteristic luminescence bands (e.g. [62]); vacancies are reported from positron-lifetime measurements to be very inhomogeneously distributed in as-grown CVD diamond [63] and to be surface-related from cathodoluminescence measurements [64].

Strain-relief processes are significant in samples implanted by ions, since the implantation generates vacancies which may become mobile and passivate (i.e. chemically combine with) the implanted species. Ab initio calculations on P-V complexes have demonstrated that they have lower energy than the isolated P and V; the P atom relaxes to the centre of the two adjacent lattice sites to give a 'semi-divacancy' structure [65]. There is no direct observation of this centre, but it has been suggested to be involved in donor-acceptor pair spectra in P-doped CVD diamond [66]. Ion-implantation doping of large atoms (Al, P) requires annealing strategies to overcome this problem of complexing with a vacancy [67].

Recently, transition metals in diamond have been recognised to act as traps for vacancies, with the 'NE4' EPR centre being identified with an Ni atom at the centre of the semi-divacancy structure [68]. These centres then act as a trap for mobile nitrogen atoms [69]; further, in both Ni- and Co-doped diamond the migration of nitrogen atoms is significantly enhanced [70].

Vacancy production should be minimised in any material used as a nuclear-particle detector; in these applications diamond appears to have significant advantages over other semiconductor materials with regard to damage produced by neutrons [71].

Finally, we note that vacancies may be the cause of the fast migration of nitrogen atoms, observed in irradiated type I diamonds [50, 51].

## K CONCLUSION

The optical properties of the two known charge states for the vacancy in diamond have been reviewed, and conversion of the vacancies between the two charge states was discussed. Electron–phonon coupling, isotope effects, production and annealing of vacancies and recent theoretical work have been described. Finally data relevant to defect engineering were considered.

## REFERENCES


[1] J.A. Baldwin [ *Phys. Rev. Lett. (USA)* vol.10 (1963) p.220-2 ]

[2] J. Isoya et al [ *Phys. Rev. B (USA)* vol.45 (1992) p.1436-9 ]

[3] J.A. van Wyk, O.D. Tucker, M.E. Newton, J.M. Baker, G.S. Woods, P. Spear [ *Phys. Rev. B (USA)* vol.52 (1995) p.12657-67 ]

[4] D.J. Twitchen, M.E. Newton, J.M. Baker, T.R. Anthony, W.F. Banholzer [ *Phys. Rev. B (USA)* vol.59 (1996) p.12900-10 ]

[5] D.C. Hunt et al [ *Phys. Rev. B (USA)* vol.61 (2000) p.3863-76 ]

[6] D.J. Twitchen, M.E. Newton, J.M. Baker, O.D. Tucker, T.R. Anthony, W.F. Banholzer [ *Phys. Rev. B (USA)* vol.54 (1996) p.6988-98 ]

[7] C.D. Clark, J. Walker [ *Proc. R. Soc. Lond. A (UK)* vol.234 (1973) p.241-57 ]

[8] G. Davies, C.M. Penchina [ *Proc. R. Soc. Lond. A (UK)* vol.338 (1974) p.359-74 ]

[9] G. Davies [ *Physica (Netherlands)* vol.273-274 (1999) p.15-23 ]

[10] G. Davies, M.F. Thomaz, M.H. Nazare, M.M. Martin, D. Shaw [ *J. Phys. C, Solid State Phys. (UK)* vol.20 (1987) p.L13-7 ]

[11] V.P. Mironov, E.F. Martinovich, V.A. Grigorov [ *Diam. Relat. Mater. (Switzerland)* vol.3 (1994) p.936-8 ]

[12] G. Davies, C.P. Foy [ *J. Phys. C, Solid State Phys. (UK)* vol.11 (1978) p.L547-53 ]

[13] C.P. Foy, G. Davies [ *J. Phys. C, Solid State Phys. (UK)* vol.13 (1980) p.L25-8 ]

[14] R.G. Farrer, L.A. Vermeulen [ *J. Phys. C, Solid State Phys. (UK)* vol.5 (1972) p.2762-8 ]

[15] A. Uedono et al [ *J. Phys., Condens. Matter (UK)* vol.11 (1999) p.4925-34 ]

[16] S. Dannefaer, D. Kerr [ *Diam. Relat. Mater. (Switzerland)* vol.7 (1998) p.339-41 ]

[17] G. Davies, E.C. Lightowlers [ *J. Phys. C, Solid State Phys. (UK)* vol.3 (1970) p.638-50 ]

[18] G. Davies [ *Nature (UK)* vol.269 (1977) p.498-500 ]

[19] D.J. Twitchen, D.C. Hunt, V. Smart, M.E. Newton, J.M. Baker [ *Diam. Relat. Mater. (Switzerland)* vol.8 (1999) p.1572-5 ]

[20] G. Davies [ *Proc. R. Soc. Lond. A (UK)* vol.336 (1974) p.507-23 ]

[21] G. Davies, S.C. Lawson, A.T. Collins, A. Mainwood, S.J. Sharp [ *Phys. Rev. B (USA)* vol.46 (1992) p.13157-70 ]

[22] R. Jones, J.P. Goss, P.R. Briddon, S. Oberg [ *Phys. Rev. B (USA)* vol.56 (1997) p.R1654-6 ]

[23] H.B. Dyer, L. du Preez [ *J. Chem. Phys. (USA)* vol.42 (1965) p.1898-906 ]

[24] G. Davies [ *Chemistry and Physics of Carbon* Eds. P.L. Walker, P.A. Thrower (Marcel Dekker Inc., New York, 1999) vol.13 p.1-144 ]

[25] S.C. Lawson, D. Fisher, D.C. Hunt, M.E. Newton [ *J. Phys., Condens. Matter (UK)* vol.10 (1998) p.6171-80 ]

[26] K. Iaboubovskii, G.J. Adriaenssens, M. Nesladek [ *J. Phys., Condens. Matter (UK)* vol.12 (2000) p.189-99 ]

[27] A.M. Stoneham [ *Solid State Commun. (USA)* vol.21 (1977) p.339-41 ]

[28] J.E. Lowther [ *Solid State Commun. (USA)* vol.20 (1976) p.933-5 ]

[29] G. Davies [ *Solid State Commun. (USA)* vol.32 (1979) p.745-7 ]

[30] G. Davies [ *J. Phys. C, Solid State Phys. (UK)* vol.15 (1982) p.L149-54 ]

[31] G. Davies [ *J. Phys. C, Solid State Phys. (UK)* vol.7 (1974) p.3797-809 ]

[32] A.T. Collins, G. Davies, H. Kanda, G.S. Woods [ *J. Phys. C, Solid State Phys. (UK)* vol.21 (1988) p.1363-76 ]

[33] S.C. Lawson [ PhD Thesis, University of London, 1992 ]

[34] G. Davies [ unpublished (1999) ]

[35] H. Holloway, K.L. Hass, M.A. Timor, T.R. Anthony, W.F. Banholzer [ *Phys. Rev. B (USA)* vol.45 (1992) p.6353 ]

[36] H.J. McSkimmin, P. Andreatch [ *J. Appl. Phys. (USA)* vol.43 (1972) p.2944-8 ]

[37] D.W. Palmer [ in *Properties and Growth of Diamond* Ed. G. Davies (INSPEC, IEE, London, UK, 1994) p.143-52 ]

[38] E.A. Burgemeister, C.A.J. Ammerlaan, G. Davies [ *J. Phys. C, Solid State Phys. (UK)* vol.13 (1980) p.L691-5 ]

[39] D.J. Twitchen, D.C. Hunt, M.E. Newton, J.M. Baker, T.R. Anthony, W.F. Banholzer [ *Physica B (Netherlands)* vol.273-274 (1999) p.628-31 ]

[40] I. Kiflawi, G. Davies, D. Fisher, H. Kanda [ *Diam. Relat. Mater. (Switzerland)* vol.8 (1999) p.1576-80 ]

[41] R. Oshima, C.Y. Nie, M. Kmatsu, H. Mori [ *Mater. Sci. Forum (Switzerland)* vol.196-201 (1995) p.1237-42 ]

[42] J. Martin, W. Grebner, W. Sigle, R. Wannemacher [ *J. Lumin. (Netherlands)* vol.83-4 (1999) p.493-7 ]

[43] A. Drabenstedt et al [ *Phys. Rev. B (USA)* vol.60 (1999) p.11503-8 ]

[44] J.W. Ager et al [ *J. Appl. Phys. (USA)* vol.76 (1994) p.4050-3 ]

[45] A. Mainwood, J. Cunningham, D. Usher [ *Mater. Sci. Forum (Switzerland)* vol.258 (1997) p.787-92 ]

[46] L. Allers, A.T. Collins, J. Hiscock [ *Diam. Relat. Mater. (Switzerland)* vol.7 (1998) p.228-32 ]

[47] D.J. Twitchen et al [ *Physica B (Netherlands)* vol.273-274 (1999) p.644-6 ]

[48] C.D. Clark, R.W. Ditchburn, H.B. Dyer [ *Proc. R. Soc. Lond. A (UK)* vol.237 (1956) p.75 ]

[49] A. Mainwood, A.M. Stoneham [ *J. Phys., Condens. Matter (UK)* vol.9 (1997) p.2453-64)

[50] A.T. Collins [ *J. Phys. C, Solid State Phys. (UK)* vol.13 (1980) p.2641-50 ]

[51] A. Mainwood [ *Phys. Rev. B (USA)* vol.49 (1994) p.7934-40 ]

[52] A.T. Collins, L. Allers, C.J.H. Wort, G.A. Scarsbrook [ *Diam. Relat. Mater. (Switzerland)* vol.3 (1994) p.932-5 ]

[53] A.M. Stoneham, A. Mainwood [ in *Proc. Conf. on Shallow-level Centers in Semiconductors* Eds. C.A.J. Ammerlann, B. Pajot (World Scientific Publishing Company, Singapore, 1997) p.165-70 ]

[54] J.E. Lowther [ *Phys. Rev. B (USA)* vol.48 (1993) p.11592-601 ]

[55] A. Mainwood [ *J. Phys. C, Solid State Phys. (UK)* vol.11 (1978) p.L449-52 ]

[56] J.E. Lowther [ *J. Phys. Chem. Solids (UK)* vol.45 (1984) p.127-31 ]

[57] L.H. Li, J.E. Lowther [ *Phys. Rev. B (USA)* vol.53 (1996) p.11277-80 ]

[58] D.P. Joubert, L. Li, J.E. Lowther [ *Solid State Commun. (USA)* vol.100 (1996) p.561-3 ]

[59] S.J. Breuer, P.R. Briddon [ *Phys. Rev. B (USA)* vol.51 (1995) p.6984-94 ]

[60] A. Zywietz, J. Furthmuller, F. Bechstedt [ *Phys. Status Solidi B (Germany)* vol.210 (1998) p.13-29 ]

[61] J.P. Goss, R. Jones, S.J. Breuer, P.R. Briddon, S. Oberg [ *Phys. Rev. Lett. (USA)* vol.77 (1996) p.3041-4 ]

[62] K. Iakoubovskii, G.J. Adriaenssens [ *Phys. Status Solidi A (Germany)* vol.172 (1999) p.123-9 ]
[63] S. Dannefaer, W. Zhu, T. Bretagnon, D. Kerr [ *Phys. Rev. B (USA)* vol.53 (1996) p.1979-84 ]
[64] L. Allers, A. Mainwood [ *Diam. Relat. Mater. (Switzerland)* vol.7 (1998) p.261-5 ]
[65] R. Jones, J.E. Lowther, J. Goss [ *Appl. Phys. Lett. (USA)* vol 69 (1996) p.2489-91 ]
[66] J. te Nijenhuis, S.M. Olsthoorn, W.J.P. van Enckevort, L.J. Giling [ *J. Appl. Phys. (USA)* vol.82 (1997) p.419-22 ]
[67] J.F. Prins [ *Diam. Films Technol.* (Japan) vol.8 (1998) p.181-94 ]
[68] V.A. Nadolinny, A.P. Yelisseyev, O.P. Yuryeva, B.N. Fergelson [ *Appl. Magn. Reson. (Austria)* vol.12 (1997) p.543-54 ].
[69] V.A. Nadolinny et al [ *J. Phys., Condens. Matter (UK)* vol.11 (1999) p.7357-76 ]
[70] D. Fisher, S.C. Lawson [ *Diam. Relat. Mater. (Switzerland)* vol.7 (1998) p.299-304 ]
[71] A. Mainwood et al [ *J. Phys. D, Appl. Phys. (UK)* vol.28 (1995) p.1279-83 ]

# A7.4 EPR measurements on the negatively charged, neutral and positively charged vacancies and the nearest neighbour divacancies in diamond

M.E. Newton, J.M. Baker and D.J. Twitchen

January 2000

## A INTRODUCTION

A vacancy is formed when a carbon atom is displaced from its lattice site, leaving four dangling bonds on its neighbouring carbon atoms. Vacancies are formed by irradiation with energetic particles, such as electrons and neutrons, or during annealing or by the motion of dislocations or slip planes. They play a significant role in the diffusion of nitrogen in diamond.

The vacancy is stable at room temperature, and its effects have been known for 90 years [1]. It took 17 years for the GR1 optical absorption, discovered in 1956 [2], which gives rise to a characteristic blue/green colour, to be firmly associated with the isolated neutral vacancy, $V^0$, with $T_d$ symmetry, requiring the measurement of the effects of applied uniaxial stress to establish the $^1E$ symmetry of the ground state [3,4]. The ND1 optical absorption was similarly shown to be due to an isolated charged vacancy, also with $T_d$ symmetry [5].

It also took a long time (about 30 years) to show that the electron paramagnetic resonance (EPR) spectra labelled S1 and S2, first observed in 1963 [6], were due to one centre, $V^-$ [7], which is correlated with the ND1 optical absorption. EPR is not observed in the ground state of $V^0$, as it has $S = 0$. It also took 34 years to show that the EPR centre labelled R4, first observed in 1964 (then labelled the e centre) [8], is a neutral nearest neighbour divacancy $(V—V)^0$ [9].

The reasons for these long delays in interpreting EPR spectra are two-fold. First, the early measurements were made on natural diamonds, and interpretation of EPR centres was confused by unknown impurity content. Secondly, by far the most valuable information about EPR centres has come from hyperfine interactions (HFI) between the unpaired electron and nuclei with magnetic moments. The low natural abundance of $^{13}C$ (1.1%), the only such isotope of carbon, has made its HFI difficult to measure. The near 100% natural abundance of $^{14}N$ is why the principal nitrogen containing centre, the P1 EPR centre, the single substitutional nitrogen atom, was understood much earlier [9]. Recent advances have been possible because of the availability of synthetic diamonds of high purity and enriched to 5–10% in $^{13}C$.

In Section B we review the information which can be deduced from the EPR spectra. Then we discuss the results for centres involving a single vacancy in Section C, and centres involving more than one vacancy in Section D. Results for complexes of vacancies and impurities are discussed in Datareview A7.3. The theory of the vacancy sites is discussed in Section E.

## B THE INFORMATION FROM EPR

The initial stage of analysis of an EPR spectrum is to describe it by a spin Hamiltonian [11]. For the centres discussed in this Datareview, this has the form:

$$H = g\mu_B\mathbf{B.S} + \mathbf{S.D.S} + (\mathbf{I.A.S} - g_N\mu_N\mathbf{B.I}) \quad (1)$$

where **S** is the effective electron spin, and **I** is the nuclear spin of any nucleus coupled to the electron. Neglecting the last two terms for the moment, there are (2S) EPR transitions corresponding to selection rules $\Delta M = 1$, between the (2S+1) values of the azimuthal quantum number M associated with **S**.

The largest term in the spin-Hamiltonian describing an EPR spectrum at the commonly used microwave frequency ($\nu$) of ~ 9 GHz and above is the Zeeman interaction $g\mu_B\mathbf{B.S}$. For S = ½ this gives a single spectral line at $B = h\nu/g\mu_B$. In diamond, g is usually very close to the free spin value $g_e = 2.0023$. Departures from $g_e$ are much smaller than in silicon because they are due to spin–orbit coupling, which is five times smaller than in Si, which introduces orbital angular momentum from excited states, which are typically five times further away in energy. The result of this is that the spectra of all defects with S = ½ are superimposed, and hence difficult to differentiate. (In principle, it would be possible to exploit a small difference in g–value by measuring at very high frequency and field **B**.) For S>½, the fine structure term **S.D.S** produces (2S) lines which are separated by ~ $D/g\mu_B$ and so are easily recognised. Also, the value of S is easily determined by counting the number of lines.

Sites with $T_d$ symmetry are special as D = 0, and all (2S) EPR transitions are superimposed on one another (and on the lines of other defects with g = 2), so that it is not possible to determine the value of S from the EPR.

Now let us consider the last two terms in EQN (1). These are not required if all neighbouring nuclei are $^{12}C$, but for those sites which have $^{13}C$ in one of the neighbouring sites they give rise to two hyperfine lines separated by $A/g\mu_B$. For an isotopic fraction x of $^{13}C$ these lines have intensity ½x. For small x there is negligible possibility of there being two $^{13}C$ neighbours, so each neighbouring carbon atom site gives rise to such a pair of hyperfine lines. A study of the separations of these lines for all directions of **B** allows the measurement of the **A**–matrix, and the simpler measurement of their relative intensities for **B** along crystal symmetry directions gives the number of equivalent neighbouring sites.

The natural abundance of $^{13}C$ (x = 0.011) is a little too low for easy measurement of the HFI, so synthetic diamonds typically with values of x between 0.05 and 0.1 have been used to facilitate measurements.

Each term **I.A.S** is usually close to axially symmetric $\{A_\parallel S_z I_z + A_\perp(S_x I_x + S_y I_y)\}$, which for the next stage of the analysis can be rewritten $A_s\mathbf{I.S} + A_p[I_z S_z - (1/3)\mathbf{I.S}]$, where $A_s$ arises from unpaired s–electrons and $A_p$ from unpaired p–electrons on the carbon atom. From the magnitudes of these parameters, the amplitude $\alpha$ of unpaired 2s and $\beta$ of unpaired 2p spin density on each nucleus may be estimated, and the direction z gives the orientation of the p-electron component [12]. The ratio $\alpha/\beta = \lambda$ may be related to the angle $\theta$ between the unpaired p component and the other three covalent bonds by $\cos^2\theta = (3+2\lambda^2)^{-1}$, and hence indicates the displacement of the carbon atom [13].

For a site with $T_d$ symmetry, these hyperfine lines are superimposed for S>½, so one cannot tell the spin S from the HFS of the EPR. A technique for measuring the separation of the hyperfine levels is required to show that S>½, because the frequencies of transitions with selection rules $\Delta M = 0$, $\Delta m = 1$ are $h\nu = AM \pm g\mu_N B$. Hence there are (S+1) such lines which gives the value of S. This may be done either by electron nuclear double resonance (ENDOR) [14] or electron spin echo envelope modulation (ESEEM) [15].

## C EPR SPECTRA OF AN ISOLATED VACANCY

### C1 The Negative Vacancy $V^-$

The EPR spectra labelled S2 and S1 [6] arise from $^{13}C$ HFS due to nearest neighbour (NN) and next nearest neighbour (NNN) $^{13}C$ atoms respectively of the negative vacancy, but S2 was initially thought to arise from a $^{14}N$ impurity [16]. The true origin of S1 and S2 was measured only by using diamonds isotopically enriched to 10% $^{13}C$ which showed clearly that S2 corresponded to hyperfine interaction with 4 equivalent (NN) $^{13}C$ atoms, and that S1 corresponded to hyperfine interaction with 12 equivalent (NNN) $^{13}C$ atoms [6]. This, coupled with the $T_d$ symmetry, is sufficient to show that the defect is centred on a substitutional site. The only other site with $T_d$ symmetry is interstitial, and has 6 NNN atoms. The spin Hamiltonian parameters for $V^-$ are given in TABLES 1 and 2.

TABLE 1 The spin, symmetry and principal values of the **g**- and **D**-matrices for centres discussed: D and A values are in MHz. The absolute signs of the components of **D** for W29 are unknown. T(K) is the temperature of measurement. In brackets after the symmetry is the number of symmetry related sites. The directions of the principal values of the matrices are indicated, either by a crystallographic direction of the unique principal value for axial symmetry, or by direction cosines in the cubic crystal axis system.

| Centre | T(K) | Spin | Symmetry (no. of sites) | **D**-matrix (direction) | | | | **g**-matrix (direction) | | | |
|---|---|---|---|---|---|---|---|---|---|---|---|
| $V^-$ | 4 | 3/2 | $T_d$ (1) | 0 | | | | 2.0027 | | | |
| | | | | | | | | 2.0027 | | | |
| | | | | | | | | 2.0027 | | | |
| $V^0$ | 5 | 2 | $T_d$ (1) | 0 | | | | 2.0033 | | | |
| | | | | | | | | 2.0033 | | | |
| | | | | | | | | 2.0033 | | | |
| NIRIM-3 ($V^+$) | 4 | ½ | $D_{2d}$ (3) | | | | | 2.0009 | [100] | | |
| | | | | | | | | 2.0024 | | | |
| | | | | | | | | 2.0024 | | | |
| $(V—V)^0$ | 33 | 1 | $C_{2h}$ (12) | +103 | 0 | −0.7071 | 0.7071 | 2.0022 | 0 | −0.7071 | 0.7071 |
| | | | | +206 | −0.8111 | 0.4136 | 0.4136 | 2.0026 | −0.7771 | 0.4450 | 0.4450 |
| | | | | −310 | 0.5850 | 0.5735 | 0.5735 | 2.0013 | 0.6293 | 0.5495 | 0.5495 |
| W29 $(V—V)^-$ | 100 | 3/2 | $C_{2h}$ (12) | +297 | 0 | −0.7071 | 0.7071 | 2.0019 | 0 | −0.7071 | 0.7071 |
| | | | | +156 | −0.785 | 0.192 | 0.192 | 2.0020 | 0.210 | 0.691 | 0.691 |
| | | | | −452 | 0.619 | 0.556 | 0.556 | 2.0024 | 0.956 | −0.207 | −0.207 |

TABLE 2 Principal values of hyperfine parameters (in MHz) for the centres discussed. In all cases the principal direction of $A_{\|}$ is along a <111> direction. The number of equivalent sites is indicated in column 2. The last column is the total unpaired spin density on the set of equivalent neighbours. No $^{13}C$ hyperfine structure has been observed for W29.

| Centre | | No. of sites | $A_{\|}/h$ | $A_{\perp}/h$ | $A_s/h$ | $A_p/h$ | α | β | λ | η |
|---|---|---|---|---|---|---|---|---|---|---|
| $V^-$ | **A(1)** | 4 | 141.5 | 81.9 | 101.8 | 19.9 | 0.027 | 0.1852 | 6.9 | 0.849 |
| | **A(2)** | 12 | 13.5 | 9.5 | 10.8 | 1.3 | 0.0029 | 0.0121 | 4.2 | 0.150 |
| $V^0$ | **A(1)** | 4 | 91.13 | 35.03 | 53.73 | 18.7 | 0.0142 | 0.1741 | 12.3 | 0.753 |
| | **A(2)** | 12 | 12.01 | 8.36 | 9.58 | 1.22 | 0.0025 | 0.0114 | 4.6 | 0.167 |
| $(V^+)$ | **A(1)** | 4 | 32.8 | 21.0 | 24.9 | 3.9 | 0.0066 | 0.0363 | 5.5 | 0.172 |
| $(V—V)^0$ | **A(1)** | 4 | 114 | 56 | 75 | 19 | 0.0199 | 0.1769 | 8.89 | 0.787 |
| | **A(2)** | 2 | 13 | 13 | 13 | 0 | 0.0034 | | | 0.0068 |

A careful study of the HFI showed that about 100% of the unpaired spin density is on the four nearest neighbours. The value of $\lambda$ suggests that there is considerable outward displacement (~ 10%, 0.015 nm) of the four NN atoms, but little displacement of the 12 NNN atoms [6].

Because of the $T_d$ symmetry, ENDOR was required to determine that the spin S = 3/2, which shows that the vacancy is negatively charged with a $^3A_2$ ground state [6]. As the intensity of the EPR absorption can be quantitatively related to the absolute concentration of $V^-$ centres, to an accuracy of ~ 10%, a comparison of EPR intensity with the optical absorbance of ND1 allowed the oscillator strength of the line to be determined [26]: see TABLE 3. This indirectly calibrated the oscillator strength of the GR1 ZPL of the neutral vacancy [26] (see TABLE 3).

TABLE 3 The correlation of EPR and optical lines.

| Centre | Symmetry | EPR | Optical | Absorp. | Lumin. | Oscillator strength $\times 10^{16}$ meV $cm^{-2}$ |
|---|---|---|---|---|---|---|
| $V^0$ | $T_d$ | not in ground state $^5A_2$, S = 2 [17] | GR1 1.673 eV | [2] | [41,42] | 1.2(3) [26] |
| $V^-$ | $T_d$ | $^4A_2$, S = 3/2 [6] | ND1 3.150 eV | [17] | X | 4.8(2) [26] |
| $V^+$ | $D_{2d}$ | NIRIM-3, S = ½ [22] | ? | | | |
| $(V—V)^0$ | $C_{2h}$ | S = 1 [32] | TH5 | [29] | X | |
| $(V—V)^-$ | $C_{2h}$ | W29, S = 3/2 [35] | ? | | | |

## C2 The Neutral Vacancy $V^0$

Although there is no EPR in the $^1E$ ground state of $V^0$, optical excitation of a diamond with hν>3.1 eV (the ND1 absorption of $V^-$) produces an EPR spectrum which is similar to S1 and S2 [17]. Similar measurements to those which showed that S = 3/2 for $V^-$ showed that the new spectrum corresponds to S = 2, and a site with 4 NN and 12 NNN atoms with similar hyperfine parameters to $V^-$. The HFI of NN atoms show that ~ 75% of the unpaired electron is located on these atoms. This, together with the substitutional $T_d$ site, confirms that the centre is also a vacancy. Even spin indicates an even number of electrons, and S = 2 requires at least four unpaired electrons. This corresponds to the $^5A_2$ excited state of $V^0$. The spin Hamiltonian parameters are given in TABLES 1 and 2. The generation of this centre by optical excitation of $V^-$ is consistent with the absence of luminescence from $V^-$ because it de–excites by emission of an electron [18].

Analysis of the values of $\lambda$ shows that, like $V^-$, there is outward relaxation of the four NN atoms by ~ 15%, 0.023 nm, and little displacement of the NNN atoms.

Lattice relaxation at a vacancy has been calculated both using Car–Parrinello molecular–dynamics [19] and using a cluster model, with local-density-pseudo-potential theory [20]. An outward displacement of the NN of 0.011 nm was predicted by the former for $V^0$, and of 0.02 nm by the latter for both $V^0$ and $V^-$, in good agreement with the measurements.

The $^5A_2$ state is not observed in diamonds which contain $V^0$ but no $V^-$, and it is observed only in a minority of diamonds which contain both $V^0$ and $V^-$ [17]; it is not known why.

No other EPR attributable to the neutral vacancy has been observed in diamonds exhibiting the GR1 absorption of the neutral vacancy. Theory predicts that the $^3T_1$ state should be fairly low lying (~ 200 meV) [21]. This state would be observable by EPR were it sufficiently populated. That it has not been observed at 300 K suggests that this state probably lies at least 100 meV above the ground state.

### C3 The Positive Vacancy $V^+$

The EPR centre NIRIM–3, produced by electron irradiation or proton implantation of synthetic boron doped (type IIb) diamonds, has been tentatively attributed to $V^+$ [22]. The centre has S = ½ and tetragonal symmetry, as one would expect for Jahn–Teller distortion of the three electron state $a_1^2t_2$; the g–values are given in TABLE 1. In boron doped diamond, the GR1 absorption of the neutral vacancy is not observed until the radiation dose is sufficient to exceed that needed to provide compensation for the boron [23]. It is therefore suggested that electrons are transferred from the vacancies produced to the boron impurities to produce $B^-$.

However, the unpaired spin density on the four NN carbon atoms of NIRIM–3 is only ~ 17%, showing either that the unpaired electron is highly delocalised, because its energy level is very close to the valence band, or that the interpretation is incorrect and the majority of the unpaired electron resides on some central substitutional impurity with zero nuclear spin.

### C4 Other Centres Involving a Single Vacancy

The GR1 absorption in diamonds which have been isochronally annealed at temperature T, has decreased by two orders of magnitude between T = 900 and 1000 K. This is interpreted as indicating that the vacancy becomes mobile around 900 K [24]. The interstitial becomes mobile at a rather lower temperature (700 K), and only some of the vacancies are lost at that temperature by recombination [25]. The amount lost in this way depends upon the concentration of interstitials, and is not more than 45%. The remainder of the vacancies are thought either to aggregate, to attach themselves to impurities, or to diffuse to the surface.

There is evidence that vacancies may coalesce with substitutional impurities. There is a great deal of evidence of such aggregations in nitrogen containing diamonds [19] (see also Datareview A7.3). W15 (EPR), H3 and H4 (optical) centres appear to be a vacancy attached to $N_s$, an A–centre and a B–centre, respectively. These defects are more stable than the divacancy, and the other possible vacancy aggregates. They are formed when $V^0$ becomes mobile. The rate of production of vacancies by irradiation in type Ib diamond [26] and in type Ia diamond [24] is higher than in type IIa. In type Ib diamonds, most isolated vacancies are in the form of $V^-$ due to charge transfer from $N_s^0$: in diamonds without other defects to confuse the issue, the concentration of $N^+$ is the same as that of $V^-$ [26].

## D EPR OF CENTRES WITH MORE THAN ONE VACANCY

### D1 The Neutral Di-Vacancy $(V—V)^0$

In irradiated type IIa diamond, there are many EPR centres which anneal in (and out again) in the temperature region above 900 K [27], which are thought therefore to involve vacancies, but the maximum concentrations of most of them are two orders of magnitude less than the initial concentration of $V^0$. However, there is one centre, R4 [7,28], also independently discovered and labelled W6 [16], which reaches its maximum concentration at about 1000 K at about 40% of the initial concentration of $V^0$. R4/W6 was first reported in 1964 [7]. The intensity of its EPR signal correlates with the TH5 optical absorption band [29], which has been shown to involve two vacancies [30]. It has been produced in low concentration by irradiation at room temperature by 2 MeV electrons (enough energy to eject two carbon atoms simultaneously), and there is some evidence that it is produced only after a threshold dose [31], and that it is not produced by low energy electrons. The concentration is greatly enhanced by annealing to ~ 1000 K. It is not produced in type Ib diamonds for two reasons: (a) $N_s$ is a competing trap for mobile vacancies [9], and (b) where a di–

vacancy is formed, $N_s$ donates an electron to form $(V—V)^-$ (see Section D2). The production rates are lower in type IaA and type IaB diamonds because of the production of H2/H3 and H4 centres respectively (Datareview A7.3).

Although R4/W6 has been tentatively associated with a divacancy for some time, it took 34 years for its identification as a nearest neighbour pair of vacancies, $(V—V)^0$, to be confirmed [9]. This again required measurement of HFI for neighbouring $^{13}C$. The divacancy has six dangling bonds on its NN carbon atoms. R4/W6 has S = 1, corresponding to an even number of unpaired electrons, taken to be $(V—V)^0$. However, this centre does not have the $D_{3d}$ symmetry one might expect for the divacancy (see FIGURE 1), but has $C_{2h}$ symmetry. The spin Hamiltonian parameters are given in TABLES 1 and 2. 80% of the unpaired electron is located on four nearest neighbour atoms out of the $C_{2h}$ plane of symmetry (a, a', d and d' in FIGURE 1). There is only a small HFI with the two in-plane atoms (b and b' in FIGURE 1). The value of $\lambda$ indicates an outward relaxation of the four out–of–plane neighbours of similar magnitude (~ 15%) to those measured for $V^0$ and $V^-$.

The **D** matrix is strongly dependent on temperature. It appears to become axially symmetric at 400 K, but it is not clear whether this is an accidental effect, or whether the symmetry really becomes $D_{3d}$. The **D** matrix is unexpectedly small, and quite non–axial at low temperatures. This, and its marked temperature dependence, may indicate competition between two large mechanisms of opposite sign: this is still an open question.

The EPR lines are very wide due to rapid spin–lattice relaxation via an Orbach process [32], indicating a low lying excited state at ~ 20 meV.

In silicon, $(V—V)^-$ or $(V—V)^+$ states have been identified, both with S = ½, but no $(V—V)^0$ states have been found, presumably because S = 0. The negative and positive divacancy states in silicon were both found to have $C_{2h}$ symmetry [33], but, unlike the neutral divacancy in diamond, the unpaired electron is primarily located in the two NN atoms in the {110} symmetry plane (b and b' in FIGURE 1). No similar states have been found in diamond, but these would have S = ½, and small g–shifts might make them difficult to recognise. A centre with S = 3/2 which may be $(V—V)^-$ is described below.

## D2 The Negatively Charged Di–Vacancy $(V—V)^-$

The EPR centre W29, which has been observed in all electron and neutron irradiated type Ib diamonds, but only in some type Ia diamonds, and which anneals in and out in the same temperature range as R4/W6, has been attributed to $(V—V)^-$ [34]. The spin Hamiltonian parameters are given in TABLE 1. That the spin of this centre is S = 3/2 has been unambiguously confirmed by observation of a $\Delta M_S = 3$ transition at low magnetic field. The assignment to a di–vacancy is indicated by features of the spectrum as well as the similar annealing behaviour to $(V—V)^0$. The **D** matrix is similar in form and symmetry to that of R4/W6, and also its line width shows a similar behaviour. In these diamonds, in addition to W29, there is a strong EPR spectrum of $V^-$ produced by irradiation, together with a decrease in the intensity of P1. The rate at which W29 is formed is greater than the rate of formation of $(N—V)^-$.

## D3 Other Centres With More Than One Vacancy

R4/W6 anneals out at 1200–1300 K. In silicon, it is known that the divacancy diffuses as a unit [33], but it is not known whether this is so in diamond, or whether it dissociates. There is a succession of new EPR centres with S = 1 which anneal in (and out) at higher temperatures, which have been tentatively attributed to strings of vacancies [28,16]. This has not been confirmed, and it is puzzling that all of the centres have $C_{2v}$ symmetry in the {110} plane. The plane of symmetry suggests that the vacancies aggregate in a {110} plane, but $C_{2v}$ symmetry implies that they correspond to odd numbers of vacancies only (even numbers would give a site of $C_{2h}$ symmetry). More work needs to be done to identify these defects and the mechanisms by which they are created.

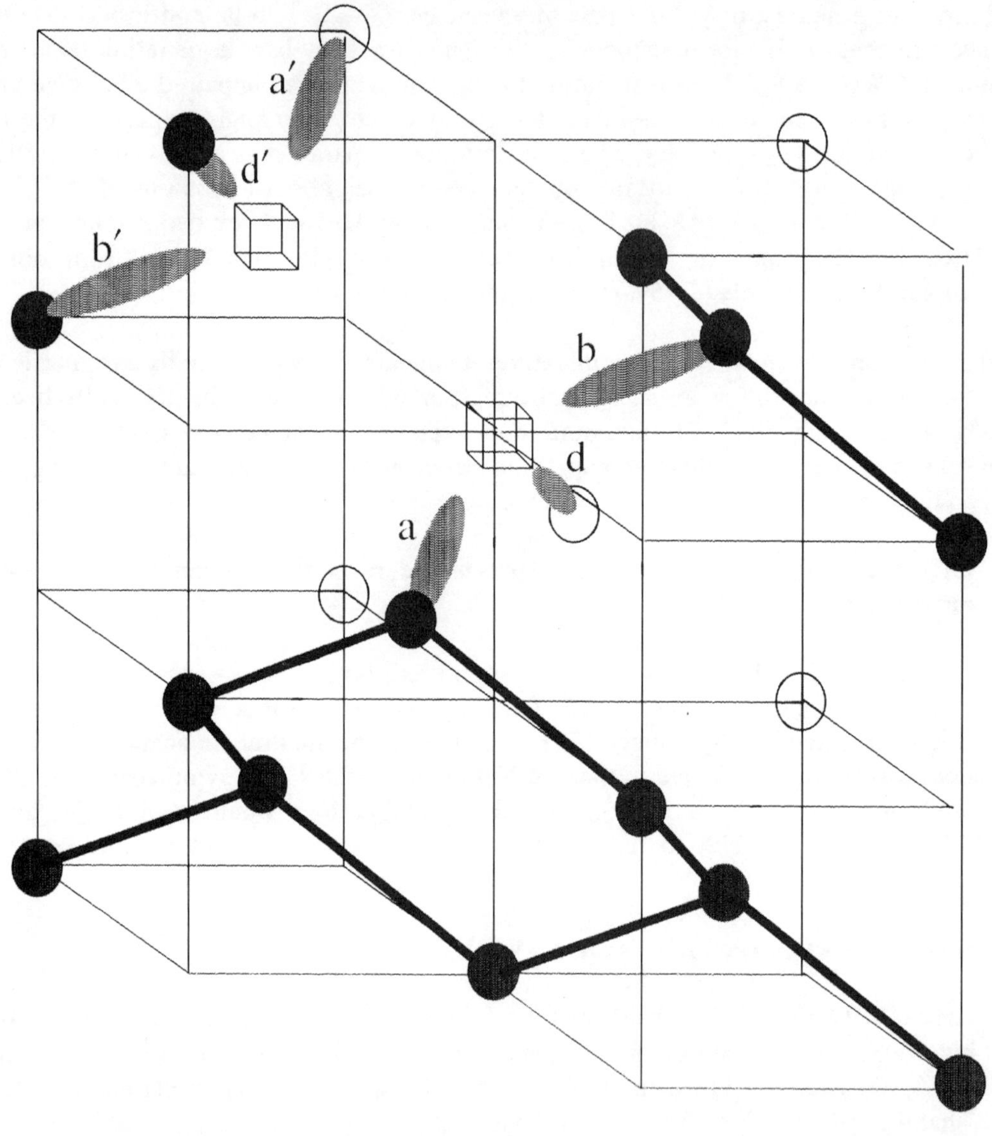


FIGURE 1 A schematic diagram of the nearest neighbours of the divacancy in silicon or diamond. The atoms are labelled following a notation similar to that used by Watkins and Corbett [33]. The plane of reflection symmetry of the site of $C_{2h}$ symmetry is taken to be the {110} plane containing atoms b and b'.

## E THEORY OF VACANCIES RELEVANT TO THE EPR MEASUREMENTS

The EPR data show that for all of the definitely identified vacancy defects, the majority of the unpaired electron spin density is in the nearest neighbour atoms to the vacancy. So to a first approximation, the ground state may be described in terms of the atomic orbitals derived from the 'dangling bonds' of the neighbouring atoms. For the $T_d$ symmetry of a single vacancy, these transform as $a_1$ and $t_2$. The calculations of Coulson and Kearsley [35] found that there were several possible ground states close in energy, that this made electron–electron correlation particularly important, and that vibronic effects are substantial. The importance of vibronic effects was discussed by Lanoo and Stoneham [36]. Mainwood and Stoneham [37] have considered the relative stability of different charge states of the vacancy, and they show that the stable charge states depend upon the position of the Fermi level $E_F$. For $E_F > E_c-2.2$ eV (type Ib diamond), $V^-$ is most stable, and for $E_F < E_v+1.0$ eV (type IIb diamond), $V^+$ is most stable. Between these limits (types Ia and IIa diamond) $V^0$ is stable. For no region is a doubly charged vacancy stable. They also derive a simple model for calculating the energy of excited states, in which identified optical transitions and ground states have been used to evaluate the relevant parameters.

This approach contrasts with that used successfully by Watkins and Corbett [33] in silicon where a one electron defect–molecule approach is used, ignoring the many–electron effects, which are taken to be smaller than Jahn–Teller effects.

The neutral vacancy has four electrons. That a $^1E$ state is lowest, which corresponds to $a_1^2t_2^2$, rather than $^5A_2$ ($a_1t_2^3$) as predicted by Hund's rules, shows the importance of correlation effects. The $^1E$ ground state suffers dynamic Jahn–Teller distortion, which splits it with an E (ground) state and an A state (at 8 meV above the ground state). The GR1 ZPL shows that a $^1T_2$ level lies at 1.673 eV. $^5A_2$ and a $^1A_1$ are predicted to lie close to this $^1T_2$ level. A $^3T_1$ level is predicted to be quite low lying, at ~ 200 meV. That it has not been observed by EPR at 300 K indicates that it lies above about 100 meV, but it might be possible to locate this state by EPR at higher temperature.

For the negative vacancy the calculations [35,37] show that the ground state is $a_1^2t_2^3$, $^4A_1$, as predicted by Hund's rules, in accordance with experiment. The first excited state, $^2T_1$, is predicted to lie at ~ 400 meV, too high for sufficient population for EPR at any temperature below 900 K when the vacancy becomes mobile, and no other state lies below 1.5 eV.

For the positive vacancy, calculations predict that $a_1^2t_2$, $^2T_2$, lies well below other levels, and it is subject to a Jahn–Teller distortion to $D_{2d}$ symmetry, giving an $a_1^2b_2$ ground state, with equal admixture of all four dangling bonds, and S = ½. The level is so low lying that the Jahn–Teller energy might push it very close to, or even below, the valence band edge.

For the neutral divacancy, in $D_{3d}$ symmetry, the one electron states are $a_{1g}$, $a_{2u}$, $e_u$ and $e_g$. Hund's rules would give a ground state with S = 3, $^7A_{2u}$, but Coulson and Larkins [38] predict a ground state $a_{1g}^2a_{2u}^2e_u^2$, $^3A_{2g}$, with no Jahn–Teller distortion. As the experimentally observed symmetry is $C_{2h}$ we approximate by starting with the Watkins and Corbett model, which in this symmetry gives two $a_g$ and two $b_u$ states. Two electrons each go into the lowest $a_g$ and $b_u$ levels, but then to get the experimentally observed result that S = 1 and the unpaired electrons primarily on the four out–of–plane NN atoms, requires one to postulate that these atoms move outwards, which causes the $a_u$ and $b_g$ levels to be displaced below the higher $a_g$ and $b_u$ levels. The observed distortion is unexpected. The increase in strain energy associated with it must be compensated by a greater reduction in the total electron energy. The divacancy in various charge states in both silicon and diamond has been discussed by Coomer et al [39] using localised density function (LDF) modelling of a large cluster. They show that the markedly different distribution of spin density in the two materials can be explained as arising from oppositely directed distortions. For the neutral divacancy in silicon this gives a state $b_u^2$ (S = 0), and in diamond $a_ub_g$ (S = 1), and for the negatively charged state in silicon $b_u^2a_g$ (S = ½), and in diamond $a_ub_gb_u$ (S = 3/2). Hence, the single electron approach appears to account for what is observed, but this must be taken cautiously in view of the known large effects of electron correlation in diamond.

## F CONCLUSION

EPR has given considerable information about the structure of the negative vacancy and the neutral divacancy in diamond. EPR in the $^5A_2$ excited state of the neutral vacancy, observed in some diamonds, confirms the $T_d$ symmetry of $V^0$. The hyperfine interaction with nearest neighbour carbon atoms around the vacancy shows that the majority of the unpaired spin density is located on these atoms, and also indicates considerable outward displacement of them. There is an EPR spectrum which might be due to a positive vacancy, but that is unconfirmed, and another which might be due to a negatively charged di-vacancy $(V—V)^-$. The unexpected $C_{2h}$ symmetry of $(V—V)^0$ still requires a complete theoretical explanation.

## REFERENCES


[1] W. Crookes [ *Diamonds* (Harper and Brothers, London, UK, 1909 ]

[2] C.D. Clark, R.W. Ditchburn, H.B. Dyer [ *Proc. R. Soc. Lond. A (UK)* vol.234 (1956) p.363 ]

[3] C.D. Clark, J. Walker [ *Proc. R. Soc. Lond. A (UK)* vol.334 (1973) p.241 ]

[4] G. Davies, C.M. Penchina [ *Proc. R. Soc. Lond. A (UK)* vol.338 (1974) p.359 ]

[5] G. Davies [ *Nature (UK)* vol.269 (1977) p.498 ]

[6] J.A. Baldwin Jr. [ *Phys. Rev. Lett. (USA)* vol.10 (1963) p.220 ]

[7] J. Isoya et al [ *Phys. Rev. B (USA)* vol.45 (1992) p.1436 ]

[8] C.D. Clark, E. Duncan, J.N. Lomer, P.W. Whippey [ *Proc. Br. Ceram. Soc. (UK)* vol.1 (1964) p.85 ]

[9] D.J. Twitchen, M.E. Newton, J.M. Baker, T.R. Anthony, W.F. Banholzer [ *Phys. Rev. B (USA)* vol.59 (1999) p.12900 ]

[10] W.V Smith, P.P. Sorokin, I.L. Gelles, G.J. Lasher [ *Phys. Rev. (USA)* vol.115 (1959) p.1546 ]

[11] A. Abragam, B. Bleaney [ *Electron Paramagnetic Resonance of Transition Ions* (Oxford Univeristy Press, UK, 1964) ]

[12] J. Morton, K. Preston [ *J. Magn. Reson. (USA)* vol.30 (1978) p.577 ]

[13] O.D. Tucker, M.E. Newton, J.M. Baker [ *Phys. Rev. B (USA)* vol.49 (1994) p.15586 ]

[14] G. Feher [ *Phys. Rev. (USA)* vol.136 (1956) p.500 ]

[15] L.G. Rowan, E.L. Hahn, W.B. Mims [ *Phys. Rev. (USA)* vol.137 (1961) p.A61 ]

[16] J.H.N. Loubser, J.A. van Wyk [ *Rep. Prog. Phys. (UK)* vol.41 (1978) p.1201 ]

[17] J.A. van Wyk, O.D. Tucker, M.E. Newton, J.M. Baker, G.S. Woods, P. Spear [ *Phys. Rev. B (USA)* vol.52 (1995) p.12658 ]

[18] G. Davies, N.B. Manson [ in *Properties and Growth of Diamond* Ed. G. Davies (INSPEC, IEE, London, UK, 1994) p.173 ]

[19] D.P. Joubet, Lin Li, J.E. Lowther [ *Solid State Commun. (USA)* vol.100 (1996) p.561 ]

[20] S.J. Breuer, P.R. Bridden [ *Phys. Rev. B (USA)* vol.51 (1995) p.6984 ]

[21] A. Mainwood, J.E. Lowther, J.A. van Wyk [ *J. Phys., Condens. Matter (UK)* vol.5 (1993) p.7929 ]

[22] J. Isoya, H. Kanda, Y. Morita [ *Proc. 4th Int. Conf. on New Diamond Science and Technology* Eds. S. Saito, N. Fujimori, O. Fukunaga, M. Kamo, K. Kobasho, M. Yoshikawa (MYU, Tokyo, Japan, 1994) p.351 ]

[23] D.W. Palmer [ in *Properties and Growth of Diamond* Ed. G. Davies (INSPEC, IEE, London, UK, 1994) p.143 ]

[24] G. Davies, S.C. Lawson, A.T. Collins, A. Mainwood, S.J. Sharp [ *Phys. Rev. B (USA)* vol.46 (1992) p.13159 ]

[25] D.C. Hunt, D.J. Twitchen, M.E. Newton, J.M. Baker, T.R. Anthony, W.F. Banholzer [ *Phys. Rev. B (USA)* vol.61 (2000) in press ]

[26] D.J. Twitchen, D.C. Hunt, V. Smart, M.E. Newton, J.M. Baker [ *Diam. Relat. Mater. (Switzerland)* vol.8 (1999) p.1572 ]

[27] C.A.J. Ammerlaan [ in *Landolt-Bornstein, Numercial Data and Functional Relationships in Science and Technology* New Series III, Eds. O. Madelung, M. Schulz (Springer-Verlag, Berlin, Germany, 1988) vol.22b p.177 ]

[28] J.N. Lomer, A.M.A. Wild [ *Radiat. Eff. (UK)* vol.17 (1973) p.37 ]

[29] C.D. Clark, R.W. Ditchburn, H.B. Dyer [ *Proc. R. Soc. Lond. A (UK)* vol.237 (1956) p.75 ]

[30] M.A. Lea-Wilson, J.N. Lomer, J.A. van Wyk [ *Philos. Mag. B (UK)* vol.72 (1995) p.81 ]

[31] C.M. Welbourn [ PhD Thesis, Reading University, UK, 1976 ]

[32] R. Orbach [ *Proc. R. Soc. Lond. A (UK)* vol.264 (1961) p.458 and p.485 ]

[33] G.D. Watkins, J.W. Corbett [ *Phys. Rev. (USA)* vol.138 (1965) p.A543 ]

[34] J.K. Kitiu, J.A. van Wyk, M.J.R. Hoch [ *Diam. Relat. Mater. (Switzerland)* vol.8 (1999) p.1569 ]

[35] C.A. Coulson, M.J. Kearsley [ *Proc. R. Soc. Lond. A (UK)* vol.241 (1957) p.433 ]

[36] M. Lanoo, A.M. Stoneham [ *J. Phys. Chem. Solids (UK)* vol.29 (1968) p.1987 ]

[37] A. Mainwood, A.M. Stoneham [ *J. Phys., Condens. Matter (UK)* vol.9 (1997) p.2453 ]

[38] C.A. Coulson, F.P. Larkins [ *J. Phys. Chem. Solids (UK)* vol.30 (1969) p.1963 ]

[39] B.J. Coomer, A. Resende, J.P. Goss, R. Jones, S. Oberg, P.R. Bridden [ *Physica B (Netherlands)* vol.273-274 (1999) p.520 ]

[40] A.T. Collins [ *J. Phys. C, Solid State Phys. (UK)* vol.14 (1981) p.289 ]

[41] G. Davies [ *J. Phys. C, Solid State Phys. (UK)* vol.15 (1982) p.L149 ]

# A7.5 The carbon interstitial and self-interstitial complexes in diamond

D.J. Twitchen, M.E. Newton, J.M. Baker and V.A. Nadolinny

January 2000

## A INTRODUCTION

The most important intrinsic defects are the vacancy and the interstitial. Although in diamond the vacancy in its neutral $V^0$ and negative $V^-$ charge states has been extensively studied by optical spectroscopy and electron-paramagnetic resonance (EPR) [1–3] (see also Datareviews A7.3 and A7.4), until very recently nothing was known about the self-interstitial.

Despite a long and intensive investigation by many workers there is still a lack of understanding of the configuration and precise role of the self-interstitial in areas such as diffusion and electronic passivation in other group IV materials. The main reason for this is thought to be associated with the mobility of the interstitial atom. In p-type silicon containing boron, for example, displacement processes occur which indicate athermal mobility (where the motion involves the positive, negative and neutral charge states) of the Si interstitials at temperatures below 2 K [4]. Recent experimental work by Watkins and Chow [5,6] which studied the athermal motion of the zinc interstitial $Zn_i^+$ in ZnSe at temperatures down to 1 K, has led to an understanding of the short lifetimes of ZnSe laser based devices.

Over the last five years a systematic study of irradiation damage in isotopically enriched single crystal diamond has resulted in the identification of the isolated interstitials—the <100>-split interstitial $[\mathbf{I}_{100}]^0$ (the R2 EPR defect), the di-<100>-split interstitial $[\mathbf{I\text{-}I}]^0$ (the R1 EPR defect) and a tri-<100>-split interstitial $I_3$ (the O3 EPR defect) complex. These defects will be the focus of this Datareview (Sections D and E). A brief summary of radiation damage processes and the theory of interstitials will be presented in Sections B and C, respectively. Identification of these defects is important for understanding processes that may occur and limit the doping of diamond using ion implantation techniques [7] and its use as a detector material [8]. This review will end with a short discussion of other extrinsic interstitials identified in diamond.

## B IRRADIATION DAMAGE IN DIAMOND

The most common process used to produce defects in diamond is irradiation with mono-energetic electrons. In addition to the displacement of target atoms, the electrons cause ionization and electronic excitation of target atoms. The latter two are in fact the predominant effects of the incoming electrons, and the main source of energy loss. As a result the penetration depth[1] is approximately 0.8 mm for 1.0 MeV electrons and 2.5 mm for 2.0 MeV [9,10]. The pattern of damage is an almost flat profile with a sharp cut off at the penetration depth [11]. Much theoretical and experimental work has been carried out to calculate the threshold energy needed to displace a carbon atom, $E_d$ [12,13]. Intuitively it can be modelled as being equal to the total binding energy $E_b$ of an atom with its neighbours, together with an additional energy $E_k$, which may be small or large compared with $E_b$, to push the atom through the gap between neighbouring atoms and onto an interstitial site. With the strength of each C–C bond in diamond deduced from the sublimation and $s^2p^2 \rightarrow sp^3$ promotion energies of 7.9 eV [12], $E_b$ is equal

[1] Although many sophisticated computer codes have been used to predict this depth, the early theoretical work by Kinchen and Pease [ *Rep. Prog. Phys. (UK)* vol.18 (1955) p.1-51 ] predicts the same results.

to 31.6 eV. Experimentally the diamond-like structure shows little directional dependence of $E_d$ [14]. $E_k$ has been modelled and can be as much as 12.7 eV [13]. Thus we would expect $E_d$ to be somewhere between 31.6 and 44.3 eV, with the lower energies being more probable. Hopping conductivity experiments made below 50 K determine energies for this process of ~36 eV [15]. Since most of the energy of the incoming electrons is lost through excitation and ionization the threshold energy for electrons to produce lattice vacancies and interstitials is measured to be 145(5) keV for electrons incident along a <100> crystallographic direction [16].

Neutron damage creates very broad features in the absorption spectrum, with the zero phonon line (ZPL) of the GR1 band (1673 meV ZPL) associated with $V^0$, often barely discernible above the phonon side band [17]. The higher mass of the neutron means that for a given kinetic energy it will transfer a greater energy to the carbon atom, such that the displaced atom often has enough energy to displace another carbon atom. This leads to multiple damage sites, creating new types of defects and large strain fields about conventional defects, resulting in the broadening of ZPLs.

Since the neutron has no charge, it does not have an ionizing effect on the lattice, unlike the electron. This ionization of the electron (or charged particle) does not contribute directly to damage centres, but it does have the secondary effect of local and global heating. The local heating can be higher than 500 K [17]. At such temperatures interstitials may be more able to migrate, either recombining with a nearby vacancy or being trapped at an alternative defect, reducing the random strain fields around the vacancy. This is an explanation of the narrower ZPLs produced by room temperature electron irradiation as compared with those of neutron irradiation [17]. This is further indicated by experiments which show that low temperature electron irradiation produces a very similar absorption spectrum to neutron irradiation at room temperature [17]. The broad ZPLs produced upon irradiation at 80 K are sharpened on heating to 750 K [17,18]. The broader spectrum is also observed after gamma-irradiation at 300 K, and again sharpened after heating to 750 K [17].

The reported production rates of the simple defects discussed in this Datareview vary significantly from study to study (see for example the review by Palmer [19]). There are at least two reasons for this. Firstly, the reported irradiations are made at different electron fluxes (number of electrons per unit area per second). If the diamond is not well heat sunk, the electron flux (usually quoted as $\mu A/cm^2$) can cause substantial bulk heating of the sample. This heating effect can vary considerably with even a small change in the flux (see [20]). Secondly, impurities can compensate the formed interstitials through either charge transfer [21], or migration and defect formation. The production rates of the defects reported here are given in TABLE 1. These rates are measured in identical IIa diamonds, where the electron energy and flux was the same in all cases.

## C THE THEORY OF THE SELF-INTERSTITIAL IN DIAMOND

There are thought to be five possible structures formed by the addition of an interstitial carbon atom to the diamond lattice – the tetrahedral, hexagonal, bond centred, <100>- and <110>-split interstitial sites [4]. Various theoretical techniques ranging from complete neglect of differential overlap (CNDO) to computer intensive ab initio density functional theory have been applied to model these interstitial sites in diamond (see Datareview A7.2). These techniques all show that the $[\mathbf{I}_{100}]^0$ is the only stable structure for the interstitial atom in diamond. This is true for the neutral, –1, +1 and +2 charge states [22–24]. The ground state of the neutral defect was determined as $^1B_1$, a singlet, and would not be detectable by EPR. The highest-frequency vibrational mode of the $[\mathbf{I}_{100}]^0$ defect has been deduced to be an infrared inactive $A_1$ mode at 2023 $cm^{-1}$, with two active modes of B symmetry at 1590 and 1600 $cm^{-1}$. These modes drop to 1984, 1560 and 1595 $cm^{-1}$ if one of the central atoms is a $^{13}C$ atom and to 1944, 1557 and 1564 $cm^{-1}$ if both of the central atoms are $^{13}C$ atoms [23]. An upper bound to the energy barrier for the motion of $[\mathbf{I}_{100}]^0$ was predicted to be 1.7 eV [23].

TABLE 1 The salient properties of the defects in diamond known to contain at least one carbon <001>-split interstitial. The production rates should be compared to that of the neutral vacancy of 1.53(10) $cm^{-1}$. The production rates were measured for type IIa diamonds irradiated at 100 K using 1.9 MeV electrons and a beam flux of 12 $\mu A\ cm^{-2}$.

| Defect label | Model [Ref] | Symmetry | Spin | Production rate ($cm^{-1}$) | Annealing temperature | Optical lines | Other information |
|---|---|---|---|---|---|---|---|
| R2 | **$[I_{100}]^0$** [18] | $D_{2d}$ ($D_2$) [18,30] | S = 1 [18,30] (excited state 50 meV above diamagnetic ground state(s)) | 1.10(10) [18,35] (At 350 K the production rate is 0.37(5)) | 670–740 K [18,35,44] | 1.685 eV, 1.859 eV, 3.99 eV [18,25,33] (These associations are from production and annealing data. $f_{1.859}$=1.1(2)×$10^{-17}$ (meV $cm^{-2}$) | Total concentration at 300 K is ~3 times measured concentration . Very wide EPR lines. Linewidth different in Ib diamonds [18]. |
| R1 | **$[I\text{-}I]^0$** [39] | $C_{2h}$ [30,39] {110} plane of reflection symmetry | S = 1 [30,39] (ground state) | 0.014(1) [20] (At 350 K the production rate is 0.14(1)) | 570–740 K [41,42,44] | --- | Spin polarization observed for optical illumination with photons of E> 1.7(1) eV and temperature <70 K [46]. |
| O3 | $I_3$ [50] | $C_2$ [50] A <100> rotation axis | S = 1 [50] (ground state) | 0.006(1) [20,50] (Not observed on irradiation at 350 K) | | --- | Indicates the existence of the **[I-C-I]** defect. |

## D THE CARBON <001>-SPLIT INTERSTITIAL

Until very recently (1997) the partial annealing of the neutral vacancy at temperatures (~700 K) below that at which it is expected to migrate (900 K) provided the most direct evidence of interstitials recombining with the vacancy [1,25–28]. Duncan [29] measured the optical absorption of a type IIa diamond at 80 K following 2 MeV electron irradiation at that temperature. The sample was then warmed to 300 K, cooled again and remeasured. He found that the GR1 band of $V^0$ is chemically stable while annealing between the irradiation temperature of 80 K and room temperature. There is no significant annihilation of vacancies by recombination with mobile self-interstitials in this temperature range.

### D1 The R2 EPR Centre

Nearly forty years after its discovery the EPR centre R2, which is produced by electron or neutron irradiation in all diamonds, has been identified as the **$[I_{100}]^0$** (shown in FIGURE 1) [18]. R2 was first observed at room temperature in a natural type IIa diamond after irradiation with $10^{18}$–$10^{20}$ 2 MeV electrons $cm^{-2}$ in 1962, then labelled the c-system [30]. It is unlike most other radiation damage centres in diamond in having very wide EPR lines (2.5–5.0 mT), and axial symmetry about <100> [28,30].

The ground state of the neutral system is non-paramagnetic (S = 0), with electrons paired off, but there is a nearby excited S = 1 state 50 meV above the ground diamagnetic state(s) which is sufficiently thermally populated at temperatures above 100 K to make observation of R2 possible

[18][2]. The linewidth of R2 is independent of concentration up to defect levels of 20 ppm, with a characteristic angular anisotropy, which has been shown to arise from a distribution of both the principal direction, and the magnitude of the axially symmetric **D**-matrix [18]. This distribution is attributed to distortion of the structure produced by some as yet unidentified motional effect, or because of neighbouring vacancies and the sensitivity of the $sp^2$ bonds at either end of the defect to internal strain. In diamonds containing ~200 ppm of nitrogen the line width of R2 is 50% greater; this is attributed to additional bulk stress associated with the incorporation of the nitrogen [18]. Measurements between 80 and 650 K showed that the magnitude of the **D**-matrix decreased linearly with increasing temperature (0.030(6) MHz $K^{-1}$) [18]. This measured temperature variation is somewhat larger in magnitude, but of the same sign, as that indicated by thermal expansion of the lattice.

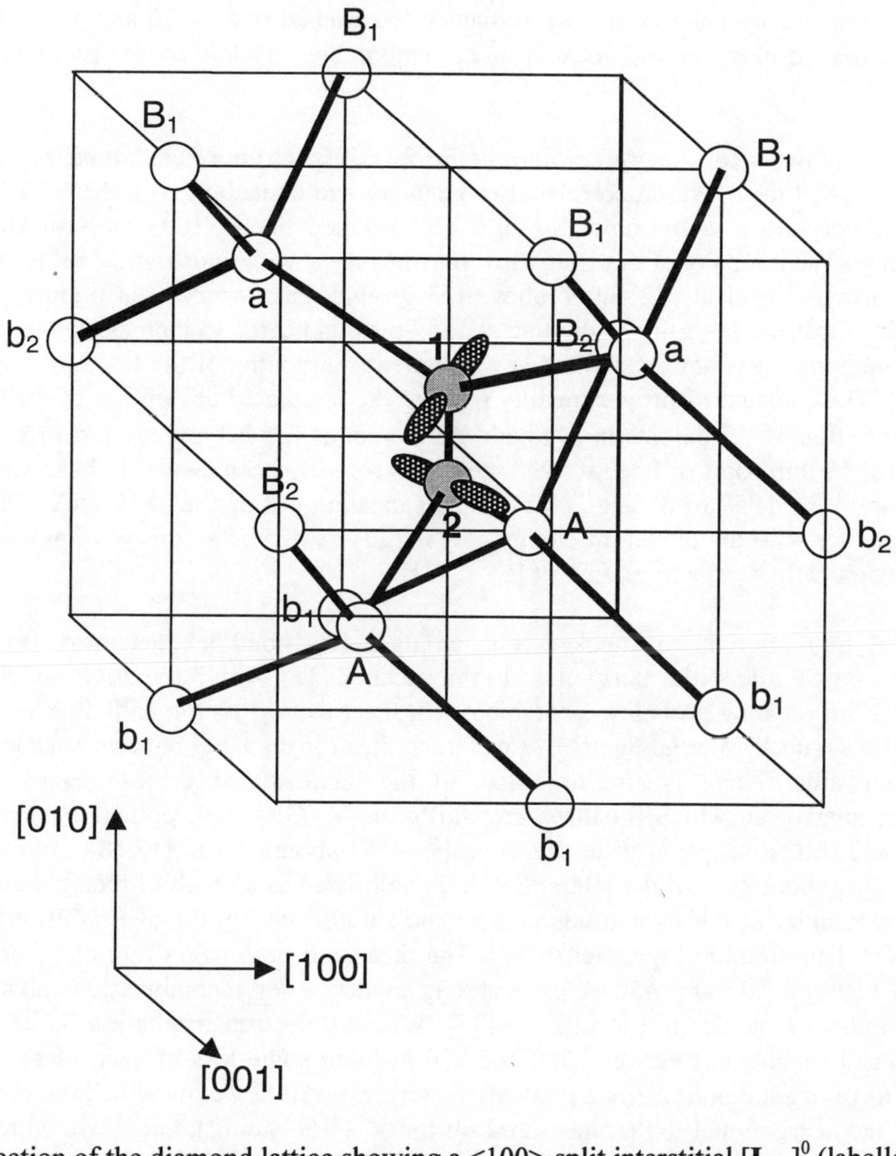


FIGURE 1 Section of the diamond lattice showing a <100>-split interstitial $[\mathbf{I}_{100}]^0$ (labelled 1 and 2) and nearby equivalent carbon neighbours (labelled by the same letter) to this interstitial. This defect has been shown to correspond to the R2 EPR centre.

[2] To take account of the population of R2 in the ground S = 0 state, the integrated concentration of R2, measured at 300 K, must be multiplied by approximately three to allow for the ground state population.

As a consequence of the large linewidth of R2 it was not possible to observe any lines symmetrically positioned on either side of the allowed $\Delta M_S = \pm 1$ central transitions which might be associated with a $^{13}C$ hyperfine interaction. It is this interaction that has been so useful in identifying EPR defects in diamond (see for example [2,39,50]). However, early work by Owen [31] found narrow lines for the $\Delta M_S = 2$ transitions, because their width is affected only in second order by the value of D. The narrow $\Delta M_S = 2$ transitions were exploited by Nadolinny et al [32] in an electron irradiated dislocation-free natural type IIa diamond, for which the typical linewidth of most EPR defects, like that of $V^-$, was 20 μT. For external field **B** exactly along <100>, parallel to the axis of the defect, the intensity of the $\Delta M_S = 2$ transition was zero. As **B** was rotated away from <100> the lines gained intensity, but also increased in width. For **B** between 1 and 1.5° from <100>, a linewidth of 20 μT allowed the resolution of $^{13}C$ hyperfine structure from three sets of equivalent carbon neighbours. Further measurements by Hunt et al [18] in diamonds enriched with 5, 10 and 50% $^{13}C$ showed that most of the unpaired electron was located in 2p nonbonding orbitals on the two equivalent carbon atoms of $[\mathbf{I}_{100}]^0$.

Walker [33] found two weak absorption lines in the optical spectrum of electron irradiated diamond at 1685 meV and 1859 meV, which correlated in intensity and annealing with the R2 EPR defect. The former line is very close to the much stronger zero-phonon line at 1673 meV of GR1, due to the neutral vacancy. The intensity of the 1685 meV line decreases with temperature below 80 K as though it originates from a level at 6(2) meV above the ground state which was thermally depopulated. Walker reported that the intensity of the line at 1859 meV appeared to increase below 80 K, by about 50% at 4 K, although it is not clear whether the evident sharpening of the line at 4 K was taken into account [18]. The constant of proportionality relating the integrated absorption of the 1859 meV line to the concentration of R2 defects in synthetic IIa diamonds has been measured to be $1.1(2) \times 10^{-17}$ meV $cm^{-2}$ [18]. A third optical line, at 3990 meV, has recently been shown to have similar annealing characteristics to the 1859 meV line [18,25]. The annealing out of the 1859 meV absorption fits to first order kinetics with an activation energy of 1.68(20) eV [25,35]—this is in agreement with that predicted theoretically for the migration of $[\mathbf{I}_{100}]^0$ [23].

The effect of irradiation is to knock carbon atoms a few bond lengths away from their lattice positions, leaving an interstitial fairly close to the vacancy [18]. Any interstitials which end up in the same chair as the vacancy probably recombine with the vacancy, so the EPR line width might arise from the strain set up by a large number of different sites for the neighbouring vacancy from the site distant $a_0$ outwards. There is also an effect of the local strain at the vacancy set up by the neighbouring interstitial, which broadens and shifts the $V^0$ GR1 ZPL optical absorption [18]. This broadening and shift disappears when R2 anneals out at about 750 K [17,18]. As the $[\mathbf{I}_{100}]^0$ defect becomes mobile, about 30% of the interstitials are annihilated as a result of recombination of $V^0$ with $[\mathbf{I}_{100}]^0$. The remainder of the interstitials escape, and simultaneously the 5RL (ZPL 4.581 eV) centre appears, which has rhombic I symmetry [34]. The measured production rate of R2 as a function of temperature between 100 and 450 K gives strong evidence for recombination-enhanced migration under the conditions of electron irradiation [35]. Without electron irradiation R2 is stable at room temperature and anneals out between 700 and 750 K (with some loss of vacancies); however under the conditions of irradiation there is a production stage at ~150 K, below which the production rate is 1.1(1) $cm^{-1}$ and nearly equal to that measured for the $V^0$ (1.5(1) $cm^{-1}$), but above which the rate falls to a third of this value [35].

R2 is observed in Ib diamonds but not until a certain threshold irradiation dose, at which $V^0$ is also observed [36–38]. In Ib diamonds, on annealing at 700 K just before R2 anneals out, the concentration of R2 increases significantly [38]. The mechanism for this is unclear.

## E CARBON <100>-SPLIT INTERSTITIAL COMPLEXES

### E1 The R1 EPR Centre

EPR measurements in electron-irradiated diamond enriched with 5% $^{13}C$ resulted in the identification of the R1 EPR centre, more than 30 years after the discovery of the centre [30]. This defect comprises two parallel nearest neighbour <100>-split interstitials $[\mathbf{I}\text{-}\mathbf{I}]^0$ [39]. In 1997 this was the first conclusive identification of an interstitial defect in diamond. Like R2, R1 has been observed in all diamonds after electron irradiation, and has also been reported to be present after room temperature phosphorus implantation [40]. EPR measurements in a synthetic IIa diamond enriched to 5% $^{13}C$ showed that the majority of the unpaired electron distribution (72%) was localized in nonbonding 2p orbitals on two separate carbon atoms [39].

R1 anneals out between 600 and 650 K [39,41,42]. The measured production rate of R1 at two different temperatures is given in TABLE 1. That R1 is observed in diamonds irradiated at 100 K [20] is conclusive evidence that the interstitial atom, under the conditions of electron irradiation, must be mobile at this temperature, since it is unlikely that R1 is a direct radiation product, but can only be formed by mobile interstitials.

Annealing studies showed that R1 annealed out without any loss of vacancies, although there was a temporary increase in the concentration of $V^-$ [42]. As R1 anneals the 3H (ZPL at 2.462 eV) absorption line grows [43], and there is a corresponding growth in the EPR centres R3 and R14 [43,44], although R3 and R14 anneal out after the 3H absorption [44]. Two different components were needed to fit the annealing of R1 [41]. One component was very fast and insufficient data were collected to calculate an activation energy; the second process had a characteristic activation energy of 0.6(1) eV. This energy is too low to be associated with bulk interstitial migration; in a recent paper this energy has been associated with the energy of the R1 defect to form $[\mathbf{I}\text{-C-}\mathbf{I}]^0$—the Humble defect [45]. This has the correct symmetry (rhombic I) and properties to be the 3H optical defect.

Two parallel <100>-split interstitials at next nearest neighbour, $[\mathbf{I}\text{-C-}\mathbf{I}]^0$, positions in the same {1 1 0} plane, results in a structure with $C_{2v}$ symmetry (rhombic I). As for R1, a reconstruction of dangling bonds of the two <100>-split interstitials is predicted for this structure, giving a defect with two non-bonding p-orbitals [24]. The calculated energy difference between this structure and R1 is negligible for the S = 1 state [24]. However, the S = 0, $[\mathbf{I}\text{-C-}\mathbf{I}]^0$ structure is calculated to be 0.4 eV lower in energy than the S = 1, $[\mathbf{I}\text{-}\mathbf{I}]^0$ defect [24]. The S = 0, $[\mathbf{I}\text{-C-}\mathbf{I}]^0$ defect would not be observable in EPR in its ground state, but is predicted to be optically active.

The R1 defect demonstrates optical spin polarization below a certain threshold temperature (which depends on the concentration of the R1 defect) when illuminated with light of energy greater than 1.7(1) eV [46]. The spin polarization arises by selective excitation/repopulation of the R1 ground state mediated by spin-orbit coupling with a higher state.

In neutron irradiated Ib diamonds, before R1 anneals out its concentration increases significantly [36,37,44], which is not observed in IIa diamonds. As with R2 the mechanism for this increase is not understood.

### E2 The Rla...f and R13 EPR Centres

In addition to the strong spectra of R1 and R2, irradiation in IIa diamonds produces a set of six satellite lines near R1 with about 0.5% the intensity of R1, which have been labelled Rla...f [44]. Four of these centres have about the same intensity, and the other two are of about half that intensity. The spin Hamiltonian describing them is the same as for R1 [39], but with a decrease of a few percent in the **D**-matrix. Rla...f are produced under the same conditions as R1 and anneal out at the same temperature, suggesting that they are related to R1. Because of their low intensity it has not been possible to observe enough of the angular variation of Rla...f to know if their symmetry is the same as R1 ($C_{2h}$) or lower. These centres have been observed in diamonds of high purity, so they are not related to impurities. An adequate, although as yet unproved, model for these centres is based on vacancies occupying neighbouring sites to R1, which cause a change in the **D**-matrix of R1 by allowing distortion of the diamond matrix which holds the R1 centre in place. To produce this distortion in an otherwise fairly rigid structure, means the vacancies must lie fairly close to the basic R1 unit. There are six possible sites for the vacancy around the R1 structure which could produce defects with the correct symmetries and relative concentrations of the measured R1a...f centres.

Another EPR centre which fits into this model is R13, which is always produced and destroyed under conditions similar to R1. It has ~1% the intensity of R1, $C_1$ symmetry and a **D**-matrix of about half the magnitude of R1 [44]. It would seem that this centre is a more severely distorted form of R1, with a vacancy within the chair of one of the interstitials in the R1 structure. No optical centres have been correlated with any of these centres, and their low intensity means further progress is likely only to be made using theoretical techniques. It is important that these defects are understood because they may provide valuable clues to the formation of more dominant defects such as R1 and R2.

## E3 The O3 EPR Centre

Recently a new EPR centre, labelled O3, was observed in diamond irradiated at 100 K [20]. A detailed investigation in a diamond with $^{13}C$ enrichment showed that the centre has a triplet S = 1 ground state and $C_2$ symmetry with a <100> rotation axis [50]. This information combined with the analysis of $^{13}C$ hyperfine couplings showed that 76% of the unpaired electronic wavefunction was localized in two non-bonding orbitals on different carbon atoms, and allowed a detailed model to be constructed for the defect [50]. The model for O3 comprises three parallel non-planar <100>-split interstitials at next-nearest neighbour positions $I_3$ (FIGURE 2). For this defect to form it seems that the defect that contains two parallel split interstitials at next nearest neighbour positions in the same {110} plane [**I-C-I**] must be present, since it is very unlikely that under the typical conditions of irradiation, a single electron could produce this or that three <001>-split interstitials could meet at the same time to form a stable complex. In neutron irradiated Ib diamonds, the concentration of O3 increases as R1 anneals out, and then itself anneals out at about 650 K [50].

First-principles density functional theory has been used to investigate the interstitial and its aggregates in diamond [24]. The theory found agreement about the stability and structure proposed for R1, O3 and R2 EPR defects.

Theory also predicted that a closed loop of four <001>-split interstitials $I_4$ was a low energy structure. The identification of O3 not only indicates the existence of [**I-C-I**] but also the likely existence of $I_4$. The addition of a fourth interstitial to O3 results in an optimized structure where all atoms are fourfold co-ordinated, and furthermore all bond lengths and angles are close to ideal. This structure has been suggested to arise in Si with little convincing evidence [47]. This $I_4$ structure has been proposed as a building block of platelets in diamonds [24].

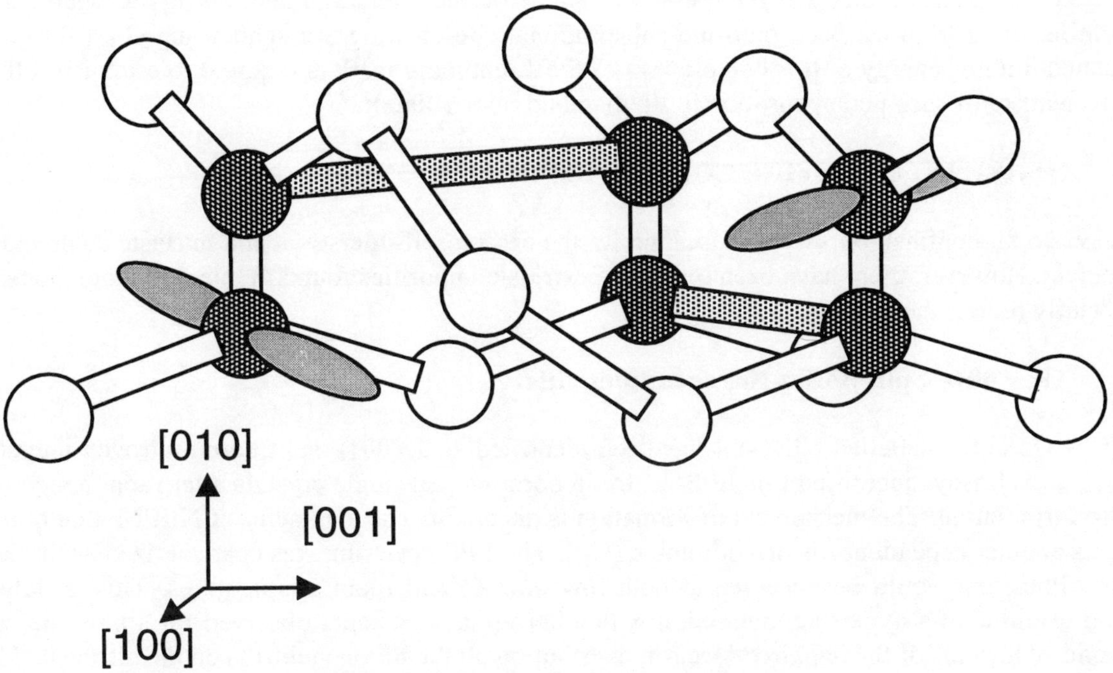


FIGURE 2 The proposed structure for the O3 EPR defect, which contains three non-planar parallel <100>-split interstitials. The $C_2$ rotation axis of this defect is along [100].

## E4 Other Possible Intrinsic Interstitial Related Defects

### E4.1 Type IIa

In the EPR of irradiated IIa diamond two defects which are observed at 300 K remain unidentified, namely R3 and R14 [44]. These both have S = 1 and $C_{1h}$ symmetry [44]. They anneal out at ~830 K, after the R2 defect, but before the vacancy anneals [44]. They are always produced on irradiation and increase in concentration as both R1 and R2 anneal out [43]. As they are created in very pure material it is likely that they involve only intrinsic defects. No definite models have been proposed.

### E4.2 Type Ib

In diamonds containing large concentrations of single substitutional nitrogen the EPR centres labelled W11, W12, W13 and W14 are formed [37]. With the exception of W11 these all have S = 3/2 ground states. W11 arises from a thermally excited S = 3/2 state; its S = 1/2 ground state has never been identified. These centres anneal below the temperature at which R2 anneals out [37,44].

### E4.3 Type IIb

On irradiation, no new EPR defects have been reported in IIb diamonds which are not observed in IIa diamonds. R1 and R2 are created but with much lower production rates than in the IIa material, and then not until a threshold dose, at which $V^0$ is also observed [38]. On irradiation, the concentration of uncompensated neutral substitutional boron (measured by the absorption maxima at 2455, 2801 and 2922 $cm^{-1}$) is significantly reduced. This reduction continues on annealing at 750 K when $[\mathbf{I}_{100}]^0$ is mobile [48]. The boron only reappears after annealing to temperatures as high as 1700 K. The origin of this compensation of the boron acceptor is not clear. Recent theoretical studies investigating boron interstitial complexes found that the boron was stable on the interstitial site, but it was energetically favourable for it to move back onto the substitutional site creating a neighbouring $[\mathbf{I}_{100}]^0$ [49]. The calculated binding energy of this complex was 3.5 eV, but more work is required to establish if this is the mechanism of boron compensation in IIb diamond on irradiation.

## F OTHER INTERSTITIAL COMPLEXES IN DIAMOND

We have so far confined ourselves to looking for the carbon self-interstitial and intrinsic complexes of this defect. However, there have been reports of extrinsic impurities found in interstitial positions. We will briefly review these cases.

### F1 The <001>-Split Boron-Nitrogen Interstitial

The S = 1/2 defect labelled NIRIM 4 has been identified as a <001>-split boron-nitrogen interstitial $[\text{B-N}]^+_{100}$ [51]. Isoya identified this EPR centre in boron-doped single crystals after room temperature electron irradiation. The mechanism of formation is uncertain. The symmetry of NIRIM 4 determined from its angular dependence is orthorhombic ($C_{2v}$). The EPR spectrum was characterized as three sets of four lines, and could be observed at both low (4.2 K) and room temperatures. This is only the second example of a hyperfine interaction with a boron nucleus being observed for a point defect in diamond[3] with 11% of the unpaired electron distribution on the boron and 61% on the nitrogen [51].

### F2 Interstitial Nitrogen

Experimental analysis of isotopic substitution data has shown that the defect responsible for the H1a absorption (1450 $cm^{-1}$), which is produced in irradiated diamond rich in A-centres and C-centres after annealing to 600 K and 950 K respectively, contains two equivalent carbon atoms and a single nitrogen atom. Theoretical modelling of the possible configurations for two carbons and one nitrogen concluded that the defect responsible was a bond-centred nitrogen interstitial atom [45]. The vibration that gives rise to the H1a absorption is a mode in which the carbon moves perpendicular to the carbon-carbon bond. Surprisingly there has been no EPR signal observed for this defect. One striking aspect of the production of the H1a centre is that for a given irradiation dose and given annealing, the H1a strength increases linearly with the local concentration of A-centres [34]. The effect could be caused by an increase in production of stable damage near the nitrogen.

### F3 Interstitial Nickel

Transition metals in the diamond lattice have been shown to occupy a series of different sites [53–56], some of which seem to depend upon the charge state of the defect [53]. Seminal EPR measurements showed that a centre observed in diamonds grown with a nickel containing solvent catalyst and nitrogen getters was an isolated $Ni^+$ at a tetrahedral interstitial site. The centre labelled NIRIM 1 had an electronic configuration $3d^9$, S = 1/2 and a g-value of 2.0112. This defect has also been associated with an optical centre at 1.40 eV. The g-values of this defect indicate that the crystal field at the tetrahedral site is dominated by the octahedrally distributed second neighbour atoms.

---

[3] The first was reported for an unidentified defect in a natural IIb diamond labelled W36, by J.H.N. Loubser and J.A. van Wyk [ *S. Afr. J. Phys. (South Africa)* vol.10 (1987) p.165 ].

## G CONCLUSION

There has been rapid and substantial progress in understanding many of the basic EPR centres created in IIa diamonds. The identification of the isolated interstitial R2 and its production conditions compared with other identified interstitial complexes like R1 and O3 represent a significant breakthrough. The understanding of the optical centres in irradiated IIa diamond is less clear and it must be a focus of continuing research. The optical information is necessary for the complete understanding of the different charge states of these primitive defects, as only certain charge states are observable in paramagnetic resonance.

There are many questions which remain about the interstitial defect processes and the defects produced in Ib and IIb diamonds. However, this should change in the near future as the knowledge of the processes in the IIa material must give valuable pointers to those which may occur in these extrinsic defect containing diamonds.

## ACKNOWLEDGEMENTS

We would like to thank Drs D.C. Hunt and A. Yelisseyev for their time in discussing some of the data presented in this review.

## REFERENCES

[1] G. Davies et al [ *Phys. Rev. B (USA)* vol.46 (1992) p.13157 ]
[2] J. Isoya et al [ *Phys. Rev. B (USA)* vol.45 (1990) p.3905 ]
[3] D.J. Twitchen, D.C. Hunt, V. Smart, M.E. Newton, J.M. Baker [ *Diam. Relat. Mater. (Switzerland)* vol.8 (1999) p.1572 ]
[4] See, for example, W. Hayes, A.M. Stoneham [ *Defects and defect processes in non-metallic solids (*John Wiley & Sons, New York, 1985) ch.6 ]
[5] K. Chow, G.D. Watkins [ *Phys. Rev. Lett. (USA)* vol.81 (1998) p.2084 ]
[6] G.D. Watkins, K. Chow [ *Physica B (Netherlands)* vol.273-274 (1999) p.861 ]
[7] J.F. Prins [ *Industrial Diamond Review (UK)* vol.1 (1996) p.22 ]
[8] L. Allers, A.S. Howard, J.F. Hassard, A. Mainwood [ *Diam. Relat. Mater. (Switzerland)* vol.6 (1997) p.353 ]
[9] J.C. Bourgoin, B. Massarani [ *Phys. Rev. B (USA)* vol.14 (1976) p.3690 ]
[10] S.C. Lawson [ private communication, 1999 ]
[11] A. Mainwood [ private communication, 1997 ]
[12] J.C. Bourgoin, M. Lanoo [ *Point Defects in Semiconductors* vol.II (SpringerVerlag, Berlin, 1986) p.228 ]
[13] W. Kohm [ *Phys. Rev. A (USA)* vol.94 (1954) p.1409 ]
[14] J. Koike, D.M. Parkin, T.E. Mitchell [ *Appl. Phys. Lett. (USA)* vol.60 (1992) p.1450 ]
[15] B. Massarani, J.C. Bourgoin [ *Phys. Rev. B (USA)* vol.14 (1976) p.3682 ]
[16] J. Steeds [ private communication, 1999 ]
[17] C.D. Clark, R.W. Ditchburn, H.B. Dyer [ *Proc. R. Soc. Lond. A (UK)* vol.237 (1956) p.75 ]
[18] D.C. Hunt et al [ *Phys. Rev. B (USA)* vol.61 (2000) in press ]
[19] D.W. Palmer [ in *Properties and Growth of Diamond* Ed. G. Davies (INSPEC, IEE, London, UK, 1994) p.143 ]
[20] D.J. Twitchen, D.C. Hunt, M.E. Newton, J.M. Baker, T.R. Anthony, W.F. Banholzer [ *Physica B (Netherlands)* vol.273-274 (1999) p.628 ]
[21] S.C. Lawson, D. Fisher, D.C. Hunt, M.E. Newton [ *J. Phys., Condens. Matter (UK)* vol.10 (1998) p.6171 ]

[22] A. Mainwood, F.P. Larkins, A.M. Stoneham [ *Solid-State Electron. (UK)* vol.21 (1978) p.1431 ]

[23] S.J. Breuer, P.R. Briddon [ *Phys. Rev. B (USA)* vol.51 (1995) p.6984 ]

[24] B.J. Coomer et al [ *Phys. Rev. Lett. (USA)* submitted (1999) ]

[25] L. Allers, A.T. Collins, J. Hiscock [ *Diam. Relat. Mater. (Switzerland)* vol.7 (1998) p.228 ]

[26] D.W. Palmer [ PhD Thesis, University of Reading, 1961 ]

[27] C.D. Clark, E.W.J. Mitchell [ *Inst. Phys. Conf. Ser. (UK)* vol.31 (1977) p.45 ]

[28] C.D. Clark, E.W.J. Mitchell [ *Radiat. Eff. (UK)* vol.9 (1971) p.219 ]

[29] P. Duncan [PhD Thesis, University of Reading, 1963 ]

[30] E.A. Faulkner, J.N. Lomer [ *Philos. Mag. B (UK)* vol.7 (1962) p.1995 ]

[31] J. Owen [ *Physical properties of diamond* Ed. R. Berman (OUP, 1965) ch.10 p.274 ]

[32] V.A. Nadolinny, E.V. Sobolev, O.P. Yurieva [ *J. Struct. Chem. (USA)* vol.626.7J (1995) p.36 ]

[33] J. Walker [ *J. Phys. C (UK)* vol.10 (1977) p.3867 ]

[34] G. Davies [ *Physica B (Netherlands)* vol.273-274 (1999) p.15 ]

[35] D.J. Twitchen et al [ *Physica B (Netherlands)* vol.273-274 (1999) p.644 ]

[36] J.A. van Wyk [ private communication, 1997 ]

[37] J.A. van Wyk [ *J. Phys., Condens. Matter (UK)* vol.6 (1994) p.801 ]

[38] D.J. Twitchen, C. Wade [ private communication, 1999 ]

[39] D.J. Twitchen, M.E. Newton, J.M. Baker, T.R. Anthony, W.F. Banholzer [ *Phys. Rev. B (USA)* vol.54 (1997) p.6988 ]

[44] J. Isoya, H. Kanda, M. Akaishi, Y. Morita, T. Ohshima [ *Diam. Relat. Mater. (Switzerland)* vol.6 (1997) p.356 ]

[41] D.J. Twitchen, M.E. Newton, J.M. Baker [ *Phys. Rev. Lett. (USA)* submitted (2000) ]

[42] D.J. Twitchen, M.E. Newton, J.M. Baker, T.R. Anthony, W.F. Banholzer [ *Mater. Sci. Forum (Switzerland)* vol.239-241 (1997) p.107 ]

[43] D.J. Twitchen [ DPhil Thesis, University of Oxford, 1997 ]

[44] C.A.J. Ammerlaan [ *Landolt-Bornstein Numerical Data and Functional Relationships in Science and Technology* New Series III, Eds. O. Madelung, M. Schulz (Springer Verlag, Berlin, Germany, 1988) vol.22b p.177 ]

[45] P. Humble [ *Proc. R. Soc. Lond. A (UK)* vol.381 (1982) p.65 ]

[46] D.J. Twitchen, M.E. Newton, J.M. Baker, W.F. Banholzer, T.R. Anthony [ *Diam. Relat. Mater. (Switzerland)* vol. 8 (1999) p.1101 ]

[47] N. Arai, S. Takeda, M. Kohyama [ *Phys. Rev. Lett. (USA)* vol.78 (1997) p.4265 ]

[48] A.T. Collins [ *Inst. Phys. Conf. Ser. (UK)* vol.31 (1977) p.346 ]

[49] J.P. Goss [ private communication, 1999 ]

[50] D.C. Hunt, D.J. Twitchen, M.E. Newton, J.M. Baker, T.R. Anthony, W.F. Banholzer [ *Phys. Rev. B (USA)* in press (2000) ]

[51] J. Isoya, H. Kanda, Y. Morita [ *Phys. Rev. B (USA)* vol.56 (1997) p.6392 ]

[52] I. Kiflawi, A. Mainwood, H. Kanda, D. Fisher [ *Phys. Rev. B (USA)* vol.54 (1996) p.16719 ]

[53] J. Isoya, H. Kanda, J.R. Norris, J. Tang, M.K. Bowman [ *Phys. Rev. B (USA)* vol.41 (1990) p.3905 ]

[54] J. Isoya, H. Kanda, Y. Uchida [ *Phys. Rev. B (USA)* vol.42 (1990) p.9843 ]

[55] D.J. Twitchen, J.M. Baker, M.E. Newton, K. Johnston [ *Phys. Rev. B (USA)* vol.61 (2000) p.9 ]

[56] V.A. Nadolinny, A.P. Yelisseyev, O.P. Yuryeva, B.N. Feigelson [ *Appl. Magn. Reson. (Austria)* vol.12 (1997) p.543 ]

CHAPTER A8

# PROPERTIES OF DEFECTS IN DIAMOND: SHALLOW STATES IN DIAMOND

# A8.1 Properties of shallow bound excitons in diamond

K. Thonke and R. Sauer

April 2000

## A INTRODUCTION

Boron is the only shallow acceptor in diamond with a binding energy of 368 meV [1–3]. It is incorporated in natural type IIb diamonds being responsible for their dark-blue to black colour but can also be introduced into synthetic high pressure-high temperature (HPHT) crystals and CVD diamond layers. Phosphorus appears to act as a shallow donor with a binding energy of 560–650 meV in HPHT and CVD diamond [4–6]. Sulphur has been claimed recently to be a shallow donor, but this evidence is not yet fully documented at present [7,8]. Bound exciton luminescence has been identified owing to boron and phosphorus.

## B BORON BOUND EXCITON

Bound exciton cathodoluminescence (CL) was studied in detail by Dean et al [9] in natural and HPHT p-type diamonds. The impurity binding the exciton was associated with the acceptor responsible for the p-conductivity of the crystals at that time believed to be aluminium. The spectrum consists of a doublet no-phonon transition at 5.368 eV and 5.356 eV (values for 100 K sample temperature, lines $D_0$' and $D_0$, respectively) yielding an exciton localisation energy of $\approx$ 55 meV relative to the free exciton threshold energy. These no-phonon lines become weakly allowed in this indirect semiconductor due to a partial relaxation of the k-selection rule, since k is no longer a good quantum number for electrons and holes localised in real space at the impurities. More intense are the wave-vector conserving transitions at 5.227 eV ($D_1$') and 5.215 eV ($D_1$) involving one TO phonon with (141 ±1) meV energy. Further weaker phonon replicas due to TA (87 ±2 meV), LO (163 ±1 meV), and (TO + $O^{\Gamma}$) (141 +165 meV) phonons and also multiple combinations of these phonons were found. The doublet splitting of $\approx$ 12 meV of all these BE lines was ascribed to the presence of different holes in the BE complex, either two equivalent $\Gamma_8$ holes or inequivalent $\Gamma_8$ and $\Gamma_7$ holes. In the first case the exchange interaction of the equivalent holes leads to stronger exciton binding yielding the BE low-energy doublet component; in the latter case the $\Gamma_7$ hole from the split-off valence band has higher energy ($\Delta_0$) and the exchange interaction is lower due to the inequivalence of the two holes yielding the BE high-energy doublet component. The original identification of aluminium as the binding acceptor was later replaced by boron [2,3].

With improved quality control of chemical vapour deposited (CVD) diamond the boron bound exciton was also observed in synthetic CVD diamond films [10–12]. Kawarada et al [12] found in boron-doped polycrystalline films a linear relationship between electrical conductivity and boron content. From the intensity ratio of the boron bound exciton line $BE_B^{TO}$ and the free exciton line $FE^{TO}$, they derived a phenomenological relation valid at T = 80 K which allows the boron doping level in the range from $10^{17}$ $cm^{-3}$ to $10^{19}$ $cm^{-3}$ to be determined from CL spectra [13].

Boron bound exciton luminescence can also be observed in boron implanted and annealed HPHT diamond. At implantation doses of $\approx 2 \times 10^{13}$ $cm^{-2}$ annealing temperatures of 1000–1200°C are needed to heal lattice damage and activate boron onto optically active substitutional lattice sites [14].

Measurements on low-strain natural type IIa or synthetic HPHT diamonds with low boron concentrations allowed resolution of a fourfold fine structure in each of the doublet components $D_0$'

and $D_0$ of the $BE^{NP}$ transition [15,16] (see FIGURE 1 and TABLE 1). FIGURE 2 shows an energy level scheme which accounts for all splitting and thermalisation features observed. In particular, the intensity ratio of the two low energy components $I_a : I_b$ equals 1 : 2.5 and is not dependent on temperature as is also the case for the intensity ratio of the two high energy components, $I_c : I_d$ [15,16]. The level scheme in FIGURE 2 reveals two transition final states of the neutral boron acceptor $A^0$ with an energy spacing of 2.0 meV. The splitting was identified [15] with the spin-orbit splitting, $\Delta_0 = 2.07$ meV, of the bound hole in the acceptor ground state as known from infrared and electronic Raman measurements [1,17]. The nature of the 3.3 meV splitting in the initial states of the bound exciton remains unclear. The large 11.4 meV splitting in the initial state, which is similarly observed for the free exciton, is tentatively explained in terms of an excitonic spin singlet-spin triplet splitting [18].

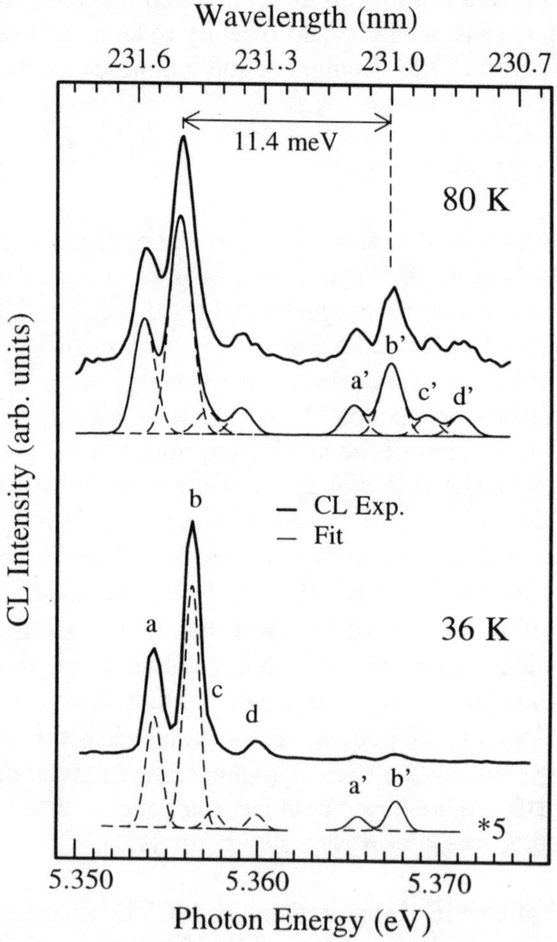


FIGURE 1 Fine structure of the boron bound exciton no-phonon line, recorded at two different temperatures. Solid line: experiment; dashed line: fit (from [15]).

TABLE 1 Photon energies of $BE^{NP}$ components at 80 K (from [18]).

| Line | hν (eV) | Line | hν (eV) |
|---|---|---|---|
| a | 5.3538 | a' | 5.3652 |
| b | 5.3558 | b' | 5.3672 |
| c | 5.3572 | c' | 5.3692 |
| d | 5.3591 | d' | 5.3711 |

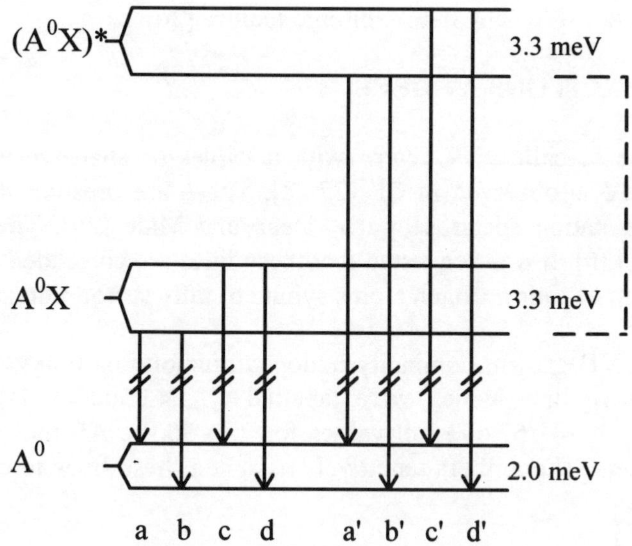


FIGURE 2 Level scheme of the boron bound exciton transitions in FIGURE 1 (from [15]).

When the boron content is increased, the BE recombination lines shift to lower energy and broaden [19], similar to the case of doped silicon [20]. For a boron concentration of ≈ 2 × $10^{20}$ $cm^{-3}$ the Mott density is reached, with a $BE^{TO}$ line energy of 5.0 eV.

## C PHOSPHORUS BOUND EXCITON

Synthetic diamond samples grown by CVD with intentional phosphorus doping and samples grown by HPHT synthesis with phosphorus as a catalyst reveal a relatively sharp new line at 5.175 eV (at T = 80 K). Based on its thermalisation behaviour relative to the free exciton line and on sample statistics, it was assigned to the TO-assisted transition of an exciton bound either to an isolated phosphorus donor or to a phosphorus-related complex [21]. This P-related exciton is bound with a localisation energy $E_{loc,X}(P)$ of ≈ 95 meV. The occurrence of this new line was accompanied by an extended spectrum of line-resolved donor-acceptor pair transitions between boron acceptors and the novel phosphorus-related donors [5]. Analysis of the spectra yielded a donor ionisation energy $E_D$ = (630 ±50) meV.

Comparing these data to the case of boron with an exciton localisation energy $E_{loc,X}(B)$ ≈ 55 meV [9] and a hole binding energy $E_A$ = 368 meV of the acceptor [1], an equivalent to Haynes' rule originally found for silicon [22] was tentatively established for diamond: the exciton localisation energy in diamond is ≈ 15% of the impurity ionisation energy [5].

## D OTHER POTENTIAL SHALLOW DOPANTS

Following theoretical predictions [23] lithium is a shallow interstitial donor, similar to the situation in silicon. Li implanted and annealed type IIa diamonds showed n-conductivity which was directly related to the dopant (rather than to the presence of damage-induced defects) but seemed to originate more probably from hopping processes than from impurity-to-band activation of carriers [24]. In CVD grown films with Li doping, a new, relatively broad transition at ≈ 5.12 eV along with a phonon replica displaced by -165 meV was observed, but could not definitely be assigned to Li [25]. For sodium, which is also expected to act as a donor [23], no CL spectra are reported to date. Samples which contain sulphur as an intentional dopant and are n-type conducting at room temperature [7] were investigated in CL, but did not show any new excitonic feature [26].

## E SHALLOW BOUND EXCITONS AT DEFECTS

In natural type I diamonds, the so-called N9 centre with a triplet of sharp zero-phonon lines at 5.252 eV, 5.262 eV and 5.277 eV is observed in CL [27,28]. These are presumably the same lines which were found earlier in excitation spectroscopy by Dean and Male [29]. The lines were also studied by Nahum and Halperin [30]. It was suggested that these lines are correlated with the nitrogen 'B aggregate', i.e. an arrangement of four nitrogen atoms symmetrically surrounding a vacancy [31].

In CL studies on hot-filament CVD grown, nominally undoped diamond particles, Robins et al [32] found a series of relatively sharp lines which were labelled C1 (4.950 eV), D1 (4.920 eV), E1 (4.832 eV), F1 (4.798 eV), and G1 (4.757 eV) (all values for T = 80 K). All these lines were most strongly observed on {100} facets. The authors tentatively assigned these lines to excitons bound to unidentified lattice defects.

## F CONCLUSION

Bound exciton cathodoluminescence owing to boron and phosphorus in diamond has been described. Lithium, sodium and sulphur are potential shallow dopants, and lines found in CL studies of nominally undoped diamond have been tentatively assigned to excitons bound to unidentified lattice defects.

## REFERENCES


[1] P.A. Crowther, P.J. Dean, W.F. Sherman [ *Phys. Rev. (USA)* vol.154 (1967) p.772 ]; A.T. Collins, E.C. Lightowlers [ *Phys. Rev. (USA)* vol.171 (1968) p.843 and references therein ]

[2] R.M. Chrenko [ *Phys. Rev. B (USA)* vol.7 (1973) p.4560 ]

[3] E.C. Lightowlers, A.T. Collins [ *J. Phys. D, Appl. Phys. (UK)* vol.9 (1976) p.951 ]

[4] S. Koizumi et al [ *Diam. Relat. Mater. (Switzerland)* vol.7 (1998) p.540 ]

[5] H. Sternschulte, K. Thonke, R. Sauer, S. Koizumi [ *Phys. Rev. B (USA)* vol.59 (1999) p.12924 ]

[6] M. Nesladek, K. Meykens, K. Haenen, L.M. Stals, T. Teraji, S. Koizumi [ *Phys. Rev. B (USA)* vol.59 (1999) p.14852 ]

[7] I. Sakaguchi et al [ *Phys. Rev. B (USA)* vol.60 (1999) p.R2139 ]

[8] R. Kalish, A. Reznik, C. Uzan-Saguy, C. Cyterman [ *Appl. Phys. Lett. (USA)* vol.76 (2000) p.757 ]

[9] P.J. Dean, E.C. Lightowlers, D.R. Wight [ *Phys. Rev. (USA)* vol.140 (1965) p.A352 ]

[10] A.T. Collins, M. Kamo, Y. Sato [ *J. Phys., Condens. Matter (UK)* vol.1 (1989) p.4029 ]

[11] W.D. Partlow, J. Ruan, W.J. Choyke, D.S. Knight [ *J. Appl. Phys. (USA)* vol.67 (1990) p.7019 ]

[12] H. Kawarada, Y. Yokota, A. Hiraki [ *Appl. Phys. Lett. (USA)* vol.57 (1990) p.1889 ]

[13] H. Kawarada, H. Matsuyama, Y. Yokota, T. Sogi, A. Yamaguchi, A Hiraki [ *Phys. Rev. B (USA)* vol.47 (1993) p.3633 ]

[14] H. Sternschulte et al [ *Appl. Phys. Lett. (USA)* vol.71 (1997) p.2668 ]

[15] H. Sternschulte et al [ *Mater. Sci. Forum (Switzerland)* vol. 258-263 (1997) p.757 ]

[16] S.J. Sharp, A.T. Collins, G. Davies, G.S. Joyce [ *J. Phys., Condens. Matter (UK)* vol.9 (1997) p.L451 ]

[17] H. Kim, R. Vogelgesang, A.K. Ramdas, S. Rodriguez, M. Grimsditch, T.R. Anthony [ *Phys. Rev. Lett. (USA)* vol.79 (1997) p.1706 and *Phys. Rev. B (USA)* vol.57 (1998) p.15315 ]

[18] R. Sauer, H. Sternschulte, S. Wahl, K. Thonke, T.R. Anthony [ *Phys. Rev. Lett. (USA)* (2000) to be published ]

[19] H. Sternschulte, T. Albrecht, K. Thonke, R. Sauer [ *Proc. 23rd Int. Conf. on the Physics of Semiconductors* Berlin, 1996, Eds. M. Scheffler, R. Zimmermann (World Scientific, Singapore, 1996) p.169 ]

[20] J. Wagner [ *Phys. Rev. B (USA)* vol.29 (1984) p.2002 ]

[21] H. Sternschulte, T. Albrecht, K. Thonke, R. Sauer, M. Grießer, M. Grasserbauer [ *Mater. Res. Soc. Symp. Proc. (USA)* vol.423 (1996) p.693 ]

[22] J.R. Haynes [ *Phys. Rev. Lett. (USA)* vol.4 (1960) p.361 ]

[23] S.A. Kajihara, A. Antonelli, J. Bernholc, R. Car [ *Phys. Rev. Lett. (USA)* vol.66 (1991) p.2010 ]

[24] S. Prawer, C. Uzan-Saguy, G. Braunstein, R. Kalish [ *Appl. Phys. Lett. (USA)* vol.63 (1993) p.2502 ]

[25] H. Sternschulte, K. Thonke, R. Sauer [ *Phys. Status Solidi A (Germany)* vol.172 (1999) p.37 ]

[26] N. Teofilov, K. Thonke, R. Sauer [ to be published ]. Samples of the group at CREST/NIRIM (Dr. Ando) were used for these studies. Hall measurements in our laboratory confirmed n-type conductivity at room temperature.

[27] D.R. Wight, P.J. Dean [ *Phys. Rev. (USA)* vol.154 (1967) p.689 ]

[28] P.A. Crowther, P.J. Dean [ *J. Phys. Chem. Solids (UK)* vol.28 (1967) p.1115 ]

[29] P.J. Dean, J.C. Male [ *Proc. R. Soc. Lond. A (UK)* vol.277 (1964) p.330 ]

[30] J. Nahum, A. Halperin [ *J. Phys. Chem. Solids (UK)* vol.23 (1962) p.345 ]

[31] A.T. Collins [ *Physica B (Netherlands)* vol.185 (1993) p.284 ]; G.S. Woods [ in *Properties and Growth of Diamond* Ed. G. Davies (INSPEC, IEE, London, UK, 1994) p.90 ]

[32] L.H. Robins, E.N. Farabaugh, A. Feldman [ *Phys. Rev. B (USA)* vol.48 (1993) p.14167 ]

# A8.2 Shallow n-type dopants in diamond: theory

A.B. Anderson

May 2000

## A INTRODUCTION

Theory is in the early stages of providing information about possible shallow n-type dopants in diamond, and comparison with experiment is often problematic because of the absence of definitive characterization of the experimental systems [1]. There are several levels of theoretical approximation that have been and continue to be used, and there are two types of donor process to consider: optical (sudden Franck–Condon ionization) and thermal (adiabatic ionization with defect site structure relaxation). For the optical case, the excited donor defect centre, D, de-excites by heating the sample (generating phonons), and the overall process may be represented at the adsorption band edge as

$$D + h\nu \rightarrow D^{+*} + e^{-}(cb) \rightarrow D^{+} + e^{-}(cb) + \text{heat} \quad (1)$$

where hν is the photon energy and the electron is excited to the bottom of the conduction band, cb. Such injections of electrons from deep levels can give rise to photocurrents [2] and with shallow n-type dopants infrared photons could inject electrons into the conduction band. In the thermal case lattice phonons interact with the donor electrons and an equilibrium is established wherein there is a certain population of the conduction band by free electrons that are essentially uncoupled from the resulting $D^{+}$. These electrons are the charge carriers for n-type conductivity and their number increases as the temperature is raised. Their number also increases exponentially with decreasing dopant ionization energies. The process in the shallow dopant case is

$$D \rightarrow D^{+} + e^{-}(cb) \quad (2)$$

Since the conduction band electrons in this process are completely free, D and $D^{+}$ are both in their relaxed structures with nuclear distribution functions determined by the zero–point vibrational wavefunctions in the low–temperature limit, with modifications at increasing temperatures due to the participation of higher energy vibrational states.

A commonly used theoretical model for semiconductor electronic properties is to rearrange electron occupations amongst the occupied and empty orbital energy levels that have been determined by a Hartree–Fock calculation on the ground electronic state for the defect model, which, preferably, is variationally optimized in its structure coordinates. Such rearrangements of occupations within the orbital energy spectrum yield energy differences between ground and excited states that are exactly the differences between the orbital energies, provided the excited state electronic wavefunctions are not variationally optimized. The demonstration of this fact is known as Koopmans' theorem, and this model is the basis of the common practice of comparing calculated initial–state energy level spectra with experimental optical and photoelectron emission spectra. It is a qualitative model, often very helpful, but subject to many sources of error, including (i) the need for relaxation of the excited or ionized state electronic wavefunction, (ii) the need for more accurate calculations on both the ground and excited state systems using electron wavefunctions of greater flexibility than provided by the single determinant (and with such wavefunctions comes the loss of the orbital picture for electronic structures provided by Koopmans' theorem), (iii) the need to add the nuclear vibrational energies to the electronic energies, and (iv) the need, in thermal excitation, EQN (2), to also optimize the structure of the excited state system.

Some of the work on n-type diamond has used parametrized Hamiltonians that gave energy level spectra that were used in the Koopmans' theorem sense to predict bandgap energy level positions. The extended Hückel (EH) and atom superposition and electron delocalization (ASED) molecular orbital procedures have been used in this way [3–7]. These methods are not improvable by (i) above because, once the Hamiltonian parameters are chosen, the spectrum is invariant to changes in orbital occupation. However, the ASED method has been used with improvement (iv) by calculating the dopant defect structure before and after ionization [4,6,7].

Density functional theory (DFT) has also been employed to predict properties of n-type donors in diamond [8,9]. Although there is no Koopmans' theorem for such calculations, the energy level spectrum is still used for comparing with experiment [8], and in recent work correction (iv), relaxation of the ionized state structure (this process also includes correction (i)), has been invoked [9].

## B CRITERIA FOR DOPANTS TO BE SHALLOW n–TYPE

According to ASED–MO energy level studies [7] of substitutional N and P placed in the vacancy site of a $C_{97}H_{78}$ cluster, it is the relative size of the valence p orbital that determines the relative closeness of the singly occupied orbital energy level to the conduction band. The different atomic electronegativity has a much smaller effect [7]. The donating electron orbital in the bandgap has antibonding (destabilizing) overlap between the donor and the surrounding carbon atoms [6]. Thus, substitutional P should be a better donor than N, S should be better than O, and Cl should be better than F; this is what is calculated [6]. Density functional calculations should give similar relative results and a comparison of the predictions of the two theoretical methods with experiment is given in TABLE 1.

TABLE 1 Calculated optical (opt.) and thermal (therm.) excitation energies (eV) from ASED cluster [6] and DFT band [8] calculations and experimental values for substitutional nitrogen [10] and (presumably) substitutional phosphorus [11–16] atoms. Negative energies mean spontaneous ionization.

| | Calculated | | | | | | Experimental | |
|---|---|---|---|---|---|---|---|---|
| | ASED | | | DFT | | | | |
| Dopant | Opt. | Therm. | | Opt. | Therm. | | Opt. | Therm. |
| N | 3.57 | 0.94 | | 0.7 | ---- | | 2.0 | 1.6–1.7 |
| P | –0.21 | –1.35 | | 0.2 | ---- | | 0.55, 0.65 | 0.46 |

Both theoretical models underestimate the donor excitation energies and both incorrectly predict the substitutional phosphorus atom to be a shallow donor and correctly predict a substitutional nitrogen atom to be a deep donor, though in the DFT case the thermal activation energy is not available for comparison, but will be less than the optical value.

All interstitial atoms except H [5] are expected to be shallow donors in diamond because of their large atomic orbital overlaps with neighbouring carbon atoms that destabilize the occupied antibonding counterpart orbitals despite the relaxation of the carbon atoms off their lattice sites. ASED calculations have predicted interstitial lithium atoms to be shallow donors and substitutional lithium atoms to be potential shallow acceptors [4]. The same will hold for the other alkali metal atoms. DFT band calculations yielded optical activation energies for substitutional Li and Na of 0.1 eV and 0.3 eV, respectively [8], and the thermal values, had they been calculated, should have been even less. There is as yet no well-defined experimentally prepared n–type lithium doped system. However, ion implantation has yielded n-type conduction activation energies of 0.10 eV [17] and 0.2 eV [18]. Lithium incorporated during diamond film growth has yielded p–type conductivity with an activation energy of 0.24 eV, which was explained as due to (unproven) boron contamination [19]. Later work showed n–type doping by lithium with an activation energy of 0.16 eV but no electrical activity was seen in samples doped by sodium or phosphorus under similar conditions [20]. Indiffusion of lithium largely quenches the band–A luminescence spectrum, attributed to vacancies, and shows no electrical activity [21]. The lack of electrical activity may be due to compensation of lithium atoms in substitutional sites (having entered vacancies) by interstitial lithium atoms [4]. Ion implantation of sodium has formed n–type diamond with a 0.13 eV activation energy [18]. As mentioned above, sodium has not been observed to incorporate in diamond during growth by chemical vapour deposition, but low energy sodium ion implantation during growth has formed p–type diamond with an activation energy of 0.044 eV for one sample and 0.082 eV for another [22]. Simply having the alkali metal atom in a substitutional site seems sufficient for producing p–type conductivity [4], though other explanations have also been proposed [16].

## C DONORS WITH TWO OR MORE ELECTRONS IN BANDGAP ORBITALS

Oxygen has smaller valence p orbitals than nitrogen so, according to the above–mentioned principles, its doubly occupied bandgap orbital should be even deeper than that of nitrogen. ASED calculations confirm this and also predict that relaxation of the $O^+$ structure still leaves it a deep thermal donor [6]. DFT calculations also predict a deep donor property [9]. Very recently, $O^+$ implantation followed by annealing has yielded n–type diamond with a thermal activation energy of 0.32 eV [14]. Whereas the proposal was made that the donors were substitutional oxygen atoms, the fact that annealing to above 600°C is expected to deactivate this electrical behaviour [14] implies that interstitial oxygen atoms may be involved. Theory also says that an atom is a shallower donor when in an interstitial site than it is in a substitutional site [4].

Sulphur has larger valence p orbitals than oxygen and is less electronegative, implying that it might be a shallower donor than oxygen. According to ASED calculations, substitutional sulphur is a much shallower thermal donor than substitutional oxygen and possibly shallower than substitutional nitrogen [6]. Recent B3LYP density functional calculations on $XC_{43}H_{42}$ and $[XC_{43}H_{42}]^+$ where X is P, S, and N, yield the result that substitutional S and N have the same thermal activation energies, and are 0.51 eV deeper than substitutional P [23]. This is comparable to the 1.2 eV difference between N and P from the experimental entries in TABLE 1, and supports the conclusion that substitutional sulphur atoms are deep donors. Recently sulphur doped diamond films were grown that demonstrated n–type conductivity with a 0.38 eV activation energy [24]. Subsequent investigation points to unintentional boron impurities as responsible for the conductivity [25].

Several other donors in mono– or di–substitutional sites were analyzed with ASED calculations: F, Cl, I, Xe, BO, BS, NO, NS and BeCl [6]. The calculated thermal activation energies were all greater than for P, and so the implication is that all are deep donors.

## D CONCLUSION

This review of the theoretical literature and comparison with the experimental literature leads one to conclude that interstitial sites are more probable donor sites for small atoms than are substitutional ones. The past theoretical work seems qualitative at best and new insight is needed to make theoretical predictions quantitative. There is a small but increasing number of well-defined experimentally prepared systems that can be used as benchmarks for theoretical models.

## ACKNOWLEDGEMENT

Unpublished results referenced herein were supported by the National Science Foundation (USA) Grant No. CHE-9816345.

## REFERENCES

[1] A.T. Collins [ *Diam. Relat. Mater. (Switzerland)* vol.8 (1999) p.1455-62 ]
[2] Yu.V. Pleskov, A.Ya. Sakharova, M.D. Krotova, L.L. Bouilov, B.V. Spitsyn [ *J. Elecroanal. Chem. (Switzerland)* vol.228 (1987) p.19-27 ]
[3] R.P. Messmer, G.D. Watkins [ *Phys. Rev. B (USA)* vol.7 (1973) p.2568 ]
[4] A.B. Anderson, S.P. Mehandru [ *Phys. Rev. B (USA)* vol.48 (1993) p.4423-7 ]
[5] S.P. Mehandru, A.B. Anderson [ *J. Mater. Res. (USA)* vol.9 (1994) p.383-95 ]
[6] A.B. Anderson, E.J. Grantscharova, J.C. Angus [ *Phys. Rev. B (USA)* vol.54 (1996) p.14341-8 ]
[7] A.B. Anderson, L.N. Kostadinov [ *J. Appl. Phys. (USA)* vol.81 (1997) p.264-7 ]
[8] S.A. Kajihara, A.Antonelli, J.Bernholc, R.Car [ *Phys. Rev. Lett. (USA)* vol.66 (1991) p.2010-3 ]
[9] S. Poykko, M. Kaukonen, M.J. Puska, R.M. Nieminen [ *Comput. Mater. Sci. (Netherlands)* vol.10 (1998) p.351-5 ]
[10] A.T. Collins, E.C. Lightowlers [ in *The Properties of Diamond* Ed. J.E. Field (Academic Press, London, 1979) p.79-105 ]
[11] S. Koizumi et al [ *Diam. Relat. Mater. (Switzerland)* vol.7 (1998) p.540-4]
[12] H. Sternschulte, K. Thonke, R. Sauer [ *Advanced Materials '98* (NIRIM, Tsukuba, Japan, 1998) p.113 ]
[13] M. Nesladek, K. Meykens, K. Haenen, L.M. Stals, T. Teruji, S. Koizumi [ *Diam. Relat. Mater. (Switzerland)* vol.8 (1999) p.882-5 ]
[14] J.F. Prins [ *Phys. Rev. B (USA)* vol.61 (2000) p.7191-4 ]
[15] M.E. Zvannt, W.E. Carlos, J.A. Freitas Jr., K.D. Jamison, R.P. Hellmer [ *Appl. Phys. Lett. (USA)* vol.65 (1994) p.2287-9 ]
[16] G. Popovici, M.A. Prelas [ *Diam. Relat. Mater. (Switzerland)* vol.4 (1995) p.1305-10 ]
[17] V.S. Vavilov, E.A. Knorova, E.B. Stepanova, E.M. Trukhan [ *Sov. Phys.-Semicond. (USA)* vol.13 (1979) p.635-8 ]
[18] S. Prawer, C. Uzan-Saguy, G. Braunstein, R. Kalish [ *Appl. Phys. Lett. (USA)* vol.63 (1993) p.2502-4 ]
[19] G.G. Fountain et al [ in *Diamond Materials* Eds. A.J. Purdes, J.C. Angus, R.F. Davis, B.M. Meyerson, K.E. Spear, M. Yoder (Electrochemical Society, 1991) p.523 ]
[20] T.H. Borst, O. Weis [ *Diam. Relat. Mater. (Switzerland)* vol.4 (1995) p.948-53 ]
[21] J. te Nijenhuis et al [ *Diam. Relat. Mater. (Switzerland)* vol.6 (1997) p.1726-32 ]
[22] K.D. Jamison, H.K. Schmidt, D. Eisenmann, R.P. Hellmer [ *Mater. Res. Soc. Symp. Proc. (USA)* vol.302 (1993) p.251-7 ]
[23] A.B. Anderson, T.V. Albu [ unpublished ]
[24] I. Sakaguchi et al [ *Phys. Rev. B (USA)* vol.60 (1999) p.R2139-41 ]
[25] R. Kalish, A. Reznik, C.U.-Saguy, C. Cytermann [ *Appl. Phys. Lett. (USA)* vol.76 (2000) p.757-9 ]

# A8.3 Passivation of acceptors in diamond: B–H interaction

J. Chevallier

December 1999

## A INTRODUCTION

When diamond is grown by the chemical vapour deposition (CVD) technique, a hydrogen-rich mixture of $CH_4/H_2$ is used as a precursor. As a consequence, large amounts of atomic hydrogen are present in the gas phase. Atomic hydrogen introduced in more conventional semiconductors (Si, GaAs, etc.) during their growth and/or their processing is known to passivate acceptor and/or donor dopants. We review here the present information and data on the boron–hydrogen interactions in p-type diamond since boron is commonly used as a p-type dopant in CVD grown diamond films and in synthetic diamond as well.

## B HYDROGEN DIFFUSION AND SOLUBILITY

Most data on the diffusion and the solubility of hydrogen in boron doped diamond come from experiments where monocrystalline (synthetic or homoepitaxially grown) diamond samples homogeneously doped with boron have been exposed to a deuterium plasma (radiofrequency at 13.56 MHz, microwave at 2.45 GHz or DC). Both the hydrogen diffusion coefficient and the hydrogen solubility strongly depend on the boron concentration and on the diffusion temperature.

Two distinct hydrogen diffusion regimes must be distinguished:

### B1 Low Temperature Diffusion Regime (T<550°C)

In this temperature range, the hydrogen diffusion profiles exhibit a quasi plateau region followed by a steep decrease of the hydrogen concentration as shown in FIGURE 1. In the plateau region, the hydrogen concentration matches the total boron concentration. The hydrogen diffusion profile shape indicates that the hydrogen diffusion is limited by the trapping of hydrogen on boron acceptors with the formation of B-H pairs: each boron atom traps one hydrogen [1]. The deuterium accumulation layer in the near surface region is attributed to deuterium trapped on plasma-induced defects.

One can define an effective diffusion coefficient $D_{eff} = L_D^2/t_0$ where $L_D$ is the plateau extent and $t_0$ the hydrogen diffusion duration. For a radiofrequency (RF) deuterium plasma exposure, the effective hydrogen diffusion coefficient at 290°C is given in TABLE 1 for the different boron concentrations of FIGURE 1.

TABLE 1 Dependence of the effective hydrogen diffusion coefficient on the boron concentration at 290°C for boron-doped diamond exposed to an RF deuterium plasma [1].

| Boron concentration ($cm^{-3}$) | $D_{eff}$ ($cm^2/s$) at 290°C |
|---|---|
| $3.00 \times 10^{20}$ | $1.6 \times 10^{-15}$ |
| $5.00 \times 10^{19}$ | $1.8 \times 10^{-14}$ |
| $2.00 \times 10^{18}$ | $9.3 \times 10^{-14}$ |

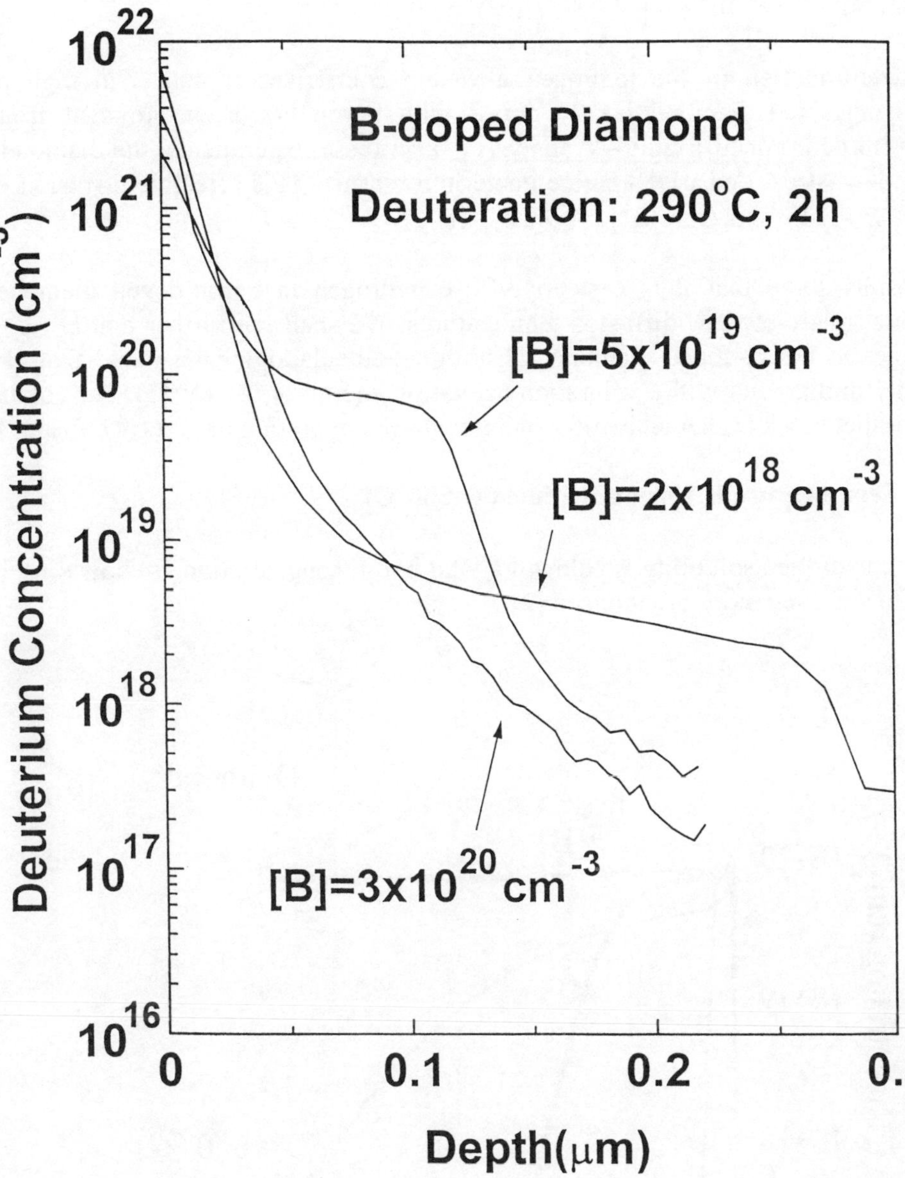


FIGURE 1 Evolution of the deuterium diffusion profiles in homoepitaxial boron-doped diamond as a function of the boron concentration. The epilayers were exposed to an RF deuterium plasma at 290°C for 2 hr [1].

The temperature dependence of the hydrogen diffusion coefficient is found to be different for an RF deuterium plasma exposure and a microwave deuterium plasma exposure:

-in a homoepitaxial diamond containing $5 \times 10^{19}$ $cm^{-3}$ boron and exposed to an RF deuterium plasma, one gets [2]:

$$D_{eff}\,(cm^2/s) = 2.5 \times 10^{-11} \exp[-0.35/kT(eV)] \quad (1)$$

-in a homoepitaxial diamond containing $8 \times 10^{19}$ $cm^{-3}$ boron and exposed to a microwave deuterium plasma, one gets [3]:

$$D_{eff}(cm^2/s) = 3.5 \times 10^{-5}\exp[-1.41/kT(eV)] \quad (2)$$

An indirect determination of the hydrogen diffusion coefficient at 400°C in high pressure/high temperature synthesized diamond lightly doped with boron has been provided from the boron passivation depth determined from C–V analysis [4]. In these experiments, the diamond sample was placed 10 cm downstream of a DC-remote deuterium plasma. This effective diffusion coefficient is $D_{eff}(400°C) = 4.7 \times 10^{-13}$ $cm^2/s$ for $[B] = 2 \times 10^{16}$ $cm^{-3}$.

The above results show that it is easy to diffuse hydrogen in boron-doped diamond by plasma exposure even at relatively low diffusion temperatures. We shall see further that $H^+$ is the dominant diffusing species in boron-doped diamond. Theoretical calculations have found that $H^+$ is a highly mobile impurity in diamond with a migration activation energy of 0.1 eV [5]. For comparison, these calculations predict much higher activation energies for the migration of $H^0$ (1.9 eV) and $H^-$ (2.5 eV).

**B2 High Temperature Diffusion Regime (T>550°C)**

In this case, the hydrogen solubility is below the total boron concentration as shown in FIGURE 2. At 700°C, this effect is even more pronounced [2].

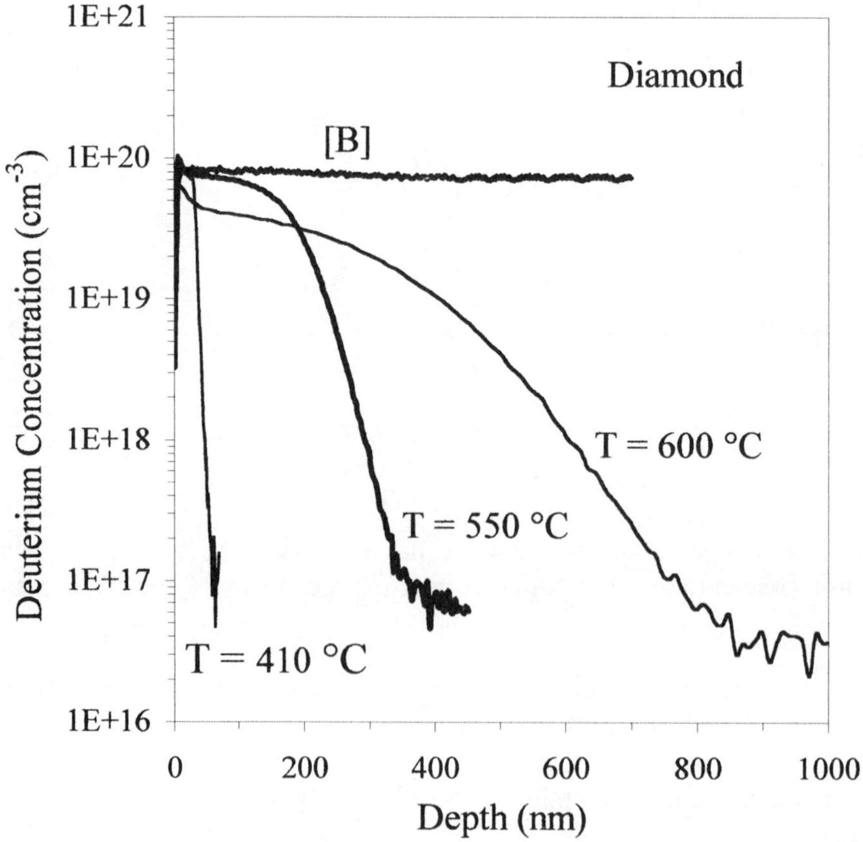


FIGURE 2 Temperature dependence of deuterium diffusion profiles in homoepitaxial boron-doped diamond exposed to a deuterium microwave plasma. For all the epilayers, the total boron concentration is $8 \times 10^{19}$ $cm^{-3}$ and the exposure duration is 2 hr [9].

As the temperature increases, the hydrogen diffusion profiles deviate from a step function and approach a complementary error function (erfc). At 700°C, the diffusion profile is correctly fitted by an erfc. For [B] = $5 \times 10^{20}$ $cm^{-3}$, the hydrogen diffusion coefficient at 700°C is $1.1 \times 10^{-12}$ $cm^2/s$ [2]. In this high temperature diffusion regime, the hydrogen transport is mainly limited by diffusion with weak B–H interactions.

Another determination of the hydrogen diffusion coefficient has been provided in type IIa natural diamond initially diffused with N, Li and O at 860°C and then co-diffused with hydrogen and boron again at 860°C [6]. In such a sample, the hydrogen diffusion coefficient is $2.4 \times 10^{-13}$ $cm^2/s$. However, this value is not representative of the hydrogen diffusion coefficient in boron doped diamond at 860°C because the hydrogen diffusion occurs in samples containing inhomogeneous distributions of different impurities.

## C ELECTRICAL PASSIVATION OF BORON ACCEPTORS

Different theoretical calculations have predicted a passivation of the electronic levels of boron acceptors by hydrogen in diamond as a result of the formation of B–H pairs [5,7].

An experimental demonstration of the electrical passivation of boron acceptors is shown in FIGURE 3 [4]. In this case, Schottky diodes have been realized on synthetic IIb diamond before and after exposure to a DC-remote deuterium plasma at 400°C for 2 hr. Before deuterium diffusion, the boron acceptor concentration is $2.7 \times 10^{16}$ $cm^{-3}$. After deuterium diffusion, the capacitance of the Schottky diodes significantly decreases. This corresponds to an increase of the depletion layer width due to a decrease of the effective acceptor density. Analysis of the C–V characteristics indicates an effective boron concentration of $5 \times 10^{13}$ $cm^{-3}$ after deuterium diffusion. This means that the passivation of boron by hydrogen in diamond is very efficient.

## D THERMAL STABILITY OF B-H PAIRS

The critical temperature of 550°C between the two hydrogen diffusion regimes is an indication that B-H pairs start to dissociate at this temperature in diamond (FIGURE 2). From reverse bias annealing experiments on deuterated boron doped diamond, it is found that B–H pairs dissociate at about 480°C [8]. For comparison, the dissociation temperature of B–H pairs in silicon is 250°C and their dissociation energy is 1.3 eV. From that, an estimated value of the dissociation energy of B–H pairs in diamond is 2 eV. This value is well above the theoretical value of 0.3 eV [5].

At the temperatures of 500–550°C, B–H pairs dissociate but the liberated hydrogen species do not out-diffuse from the diamond sample. Deuterium out-diffusion experiments, performed by heating the samples in a linear temperature ramp (20 K/min) while measuring the mass 4 ($D_2$) signal of a quadrupole mass spectrometer, show that the out-diffusion occurs at 880°C for microwave plasma exposed diamond as seen in FIGURE 4. Moreover, the effusion peak temperature does not depend on the hydrogen diffusion depth which implies that the hydrogen diffusion in diamond is not the limiting process for hydrogen out-diffusion. Since neither the hydrogen diffusion nor the B–H complex dissociation is the limiting process, it is concluded that the diamond surface barrier limits the hydrogen out-diffusion [9].

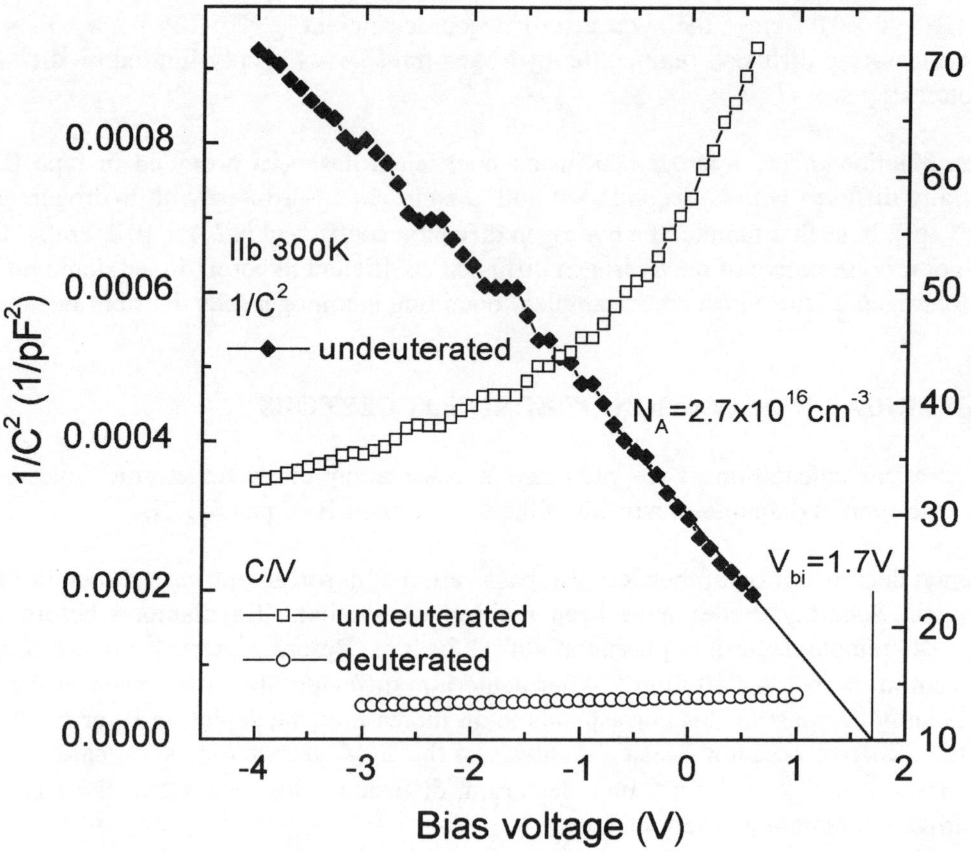


FIGURE 3 Capacitance-voltage characteristics of a deuterated (open circles) and an undeuterated (open squares) Schottky diode on synthetic type IIb diamond. The inverse squared capacitance data are shown only for the undeuterated diode (solid diamond). For the deuterated diode, the analysis of the C–V characteristics must take into account the presence of the efficiently passivated region in order to extract the effective boron concentration in this region [4].

## E THE HYDROGEN CHARGE STATE IN BORON DOPED DIAMOND

In reverse bias annealing experiments at 480°C on deuterated boron-doped diamond Schottky diodes, changes of the C–V characteristics are due to the B–H pair dissociation in the high electric field region below the surface and to a drift of hydrogen deeper in the sample where new B–H complexes are formed [8]. This means that hydrogen diffuses in the positive charge state $H^+$ in boron-doped diamond.

Calculations have also concluded that hydrogen acts as a mid-gap donor $H^{0/+}$ in diamond, hence with a level above the fundamental acceptor level of boron at $E_V$+0.37 eV [5]. As a consequence, hydrogen atoms ionize into $H^+$ by compensating the free holes and the neutral boron acceptors available at the diffusion temperature. If the hydrogen concentration is equal to or above the boron concentration, all the boron acceptors are ionized as $B^-$ and, at T<550°C, $H^+$ and $B^-$ form neutral complexes by coulombic attraction. In boron-doped diamond, the decrease of the neutral boron concentration under hydrogenation has been detected by infrared absorption [1].

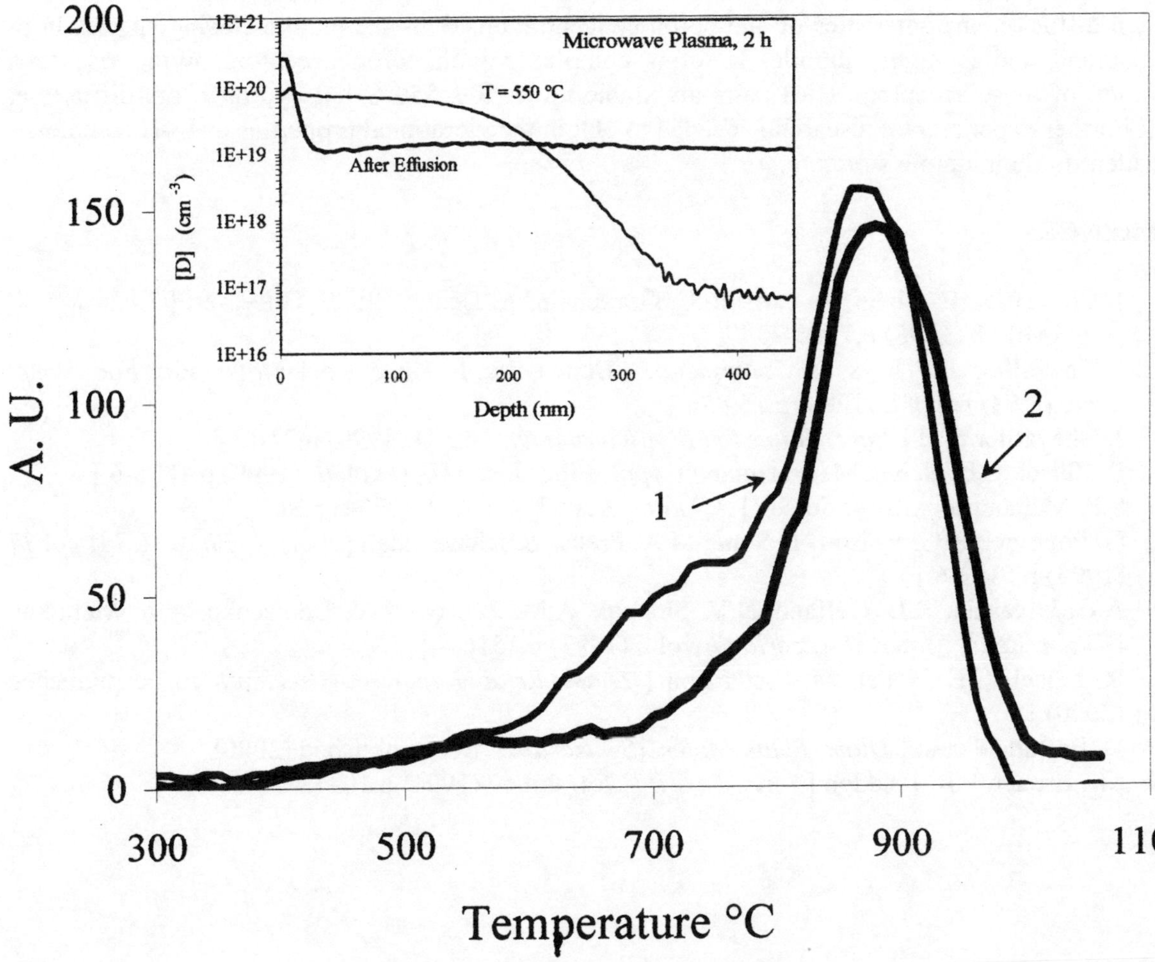


FIGURE 4 Deuterium out-diffusion experiments for homoepitaxial boron-doped diamond with 8 × $10^{19}$ $cm^{-3}$ boron concentration after exposure to a deuterium microwave plasma for 2 hr at 600°C (1) and 450°C (2). The shoulder at 700°C for sample (1) is probably due to deuterium trapped on plasma-induced defects. The D concentration profiles of a 550°C diffused sample before and after effusion are shown in the insert [9].

## F THEORETICAL ATOMIC STRUCTURES OF B–H PAIRS

So far, the local vibrational mode of the B–H complexes in diamond has not been experimentally observed. Three models have been proposed by theorists for the description of B–H complexes in diamond. First principles local-density functional theory predicts a hydrogen atom bonded to boron with the B–H bond along a <100> direction [10]. A semi-empirical molecular orbital calculation predicts that hydrogen forms a puckered bridged bond between B and one of its four C nearest neighbours [5]. Differing from these two models which exclude a trigonal symmetry, molecular-orbital linear combination of atomic orbitals calculations conclude that hydrogen is sitting in a tetrahedral interstitial position nearest to the boron atom [7].

## G CONCLUSION

Hydrogen-boron interactions in diamond are demonstrated through a strong dependence of the hydrogen diffusion characteristics on the boron concentration. $H^+$ is the main diffusing species in p-type diamond and is highly mobile. It forms complexes with boron acceptors giving rise to a passivation of these acceptors. B–H pairs are stable up to 500–550°C but hydrogen out-diffuses at 880°C. Further experimental research is needed to obtain the vibrational properties of B–H complexes and to identify their atomic structure.

## REFERENCES


[1] J. Chevallier, B. Theys, A. Lusson, C. Grattepain, A. Deneuville, E. Gheeraert [ *Phys. Rev. B (USA)* vol.58 (1998) p.7966-70 ]

[2] J. Chevallier, B. Theys, C. Grattepain, A. Deneuville, E. Gheeraert [ *Mater. Res. Soc. Symp. Proc. (USA)* vol.510 (1998) p.169-74 ]

[3] J. Chevallier et al [ *Phys. Status Solidi A (Germany)* vol.174 (1999) p.73-81 ]

[4] R. Zeisel, C.E. Nebel, M. Stutzmann [ *Appl. Phys. Lett. (USA)* vol.74 (1999) p.1875-6 ]

[5] S.P. Mehandru, A.B. Anderson [ *J. Mater. Res. (USA)* vol.9 (1994) p.383-95 ]

[6] G. Popovici, R.G. Wilson, T. Sung, M.A. Prelas, S. Khasawinah [ *J. Appl. Phys. (USA)* vol.77 (1995) p.5103-6 ]

[7] A.G. Ulyashin, R.B. Gelfand, N.V. Shlopak, A.M. Zaitsev, A.V. Denisenko, A.A. Melnikov [ *Diam. Relat. Mater. (Switzerland)* vol.2 (1993) p.1516-8 ]

[8] R. Zeisel, C.E. Nebel, M. Stutzmann [ *Diam. Relat. Mater. (Switzerland)* to be published (2000) ]

[9] D. Ballutaud et al [ *Diam. Relat. Mater. (Switzerland)* to be published (2000) ]

[10] S.J. Breuer, P.R. Briddon [ *Phys. Rev. B (USA)* vol.49 (1994) p.10332-6 ]

# A8.4 Photoconductivity in diamond

E. Pereira

April 2000

## A INTRODUCTION

Photoconductivity of bulk diamond has been studied, for all types of natural diamonds, as well as irradiated diamonds. Also in recent years photoconductivity of diamond films has been studied. Among the best studied defects that originate photoconductivity are the neutral vacancy (GR system), the negatively charged vacancy (ND1) and the boron containing IIb diamonds.

## B VACANCY ASSOCIATED PHOTOCONDUCTIVITY

The neutral vacancy has an absorption spectrum to several excited states (GR1–GR8). While the GR1 zero phonon line occurs at 1.673 eV, the GR2–GR8 lines occur between 2.881 and 3.006 eV. Luminescence occurs only from GR1, while the excited states associated with the other transitions give photoconduction [1]. In the photoconduction spectrum the GR1 originates a negative signal, indicating that the photons absorbed into GR1 do not originate ionization. By photo Hall measurements it was established that the photoconduction of the GR system is due to holes [2].

The negatively charged vacancy (ND1) is observed in absorption but not in luminescence. Its zero phonon line occurs at 3.149 eV and presents a well structured side band. The photocurrent is due to the emission of the electron [1].

## C SEMICONDUCTING DIAMOND

In diamond boron acts as an acceptor with a level 370 meV above the valence band. Two interesting phenomena occur in this photoconductivity: the observation of oscillations above the acceptor ionization energy and photothermal ionization and tunnelling.

In absorption, above the ionization edge, at least three series of peaks separated by 158 meV, and with a decrease of about one order of magnitude from one series to the next, are observed. In the photoconductivity spectrum temperature independent minima are observed at 452, 469 and 510 meV and temperature dependent minima at 375.8, 540.5, 702.1 and 865.8 meV. The temperature independent features are attributed to Fano resonances, while the temperature dependent minima are due to fast de-excitation to near ionization states of the acceptor [3].

Sharp maxima in the photoconductivity spectrum are observed at 77 K below the ionization energy of the acceptor and at energies where absorption occurs. This is interpreted as due to optical absorption to excited states of the acceptor followed by thermal excitation to the valence band [4]. This photoconductivity decreases with temperature but at 4 K in some diamonds a weak broad peak is observed attributed to tunnelling of holes from the excited states of the acceptor to neighbouring ionised centres [4].

## D DIAMOND FILMS

In diamond films there have been in recent years several studies on photoconductivity. The analysis is complicated by the fact that the films are polycrystalline and at the interfaces there are non-diamond phases. Thus most of the studies have shown broad structureless bands in a wide range of photon energies, very sensitive to the growth conditions and the presence of the boron acceptor. However, some features, such as the ND1, are common with natural diamonds, although the structure is not so well defined as in bulk diamond. Also a dip in the spectrum can be observed in the region of the GR1 absorption. The presence of vacancies on the films is to be expected, according to the thermodynamics of the growth process. Photoconductivity originating in the bulk of the diamond films, grown by the hot filament method on silicon wafers, has been reported with a threshold of 1 eV. A contribution from the contacts was ruled out. Up to 1.9 eV the square root of the photoyield is linear with photon energy, but it levels off above this energy. Above 3 eV a fast increase of photocurrent is observed. The origin of the free carriers was suggested to be un-ionized acceptors [5]. On films grown by MPCVD on Si and using Ti contacts, bands have been found around 0.5, 1.1, 1.4 and 1.9 eV [6]. Photoconductivity of undoped and boron doped films is similar, except when the boron doped film is observed without applied bias. In this situation the peak at 1.1 eV is absent. This situation is attributed to electron filling of the acceptors responsible for the photocurrent, due to hole depletion in the region close to the Schottky contact. With a bias voltage, the photocurrent originates away from the contact, and thus the 1.1 eV peak is observed. The 1.4 eV and 1.9 eV bands are attributed to deeper states [7]. In HFCVD diamond films on silicon, features observed at 2.2 and 3.6 eV have been attributed to the photoionization of single substitutional nitrogen and intrinsic defects [8]. A study of photoconductivity of free-standing films, grown on Si or Cu substrates, and their temperature dependence has shown that bands at 1.8–1.9, 2.6–2.7 and 3.1–3.2 eV are observed. In both types of film the activation energy is similar for the bands. The lower energy bands have activation energies around 90 meV, while the band at 3.1–3.2 eV has an activation energy of 140 meV. In the Cu grown film, with smaller grains and more non-diamond phases, extra bands at 2.9 and 3.9 eV can be observed, while in the Si grown film a band at 2.2 eV is observed [9]. The extra substrate dependent bands are attributed to defects at the grain boundaries, while the others originate in the bulk of diamond crystallites.

## E CONCLUSION

The photoconductivity of bulk diamond is rather well understood. However, the behaviour of diamond films is much more complex. Features are dependent on growth process and difficulties are increased due to the different film morphologies and the contact behaviour. However, some understanding of the films is emerging, mainly triggered by the potential applications of diamond films as photon detectors in a wide energy range.

## REFERENCES


[1] R.G. Farrer, L.A. Vermeulen [ *J. Phys. C (UK)* vol.5 (1972) p.2762-8 ]

[2] C.C. Clark, B.J. Parsons, L.A. Vermeulen [ *J. Phys. C (UK)* vol.12 (1979) p.2597-610 ]

[3] G. Davies, R. Stedman [ *J. Phys. C (UK)* vol.20 (1987) p.2119-27 ]

[4] A.T. Collins, E.C. Lightowlers [ *Phys. Rev. (USA)* vol.171 (1968) p.843-55 ]

[5] K. Okumura, J. Mort, M.A. Machonkin [ *Philos. Mag. Lett. (UK)* vol.65 (1992) p.105-12 ]

[6] P. Gonon, A. Deneuville, E. Gheeraert, F. Fontaine [ *Diam. Relat. Mater. (Switzerland)* vol.3 (1994) p.836-9 ]

[7] P. Gonon, A. Deneuville, F. Fontaine, E. Gheeraert [ *J. Appl. Phys. (USA)* vol.78 (1995) p.6633-8 ]

[8] M.C. Rossi, S. Salvatori, F. Galluzzi [ *Diam. Relat. Mater. (Switzerland)* vol.6 (1997) p.712-6 ]

[9] L. Pereira, E. Pereira, H. Gomes [ *Diam. Relat. Mater. (Switzerland)* to be published (2000) ]

CHAPTER B1

# SINGLE CRYSTAL AND MICROCRYSTALLINE DIAMOND FILMS: GROWTH ISSUES

# B1.1 High temperature high pressure synthesis of single crystal diamond

H. Kanda and T. Sekine

February 2000

## A INTRODUCTION

High pressure diamond is synthesized at conditions where diamond is thermodynamically more stable than graphite. Knowledge of the phase diagram of carbon is, therefore, important for diamond synthesis. However, the diamond stable condition is not sufficient for the synthesis. Kinetic considerations are also necessary. The activation energy of direct conversion (solid–solid transformation) from graphite to diamond is very large, and very high temperatures and pressures are required. In order to decrease the activation energy, so-called catalysts are employed. The conversion from graphite to diamond with/without the catalysts is summarized in this Datareview.

Synthetic diamond has three forms, i.e. single crystal, polycrystalline aggregate, and powder. The characteristics of only the single crystal are summarised here, because the other two forms are rarely used in electronic applications.

## B THE PHASE DIAGRAM OF CARBON

FIGURE 1 is a phase diagram of carbon showing several phase boundaries [1,2]. The boundary between diamond and graphite has been studied in most detail. The boundary expressed by $P(GPa) = 1.94 + 0.0025T(°C)$ is the most reliable to date [3]. The relationship was determined based on experiments on growth and dissolution of both diamond and graphite, accounting for the pressure effect on EMF of a thermocouple.

## C DIRECT CONVERSION

There are two ways to synthesize diamond without the help of any catalysts, i.e. static and dynamic compression.

In static compression, 8 to 20 GPa of pressure and 1000 to 3000°C are required [1]. In order to generate such high pressure, a belt-type apparatus, diamond anvil cell and multi-anvil apparatus are used. Electric current or a laser is applied for heating the sample. Carbon sources used are graphite, amorphous carbon, glassy carbon and fullerite $C_{60}$. The diamond synthesized is less than 20 μm in size. Hexagonal diamond (lonsdaleite) has been formed as well as cubic diamond.

Conversion to diamond has been detected even at room temperature using in-situ X-ray diffraction, but the diamond is not quenchable [4]. When $C_{60}$ was used as the carbon source, diamond was recovered after high pressure compression of 20 GPa at room temperature [5].

Dynamic compression techniques have produced diamond at pressures of 7 to 150 GPa for diamond synthesis [1]. The temperature is variable depending on the initial sample density. A variety of carbon sources, such as $C_{60}$ and nanotubes, as well as graphite have been employed [1,6]. In most cases, the duration of loading is of the order of a microsecond, and the diamond formed is quite small, less than 100 nm grain size.

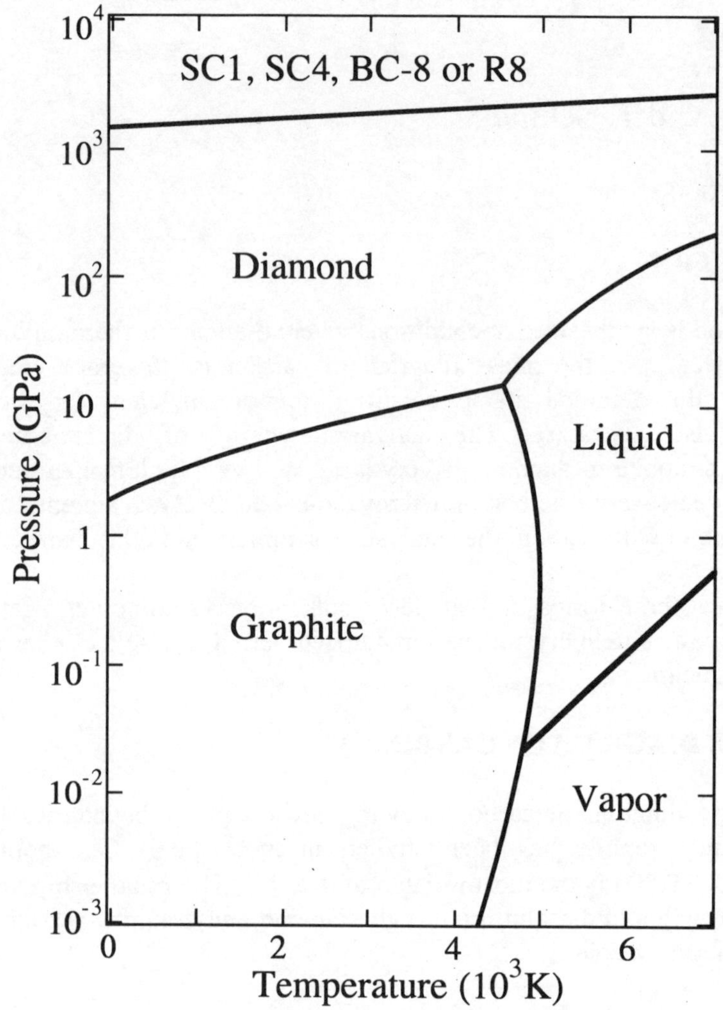


FIGURE 1 Pressure–temperature phase diagram of carbon.

The threshold pressure for the conversion to diamond depends on the nature of the initial carbon. The formation zone for the hexagonal diamond is located at relatively lower temperatures than that for cubic diamond. Well-crystalline graphite enhances the formation of hexagonal diamond. This connection is consistent with the martensitic nature for the graphite–hexagonal diamond transformation. In order to recover the hexagonal diamond irreversibly, it is necessary to heat the compressed sample to about 1000°C [7].

There are many proposals which seek to explain the transformation mechanism of graphite–cubic diamond: (i) by means of the presence of rhombohedral graphite in natural graphite [8]—this is now not plausible because rhombohedral graphite is rather rare; (ii) via liquid carbon [9]; (iii) via pressure-assisted diffusion in highly defective lattices [10]; (iv) through the solid-vapour-liquid-solid sequence [11]; (v) through intermediate phases possibly such as hexagonal diamond; (vi) a martensitic transformation from suitably oriented graphite [12,13].

## D CATALYTIC CONVERSION

### D1 Catalysts

There are many kinds of materials that assist conversion of graphite to diamond at lower temperature and pressure conditions. The materials have been called catalysts or catalyst-solvents.

The catalysts are shown with comments in the following:

(i) Conventional catalysts [14]: Transition metals of group VIII elements (Fe, Co, Ni, Ru, Rh, Pd, Os, Ir, Pt), Mn, Cr and Ta.

Ni, Co, Fe and alloys containing them as main components are popularly used for production of diamond. More than 10 at % of carbon dissolves into the molten metals, and they are essentially the solvents of carbon.

(ii) Cooperative catalysts [15]: Alloys of carbide forming elements (Ti, Zr, Hf, V, Nb, Mo, W) and inert elements (Cu, Ag, Au).

The elements of each group do not have the catalytic effect. The catalytic effect of the alloy is interpreted to be due to the same electronic configuration as that of the conventional catalysts. The alloys are, therefore, in the same category as the conventional catalysts.

(iii) Mg [16].

This plays the role of the solvent of carbon essentially at high temperatures where its carbide decomposes, and may be in the same category as the conventional catalysts, although this is not a transition metal.

(iv) Oxygen-containing materials [17–20]: Carbonates ($CaCO_3$, etc.), hydroxides ($Ca(OH)_2$, etc.), sulphates ($MgSO_4$, etc.), $H_2O$, kimberite.

Any compounds containing oxygen may be effective as catalysts except thermally stable oxides. Oxygen liberated from the compounds may play an essential role for the diamond synthesis. The catalysts are likely to have a close correlation with growth of natural diamond.

(v) Inert elements [21,22]: P, Cu, Zn, Ge, Sn, Sb.

These elements are known to show little reactivity with carbon, and they do not work as catalysts even if they melt, in contrast to the conventional catalysts. Much higher temperatures than their melting temperature are required for diamond synthesis. The rate of conversion to diamond is low.

(vi) Hydrides [23]: LiH, $CaH_2$.

Micrometre sized crystals have been formed.

### D2 P–T Condition

A pressure–temperature region where diamond grows is bounded by two lines as shown in FIGURE 2(a) [24,25], i.e. the diamond–graphite equilibrium line and a melting line where a liquid phase is formed in the catalyst-carbon system. This region is well fit for the conventional catalysts. The melting line is lower than the melting temperature of pure metals, if the metal has a eutectic relation with carbon [26]. On the other hand, the melting line is higher than the melting temperature of pure metals if the metals form stable carbides. Ni and Co have a eutectic relation with carbon, whereas Fe and Mn form stable carbides. In order to form the liquid phase at low temperatures, alloys of the metals are employed. 5 to 6 GPa and 1200 to 1500°C are the typical conditions for commercial production using the conventional catalysts [27]. However, there is a report that diamond grows from solid catalyst metal, although the yield of diamond is low [28].

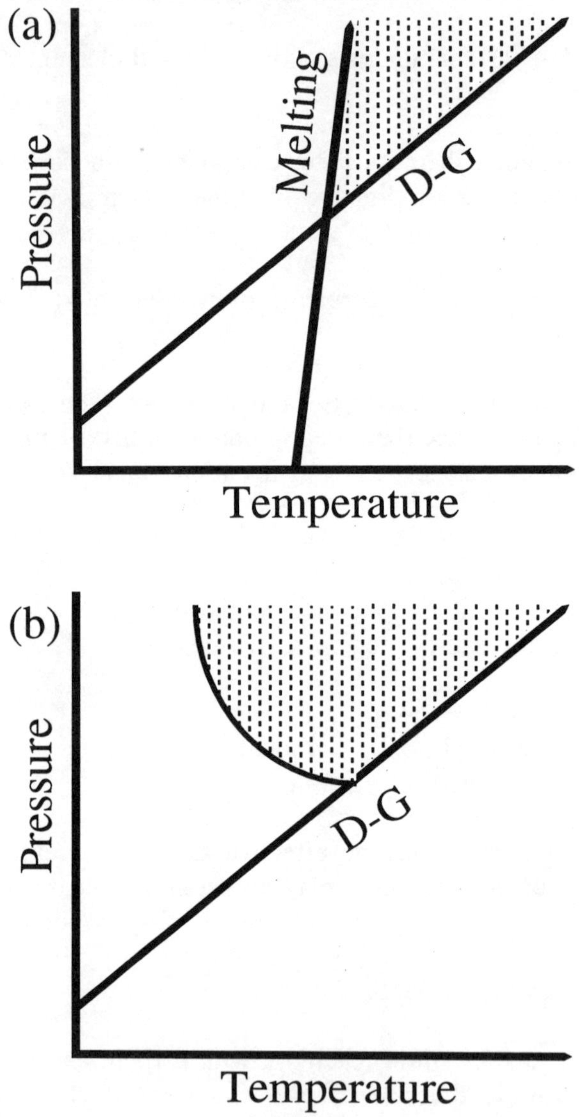


FIGURE 2 Schematic of diamond growth regions from (a) the conventional catalysts and (b) non-metallic catalysts. Diamond grows in the shaded regions.

To be more exact, the growth region does not fit to the equilibrium line; some excess pressure is required [29]. The deviation is prominent for the non-conventional catalysts as shown in FIGURE 2(b) [30]. Higher pressure is required to grow diamond at lower temperature. The presence of a negative slope at the low temperature range of the growth region indicates that diamond does not grow even if the catalyst is in the liquid phase. The minimum temperature for growth is higher than the melting temperature of the catalyst at a given pressure. C-H-O fluid, typically $H_2O$, is expected to be in the liquid phase even at very low temperatures, but it has been reported that 1150°C is a minimum temperature for the growth [31]. The minimum pressure and temperature for the growth is lowered by employing a seed crystal, indicating that the nucleation barrier is considerable for formation of diamond.

### D3 Mechanism

The conventional catalysts essentially play the role of the solvent of carbon at the P–T condition for diamond growth, and it is reasonably understood that diamond grows from solutions which are supersaturated with carbon [25]. There are two ways to form the supersaturated solution. (i) The chemical potential of graphite is higher than that of diamond in a diamond-stable condition, and solution saturated with graphite is supersaturated for diamond. The solubility difference between graphite and diamond produces continuous conversion from graphite to diamond. This mechanism is applied for the production of diamond grits up to 1 mm. (ii) If there is a temperature gradient in the catalyst solution, carbon dissolved in a higher temperature region is transported to a cooler region, where the solution becomes supersaturated [25,26]. This is because the carbon solubility increases with temperature. If a seed crystal is placed at the cooler region, the carbon precipitates on the seed, and a large crystal grows. This method is termed the temperature gradient method.

As for the non-conventional catalysts such as carbonates and copper, the catalytic effect has not been studied systematically. As described in the preceding section, their liquid phase is not always a solution from which diamond precipitates.

There are reports presenting diamond growth at much lower pressure and temperature conditions. One report says that the diamond has grown at 140 MPa and 800°C which is in a graphite-stable region in a system of Ni, $H_2O$ and graphite [32]. The growth mechanism may be completely different from the catalytic effect described above.

## E SINGLE CRYSTAL DIAMOND

The temperature gradient method is used to grow large crystals above 1 mm. Parameters which influence the characteristics of the grown crystals are growth temperature, temperature gradient, growing time, composition of the catalysts, and presence of impurities in the growth cell. The characteristics of the grown crystals are:

(i) Size

The world's largest crystal is 34.80 carats [33], and up to 10 mm crystals are commercially available [34]. Growth rates depend on growth temperature and the temperature gradient. They are about 10 mg/hr or less for high quality crystals [35]. The growth rate must be kept lower to grow colourless crystals than for the growth of yellow crystals [36].

(ii) Morphology

A single crystal polyhedron consists of {111}, {100}, {113}, {110} and {115} [26,37]. The former two appear as major surfaces. {113} and {110} are commonly seen as truncations of the former two. The morphology depends on the catalysts [38] and the growth temperature [39]. On the growth surfaces fine patterns reflected from the solidified texture of the metal catalyst are observed. They are dendritic or striated, depending on the composition of the metal catalyst [40].

(iii) Defects

Defects such as inclusions, stacking faults and dislocations are commonly present. Bulk inclusions seen in diamond are solidified metal used as the catalyst [41]. Micrometre sized particles such as metal carbides dispersed in the crystals are also detected if dark field illumination is applied [42]. Dislocations originate from the seed crystal and extend along the growth direction. Crystals free from such defects have recently been produced using a high quality diamond as a seed crystal [43].

## F ATOMIC IMPURITIES RELATED TO OPTICAL PROPERTIES

Species of atomic impurities incorporated into the diamond lattice are limited: H, B, N, Si, P, Ni and Co.

(i) Hydrogen is not common in high pressure diamond. A hydrogen related peak at 3107 $cm^{-1}$ has been found in some of the crystals after heat treatment above 2100°C [44], but not in all of the high pressure diamond.

(ii) Boron is readily incorporated to produce a blue colour, if boron-containing materials such as boron, boron nitride and borides are present in the growth cell. The blue colour becomes deeper with increasing boron additives. However, a yellow colour due to nitrogen persists until boron concentrations exceed the nitrogen concentrations, even if boron is doped [45]. Few reports have been published on concentration of boron, although 270 ppm of boron has been documented [46].

(iii) Nitrogen is present in crystals with a wide range of concentrations from near zero to 800 ppm [26,47,48]. The concentrations depend on the composition of the alloy used as the catalyst and the growth temperature. The diamonds grown from typical metal catalysts such as Ni, Co and Fe contain 50 to 300 ppm of nitrogen. By adding nitrogen getters, Ti, Zr or Al, to the catalysts, the nitrogen concentrations are reduced to undetectable levels, i.e. type IIa. On the other hand, it is difficult to dope more than 1000 ppm of nitrogen, which is commonly found in natural diamonds.

The concentrations decrease with increasing growth temperatures when Ni and Co are used, although the temperature effect is weak [39,49]. On the other hand, they increase with an increase of growth temperature, when metal catalysts containing the nitrogen getters are used, i.e. the gettering effect decreases with increasing temperature.

The nitrogen is incorporated as a single substitutional form, but nitrogen pairs are found in crystals grown at higher temperatures [50]. The pairs are believed to be formed by migration of nitrogen during growth, although there is a report suggesting that the pairs are captured directly during growth [51].

(iv) Nickel impurity is incorporated when nickel-containing metal is used as the catalyst and produces a number of characteristic absorption and luminescence bands. The bands depend on the composition of metallic catalysts and the temperatures of heat treatment, i.e. concentrations and structures of nitrogen impurities [52–55]. With decreasing nitrogen concentrations, a brownish-yellow colour, which is seen in a crystal grown from pure nickel, changes to green and then to brown. With heat treatment, the brownish-yellow and green colours change to deep brown. Electron spin resonance (ESR) studies have revealed that substitutional $Ni^{-}$ and interstitial $Ni^{+}$ are present in nitrogen-rich and nitrogen-poor crystals, respectively [56,57]. Heat treatment produces complexes consisting of Ni, N and vacancies [58].

(v) Cobalt impurity has also been reported to be incorporated, when cobalt-containing metal is used as the catalyst. The cobalt impurity does not produce strong absorption bands. However, strong yellow luminescence is seen in crystals after heat treatment [59]. When nitrogen getter is added to cobalt metal used as the catalyst, the grown crystal does not give absorption or luminescence bands in contrast to the case of nickel. 20 ppm of Co has been detected by the X-ray fluorescence technique [60].

(vi) Silicon impurity is commonly found in chemical vapour deposition (CVD) diamond and a peak at 1.68 eV is seen in both absorption and luminescence spectra. This has also been detected even in high pressure diamond grown from silicon-containing alloys [61].

(vii) Phosphorus has been known to be incorporated in CVD diamond. In contrast, phosphorus impurity has been little studied for high pressure diamond. As described above, phosphorus plays the role of catalyst for diamond synthesis [21]. However, no characteristic optical centre has been found. Phosphorus impurity accompanied by nitrogen has been detected from a crystal grown from phosphorus using an ESR technique [62].

(viii) Distribution of the impurities is not uniform. A growth sector dependence is clearly seen. For nitrogen, the order of its concentrations is {111} > {100} > {113} = {110} [45], although the order is reverse, i.e. {100} > {111}, in crystals grown at low temperatures [39]. A blue colour caused by boron impurity also depends on the growth sector. The depth of the blue colour is in the order of {111} > {113} > {110} > {100} [45]. However, when the amount of boron additive is small, {111} shows a yellow colour, whereas {113} and {110} become blue. Ni and Co impurities are restricted to {111} sectors [52,63].

Coloured striations parallel to growth surfaces are seen in a cross section of diamond, which is due to a change of growth conditions during growth. With a sudden decrease of temperature during a growth run, the blue colour has changed to yellow in a boron-doped diamond [49]. Deep coloured striations are formed in a Ni-containing diamond with the same temperature change.

Aggregation rates of nitrogen from an isolated form to pairs are not uniform. They are higher in {111} sectors than in {100} [39]. Impurities such as Ni and Co enhance the aggregation rates of nitrogen [64].

## G CONCLUSION

This Datareview focused on single crystal diamond. The growth technique has been well developed to produce high purity crystals with good crystal perfection. However, crystal size is limited, and impurity doping is also limited.

## REFERENCES


[1] H. Kanda, T. Sekine [ in *Properties and Growth of Diamond* Ed. G. Davies (INSPEC, IEE, London, UK, 1994) p.404-8 ]

[2] F.P. Bundy, W.A. Bassett, M.S. Weathers, R.J. Hemley, H.K. Mao, A.F. Goncharov [ *Carbon (UK)* vol.34 (1996) p.141-53 ]

[3] C.S. Kennedy, G.C. Kennedy [ *J. Geophys. Res. (USA)* vol.81 (1976) p.2467-9 ]

[4] T. Yagi, W. Utsumi, M. Yamanaka, T. Kikegawa, O. Shimomura [ *Phys. Rev. B (USA)* vol.46 (1992) p.6031-9 ]

[5] M.N. Regueiro, P. Monceau, J.-L. Hodeau [ *Nature (UK)* vol.355 (1992) p.237-9 ]

[6] Y.Q. Zhu, T. Sekine, T. Kobayashi, E. Takazawa, M. Terrones, H. Terrones [ *Chem. Phys. Lett. (Netherlands)* vol.287 (1998) p.689-93 ]

[7] F.P. Bundy, J.S. Kasper [ *J. Chem. Phys. (USA)* vol.46 (1967) p.3437-46 ]

[8] P.S. DeCarli, J.C. Jamieson [ *Science (USA)* vol.133 (1961) p.1821 ]

[9] V.I. Trefilov, G.I. Savvakin, V.V. Skorokhod, Yu.M. Solonin, A.F. Khreinko [ *Sov. Phys.-Dokl. (USA)* vol.23 (1978) p.269-71 ]

[10] D.G. Morris [ *J. Appl. Phys. (USA)* vol.51 (1980) p.2059-65 ]

[11] J. Kleiman, R.B. Heimann, D. Hawken, N.M. Salansky [ *J. Appl. Phys. (USA)* vol.56 (1984) p.1440-54 ]

[12] E.J. Wheeler, D. Lewis [ *Mater. Res. Bull. (USA)* vol.10 (1975) p.687-94 ]

[13] D.J. Erskine, W.J. Nellis [ *J. Appl. Phys. (USA)* vol.71 (1992) p.4882-6 ]

[14] H.P. Bovenkerk, F.P. Bundy, H.T. Hall, H.M. Strong, R.H. Wentorf Jr. [ *Nature (UK)* vol.184 (1959) p.1094-8 ]

[15] M. Wakatsuki [ *Jpn. J. Appl. Phys. (Japan)* vol.5 (1966) p.337 ]

[16] N.V. Novikov, A.A. Shulzhenko [ in *Science and Technology of New Diamond* Eds. S. Saito, O. Fukunaga, M. Yoshikawa (KTK, Tokyo, 1990) p.217-9 ]

[17] M. Akaishi, H. Kanda, S. Yamaoka [ *J. Cryst. Growth (Netherlands)* vol.104 (1990) p.578-81 ]

[18] M. Akaishi, H. Kanda, S. Yamaoka [ *Jpn. J. Appl. Phys. (Japan)* vol.29 (1990) p.L1172-4 ]

[19] M. Arima, K. Nakayama, M. Akaishi, S. Yamaoka, H. Kanda [ *Geology (USA)* vol.21 (1993) p.968-71 ]

[20] S.M. Hong, M. Akaishi, S. Yamaoka [ *J. Cryst. Growth (Netherlands)* vol.200 (1999) p.326-8 ]

[21] M. Akaishi, H. Kanda, S. Yamaoka [ *Science (USA)* vol.259 (1993) p.1592-4 ]

[22] H. Kanda, M. Akaishi, S. Yamaoka [ *Appl. Phys. Lett. (USA)* vol.65 (1994) p.784-6 ]

[23] L.F. Kulikova M.D. Shalimor, V.N. Slesaren, E.N. Yakovlev [ *Sov. J. Superhard Mater. (USA)* vol.10 (1988) p.1-3 ]

[24] F.P. Bundy, H.P. Bovenkerk, H.M. Strong, R.H. Wentorf Jr. [ *J. Chem. Phys. (USA)* vol.35 (1961) p.383-91 ]

[25] F.P. Bundy, H.M. Strong, R.H. Wentorf Jr. [ in *Chemistry and Physics of Carbon* vol.10 Eds. P.L. Walker Jr., P.A. Thrower (Marcel Dekker, New York, 1973) p.213-63 ]

[26] H.M. Strong, R.M. Chrenko [ *J. Phys. Chem. (USA)* vol.75 (1971) p.1838-43 ]

[27] O. Fukunaga, Y.S. Ko, M. Konoue, N. Ohashi, T. Tsurumi [ *Diam. Relat. Mater. (Switzerland)* vol.8 (1999) p.2036-42 ]

[28] S. Hosomi [ *Mater. Res. Bull. (USA)* vol.19 (1984) p.479-86 ]

[29] H.M. Strong, R.E. Hanneman [ *J. Chem. Phys. (USA)* vol.46 (1967) p.3668-76 ]

[30] M. Akaishi [ *Diam. Relat. Mater. (Switzerland)* vol. 2 (1993) p.183-9 ]

[31] Yu.A. Palyanov, A.G. Sokol, Yu.M. Borzdov, A.F. Khokhryakov, N.V. Sobolev [ *Nature (UK)* vol.400 (1999) p.417-8 ]

[32] X.-Z. Zhao, R. Roy, K.A. Cherian, A. Badzian [ *Nature (UK)* vol.385 (1997) p.513-4 ]

[33] De Beers press release (1992)

[34] S. Satoh, H. Sumiya [ *Rev. High Press. Sci. Technol. (Japan)* vol.2 (1993) p.315-20 ]

[35] S. Vagarali, M. Lee, R.C. DeVries [ *J. Hard Mater. (UK)* vol.1 (1990) p.233-45 ]

[36] R.H. Wentorf Jr. [ *J. Phys. Chem. (USA)* vol.75 (1971) p.1833-7 ]

[37] M.-L.T. Rooney [ *J. Cryst. Growth (Netherlands)* vol.116 (1992) p.15-21 ]

[38] H. Kanda, T. Ohsawa, O. Fukunaga, I. Sunagawa [ *J. Cryst. Growth (Netherlands)* vol.94 (1989) p.115-24 ]

[39] S. Satoh, H. Sumiya, K. Tsuji, S. Yazu [ in *Science and Technology of New Diamond* Eds. S. Saito, O. Fukunaga, M. Yoshikawa (KTK, Tokyo, 1990) p.351-5 ]

[40] H. Kanda, M. Akaishi, N. Setaka, S. Yamaoka, O. Fukunaga [ *J. Mater. Sci. (UK)* vol.15 (1980) p.2743-8 ]

[41] J.E. Shigley, E. Fritsch, C.M. Stockton J.I. Koivula, C.W. Fryer, R.E. Kane [ *Gem Gemol. (USA)* vol.22 no.4 (1986) p.192-208 ]

[42] J.E. Shigley et al [ *Gem Gemol. (USA)* vol.23 (1987) p.187-206 ]

[43] H. Sumiya, N. Toda, Y. Nishibayashi, S. Satoh [ *J. Cryst. Growth (Netherlands)* vol.178 (1997) p.485-94 ]

[44] I. Kiflawi, D. Fisher, H. Kanda, G. Sittas [ *Diam. Relat. Mater. (Switzerland)* vol.5 (1996) p.1516-8 ]

[45] R.C. Burns et al [ *J. Cryst. Growth (Netherlands)* vol.104 (1990) p.257-79 ]

[46] R.M. Chrenko [ *Nature (Phys. Sci.) (UK)* vol.229 (1971) p.165-7 ]

[47] R.M. Chrenko, H.M. Strong, R.E. Tuft [ *Philos. Mag. (UK)* vol.23 (1971) p.313-8 ]

[48] A.T. Collins, S.C. Lawson [ *Philos. Mag. Lett. (UK)* vol.60 (1989) p.117-22 ]

[49] H. Kanda, S.C. Lawson [ *Indust. Diam. Rev. (UK)* vol.55 (1995) p.56-61 ]

[50] H. Kanda, S. Yamaoka [ *Diam. Relat. Mater. ( Switzerland)* vol.2 (1993) p.1420-3 ]

[51] A.P. Yelisseyev, V.A. Nadolinny, B. Feigelson, S. Terentyev, S. Nosukhin [ *Diam. Relat. Mater. ( Switzerland)* vol.5 (1996) p.1113-7 ]

[52] A.T. Collins, H. Kanda, R.C. Burns [ *Philos. Mag. B (UK)* vol.61 (1990) p.797-810 ]

[53] S.C. Lawson, H. Kanda, M. Sekita [ *Philos. Mag. B (UK)* vol.68 (1993) p. 39-46 ]

[54] S.C. Lawson, H. Kanda [ *J. Appl. Phys. (USA)* vol.73 (1993) p.3967-73 ]

[55] H. Kanda, K. Watanabe [ *Diam. Relat. Mater. (Switzerland)* vol.8 (1999) p.1463-9 ]

[56] J. Isoya, H. Kanda, J.R. Norris, J. Tang, M.K. Bowman [ *Phys. Rev. B (USA)* vol.41 (1990) p.3905-13 ]

[57] J. Isoya, H. Kanda, Y. Uchida [ *Phys. Rev.B (USA)* vol.42 (1990) p.9843-52 ]

[58] V.A. Nadolinny, A.P. Yelisseyev [ *Diam. Relat. Mater. (Switzerland)* vol.3 (1994) p.1196-200 ]

[59] S.C. Lawson, H. Kanda, K. Watanabe, I. Kiflawi, Y. Sato, A.T. Collins [ *J. Appl. Phys (USA)* vol.79 (1996) p.4348-57 ]

[60] X. Jia, S. Hayakawa, W. Li, Y. Gohshi, M. Wakatsuki [ *Diam. Relat. Mater. (Switzerland)* vol.8 (1999) p.1895-9 ]

[61] G. Sittas, H. Kanda, I. Kiflawi, P.M. Spear [ *Diam. Relat. Mater. (Switzerland)* vol.5 (1996) p.866-9 ]

[62] J. Isoya, H. Kanda, M. Akaishi, Y. Morita, T. Ohshima [ *Diam. Relat. Mater. (Switzerland)* vol.6 (1997) p.356-60 ]

[63] M. Wakatsuki, S. Hayakawa, A. Aoki, Y. Gohshi, A. Iida [ in *New Diamond Science and Technology* Eds. R. Messier, J.T. Glass, J.E. Butler, R. Roy (Mater. Res. Soc., Pittsburgh, 1991) p.143-8 ]

[64] D. Fisher, S.C. Lawson [ *Diam. Relat. Mater. (Switzerland)* vol.7 (1998) p.299-304]

# B1.2 Thermodynamic and kinetic issues of metastable diamond growth

J.C. Angus

May 2000

## A INTRODUCTION

Diamond synthesis can be divided into two broad categories: processes operating at high pressure where diamond is the stable phase and low-pressure processes where diamond is metastable. In this Datareview the thermodynamic and kinetic limitations of diamond growth when diamond is the metastable phase are discussed. Thermodynamics, by its nature, cannot describe the details of molecular mechanisms or the rates at which chemical transformations take place. It can, however, provide a fundamental conceptual framework, within which proposed chemical mechanisms must fall.

## B THERMOCHEMISTRY OF DIAMOND AND GRAPHITE

Carbon exists in two primary crystal forms: diamond and graphite. At 1 bar and 298 K, graphite is the stable phase and diamond is metastable. The basic thermodynamic properties of diamond and graphite at 298 K and 1 bar are given in TABLE 1.

TABLE 1 Thermodynamic properties of diamond and graphite [1].

| Property, 298.15 K, 1 bar | Graphite, $C_{(g)}$ | Diamond, $C_{(d)}$ |
|---|---|---|
| Enthalpy, $\Delta_f H^0$ | 0 | 1.895 kJ/mole |
| Gibbs free energy, $\Delta_f G^0$ | 0 | 2.900 kJ/mole |
| Entropy, $S^0$ | 5.740 J/mole K | 2.377 J/mole K |
| Specific volume, $V^0$ | 5.315 cm$^3$/mole | 3.417 cm$^3$/mole |
| Heat capacity, $C_p$ | 8.527 J/mole K | 6.113 J/mole K |

The relative stability of graphite and diamond is determined by the difference in their Gibbs free energies (chemical potentials), $\mu_d - \mu_g$, which is determined by integration of the Gibbs–Duhem equation over an appropriate path from 298 K and 1 bar:

$$(\mu_d - \mu_g) = (\mu_d^0 - \mu_g^0) + \int_{T^0, P^0}^{T, P} \left[ -(S_d - S_g)dT + (V_d - V_g)dP \right]$$

where $\mu_d^0 - \mu_g^0 = \Delta_f G_d^0 - \Delta_f G_g^0 =$ 2.900 kJ/mole at 298.15 K and 1 bar. The diamond-stable region is determined by the condition $\mu_d - \mu_g < 0$ and the graphite-stable region by the condition $\mu_d - \mu_g > 0$. Values of T, P for which $\mu_d - \mu_g = 0$ fix the diamond/graphite equilibrium line. The principles behind this calculation have been known for many years [2–5] and details have been described elsewhere [6–10]. The equilibrium pressures are high, 16 000 bar at 298 K and 64 000 bar at 2000 K [8]. The principal difficulties in calculating the equilibrium line are believed to be in estimating the heat capacities, compressibilities and thermal expansion coefficients of graphite and diamond in the high temperature/high pressure region. However, a recent re-visitation of this question by Berman [11] indicates that other factors may also lead to uncertainties in the calculations.

The difference in free energy between diamond and graphite is not large compared to the energy changes of other processes. For example, at 1 bar, for a growth temperature of 1100 K, $\mu_d - \mu_g =$ 6.41 kJ/mole [7], which is only 0.066 eV/carbon atom. For comparison, at 1100 K, kT = 0.095 eV/carbon atom. In other words, at growth temperatures, the thermal energies of atoms are greater than the difference in free energies of carbon atoms in the two bulk phases. This means the activation barrier between the two carbon phases is often the important factor in metastable diamond growth rather than the intrinsic difference in energies of the two phases.

Since $V_d < V_g$ and $S_d < S_g$, increasing temperature and pressure influence $\mu_d - \mu_g$ in opposite directions. Increasing the temperature at constant pressure enhances the relative stability of graphite, whereas increasing pressure enhances the relative stability of diamond. It follows that elevated temperatures are used in diamond synthesis, not for thermodynamic advantage, but for chemical kinetic reasons. At higher temperatures, the facile attachment kinetics and increased surface diffusion rates aid in obtaining high quality crystals in both the high-pressure and low-pressure syntheses.

## C DIAMOND SYNTHESIS AT HIGH PRESSURES

At 1700 K, static pressures in excess of 50 kbar are required to reach the diamond-stable region. However, graphite held at these conditions does not transform to diamond at measurable rates. The discovery that the activation barrier for the direct solid-state transition to diamond could be circumvented by dissolving the graphite in a transition metal 'solvent-catalyst' was the key for the successful synthesis of diamond at high static pressures [12]. When held above the metal/eutectic temperature, the molten metal dissolves the graphite (or other forms of non-diamond carbon); diamond subsequently crystallises from the liquid metal solution. Metal solvents are not required; alkaline earth and alkaline metal sulphates and carbonates have also been successfully used as diamond growing media [13]. Recently, low pressure processes have been reported that use molten metals as a catalyst; see Section G.

## D GRAPHITISATION OF DIAMOND

Diamond can, in principle, spontaneously transform into graphite at any point within the graphite-stable region. The kinetic barrier inhibiting this transition is a necessary condition for extension of a diamond crystal at T, P where graphite is the stable phase. The temperature at which graphitisation of diamond occurs depends on the perfection of the diamond and the nature of the ambient. For example, graphitisation is catalysed by oxygen and tends to initiate at surfaces and at internal flaws where the increase in volume accompanying the transformation can be more easily accommodated [14,15]. Clean diamond surfaces in vacuum do not graphitise at measurable rates below 1800 K [16]. Most low-pressure diamond growth processes take place at temperatures no greater than 1300 K, well below the temperature where appreciable graphitisation occurs.

Theoretical studies indicate that (111) diamond surfaces preferentially graphitise near a step [17,18]. Other recent computational studies have examined details of the diamond to graphite transition [19–25]. Of great interest are recent experimental studies on the graphitisation of the surfaces of small diamonds to give fullerene-like shells on the surface [26,27], similar to the domed structures predicted by first-principles molecular dynamic studies [25]. Moreover, the (0001) planes of the graphite are parallel to the (111) planes of the original diamond. Transmission electron microscopy of the interface shows that three diamond (111) bi-layers are matched with two graphite (0001) sheets. This is the low energy diamond/graphite interface that was proposed by Lambrecht et al [24] and also studied by DeVita et al [25]. The orientation is the same as observed for graphite flakes included in diamond grown at high pressure, in diamond grown on graphite substrates by chemical vapour deposition [28], and in graphitised diamond [14,15]. This intimate physical connection between the graphite and diamond that is found in many situations has been postulated to play a role in diamond nucleation [24, 25]. Another study of interest is by Reznik et al [29] who annealed diamond damaged by ion implantation and observed the factors influencing its reversion to diamond or conversion to graphite.

## E CHEMICAL VAPOUR DEPOSITION OF DIAMOND AT LOW PRESSURE

In the graphite-stable region of the phase diagram, the direct transition of graphite, $C_{(g)}$, to diamond, $C_{(d)}$, is thermodynamically forbidden because $\mu_d - \mu_g > 0$.

$$C_{(g)} = C_{(d)} \qquad \Delta G = \mu_d - \mu_g = +6.41 \text{ kJ/mole (1100 K, 1 bar [7])}$$

This thermodynamic limitation for the direct conversion is circumvented by using an indirect process starting with reactants at high chemical potential. Atomic hydrogen serves as the active reagent in virtually all reported low-pressure syntheses, although there is no thermodynamic reason why other chemistries could not work. The atomic hydrogen, which has an extremely high chemical potential, participates in a number of processes, all of which are thermodynamically allowed, i.e. proceed with a net decrease in Gibbs free energy. At the end of these processes, diamond is formed, and the atomic hydrogen has either returned to its molecular form, $H_2$, or appears in a stable hydrocarbon-containing species. The atomic hydrogen must participate in the elementary reactions that lead to diamond formation. Simple recombination of 2H to $H_2$ by itself dissipates the available free energy supplied by the plasma or hot filament. The process can be illustrated with a small set of elementary reactions that are believed to take place during diamond formation, as shown below. (The complete growth mechanism is more complex and is reviewed elsewhere in this volume [30].)

Generation of free radicals in the gas phase:

$$H + CH_4 = \dot{C}H_3 + H_2$$

Addition of hydrogen to a bare diamond surface:

$$H + C_{(d)} = C_{(d)}H$$

Abstraction of hydrogen from a hydrogenated surface to form a surface free radical:

$$H + C_{(d)}H = C_{(d)} + H_2$$

Attachment of a methyl radical to the diamond surface:

$$CH_3 + C_{(d)} = C_{(d)}CH_3$$

The sum of the above elementary reactions results in the attachment of one carbon atom to the diamond surface in the form of a methyl group.

$$CH_4 + 3H + C_{(d)} = C_{(d)}CH_3 + 2H_2 \qquad \Delta G < 0$$

Further abstraction reactions and rearrangements cause the attached carbon atom to become fully incorporated into the growing diamond crystal [30]. The net result is the conversion of stable gas phase species, e.g. $CH_4$, to diamond by a series of steps, each of which is favoured thermodynamically. The diamond crystal does not graphitise because of the large activation barrier to graphitisation described earlier, and remains trapped in a metastable state.

## F DIAMOND NUCLEATION

The reactions in Section E describe the extension of an existing diamond crystal. They do not explain why graphite is not formed during low-pressure diamond growth, nor do they describe how new crystals of diamond nucleate from the vapour phase. These questions can only be answered by considering molecular mechanisms and the rates of competing chemical processes and cannot be fully explained by thermodynamic arguments alone. Thermodynamics does, however, provide some insight into the problem.

Nucleation of new crystals may be accomplished on small diamond debris left over from diamond polishing (not truly a nucleation process) or by nucleation on other solids with similar structure, such as silicon carbide, or on graphite itself. It has also been well established experimentally that diamond can nucleate in the absence of these centres, but there is as yet no widely accepted mechanism for this situation.

In the diamond-growing environment there is a careful balance of a flux of carbon to the surface in the form of hydrocarbon radical species and a very high flux of atomic hydrogen. The fluxes are chosen so that there is a slight carbon supersaturation in the region of the substrate. In this environment, high molecular weight condensed ring aromatics and solid carbons that form by gas phase pyrolysis are hydrogenated to produce saturated $sp^3$ structures. Badziag et al [31] first pointed out that small, fully hydrogenated diamond nuclei have a lower energy than fully hydrogenated graphite nuclei. The argument is based on the fact that the hydrogenation of an aromatic bond by atomic hydrogen is exothermic. Angus extended this argument by considering the free energy changes for reactions involving a series of molecular analogues of graphite and diamond nuclei [32]. For example, the hydrogenation of phenanthrene to produce diamantane, the 14 carbon molecular analogue of diamond, is strongly favoured thermodynamically.

$$C_{14}H_{10} + 10H = C_{14}H_{20} \qquad \Delta G = -166 \text{ kJ/mole C [32]}$$

The further evolution of these precursors depends on the details of their structure. One path is further hydrogenation and ultimate gasification by the atomic hydrogen. This process has long been posited as a key process permitting the chemical vapour deposition of diamond. Theoretical studies show that during hydrogenation the prism faces of graphite precursors are reactive to atomic hydrogen and become hydrogenated [33,25]. These prism faces also provide a low energy, coherent interface with diamond [24,25] and are therefore favourable sites for nucleation of small diamond crystals. Moreover, some of these diamond nuclei will be twinned [24], which enhances growth rates by providing a re-entrant corner for easy atom attachment [34]. The twinned crystals will have a kinetic advantage over un-twinned crystals and may grow more easily to macroscopic size. Structures without a kinetic advantage for growth may be gasified.

## G HYBRID METAL SOLVENT PROCESSES

An interesting modification to the chemical vapour deposition of diamond has been described in which a mixture of molten metal and graphite is treated with atomic hydrogen from a plasma source

[35–38]. The molten metal is typically iron or nickel. Mallika et al [39] have recently summarised the process. The diamond appears to nucleate and grow both on and within the melt. From the thermodynamic standpoint, this is a variant of the conventional process using $CH_4$ as the carbon source. The thermodynamic potential driving the process is again the very high chemical potential of the atomic hydrogen. It is believed that the metal serves a catalytic function, similar to that in the high-pressure process, by dissolving the graphite. The mechanism for converting the dissolved carbon to diamond is not understood, however, and is controversial [40].

Because the process is at low pressure, the temperature of the metal/carbon eutectic is not raised, as it is in the high-pressure synthesis of diamond using a transition metal catalyst. Instead, the eutectic temperature is lowered because of the increased level of dissolved hydrogen caused by the extremely high chemical potential of the atomic hydrogen. This effect was first noticed by Sato et al [41] and Yang et al [42,43] while attempting to grow diamond on nickel substrates from a microwave plasma. Liquid phases were observed several hundred degrees below the Ni/C eutectic temperature of 1591 K.

## H CONCLUSION

Thermodynamic and kinetic limitations of the growth of metastable diamond have been discussed. Diamond can be synthesised at high pressures or at low pressures, the latter via CVD. Theoretical treatments of nucleation are still incomplete.

## REFERENCES


[1] D.D. Wagman et al [ *J. Phys. Chem. Ref. Data (USA)* vol.11 supplement 2 (1982) p.83 ]

[2] G.N. Lewis, M. Randall [ *J. Am. Chem. Soc. (USA)* vol.37 (1915) p.458-70 ]

[3] F. Simon [ *Handb. Phys. (Germany)* vol.X (1926) p.350-404 ]

[4] O.I. Leipunskii [ *Usp. Khim. (USSR)* vol.8 (1939) p.1519-34 ]

[5] F.D. Rossini, R.S. Jessup [ *J. Res. Natl. Bur. Stand. (USA)* vol.21 (1938) p.491 ]

[6] R. Berman, F. Simon [ *Z. Elektrochem. (Germany)* vol.59 (1955) p.333-8 ]

[7] R. Berman [ in *Physical Properties of Diamond* Ed. R. Berman (Clarendon Press, 1965) ch.14 p.371-93 ]

[8] R. Berman [ in *The Properties of Diamond* Ed. J. Field (Academic Press, 1979) ch.1 p.4-22 ]

[9] G. Davies [ *Diamond* (Adam Hilger Ltd., 1984) ch.5 p.67-88 ]

[10] R. Berman [ in *Properties and Growth of Diamond* Ed. G. Davies (INSPEC, IEE, London, UK, 1994) p.30-2 ]

[11] R. Berman [ *Solid State Commun. (USA)* vol.99 (1996) p.35-7 ]

[12] F.P. Bundy, H.T. Hall, H.M. Strong, R.H. Wentorf [ *Nature (UK)* vol.176 (1955) p.51-5 ]

[13] M. Akaishi, H. Kanda, S. Yamaoka [ *Jpn. J. Appl. Phys. (Japan)* vol.29 (1990) p.L1172-74 ]

[14] T. Evans, P. James [ *Proc. R. Soc. Lond. A (UK)* vol.277 (1964) p.260-9 ]

[15] G. Davies, T. Evans [ *Proc. R. Soc. Lond. A (UK)* vol.328 (1972) p.413-27 ]

[16] T. Evans [ in *The Properties of Diamond* Ed. J. Field (Academic Press, 1979) ch.13 p.408-19 ]

[17] B.N. Davidson, W.E. Pickett [ *Phys. Rev. B (USA)* vol.49 (1994) p.14770-3 ]

[18] G. Kern, J. Hafner [ *Phys. Rev. B (USA)* vol.58 (1998) p.13167-75 ]

[19] S. Scandalo, M. Bernasconi, G.L. Chiarotti, P. Focher, E. Tosatti [ *Phys. Rev. Lett. (USA)* vol.74 (1995) p.4015-8 ]

[20] G. Jungnickel, C.D. Latham, M.I. Heggie, Th. Frauenheim [ *Diam. Relat. Mater. (Switzerland)* vol.5 (1996) p.102-7 ]

[21] G. Jungnickel et al [ *Phys. Status Solidi (Germany)* vol.154 (1996) p.109-25 ]

[22] A.T. Balaban, D.J. Klein [ *Carbon (UK)* vol.35 (1997) p.247-51 ]

[23] A.V. Petukhov, D. Passerone, F. Ercolessi, E. Tosatti, A. Fasolino [ *Phys. Rev. B (USA)* vol.61 (2000) p.R10590-3 ]

[24] W.R.L. Lambrecht, C.H. Lee, B. Segall, J.C. Angus, Z. Li, M. Sunkara [ *Nature (UK)* vol.364 (1993) p.607-10 ]

[25] A. De Vita, G. Galli, A. Canning, R. Car [ *Nature (UK)* vol.379 (1996) p.523-6 ]

[26] V.L. Kuznetsov, A.L. Chuvilin, Y.V. Butenko, S.V. Stankus, R.A. Khairulin, A.K. Gutakovskii [ *Chem. Phys. Lett. (Netherlands)* vol.289 (1998) p.353-60 ]

[27] V. Kuznetsov, I.L. Zilberberg, Yu.V. Butenko, A.L. Chuvilin, B. Segall [ *J. Appl. Phys. (USA)* vol.86 (1999) p.863-70 ]

[28] Z. Li, L. Wang, T. Suzuki, A. Argoitia, P. Pirouz, J.C. Angus [ *J. Appl. Phys. (USA)* vol.73 (1993) p.711-5 ]

[29] A. Reznik, V. Richter, R. Kalish [ *Diam. Relat. Mater. (Switzerland)* vol.7 (1998) p.317-21 ]

[30] J.E. Butler, D.G. Goodwin [ Datareview in this book: *B1.3 CVD growth of diamond* ]

[31] P. Badziag, W.S. Verwoerd, W.P. Ellis, N.R. Greiner [ *Nature (UK)* vol.342 (1990) p.244-5 ]

[32] J.C. Angus [ in *Proc. Int. School of Physics "Enrico Fermi" Course CXXXV* Eds. A. Paoleti, A. Tucciarone (IOS Press, Amsterdam, 1997) p.9-28 ]

[33] S.P. Mehandru, A.B. Anderson, J.C. Angus [ *J. Phys. Chem. (USA)* vol.96 (1992) p.10978-82 ]

[34] J.C. Angus, M. Sunkara, S. Sahaida, J.T. Glass [ *J. Mater. Res. (USA)* vol.7 (1992) p.3001-9 ]

[35] R. Roy, H.S. Dewan, P. Ravindranathan [ *Mater. Res. Bull. (USA)* vol.28 (1993) p.861-6 ]

[36] R. Roy, H.S. Dewan, P. Ravindranathan [ *J. Mater. Chem. (UK)* vol.3 (1993) p.685-6 ]

[37] R. Roy et al [ *Mater. Lett. (Netherlands)* vol.25 (1995) p.191-3 ]

[38] R. Roy, K. Cherian [ *Mater. Technol. (USA)* vol.11 (1996) p.6-9 ]

[39] K. Mallika, R.C. DeVries, R. Komanduri [ *Thin Solid Films (Switzerland)* vol.339 (1999) p.19-33 ]

[40] P.K. Bachmann [ *Adv. Mater. (Germany)* vol.5 (1993) p.137-40 ]

[41] Y. Sato, H. Fujita, T. Ando, T. Tanaka, M. Kamo [ *Philos. Trans. R. Soc. Lond. A (UK)* vol.342 (1993) p.225-31 ]

[42] P.C. Yang, W. Zhu, J.T. Glass [ *J. Mater. Res. (USA)* vol.8 (1993) p.1773-6 ]

[43] P.C. Yang et al [ *J. Cryst. Growth (Netherlands)* vol.187 (1998) p.81-8 ]

# B1.3 CVD growth of diamond

J.E. Butler and D.G. Goodwin

May 2000

## A INTRODUCTION

The growth of diamond from the gas phase by chemical vapour deposition (CVD) occurred as early as the mid 1950s, with a continued low level of research in the 1960s and early 1970s because of extremely low growth rates and scepticism in the scientific community [1–4]. Key papers demonstrating the CVD of diamond appeared in the late 1970s, but it was not until the mid 1980s that the progress had proceeded to the point where significant scientific and industrial interest was generated. By the mid 1980s, it was accepted that small crystals (~ micrometres in size) and continuous polycrystalline thin films (~ tens of micrometres thick) could be grown by CVD. A typical CVD deposition of diamond is from a flowing gas mixture comprised of a small amount of methane (typically less than 1% atomic) in hydrogen at between 10 and 50 torr, activated with a hot filament (~ 2200°C) or plasma near a substrate heated to between 700 and 1000°C [5]. A schematic of a generic diamond CVD process is shown in FIGURE 1 [6].

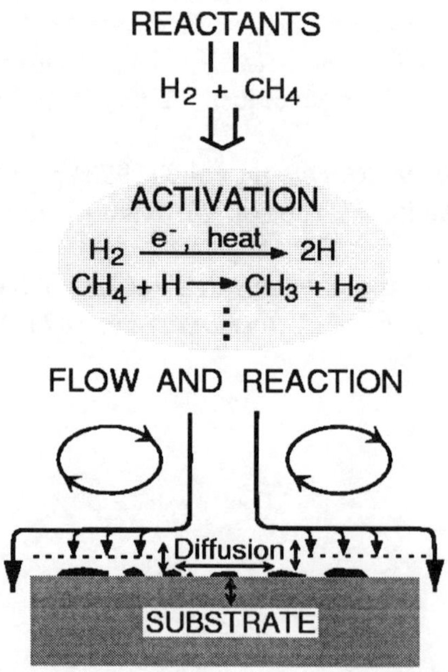


FIGURE 1 Schematic showing the principal elements in the complex diamond CVD process: flow or reactants into the reactor, activation of the reactants by thermal and plasma processes, reaction and transport of the species to the growing surface, and surface chemical processes depositing diamond and other forms of carbon (from [6]).

In the late 1980s, the first studies of the chemistry of the process began to appear using optical emission, in situ laser diagnostics, mass spectrometry, and other diagnostic techniques [7]. These techniques demonstrated the presence of methyl radicals, acetylene, CH, $C_2$, and atomic hydrogen near the growth surface, and led to computational studies of the gaseous and surface chemistries [8]. Intense discussion raged over the predominant growth species: methyl radical versus acetylene. Subsequent experiments have demonstrated that under certain conditions, most of the deposited diamond results from methyl (or $C_1$ species) and that acetylene (or $C_2$ species) can also be important to growth under other conditions. No single surface mechanism or growth model has been sufficient to describe all the observations, and the community has come to understand that many reactions (models of growth) are often occurring simultaneously due to the complex chemical nature of the growth surface.

## B GROWTH MECHANISM

Generic growth models which ignore the specific growth site stereochemistry have been extremely successful in describing, both qualitatively and quantitatively, a wide variety of growth environments [8,9]. In these models, atomic hydrogen is the key chemical reactant driving both the gaseous chemistries generating a mix of hydrocarbon free radicals and molecules in the gas phase, and the surface chemistry, and influencing the deposition rate and the material quality. The importance of atomic hydrogen for CVD diamond growth is seen by considering a generic set of diamond surface reactions.

$$C_D\,H + H\bullet \rightarrow C_D\bullet + H_2 \tag{1}$$

$$C_D\bullet + H\bullet \rightarrow C_D\,H \tag{2}$$

$$C_D\bullet + C_xH_y \rightarrow C_D - C_xH_y \tag{3}$$

Reaction (1) represents activation of a surface site by removal of a surface hydrogen atom from a carbon atom on the diamond surface ($C_D$). An activated surface site is required for hydrocarbon addition. Either a radical (e.g. $CH_3\bullet$) or unsaturated molecule (e.g. $C_2H_2$) may add to a radical site. Under conditions of diamond CVD, reactions (1)–(3) are fast and the reactions are in steady-state. Assuming that the H abstraction rate $k_1[C_DH][H\bullet]$ is much greater than the unimolecular desorption rate $k_{-3}[C_D\text{-}C_xH_y]$, and applying steady-state to $[C_D]$ results in EQN (4) for the fraction of open sites

$$\frac{[C_D\bullet]}{[C_D\,H]} = \frac{k_1\,[H\bullet]}{k_{-1}[H_2] + k_2[H\bullet] + k_3\,[C_xH_y]} \tag{4}$$

Values for the rate coefficients in EQN (4) are given in TABLE 1. Examination of the terms in EQN (4) for conditions typical of diamond CVD shows that $k_{-1}$ is small ($k_{-1}$ is calculated from $k_1$ and the equilibrium constant for reaction (1)) so $k_{-1}[H_2]$ is much less than $k_2[H\bullet]$. The concentration of reactive hydrocarbons $[C_xH_y]$ is usually less than $[H\bullet]$ so $k_3[C_xH_y]$ is less than $k_2[H\bullet]$. Thus EQN (4) can be rewritten in terms of $k_1$ and $k_2$ only,

$$\frac{[C_D\bullet]}{[C_D\,H]} = \frac{k_1}{k_2} \tag{5}$$

In the limit of small hydrocarbon and high atomic hydrogen mole fractions the fraction of open sites depends only on temperature. The fraction of open sites as a function of temperature is shown in FIGURE 2 [9].

The surface mediated conversion of two gaseous hydrogen atoms to one hydrogen molecule, reactions (1) and (2), generates a significant amount of surface heating and is a significant contribution to heat flow in the diamond CVD process.

TABLE 1 Rate coefficients used in the site fraction and quality calculations (from [9]). The units of A are given in terms of moles, cubic centimetres and seconds. The units of $E_a$ are kcal/mole. The rate coefficient is expressed as $AT^n\exp(-E_a/RT)$.

| Rate coefficient | A | n | $E_a$ | Ref |
|---|---|---|---|---|
| $k_1$ | $5.75 \times 10^{12}$ | 0 | 5.1 | a |
| $k_2$ | $1 \times 10^{14}$ | 0 | 0 | b |
| $k_3$ | $1 \times 10^{13}$ | 0 | 0 | b |
| $k_{14}$ | $1.12 \times 10^{16}$ | −1 | 48 | c |

[a] By analogy with hydrogen abstraction from i-$C_4H_{10}$, J.A. Kerr, S. Moss [ in *CRC Handbook of Bimolecular and Termolecular Gas Reactions* vol.1 (CRC Press, Boca Raton, FL, 1981) ].
[b] S. Harris [ *Appl. Phys. Lett. (USA)* vol.56 (1990) p.2298-300 ].
[c] Estimated by analogy with the reaction $CH_3$-CH=CH$^{\cdot}$ → $CH_3^{\cdot}$ + $C_2H_2$, calculated from the equilibrium constant and assuming a reverse rate coefficient of $1 \times 10^{13}$ $cm^3$ $mole^{-1}$ $s^{-1}$, S.W. Benson [ *Thermochemical Kinetics* (Wiley and Sons, NY, 1976) ].

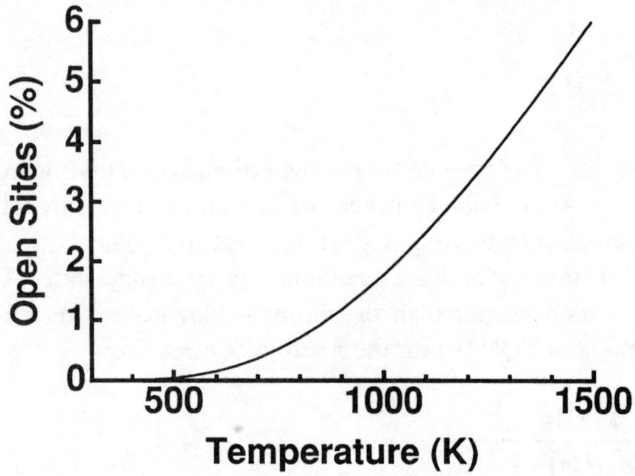


FIGURE 2 The fraction of open (hydrogen vacancy) sites on the hydrogenated surface versus temperature is computed using EQN (4) with the values in TABLE 1 (from [9]).

The rate of adding carbon to the surface, reaction (3), is thus independent of atomic hydrogen concentration in the limit of high atomic hydrogen flux at the surface. The reverse of the carbon addition, $k_{-3}$, is the unimolecular desorption of the hydrocarbon species and is important at higher temperatures. However, when the hydrocarbon species has more than a single bond to the lattice carbons, the unimolecular desorption process is generally negligible.

Upon addition of a hydrocarbon radical or molecule to the surface, the newly added carbon has one bond to the lattice. To be fully incorporated, at least two more bonds must be made (assuming the fourth bond will be to the next layer to be deposited). Since the hydrocarbons tend to be hydrogenated, incorporation must involve removal of hydrogen. In most CVD diamond processes atomic hydrogen is the primary reagent for this process, analogous to reaction (1). Hence, we shall define growth of the lattice as the formation of at least two bonds between the hydrocarbon adsorbates and the diamond lattice, as opposed to the process we have considered thus far, namely the addition of a carbon species to the surface.

Now let us consider the kinetics of this chemisorbed hydrocarbon species and an adjacent hydrogen terminated surface site, $\underline{CH}...HC_D$

$$\underline{CH}...HC_D + H\bullet \rightarrow \underline{C}\bullet...HC_D + H_2 \tag{6}$$

$$\underline{C}\bullet...HC_D + H\bullet \rightarrow \underline{CH}...HC_D \tag{7}$$

$$\underline{C}\bullet...HC_D + H\bullet \rightarrow \underline{C}\bullet...\bullet C_D \tag{8}$$

$$\underline{C}\bullet...\bullet C_D \rightarrow \underline{C}-C_D \tag{9}$$

Using similar assumptions to those above and noting the reactions (6) and (8) are the same as reaction (1), and reaction (7) is the same as reaction (2), we consider steady state on the $\underline{C}\bullet...HC_D$ species to obtain EQN (10), which is independent of atomic hydrogen concentration analogous to EQN (5).

$$\frac{[\underline{C}\bullet...HC_D]}{[\underline{CH}...HC_D]} = \frac{k_1}{k_1 + k_2} \tag{10}$$

Incorporation of carbon into the diamond lattice is represented by the sum of EQNS (8) and (9), and shows that the rate of lattice incorporation depends directly on the atomic hydrogen concentration.

$$\underline{C}\bullet...HC_D + H\bullet \rightarrow \underline{C}-C_D + H_2 \tag{11}$$

## C ALTERNATIVE CHEMISTRIES

Oxygen and halogen species have also proven to be important in certain deposition chemistries, with the growth in atmospheric combustion flames startling the community in 1988 [10,11]. Diamond deposition was observed when flames fuelled by oxygen and acetylene were run slightly fuel-rich and a substrate cooled to 800 to 1100°C was placed just outside the inner or premixed flame front [11]. Other hydrocarbon fuels were found to work as well, including ethylene [12], propylene [13], and methyl acetylene [14].

As experience with many different activation schemes (plasmas, filaments, flames), reactor designs, and chemical recipes developed in the early 1990s, two important empirical patterns developed: the H-C-O 'phase diagram' [15], and the 'alpha parameter' [16] describing the morphology of individual crystals or the polycrystalline texture.

The H-C-O 'phase diagram' showed that hydrogen must be present and that the carbon-to-oxygen ratio in the reactants must equal or exceed unity for successful diamond deposition to occur. The addition of oxygen enabled the use of significantly higher carbon reactant concentrations. This 'phase' diagram has been a useful tool for predicting appropriate reactant ratios, and its validity is derived from the complex chemistry and long residence times of the gases in most reactors which leads to essentially all the oxygen species reacting to form the chemically stable and unreactive CO molecule.

The morphology of most CVD diamond deposits displays cubo-octahedral habits, i.e. macro facets of cubic (100) or octahedral (111) crystallographic orientation. The 'alpha parameter' is directly related to the ratio of the growth velocities of these faces, and hence expresses the continuous ranges of habits from cubic to octahedral. The other low index face (110) grows significantly faster than either

(100) or (111) and hence is not stable. Growth under the appropriate alpha conditions has led to extremely smooth (100) textured films.

## D DIAMOND SURFACE STRUCTURE AND CHEMISTRY

In parallel with the developments in the CVD of diamond, our knowledge of the surface chemistry and structure has progressed greatly [17,18]. Diamond has the highest atomic bulk and surface densities of any material. Hence most atomic and molecular species are too large to terminate the bonds at the surface of the bulk. Even hydrogen is sterically hindered on the unreconstructed (100) surface. Given the small size of hydrogen and the strength of the C–H bond, it is not surprising that most diamond is usually hydrogen-terminated, and hence acts chemically like an alkane surface. Less than monolayer amounts of oxygen, halogen, and other species have been observed bonded to the surface and can significantly modify the electrical properties of the surface.

The large flux of atomic hydrogen bombarding the surface limits the size and residence times of hydrocarbon adsorbates by beta-scission reactions [9]. As an example of the removal of surface species by beta-scission, two possible ways to attack a surface-bonded ethyl group are shown in FIGURE 3 [8]. An H atom is first abstracted by gas-phase H to create a radical site (an unpaired electron). In (a), a methyl hydrogen is abstracted, while in (b) the hydrogen on the diamond surface carbon is abstracted. Once this has occurred, the bond once-removed (beta) from the radical site may now break, allowing double bond formation to the carbon from which the hydrogen was abstracted. This process results in scission of all or part of the adsorbate from the lattice. Similar beta-scission processes will occur for other multi-carbon surface species. The beta-scission process plays an important role in diamond growth, since it is believed to be an important mechanism by which H is able to etch non-diamond sp$^2$ carbon [9]. This problem is a severe one for many proposed growth mechanisms which proceed through intermediate surface species susceptible to removal by beta-scission [19].

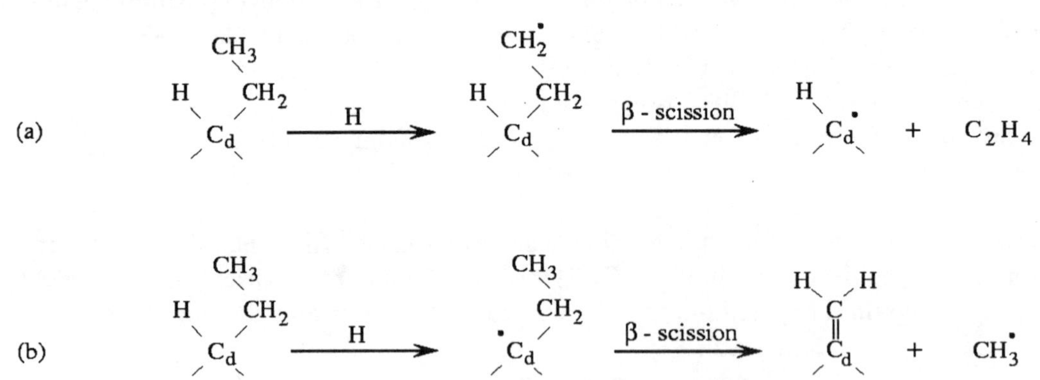


FIGURE 3 Illustration of two beta-scission reaction processes to attack a surface-bonded ethyl radical (from [8]).

A consequence of beta-scission reactions, as with thermal desorption, is that it is improbable to find hydrocarbons with chain lengths longer than one carbon adsorbed on the surface, or even two hydrocarbons which adsorb nearby and then react. The steady-state surface coverage of these vulnerable intermediates may be so small that the probability of two being simultaneously adjacent is negligible. These arguments suggest that the possible chemical pathways through which carbon is added to the lattice during diamond CVD are constrained in ways not encountered in other, 'more traditional' CVD environments, or in film growth by MBE. The small open site fraction and the problem of steric hindrance limit the number of sites at which hydrocarbon species may bond. The fact that many surface species are kinetically unstable with respect to beta-scission means that these species must be incorporated into the lattice very quickly, through paths which do not require waiting for a second species to bond nearby, such as reaction at surface steps or edges of plateaus on low index surfaces. Only relatively infrequent events, such as nucleating a new atomic layer on a low index surface, could conceivably proceed through surface species subject to removal by beta-scission. Recent kinetic Monte Carlo modelling of homoepitaxial diamond growth illustrates quite clearly how improbable nucleation of the next atomic layer events control the overall growth rate and morphology of the faceted growth surfaces [20].

## E SITE SPECIFIC MODELS

### E1 Addition to the (100) Surface

The (100) surface is unique in that addition of only one carbon is required to form two bonds to the surface. The (100) faces of CVD diamond often appear atomically smooth [42,43], and (100) growth sectors have low defect density [44]. For these reasons, growth on the (100) surface has received the most theoretical attention. The structure of the hydrogenated diamond (100) surface is generally accepted as a 2×1 mono-hydride reconstruction, because the fully hydrogen-terminated (dihydride) surface is sterically unstable [21,22]. Scanning tunnelling microscopy (STM) and atomic force microscopy (AFM) images of as-grown (100) CVD diamond surfaces clearly show a (2×1) pattern of dimer rows, consistent with the (2×1):1H surface [23–25]. Low energy electron diffraction (LEED) and high resolution electron energy loss spectroscopy (HREELS) measurements also indicate a (2×1) reconstruction [26].

The first mechanism proposed for methyl addition to the (100) surface was for the unreconstructed dihydride (100) surface using as a model of the surface the molecule bicyclo[3.3.1] nonane (BCN) [27]. A set of reactions was proposed to add a methyl group to the central opposing-hydrogen site (the 'HH site'), creating the molecule adamantane. The mechanism consists of abstraction of one of the hydrogens of the HH site, methyl addition, abstraction of either a methyl hydrogen or the other HH hydrogen, and finally one more H-abstraction to create the bridging methylene group. Rate constants for each step were estimated based on known values for similar gas-phase reactions, and thermochemistry was estimated using group additivity methods. It was shown that this mechanism, with no adjustable parameters, resulted in a conversion from BCN to adamantane at a rate which, if carried out simultaneously everywhere on the (100) surface, would result in a diamond growth rate of 0.06–0.6 μm/hr, for typical hot-filament conditions which experimentally result in growth at 0.1–1.0 μm/hr. Harris [27] states, however, that the near-perfect agreement is fortuitous, and estimates an uncertainty in growth rate of two orders of magnitude. In this work, no account was taken of the extreme strain which would result if this mechanism were actually carried out on the (100)-(1×1):2H surface, rather than on BCN. A prohibitive barrier of 80 kcal/mole was estimated for this mechanism on the (100)-(1×1):2H surface [28]. However, as discussed below, the basic HH site topology occurs commonly at steps on other surfaces or between dimer rows on the (100)-(2×1):1H surface. All of these HH-like sites have much lower strain than on the (1×1):2H surface.

Once it became clear that the relevant (100) surface was the (2×1):1H surface, rather than the (1×1):2H surface, several groups focused attention on developing mechanisms for methyl addition to this surface. Since the dimer bonds are part of 5-membered rings, a mechanism is required to open these rings and insert a carbon, in order to form the 6-membered rings of $sp^3$ diamond. A mechanism (GDSB) [29] with a low energy barrier for insertion of carbon into the dimer bond is shown schematically in FIGURE 4 [8]. The dimer is opened by abstraction of a surface hydrogen, addition of $CH_3$, abstraction of a methyl hydrogen, followed by a beta-scission reaction leading to a methylidene intermediate and a radical site.

This intermediate then reacts rapidly with the adjacent radical [30], creating a bridging methylene group and propagating the radical site. Hydrogen-free dimer sites (which are predicted to form highly strained, reactive beta-bonds) may also be an important site type where $CH_3$ may chemisorb [31]. In this case, an H migration would lead to the fourth structure in FIGURE 4. A similar mechanism has also been proposed where the fourth structure in FIGURE 4 is transformed directly into the final bridging-carbon structure though a triangular transition state [28]. However, the activation energy for this transition state is calculated to be 55 kcal/mole, indicating it will be a less important pathway than the beta-scission pathway in the GDSB mechanism. The GDSB mechanism is now widely believed to be a principal means for opening and inserting carbon into dimer bonds, and has been incorporated as a sub-mechanism into more extensive mechanisms discussed below. If a carbon is inserted into every dimer on the (2×1) surface, 1/2 of a monolayer results. To complete the new surface layer, carbon must be inserted between every pair of dimers on adjacent rows, thus bridging the 'troughs' between dimer rows. A mechanism alternating the GDSB mechanism with a modified HH mechanism to insert into dimers and to bridge between dimers has also been proposed [32]. A detailed thermochemical analysis of this proposal was conducted, using the MM3 molecular mechanics program to obtain accurate estimates of the free energy change for each step [32]. Several points emerged from this analysis. It was found that once the GDSB mechanism inserts a methylene group into one dimer, it cannot insert one into the dimer on the adjacent row, due to steric hindrance. Also, the distance between two dimerized carbons on adjacent rows is too great for a methylene group to bridge them. However, once the GDSB mechanism acts on one dimer, the distance to the nearest carbon atom on the adjacent dimer is decreased to the point that a methylene group may bridge the trough with little strain. Thus, once the GDSB mechanism acts, the HH mechanism becomes enabled to bridge the trough. An additional result from this study was the prediction that the (modified) HH mechanism proceeds preferentially through a path in which both 'HH' hydrogens are missing (a diradical site). That diradical sites are important chemisorption sites is in accord with many previous conclusions [28,31,33]. Thermochemical analysis [32] predicted that the trough-bridging (HH) portion of the mechanism is rate-limiting, and the resulting growth rate is found to be very close to that predicted by the original HH mechanism [27].

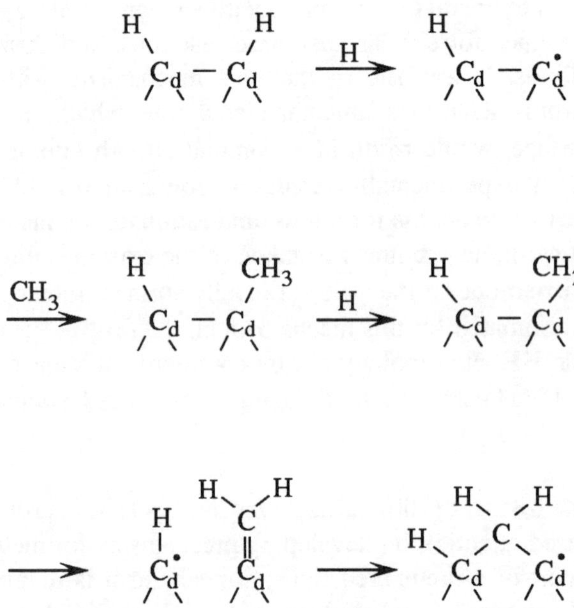

FIGURE 4 The GDSB mechanism for dimer opening and carbon insertion (from [8]) [29].

The above (100) growth mechanisms [29,32] do not explicitly address the question of why (100) surfaces have a smooth morphology. If carbon is able to incorporate everywhere on the surface through these mechanisms, a rough surface would result unless some form of surface migration or reverse etching reaction were operative. This point was discussed and it was suggested that the incorporation may occur preferentially at steps, due to lower steric hindrance at these locations [32]. A mechanism for growth from methyl radicals was proposed which explicitly addresses the issue of smooth morphology [34]. In this mechanism, preferential addition of methyl groups occurs at the end of dimer rows on a (100)-(2×1) monohydride surface with a type B single step. This mechanism makes use of the GDSB sub-mechanism to open the dimer on the lower level immediately in front of the propagating dimer row; this is followed by a dimer shift reaction, which allows formation of a methylene bridge at the end of the dimer row. MM3 was used to calculate the enthalpy of reaction for each step in this mechanism, but did not evaluate free energy changes [34]. The most endothermic step (+6 kcal/mole) was the dimer shift step. No estimates of growth rate were made in this study [34]. More recently, kinetic Monte Carlo modelling of the growth on diamond (100) using the GDSB mechanism predicted rough surfaces with a growth rate 10 times higher than experiment, while the same KMC results were in agreement with experiment for growth on (111) and (110) [35]. The energetics of the GDSB mechanism were then re-examined using a semi-empirical tight binding method for computing the forces within a large slab of the (100) diamond surface five layers thick, and then using density functional theory to treat 20 atoms in the reaction zone, with the result that the reverse of the GDSB methyl insertion into the dimer bonds (e.g. etching of $CH_3$ from the (100) surface by atomic hydrogen) is far more rapid than previously believed [35]. Addition of the etching reaction brought the rate and morphology of the KMC results into agreement with experiment [35].

Proposals have also been advanced for incorporation of other C 1 species on the (100) surface. Since the above methyl mechanisms all involve an H abstraction to form $CH_2$, direct addition of $CH_2$ could clearly occur through these mechanisms for any environment in which significant $CH_2$ is present. Also, atomic carbon would be expected to be quite reactive with the surface, and could be expected to add via mechanisms similar to these or by more direct routes [28,36]. A mechanism for adding CO to (100) was proposed, but a large barrier (51 kcal/mole) was predicted [28]. In any case, there is no experimental evidence for thermal CO contributing to diamond growth. For example, the diamond growth rate in oxyacetylene torches at 1 atm tracks the hydrocarbon concentration at the surface (which increases dramatically for $C_2H_2/O_2 > 1$), not the CO concentration, which is substantial even for $C_2H_2/O_2 > 1$ [37].

### F2 Addition to the (110) Surface

On the flat (110) surface, two carbons must be added to form a stable structure. For this reason, like (111), it has received less attention theoretically than the (100) surface for methyl addition. The (110) surface never appears in CVD diamond films, indicating it has a high growth rate and therefore grows out. Indeed, the growth rate measured on single-crystal diamond is (slightly) higher for the (110) face than for either the (111) or (100) face. Nevertheless, (110) faces are typically rough [38], and therefore a nominally (110) surface may be microfaceted into (111) [39] and (100) regions, and the high growth rate may be a reflection of a high step density of (100)-like sites.

A mechanism for methyl addition to (110) was presented which begins with sequential addition of two methyl radicals to form an ethyl ($C_2H_5$) species [9]. This species may be removed from the surface by thermal desorption or H-abstraction followed by beta-elimination of ethylene. However, an alternative path is abstraction of a neighbouring H, allowing formation of a bridging $C_2H_4$ structure. Once this 'growth site' is formed, growth may be continued by adding single methyl radicals to this site. The local topology around the growth site is that of the HH site (or a dimer site if the two hydrogens are abstracted). The propagation can therefore occur through either the HH or GDSB mechanism, or variations on these.

## E3 Addition to the (111) Surface

One of the first diamond growth mechanisms proposed was one from methyl cations [40]. In this mechanism, the diamond (111) surface is assumed to first become covered with a complete monolayer of adsorbed $CH_3$ groups; a methyl cation $CH_3^+$ then attacks three adjacent surface methyl groups, and, in a sequence of steps, binds the methyl groups together, with the evolution of three $H_2$ molecules. One difficulty with this mechanism is the very low concentration of ionized species such as $CH_3^+$ found in thermal (hot-filament or combustion) environments. Also, the energetics of bonding three methyl groups at adjacent positions on the (111) surface was calculated to be energetically unstable, due to the severe steric hindrance between adjacent groups [41]. Consequently, the methyl cation mechanism [40] must be regarded as improbable.

No other mechanisms to add methyl (or any other C 1 species) have been proposed for the flat (111) surface. The essential difficulty is that addition of 3 carbons is required on this surface before a second, stabilizing bond to the surface can be made. In any case, (111) surfaces of CVD diamond appear to be rough on an atomic scale [42,43], and (111) growth sectors are highly defective [44]. This may indicate that growth on (111) occurs at step edges, local at sites with (100) or (110) character, after infrequent nucleation of the next layer atomic layer by island formation or at re-entrant corners formed by twinning or multiple crystal intersection.

## F CONCLUSION

Current models of diamond CVD are in reasonable agreement with experimental observations of the chemical composition of the reaction environment, growth rates, temperature regimes, and reactant stoichiometry. The picture that emerges is one where gaseous atomic hydrogen (or similar reactive species like halogens, etc.) drive both the gaseous and surface processes. Carbon incorporation occurs primarily from single carbon species, e.g. the $CH_3$ radical, by incorporation at reactive sites on step edges, with the overall growth rate and morphology limited by infrequent nucleation of next layer islands or steps.

## ACKNOWLEDGEMENTS

The authors would like to thank H. Windischmann, R.L. Woodin and S.J. Harris for helpful discussions and collaborations. The Office of Naval Research/Naval Research Laboratory provided partial support for this work.

## REFERENCES

[1] J.C. Angus [ in *Synthetic Diamond* Eds. K.E. Spear, J.P. Dismukes (John Wiley & Sons, New York, 1994) ch.2 p.21-39 ]

[2] D.V. Fedoseev [ in *Synthetic Diamond* Eds. K.E. Spear, J.P. Dismukes (John Wiley & Sons, New York, 1994) ch.3 p.41-56 ]

[3] N. Setaka [ in *Synthetic Diamond* Eds. K.E. Spear, J.P. Dismukes (John Wiley & Sons, New York, 1994) ch.4 p.57-90 ]

[4] J.E. Butler, H. Windischmann [ *MRS Bull. (USA)* vol.23 (1998) p.22-7 ]

[5] See multiple papers on the hot filament, microwave plasma, and DC arcjet plasma driven deposition of diamond films in: S. Saito, O. Fukunaga, M. Yoshikawa (Eds.) [ *Proc. 1st Int. Conf. on New Diamond Science and Technology* (KTK Scientific/Terra Scientific, Tokyo, 1988) ]; A.R. Badzian, R.C. DeVries [ *Mater. Res. Bull. (USA)* vol.23 (1988) p.385 ]

[6] P.E. Pehrsson, F.G. Celii, J.E. Butler [ in *Diamond Films and Coatings* Ed. R.F. Davis (Noyes Publications, Park Ridge, NJ, 1993) ch.3 p.68-146 ]

[7] H.R. Thorsheim, J.E. Butler [ in *Synthetic Diamond* Eds. K.E. Spear, J.P. Dismukes (John Wiley & Sons, New York, 1994) ch.7 p.193-242 ]

[8] D.G. Goodwin, J.E. Butler [ in *Handbook of Industrial Diamonds and Diamond Films* Eds. M.A. Prelas, G. Popovici, L.K. Bigelow (Marcel Dekker, New York, 1998) ch.11 p.527-81 ]

[9] J.E. Butler, R.L. Woodin [ *Philos. Trans. R. Soc. Lond. (UK)* vol.342 (1993) p.15-30 ]

[10] Y. Hirose [ in *1st Int. Conf. on New Diamond Science and Technology* (New Diamond Forum, Japan, 1988) p.38 ]

[11] L.M. Hanssen, W.A. Carrington, J.E. Butler, K.A. Snail [ *Mater. Lett. (Netherlands)* vol.7 (1988) p.289-92 ]

[12] W.A. Carrington, L.M. Hanssen, K.A. Snail, D.B. Oakes, J.E. Butler [ *Metall. Trans. A (USA)* vol.20 (1989) p.1282-4 ]

[13] H.S. Shin, D.G. Goodwin [ *Appl. Phys. Lett. (USA)* vol.66 (1995) p.2909 ]

[14] S.J. Harris, H.S. Shin, D.G. Goodwin [ *Appl. Phys. Lett. (USA)* vol.66 (1995) p.891 ]

[15] P.K. Bachmann [ in *Handbook of Industrial Diamonds and Diamond Films* Eds. M.A. Prelas, G. Popovici, L.K. Bigelow (Marcel Dekker, New York, 1998) ch.21 p.821-50 ]

[16] R.E. Clausing [ in *Handbook of Industrial Diamonds and Diamond Films* Eds. M.A. Prelas, G. Popovici, L.K. Bigelow (Marcel Dekker, New York, 1998) ch.2 p.19-47 ]

[17] J.E. Butler, B.D. Thoms, M. McGonigal, J.N. Russell Jr., P.E. Pehrsson [ in *Wide BandGap Electronic Materials* NATO Meeting/NATO ASI Series Volume 3/1, *Proc. NATO Advanced Workshop on Wide BandGap Semiconductors* Minsk, Belarus, May 4-6, 1994, Eds. M.A. Prelas, P. Gielisse, G. Popovici, B.V. Spitsyn, T. Stacy (Kluwer Academic Publishers, Dordrecht, Netherlands, 1995) p.105-14 ]

[18] M.P. D'Evelyn [ in *Handbook of Industrial Diamonds and Diamond Films* Eds. M.A. Prelas, G. Popovici, L.K. Bigelow (Marcel Dekker, New York, 1998) ch.4 p.89-146 ]

[19] S. Skokov, C.S. Carmer, B. Weiner, M. Frenklach [ *Phys. Rev. B (USA)* vol.49 (1994) p.5662-71 ]

[20] C.C Battaile, D.J. Srolovitz, J.E. Butler [ *Diam. Relat. Mater. (Switzerland)* vol.6 (1997) p.1198-206 ]

[21] Y.L. Yang, M.P. D'Evelyn [ *J. Vac. Sci. Technol. A (USA)* vol.10 (1992) p.978-84 ]

[22] S.P. Mehandru, A.B. Anderson [ *Surf. Sci. (Netherlands)* vol. 248 (1991) p.369-81 ]

[23] T. Tsuno, T. Imai, Y. Nishibayashi, K. Hamada, N. Fujimori [ *Jpn. J. Appl. Phys. (Japan)* vol.30 (1991) p.1063-6 ]

[24] H.G. Busmann, H. Sprang, I.V. Hertel, W. Zimmermann-Edling, H.J. Guntherodt [ *Appl. Phys. Lett. (USA)* vol.59 (1991) p.295-7 ]

[25] L.F. Sutcu, C.J. Chu, M.S. Thompson, R.H. Hauge, J.L. Margrave, M.P. D'Evelyn [ *J. Appl. Phys. (USA)* vol.71 (1992) p.5930-40 ]

[26] B.D. Thoms, J.E. Butler [ *Surf. Sci. (Netherlands)* vol. 328 (1995) p.291-301 ]

[27] S.J. Harris [ *Appl. Phys. Lett. (USA)* vol.56 (1990) p.2298-300 ]

[28] D. Huang, M. Frenklach [ *J. Phys. Chem. (USA)* vol.96 (1992) p.1868-75 ]

[29] B.J. Garrison, E.J. Dawnkaski, D. Srivastava, D.W. Brenner [ *Science (USA)* vol.255 (1992) p.835-8 ]

[30] C.B. Musgrave, S.J. Harris, W.A. Goddard III [ *Chem. Phys. Lett. (Netherlands)* vol.247 (1995) p.359-65 ]

[31] E.J. Dawnkaski, D. Srivastava, B.J. Garrison [ *Chem. Phys. Lett. (Netherlands)* vol.232 (1995) p.524-30 ]

[32] S.J. Harris, D.G. Goodwin [ *J. Phys. Chem. (USA)* vol.97 (1993) p.23-8 ]

[33] S. Skokov, B. Weimer, M. Frenklach [ *J. Phys. Chem. (USA)* vol.98 (1994) p.8-11 ]

[34] M. Zhu, R.H. Hauge, J.L. Margrave, M.P. D'Evelyn [in A.J. Purdes et al (1993) ]

[35] C.C. Battaile et al [ *J. Chem. Phys. (USA)* vol.111 (1999) p.4291-9 ]

[36] B.W. Yu, S.L. Girshick [ *J. Appl. Phys. (USA)* vol.75 (1994) p.3914-23 ]
[37] D.G. Goodwin [ *Appl. Phys. Lett. (USA)* vol.59 (1991) p.277-9 ]
[38] C.J. Chu, M.P. D'Evelyn, R.H. Hauge, J.L. Margrave [ *J. Appl. Phys. (USA)* vol.70 (1991) p.1695-705 ]
[39] T.W. Mercer, J.N. Russell, P.E. Pehrsson [ *Surf. Sci. Lett. (Netherlands)* vol.392 (1998) p.L21-6 ]
[40] M. Tsuda, M. Nakajima, S. Oikawa [ *J. Am. Chem. Soc. (USA)* vol.108 (1986) p.5780-3 ]
[41] S.M. Valone [ in *Processing Science of Advanced Ceramics* Eds. I.A. Aksay, G.L. McVay D.R. Ulrich (Materials Research Society, 1989) p.227-34 ]
[42] M.W. Geis [ in *Diamond, Silicon Carbide and Related Wide Bandgap Semiconductors* Eds. J.T. Glass, R. Messier N. Fujimori (Materials Research Society, 1990) p.15-22 ]
[43] L.F. Sutcu, M.S. Thomson, C.J. Chu, R.H. Hauge, J.L. Margrave, M.P. D'Evelyn [ *Appl. Phys. Lett. (USA)* vol.60 (1992) p.1685-7 ]
[44] Z.L. Wang, J. Bentley, R.E. Clausing, L. Heatherly, L.L. Horton [ *J. Mater. Res. (USA)* vol.9 (1994) p.1552-65 ]

# B1.4 Nucleation of polycrystalline diamond films

I.J. Ford and M.D. Whitfield

March 2000

## A INTRODUCTION

In general, the nucleation of diamond on non-diamond substrates is difficult and polycrystalline films invariably result. This arises from a combination of the high surface energy of diamond, low precursor sticking probability and strong competition from other non-diamond carbon phases. However, control over the nucleation of polycrystalline diamond films is desirable as this would allow the optimisation of several key properties. First among these is the alignment of the crystals, giving improved optical and electronic behaviour. Secondly, control over nucleation might improve the adhesion of the films to substrates. And in general, a high nucleation density is important for producing a uniform coating and a smoother surface, and in reducing the deposition time for a given thickness of film. Much attention has been directed towards understanding how diamond nucleates on substrates in CVD processes, and how the density and orientation of the crystals might be controlled. In particular, prior surface treatments have been investigated.

## B SURFACE TREATMENTS

The simplest surface treatment is to do nothing except clean the substrate thoroughly, which normally leads to randomly oriented polycrystalline films with a low nucleation density. Heteroepitaxial diamond growth has been demonstrated using cubic boron nitride, either as a single crystal [1,2], or as a polycrystalline film [3]. The mismatch in lattice parameter between c-BN and diamond is less than 2%, and the two materials also have similar thermal expansivities, which is important in avoiding distortion on cooling from the deposition temperature. Unfortunately, c-BN is an expensive substrate which is not available in a large area form.

Cheaper materials with similar potential for epitaxy include β-SiC, although the lattice parameter mismatch is as much as 18% [4]. This material is often formed at silicon/diamond interfaces. For silicon, the mismatch in lattice parameter is 34% and the potential for epitaxial growth lower. Of course, the quality of the bonding across the interface is as important in determining whether diamond will grow on a substrate. But it is because the nucleation of diamond films on silicon is so difficult that various surface preparation treatments have been developed; these include scratching, seeding and electrical biasing.

### B1 Scratching

Atomically smooth, single crystal silicon wafers stubbornly resist the nucleation of diamond crystals on their surfaces. However, scratching the surface with diamond powders can increase the nucleation density immensely (typically from about $10^4$ to about $10^8$ per $cm^2$ [5,6]). The nucleation enhancement is approximately inversely proportional to the abrasive particle size and the nucleated crystals tend to lie along the scratch marks. Either the marks created by scratching are more favourable for the formation of an embryonic crystal of diamond, or the nucleation occurs homoepitaxially on diamond particles left embedded in the surface following the treatment. There is some evidence that diamond crystals nucleate preferentially on the sharp shoulders of the scratches rather than in the troughs [7], which tends to support the first option. Furthermore, abrasion with particles other than diamond, such as c-BN [8,9], can enhance the nucleation density. The nucleated crystals display a random alignment with the underlying substrate. But scratching the surface is messy and inconvenient, and the scratch coverage of the surface can lack uniformity. Also, the interface will not be smooth, which is an important consideration for optical or electronic properties, and the orientation of the nucleated crystals cannot be controlled.

## B2 Interlayers (Seeding)

A thin interlayer might ease the lattice parameter mismatch between the diamond phase and the underlying substrate. Treatment with diamond powder has been used for enhancing nucleation on single crystal and polycrystalline nickel substrates [10,11]. Although initially investigated as a scratch treatment, the powder was found to be just as effective when simply sprinkled onto the surface. The effectiveness seems to derive from the carbon seed material on the substrate, since non-diamond carbon powders, including graphite and $C_{60}$, also enhance nucleation [12]. Provision of a thin carbon-rich layer seems to work for other substrate materials including silicon [13]. Other successful seed materials have included hydrocarbon oils and amorphous hydrocarbon films. However, most of these methods are poorly characterised and few general conclusions can be drawn. The usual view is that graphite is detrimental to diamond formation, and the role of hydrogen in the gaseous mixture is to etch it away in preference to the diamond phase [14]. However, it has also been suggested that the hydrogenation and resulting carbonaceous bridging of graphitic planes can allow diamond to nucleate from the edges of graphitic plates [15], in which case a graphitic precursor may be key to nucleation. An interlayer often develops during CVD, as a result of chemical reactions, inter-diffusion or disordered material deposition. Various inter-layers on silicon have been seen, including β-SiC [16] and amorphous silicon [17] or carbon. The formation of interfacial layers probably occurs because of the unsuitability of bare silicon as a substrate.

Copper is closely lattice matched to diamond, and diamond can be grown on it without too much difficulty, but a study of deposition using hot filament CVD has indicated that an amorphous carbon interlayer about 10 nm thick forms at the interface during growth [18], in spite of the similarity in lattice parameter. The microcrystals of diamond seem to grow out of this layer.

## B3 Bias-Enhancement

An alternative surface treatment is bias enhanced nucleation (BEN) in which a negative bias of ~100–250 V is applied to the substrate for a short time during nucleation. BEN can be carried out in-situ, controlled by varying the voltage and application time, and causes minimal substrate damage. Progress can be monitored through the temporal development of the bias current or by simple laser diagnostics. Typical nucleation densities of $\sim 10^{11}$ cm$^{-2}$ are obtained on pristine silicon compared to values of $\sim 10^{5}$ cm$^{-2}$ and $10^{9}$ cm$^{-2}$ for untreated and scratched silicon, respectively. Theoretical estimates, based on critical nuclei considerations, have placed an upper limit of $10^{14}$ cm$^{-2}$ on nucleation densities achievable using BEN [19]. Recent reviews have been provided [20,21].

During BEN the methane fraction is often increased to ~3–5%, substrate temperature increased to ~850–900°C and the reactor pressure reduced to ~15–20 torr to obtain a more uniform plasma. However, it is typically necessary to optimise the bias parameters for a given reactor and conditions. BEN can also give rise to a limited form of 'epitaxy' in which the diamond nuclei show a common orientation with respect to the substrate but the extent is critically dependent on the exact bias parameters. 'Epitaxy' is most commonly observed on silicon, but has also been seen on silicon carbide and iridium. When BEN is combined with (100) textured growth, highly oriented diamond (HOD) films, with improved electrical characteristics [22], can be produced. The effects of BEN are most likely due to a combination of the enhanced ion bombardment of the substrate brought about by the strong electric field in the 'sheath' region above the substrate together with some modification of the plasma chemistry in this region.

BEN was first reported in hot filament reactors using biased filaments or additional electrodes between the hot-filament and the substrate [23–25]. A threshold for nucleation enhancement of around –100 V was observed. Explanations included: sheath effects; enhanced near surface atomic hydrogen concentration which could reduce $sp^2$ formation; and enhanced active species generation near to the substrate by accelerated $H^+$ ions. A systematic study of the effects of biasing in a microwave CVD system was reported by Stoner et al [26]. A negative bias of 250 V relative to earth was applied to the substrate in a 2% methane–hydrogen plasma at a total pressure of 2 kPa, producing a current approximately equal to 15 mA $cm^{-2}$ at the surface. Surface analysis revealed that a 6–10 nm layer of amorphous SiC was formed quite rapidly by a reaction between the substrate and a hydrogenated, amorphous carbonaceous layer present on the surface prior to biasing. After about an hour, a carbonaceous layer with a thickness of up to 1 nm reappears at the surface, due either to the condensation of carbon species from the vapour (accelerated by the bias), or by excess sublimation of Si from the SiC. Out of this layer diamond eventually emerges during the bias stage, when the SiC layer also shrinks. Thus, the process is fundamentally similar to the deliberate provision of a carbonaceous layer on the surface, as described in the previous section. The SiC layer is a necessary precursor to diamond nucleation, but is not the medium from which the diamonds emerge. Stoner et al [26] suggest that conditions do not favour diamond nuclei formation in the carbonaceous material at the surface until the SiC layer is thick enough to suppress silicon diffusion through it.

A similarly detailed study of diamond deposition on (100) silicon was carried out by Sheldon et al [27]. The substrate was at –200 V relative to earth and 865°C, and the $CH_4$ concentration was 1.6% with a total pressure of 5 kPa. Again, a bias stage was conducted followed by an unbiased growth stage. In this case, an amorphous carbon layer was seen to deposit, at the very slow temperature-independent rate of about 0.5 Å/min, upon which diamond then formed. The layer was assumed to form by the deposition of carbonaceous ions from the plasma. The ion energies were estimated to be about 2–20 eV, which is too small to penetrate the surface deeply. SiC was not seen in significant amounts in these studies, in contrast to the work of Stoner et al [26]. The carbonaceous medium for diamond nucleation forms without the prior generation of a SiC layer. This may be due to overall lower ion energies in the work of Sheldon et al, produced by a lower bias voltage and a higher system pressure (2 kPa).

Wolter et al [28] subsequently demonstrated the importance of carbide formation by investigating the effect of BEN on a range of carbide and non-carbide forming substrates. Only small enhancements in nucleation density were found for copper [29], which does not form a carbide but which is a better lattice match to diamond. The degree of nucleation enhancement at fixed bias conditions increased with the carbon affinity of the substrate, although silicon proved to be an exception which the authors suggested may be due to crystal structure effects.

The first BEN oriented nuclei were reported on β-SiC substrates [30,31]. Up to ~50% of the nuclei exhibited preferential orientation with respect to the underlying substrate. On silicon, 30% orientation was achieved following a newly introduced pre-bias carburisation stage [32,33]. The authors suggested that the carburisation stage quickly formed an epitaxial SiC layer which reduced the nucleation time thus minimising bias damage to the oriented nuclei. Subsequently it has been shown that a carburisation layer is helpful but not essential to the generation of BEN oriented nuclei on silicon [34]. For carefully optimised bias conditions combined with (100) textured growth, highly oriented diamond films (HODs) with nearly 100% (100) grains can be achieved [33]. Clearly, the ordered SiC can influence the subsequent development of diamond nuclei in a way that the amorphous layer observed in early experiments [26] could not. Textured growth following a bias pretreatment has also been seen on α-SiC single crystals. Currently the best films achieve HOD areas greater than 30 $cm^2$ [35] compared to only a few $mm^2$ for early films.

Models of conventional nucleation (i.e. not BEN) on non-diamond substrates have been summarised [21] and can be viewed as nucleation on: intermediate diamond-like amorphous carbon [18], metal carbide [37] or graphite [15], formed during the incubation period. The interlayer thickness ranges from a few angstroms for platinum substrates [38] to a few micrometres for molybdenum [39]. All the models assume the existence or formation of 'nucleation sites' generated from carbonaceous precursors. Possibilities include: hydrogenated $sp^2$ bonded carbon structures and transformation of rhombohedral graphite sheets into (111) diamond-like structures. The exact details remain unclear and therefore any models purporting to explain the more complex process of bias nucleation must be considered as rather more speculative. Many plausible explanations for BEN have been given but no definitive choice can yet be made; recent experiments seem to claim a key enabling role for energetic ions accelerated across the bias sheath to the substrate [40,41]. However, doubts have subsequently been articulated regarding these claims [42,20]. Gerber et al [43] proposed a sub-plantation model to account for the nucleation of diamond in an amorphous carbon layer which requires ion energies of around 80–100 eV to generate the nucleation sites. Recently, difficult to perform ion extraction experiments have indicated that at least some carbonaceous ions with the necessary energies to support this model are incident on the substrate during biasing [44]. However, attempts to nucleate diamond film using ion beams of a similar energy and species content have generally been unsuccessful and this suggests that the role of the bias modified plasma chemistry at and directly above the substrate cannot be ignored. Especially for the nucleation of oriented nuclei it has been proposed that the DC plasma which forms just above the substrate may play a significant role in modifying the plasma chemistry in a way which favours the creation and survival of this type of nucleus [45]. The presence of a DC plasma during bias is common to both hot filament and microwave CVD and its properties have been reported in a number of recent measurements [41,46].

Once diamond has nucleated satisfactorily, bias conditions hinder growth. The rate of diamond deposition is increased by more than two orders of magnitude once the bias is turned off. Thus, while ion bombardment tends to favour the emergence of ordered clusters from a disordered phase, the same conditions interfere with the orderly construction of an $sp^3$ bonded crystal, and are not favourable for diamond growth. It may be that the ion bombardment simply accelerates the supply of carbon, which might favour the nucleation process in the early stages of deposition, but is generally detrimental to ordered growth. Indeed, a recent study has indicated that there is a limit to the degree of epitaxy which can ultimately be achieved using bias that arises from damage generated during application of the bias itself [47].

## C INTERPRETATION

It seems that in the majority of cases diamond grows out of an initially disordered carbonaceous medium which exists, or accumulates, on the surface of the substrate. This occurs for particular fluxes of various species, and under certain conditions of temperature and pressure. The apparent necessity for hydrogen to be in the gas mixture suggests that it is the presence of this species, probably in atomic form, which upsets the usual preference for the growth of graphite. It is well established, in fact, that graphite is etched faster than diamond in a hydrogen plasma [48], so that it is conceivable that carbon which happens to bond to the surface with $sp^2$ hybridisation is removed faster than $sp^3$ carbon, leading to a build up of the latter on the surface [14]. Only those atoms which are the most strongly bound to the surface survive attack by hydrogen atoms, and these tend to be bonded with four-fold coordination.

The key parameter is the lifetime of a carbon atom bound to the surface, or within a thin layer on the surface. It would appear that surface treatments affect the local environment of potential nucleation sites in such a way as to favour the formation of diamond carbon. Attack by reactive chemical species such as atomic hydrogen, or structural disruption due to ion-bombardment, also affect the stability, since the chemical composition of the gas mixture, the reactor conditions and the bias voltage (and hence the ion current) influence the nucleation rate.

So deposition and removal processes are occurring simultaneously, and only the most strongly bound atoms survive until they are themselves covered over by new material. The character of the deposited material, ranging from soft amorphous carbon to nearly defect-free diamond, is determined by a competition between various deposition and removal processes. The growth of diamond crystals aligned with the underlying substrate lattice is also likely to be due to the particular stability of suitable embryonic structures.

All this applies to both the nucleation and growth of a diamond film. For nucleation, however, the starting configuration of the surface can vary widely, and the possible processes are more numerous than for the case of growth of an established film. Nucleation requires an incubation period, indicating that the deposition rate of material during the initial formation of stable nuclei is much lower than the rate during growth. Furthermore, the optimal conditions for nucleation and growth can differ, as shown by substrate biasing studies.

## D NUCLEATION MODELLING

The classical theory of heterogeneous nucleation of condensed phases at favourable locations in a system, including at the surfaces of foreign bodies, is well developed. A free energy cost is associated with the phase change, depending on the size of the nucleus formed. This typically involves contributions proportional to the volume and surface of the nucleus of the new phase, plus smaller corrections. Nuclei greater than a critical size are stable, while smaller ones exist only temporarily. Populations of nuclei of the new phase are formed due to random accretion and decay. The number of critical nuclei formed, and hence the nucleation rate of particles of the new phase, depend sensitively on the likely size of energy fluctuations, and hence on the temperature.

The main problem with classical heterogeneous nucleation theory is that the free energy is estimated from bulk material properties, and takes no account of the atomic scale, often assuming, for example, that the cluster is spherical. Nevertheless, it has been applied to the formation of diamond crystallites [49], and the related theory of homogeneous nucleation has been applied in the case of the formation of diamond particles in suspension in the gas [50]. Atomistic versions of nucleation theory have been developed, but they are not particularly successful.

Nucleation is normally associated with thermodynamic fluctuations. An alternative view of the process is that the populations of various sizes of nuclei evolve according to rate equations involving growth and decay coefficients. The latter depend on temperature and mixture composition. It is often easier to include additional mechanisms into models of this type, rather than to consider the thermodynamic fluctuations of a more complex system. The approaches are, however, essentially equivalent.

Rate equation modelling of diamond nucleation was developed some time ago [51–53]. In [52] a model of growth based on Frenklach and Wang's reaction scheme involving ethyne attachment [54] was examined. The treatment describes the explicit evolution of surface densities of three species: available nucleation sites; subcritical crystal embryos; and stable, growing particles. The rate coefficients for each conversion process are evaluated either using experimental rate coefficients for the reactions used in the model, or by comparing the model calculations with observed nucleation kinetics. The data support the assumption that the nucleation of particles is a two-step process, going from nucleation site to particle via an unstable intermediary, the subcritical embryos. This is the simplest possible nucleation model. Once formed, the particle radii are assumed to grow linearly with time, and the observed size distributions of crystals can be compared with the model.

## E CONCLUSION

A number of surface treatments which promote diamond nucleation on various materials have been identified, and experimental investigations have begun to determine the microstructures which are generated in each case, and why they are effective. One can scratch the surface with diamond particles, or to deposit a thin layer of carbonaceous material prior to film deposition. The most effective treatment, however, is to bombard the surface with positive ions, by negatively biasing the substrate in a plasma-driven reactor. The resulting modification of the surface, or the acceleration of the reactions taking place upon it, leads to the highest nucleation densities.

An ordered crystalline structure emerges from disorder within a thin layer of deposited material on the substrate, at particular choices of bombarding species and fluxes. This might be interpreted to mean that only material deposited in special locations, with a particular geometrical alignment with the immediate neighbourhood on the surface, is likely to survive long enough to be incorporated into the film. For correlations in crystallographic alignment to appear, the substrate lattice beneath the thin carbonaceous layer must also have a role to play.

The ultimate goal of research into diamond crystal nucleation would be to account for the success of the various enhancement methods, to optimise them, and possibly to identify new methods, particularly those which might favour the mutual alignment of the crystal axes of the nucleated particles. The ultimate goal would be to enable heteroepitaxial growth of diamond films to be achieved on large areas of substrate.

## REFERENCES


[1] S. Koizumi, T. Murakami, T. Inuzuka, K. Suzuki [ *Appl. Phys. Lett. (USA)* vol.57 (1990) p.563 ]

[2] L. Wang, P. Pirouz [ *Appl. Phys. Lett. (USA)* vol.63 (1993) p.1336 ]

[3] T.A. Friedmann et al [ *Appl. Phys. Lett. (USA)* vol.63 (1993) p.1342 ]

[4] W. Zhu et al [ *Phys. Rev. B (USA)* vol.47 (1993) p.6529 ]

[5] S. Matsumoto, Y. Sato, M. Kamo, N. Setaka [ *Jpn. J. Appl. Phys. (Japan)* vol.21 (1982) p.L183 ]

[6] R.A. Bauer, N.M. Sbrockey, W.E. Brower Jr. [ *J. Mater. Res. (USA)* vol.8 (1993) p.2858 ]

[7] P.A. Dennig, H. Shiomi, D.A. Stevenson [ *Thin Solid Films (Switzerland)* vol.212 (1992) p.63 ]

[8] K. Suzuki, A. Sawabe, H. Yasuda, T. Inuzuka [ *Appl. Phys. Lett. (USA)* vol.50 (1987) p.728 ]

[9] A. Sawabe, T. Inuzuka [ *Thin Solid Films (Switzerland)* vol.137 (1986) p.89 ]

[10] P.C. Yang, W. Zhu, J.T. Glass [ *J. Mater. Res. (USA)* vol.8 (1993) p.1773 ]

[11] W. Zhu, P.C. Yang, J.T. Glass [ *Appl. Phys. Lett. (USA)* vol.63 (1993) p.1640 ]

[12] P.C. Yang, W. Zhu, J.T. Glass [ *J. Mater. Res. (USA)* vol.9 (1994) p.1063 ]

[13] K.V. Ravi, C.A. Koch [ *Appl. Phys. Lett. (USA)* vol.57 (1990) p.348 ]

[14] B.V. Spitsyn, L.L. Bouilov, B.V. Deryagin [ *J. Cryst. Growth (Netherlands)* vol.52 (1981) p.219 ]

[15] W.R.L. Lambrecht, C.H. Lee, B. Segall, J.C. Angus, Z. Li, M. Sunkara [ *Nature (UK)* vol.364 (1993) p.607 ]

[16] B.E. Williams, J.T. Glass [ *J. Mater. Res. (USA)* vol.4 (1989) p.373 ]

[17] N. Jiang, Z. Zhang, B.W. Sun, D. Shi [ *Appl. Phys. Lett. (USA)* vol.63 (1993) p.328 ]

[18] J. Singh [ *Acta Metall. Mater. (USA)* vol.42 (1994) p.3197 ]

[19] S. Yugo, K. Semoto, T. Kimura [ *Diam. Relat. Mater. (Switzerland)* vol.5 (1995) p.25 ]

[20] S.P. Bozeman, B.R. Stoner, J.T. Glass [ in *Handbook of Industrial Diamonds and Diamond Films* Eds. M.A. Prelas, G. Popovici, L.K. Bigelow (Marcel Dekker Inc., New York, USA, 1998) ]

[21] H. Liu, D.S. Dandy [ *Diam. Relat Mater. (Switzerland)* vol.4 (1995) p.1173 ]

[22] B.R. Stoner, C.-T. Kao, D.M. Malta, R.C. Glass [ *Appl. Phys. Lett. (USA)* vol.62 (1993) p.2347 ]

[23] S. Yugo, T. Kanai, T. Kimura, T. Muto [ *Appl. Phys. Lett. (USA)* vol.58 (1991) p.1036 ]

[24] Y.H. Lee, P.D. Richard, P.K. Bachmann, J.T. Glass [ *Appl. Phys Lett. (USA)* vol.56 (1990) p.620 ]

[25] J.J. Chang, T.D. Mantei [ *J. Appl. Phys. (USA)* vol.71 (1992) p.5742 ]

[26] B.R. Stoner, G.-H.M. Ma, S.D. Wolter, J.T. Glass [ *Phys. Rev. B (USA)* vol.45 (1992) p.11067 ]

[27] B.W. Sheldon, R. Csencsits, J. Rankin, R.E. Boekenhauer, Y. Shigesato [ *J. Appl. Phys. (USA)* vol.75 (1994) p.5001 ]

[28] S.D. Wolter, J.T. Glass, B.R. Stoner [ *J. Appl. Phys. (USA)* vol.77 (1995) p.5119 ]

[29] S. Ojika, S. Yamashita, K. Kataoka, T. Ishikura, A. Yamaguchi, H. Kawarada [ *Jpn. J. Appl. Phys. (Japan)* vol.32 (1993) p.L200 ]

[30] B.R. Stoner, J.T. Glass [ *Appl. Phys. Lett. (USA)* vol.60 (1992) p.698 ]

[31] R. Kohl, C. Wild, N. Herres, P. Koidl, B.R. Stoner, J.T. Glass [ *Appl. Phys. Lett. (USA)* vol.63 (1993) p.1792

[32] S.D. Wolter et al [ *Appl. Phys. Lett. (USA)* vol.62 (1993) p.1215 ]

[33] B.R. Stoner, S.R. Sahaida, J.P. Bade, P. Southworth, P.J. Ellis [ *J. Mater. Res. (USA)* vol.8 (1993) p.1334 ]

[34] M.G. Jubber, D.K. Milne [ *Phys. Status Solidi A (Germany)* vol.154 (1996) p.185 ]

[35] A. Floter et al [ *Diam. Relat. Mater. (Switzerland)* vol.7 (1998) p.283 ]

[36] J. Singh [ *J. Mater. Sci. (UK)* vol.29 (1994) p.2761 ]

[37] B. Lux, R. Haubner [ in *Diamond and Diamond-like Films and Coatings* Eds. R.E. Clausing, L.L. Horton, J.C. Angus, P. Koidl (Plenum Press, New York, 1991) p.579 ]

[38] D.N. Belton, S.J. Schmieg [ *Thin Solid Films (Switzerland)* vol.212 (1992) p.68 ]

[39] P. Bou, R. Vandenbulcke, R. Herbin, F. Hillion [ *J. Mater. Res. (USA)* vol.7 (1992) p.2151 ]

[40] S.P. McGinnis, M.A. Kelly, S.B. Hagstrom [ *Appl. Phys. Lett. (USA)* vol.66 (1995) p.311 ]

[41] M. Schreck, B. Stritzker [ *Phys. Status Solidi A (Germany)* vol.154 (1996) p.197 ]

[42] W. Kulisch, L. Ackermann, B. Sobisch [ *Phys. Status Solidi A (Germany)* vol.154 (1996) p.155 ]

[43] J. Gerber, M. Weiler, O. Sohr, K. Jung, H. Ehrardt [ *Diam. Relat. Mater. (Switzerland)* vol.3 (1993) p.506 ]

[44] S. Katai, Z. Tass, G. Hars, P. Deak [ *J. Appl. Phys. (USA)* vol.86 (1999) p.5549 ]

[45] R. Stockel, K. Janischowsky, M. Stammler, L. Ley, M. Albrecht, H.P. Strunk [ *Diam. Relat. Mater. (Switzerland)* vol.7 (1998) p.147 ]

[46] M.D. Whitfield, R.B. Jackman, J.S. Foord [ *Vacuum (UK)* vol.56 (2000) p.15 ]

[47] K.H. Thurer, M. Schreck, B. Stritzker [ *Phys. Rev. B (USA)* vol.57 (1998) p.15454 ]

[48] N. Setaka [ *J. Mater. Res. (USA)* vol.4 (1989) p.664 ]

[49] B.V. Deryagin, D.V. Fedoseev [ *Surf. Coat. Technol. (Switzerland)* vol.38 (1989) p.131 ]

[50] W. Howard et al [ *J. Appl. Phys. (USA)* vol.68 (1990) p.1247 ]

[51] M. Tomellini, R. Polini, V. Sessa [ *J. Appl. Phys. (USA)* vol.70 (1991) p.7573 ]

[52] E. Molinari, R. Polini, M.L. Terranova, P. Ascarelli, S. Fontana [ *J. Mater. Res. (USA)* vol.7 (1992) p.1778 ]

[53] M. Tomellini [ *J. Mater. Res. (USA)* vol.8 (1993) p.1596 ]
[54] M. Frenklach, H. Wang [ *Phys. Rev. B (USA)* vol.45 (1991) p.1520 ]

# B1.5 Effect of B, N and P on low-temperature diamond growth

R. Haubner and B. Lux

April 2000

## A INTRODUCTION

For many years, additions of boron, nitrogen and phosphorus during low-pressure diamond deposition have been applied in order to obtain specific physical properties of diamond (e.g. semiconductivity) or to improve diamond growth conditions [1].

Boron-doped diamond was frequently investigated [2–6] while only a few publications describe the effects of phosphorus addition [7–10]. The influence of nitrogen additions was investigated over a wide range of concentrations. Small nitrogen additions can be introduced unintentionally during diamond chemical vapour deposition (CVD) processes due to leaks or air residues in the reaction gases. These contaminations cause not only changes in crystal growth and orientation [11,12] of the diamond crystals; small and medium nitrogen additions influence also electrical [13], and thermal [14] properties, as well as morphology and phase purity of the diamond [15,16].

The influence of impurities is complex as many parameters must be considered, in particular the diamond deposition method, the deposition parameters, and impurity interactions in many different ways depending on the chemistry at the conditions present.

## B BORON ADDITION

Boron addition during low-pressure diamond deposition has been used for a long time. During the early years the boron was used to increase diamond growth rates and to improve crystallinity [2–5]. Later, boron admixture during CVD diamond growth became a favoured research area, as the semiconducting technology needed boron-doped diamond [17].

For incorporation of boron into low-pressure diamonds various methods and precursors were used:

- The simplest way is to use h-BN as a substrate holder. The atomic hydrogen reacts with the h-BN forming $BH_3$ which is incorporated into the diamond during its growth [18].
- Pure boron sticks can be put into the hydrogen plasma to evaporate $BH_3$. The advantage of this method is that boron doping is limited to the exposure time and thus very thin boron-doped diamond layers can be deposited in between undoped layers [19].
- Gaseous boron precursors can be added, such as borane ($BH_3$) [2], $B(CH_3)_3$ and $B(C_2H_5)_3$ [1]. However, thereby boron contaminates the whole reactor. Thus, an alternation from doped to undoped layers is no longer possible by switching on and off the B-precursor flow.

The deposition methods usually attain gas activation by microwave plasma or a hot-filament which allow easy cleaning of the reactor from the boron contamination. Using a hot-filament the reactions of the metallic filament materials with boron (formation of borides) has to be taken into account.

## B1 Influences of Boron Concentrations and Deposition Parameters

Independently of the boron precursor and the deposition method used, an improved diamond quality is commonly found for low B-concentrations in the gas phase (< 500 ppm B/C) [20–22]; the crystal habit becomes better pronounced and slightly increased crystal sizes are observed.

For medium B-concentrations in the gas phase (1000–5000 ppm B/C) the growth rate drops to values comparable to those for undoped layers and stays constant over a large range. The morphology remains fairly constant but exhibits steadily decreasing crystal size. Analytical measurements on B-containing diamond reveal inhomogeneous boron distribution [23].

For high boron concentrations in the gas phase (> 5000 ppm B/C) the growth rate drops and the morphology becomes unfaceted. At extreme addition levels needle-like features appear (FIGURE 1(a) and (b)).

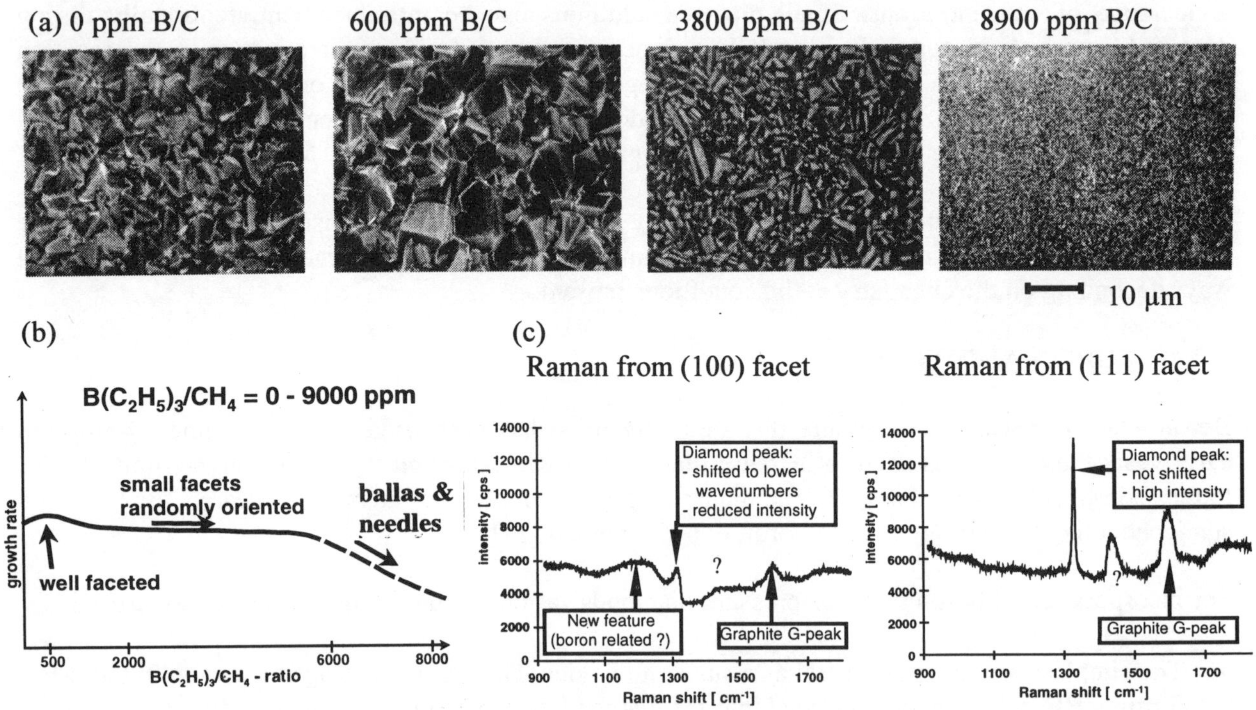


FIGURE 1 Growth characteristics of boron addition during diamond deposition: (a) morphology of selected samples, (b) schematic drawing of growth rate dependence, (c) Raman spectra on (100) and (111) facets showing inhomogeneous boron distribution (1% $CH_4/H_2$, $T_{sub}$=820°C, $T_{fil}$=2450°C, 15 torr).

## B2 Inhomogeneous Boron Incorporation

Optical microscopy and scanning electron microscopy (SEM) [24] of polished diamond samples reveal the inhomogeneities in boron distribution. The different regions with high and low boron contents were characterised by Raman, secondary ion mass spectrometry (SIMS), transmission electron microscopy (TEM) and electron energy loss spectroscopy (EELS).

- Boron is incorporated at up to 3 at.% in the {111} and less than 0.3 at.% in the {100} grown sectors [25].

- The Raman spectrum obtained from a (100) facet is very similar to a normal diamond spectrum, while the spectrum from a (111) facet shows strongly reduced Raman intensities and the presence of disordered graphite and B-doped microcrystalline diamonds [23].
- By TEM and EELS so called 'square defects' were obtained. The defect-free interior of the square is connected to growth on a (100) face and EELS indicates a low boron content while outside the square many lattice defects and higher boron contents were detected [23,26,27].
- Some graphite lamellae could also be detected in the boron-doped diamonds [23]. It seems that they are caused by small amounts of nitrogen [28].

Secondary ion mass spectrometry [25], nuclear reaction analysis (NRA) [29] and elastic recoil detection (ERD) [30] were used to quantify the boron contents in the diamond layers. The upper limit of boron incorporation is given at about 3 at.% and is accompanied by morphological deterioration including the onset of graphite formation.

### B3 Influence of Deposition Parameters

It is not possible to give a general statement about the influences of the various deposition parameters. Diamond growth (deposition rate and phase purity) in the presence of boron is quite sensitive to surface temperature (the diamond deposition window narrows) and gas activation, more than for undoped diamond growth [1,31,32].

In general, the B contents in the diamond layers were found to be higher than was the gas phase B/C ratio used.

## C NITROGEN ADDITION

The influences of $N_2$ additions on diamond growth were examined with various deposition methods [33]. Two types of experiment can be distinguished:

- Low nitrogen contents originating from the process gas impurities and the equipment leakages.
- Higher nitrogen concentrations ($N_2/CH_4 = 0.05–10$) to study chemistry of nitrogen additions.

### C1 The Influences of Nitrogen Concentration

In many diamond CVD processes unintentional small additions of nitrogen occur due to leaks or residues in the reaction gases (commonly found in $H_2$ and $CH_4$). These contaminations are a problem, as they influence the crystal growth and orientation drastically [34]. Furthermore, small and medium nitrogen additions change the outstanding electrical [13,35] and thermal [14] properties of diamond layers as well as their morphology and phase purity [15,16].

Thermodynamic equilibrium calculations showed that HCN seems to be the most important compound involved. Examinations of diamond morphology, crystal quality, orientation and growth rate for a large range of $N_2$-concentrations are used to discuss the mechanisms and possibilities of the nitrogen-induced effects (FIGURE 2).

For particular growth conditions, the diamond growth rate passed through two maxima with increasing $N_2$ addition. Depending upon the amount of added $N_2$, the results can be separated into four domains which are based on two fundamental mechanisms (FIGURE 2):

- The first mechanism is the abstraction of adsorbed hydrogen atoms caused by CN and HCN. At low $N_2$ additions ($N_2/CH_4 = 0.05–0.1$) it resulted in improved diamond phase purity and increased growth rates due to a reduction of carbon supersaturation [36]. At higher $N_2$ concentrations ($N_2/CH_4$ = 0.2–0.4) the enhanced hydrogen abstraction probably led to the start of a surface reconstruction which caused a lower diamond phase purity and decreased growth rates.

- A second mechanism probably caused by modified carbon attachment kinetics started with even higher $N_2$ concentrations. Again, with increasing $N_2$ additions the growth rate passed through a maximum (at $N_2/CH_4$ = 1), and the growth of the diamond phase was first improved (up to $N_2/CH_4$ = 2) and then deteriorated.

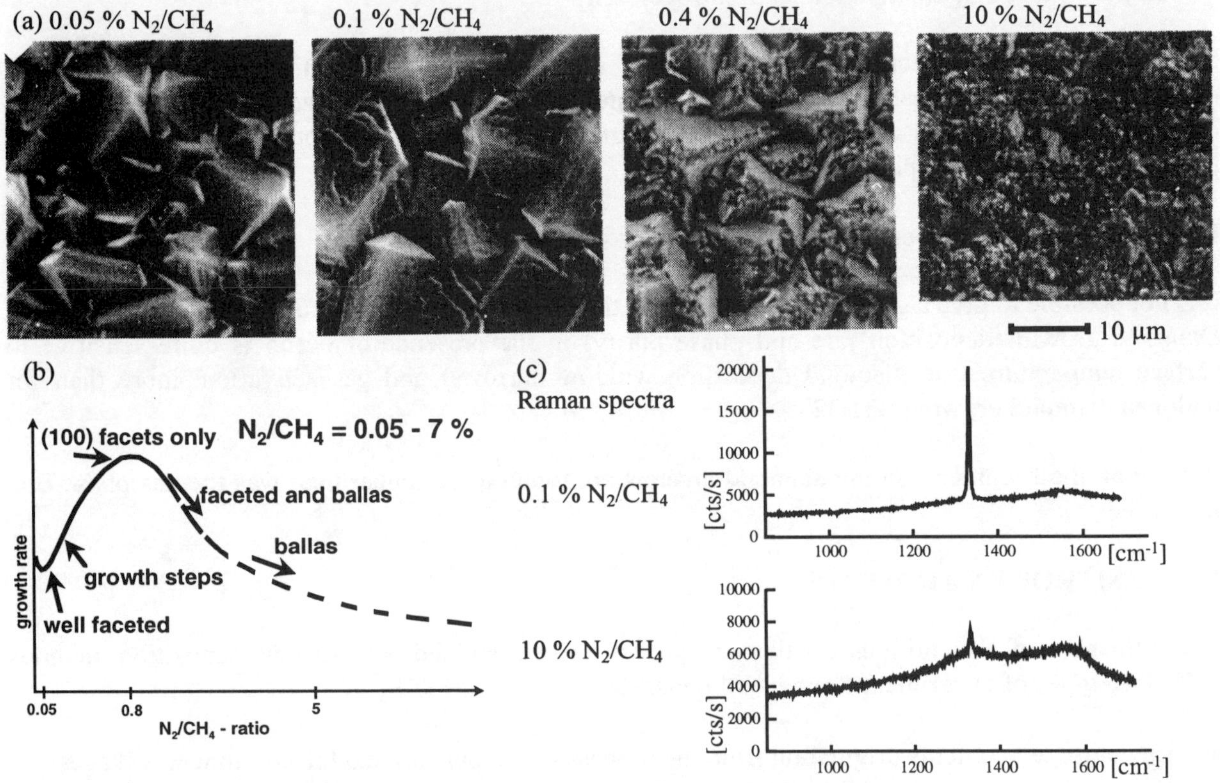


FIGURE 2 Growth characteristics of nitrogen addition during diamond deposition: (a) morphology of selected samples, (b) schematic drawing of growth rate dependence, (c) selected Raman spectra (1% $CH_4/H_2$, $T_{sub}$=820°C, $T_{fil}$=2450°C, 15 torr).

## C2 Influences of the Deposition Parameters

A strong decrease in diamond growth rate with increasing substrate temperature indicates that surface reactions of nitrogen species are involved during deposition. Variation of the gas pressure and filament temperature show a slight influence during diamond nucleation, but during crystal growth (i.e. longer deposition) no changes compared to the pure systems are observed. This indicates that the nitrogen gas phase reactions did not determine the diamond crystal growth [1].

The diamond deposition method used also has consequences for the nitrogen additions. It seems that high power density increases the formation of HCN and so the effects become more intense. In the case of hot-filament activation several % $N_2$ can be added to the gas before ballas deposition occurs while in microwave CVD only about 0.5% $N_2$ can be tolerated [15,37].

## D PHOSPHORUS ADDITION

Phosphorus addition was mainly used to produce n-type doping in the diamond. Considering only geometrical and energetic factors, an easy introduction of phosphorus into the diamond lattice seems unlikely since the covalent radius of phosphorus is larger than that of carbon [38]. Nevertheless, some reports on the formation of n-type CVD diamond films using phosphorus-containing doping sources are found in the literature [8–10]. A maximum phosphorus content of 1 wt.% in diamond films was reported [39].

The observed changes in diamond growth rate, morphology and crystallinity (FIGURE 3) are attributed either to surface or to gas phase reactions as well as to a reduction of the carbon supersaturation and defect-induced growth [7,40].

Phosphorus additions increased the diamond growth rate at lower P concentrations (330–1320 ppm $PH_3/CH_4$) and decreased it at higher ones (2640–5280 ppm $PH_3/CH_4$). Thus, the morphology changed from (111)- via (100)-faceted to unfaceted or ballas-like diamond.

With P addition (660 ppm $PH_3/CH_4$) the best deposition conditions with respect to growth rate and diamond crystal quality were obtained at low substrate temperatures (820°C), higher pressures (50–75 torr) and moderate filament temperatures (~2300°C). All diamond layers deposited with phosphine addition contained phosphorus, as confirmed qualitatively by SIMS measurements [29].

At medium P additions (e.g. 2640 ppm $PH_3/CH_4$) inhomogeneous layers were obtained. The central zone (the area with optimal diamond deposition conditions) of the layer revealed a distinctly decreased diamond peak and ballas morphology. At the edges of the sample faceted diamond was observed. This can be explained by increased HCP formation in the centre and reduced HCP concentrations at the edges [41].

The mechanism for the influence of phosphorus was shown to be a changed carbon supersaturation or atomic hydrogen coverage due to surface reactions at the growing surface rather than an altered gas activation with regard to atomic hydrogen and carbon growth species. It is suggested that the reactions occurred between phosphorus species such as methinophosphide (HCP), and adsorbed hydrogen and carbon. Thus the carbon attachment kinetics were modified, which is fundamental for the diamond growth features.

## E COMPARISON OF B, N AND P ADDITIONS

Thermodynamic calculations can help to understand the reactions involved. The important species are HCP and P for P-addition, HCN and CN in the case of $N_2$, and B, BH and $BH_2$ for boron. It could be shown by variation of the deposition parameters, that formation of the relevant species decreases diamond quality (ballas or non-diamond deposits) [42]. For example some calculations for phosphorus and nitrogen addition and variation of filament temperature are shown in FIGURE 4.

The added impurities can affect both the gas activation and gas phase recombination of the atomic hydrogen, as well as the surface mobility of the carbon species and the diamond crystal growth kinetics. Because of this the observed results can be different for the three elements (FIGURE 5).

The high amounts of nitrogen which can be added to the deposition can be explained as follows. In the case of the hot-filament method the thermal activation of the nitrogen is comparatively low and molecular nitrogen has only a small influence on diamond deposition [16]. The microwave method seems to activate the nitrogen more effectively, which allows only lower nitrogen additions [37].

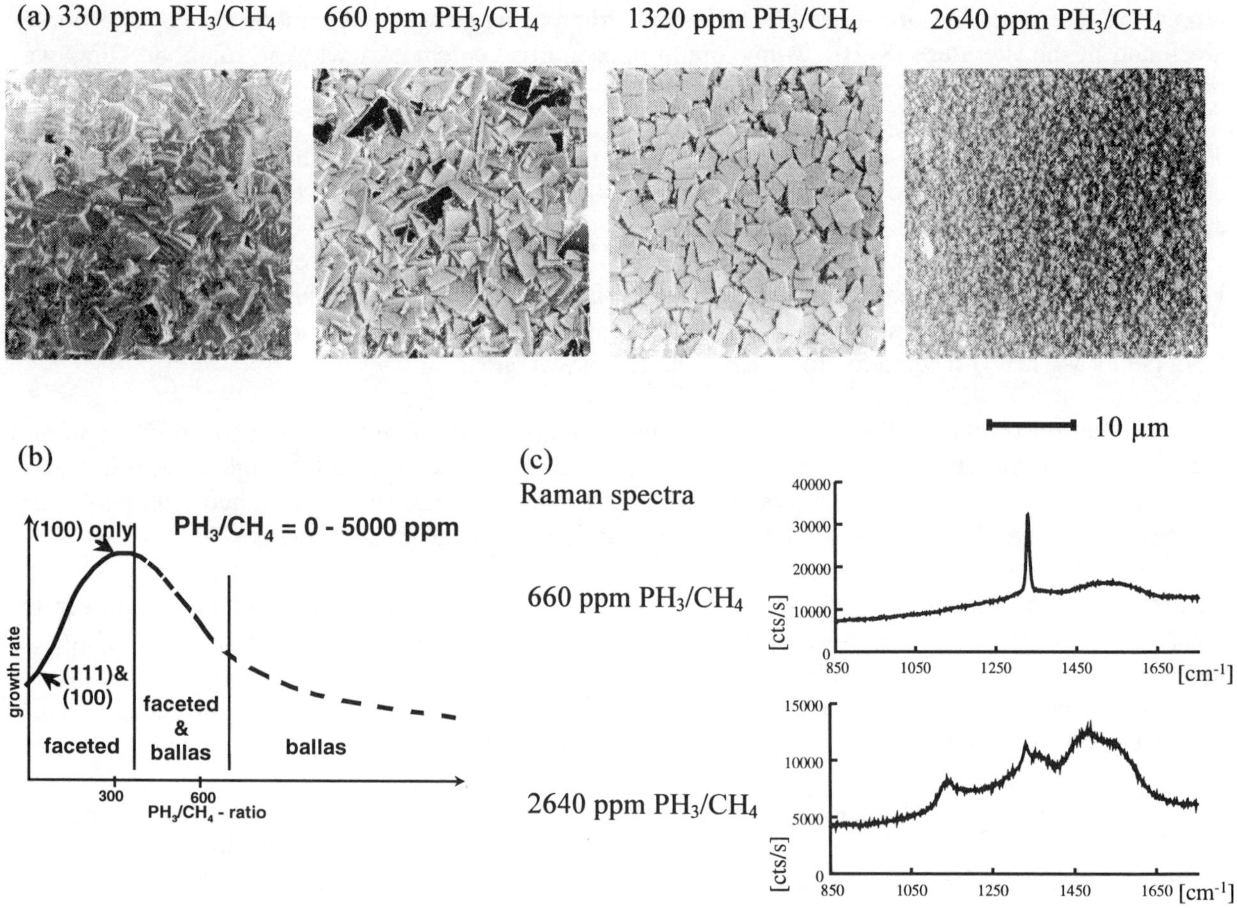


FIGURE 3 Growth characteristics of phosphorus addition during diamond deposition: (a) morphology of selected samples, (b) schematic drawing of growth rate dependence, (c) selected Raman spectra (1% $CH_4/H_2$, $T_{sub}$=820°C, $T_{fil}$=2450°C, 15 torr).

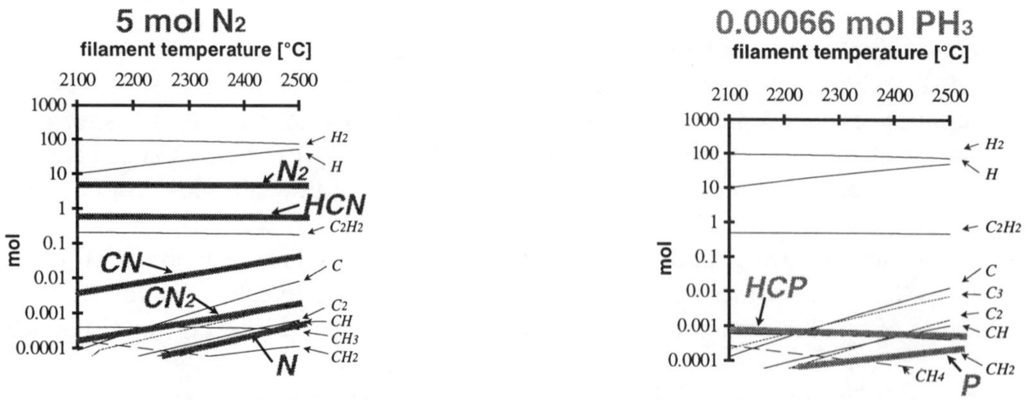


FIGURE 4 Thermodynamic calculations for nitrogen and phosphorus additions.

## F SIMULTANEOUS ADDITION OF BORON AND NITROGEN

Simultaneous addition of boron and nitrogen has many influences on CVD diamond deposition. Graphite and diamond in large amounts and various ratios can coexist [28]. These results can explain the formation of graphite during diamond doping caused by nitrogen impurities [24,26].

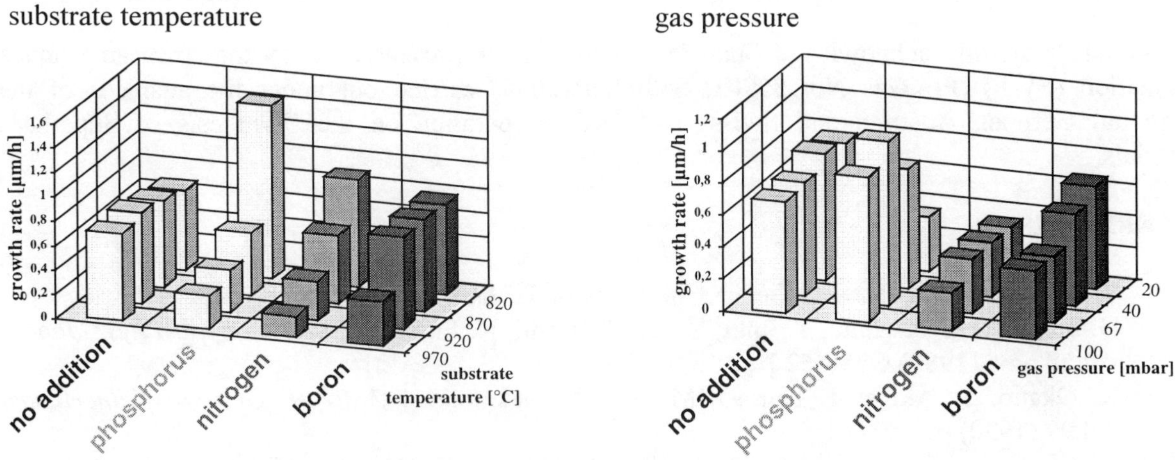


FIGURE 5 Comparison of the diamond growth rates with B, N and P addition and variation of the process parameters substrate temperature and gas pressure ($PH_3/CH_4$=0.00066, $N_2/CH_4$=5, $B(C_2H_5)_3/CH_4$=0.0035, 15 h, hot-filament, 2500°C, 20 mbar, 500 sccm $H_2$, 1% $CH_4$).

At constant boron and increasing nitrogen content the growth rate reveals a distinct minimum. Nitrogen can, depending on boron concentration and substrate temperatures, either deteriorate or improve the crystallinity of the diamond deposition.

Within the investigated concentration ranges neither h-BN nor the formation of B/C/N phases were observed; nitrogen incorporation in the mixed layers was not detectable by SIMS (<200 ppm) [28].

The mechanisms of the simultaneous B and N influences, even at low levels, on the diamond growth are important, but not yet fully understood.

## G CONCLUSION

By investigating diamond deposition at various deposition conditions and the single additions of phosphorus, nitrogen and boron the chemical interactions of the impurities during diamond deposition could be explained.

Boron addition shows no significant maximum in diamond growth and well faceted diamond growth is possible up to high boron concentrations. Boron is incorporated into the diamond but surface reactions seem to have less influence on diamond morphology. In the case of boron, gas phase reactions on the hot-filament seem to be strongly involved.

Nitrogen addition causes the formation of HCN or CN during diamond deposition. The formation of these compounds can decrease the carbon saturation in the system and thus improves diamond quality. CN radicals can also take the role of atomic hydrogen and help to stabilize the diamond surface by preferential etching. But, similar to the phosphorus additions, surface reactions of the nitrogen species influence the diamond deposition.

In the case of phosphorus additions defect-induced growth and surface reactions of phosphorus compounds (HCP) formed during gas activation mainly influence the diamond growth.

## ACKNOWLEDGEMENT

The authors gratefully acknowledge financial support of the present work by the Austrian Science Foundation (FWF) (Projects No. S5901, S5907, P12055) carried out under the auspices of the trinational German, Austrian and Swiss 'D-A-CH' cooperation on the 'Synthesis of Superhard Materials'.


## REFERENCES


[1] S. Bohr [ Doctoral Thesis, Vienna University of Technology, 1995 ]

[2] H. Shiomi, H. Nakahata, T. Imai, Y. Nishibayashi, N. Fujimori [ *Jpn. J. Appl. Phys. (Japan)* vol.28 no.5 (1989) p.758-62 ]

[3] K. Okano, Y. Akiba, T. Kurosu, M. Iida, T. Nakamura [ *J. Cryst. Growth (Netherlands)* vol.99 (1990) p.1192-5 ]

[4] E. Gheeraert, P. Gonon, A. Deneuville, L. Abello, G. Lucazeau [ *Diam. Relat. Mater. (Switzerland)* vol.2 (1993) p.742-5 ]

[5] J. Mort et al [ *Appl. Phys. Lett. (USA)* vol.55 no.11 (1989) p.1121-3 ]

[6] X.H. Wang, G.-H.M. Ma, J.T. Glass, L. Bergmann, K.F. Turner, R.J. Nemanich [ *Diam. Relat. Mater. (Netherlands)* vol.1 (1992) p.828-35 ]

[7] S. Bohr, R. Haubner, B. Lux [ *Diam. Relat. Mater. (Switzerland)* vol.4 (1995) p.133-44 ]

[8] N. Fujimori, T. Imai, H. Nakahata, H. Shiomi, Y. Nishibayashi [ *Mater. Res. Soc. Symp. Proc. (USA)* vol.162 (1990) p.23-33 ]

[9] K. Okano et al [ *Appl. Phys. A (Germany)* vol.51 (1990) p.344-6 ]

[10] M. Kamo, H. Yurimoto, T. Ando, Y. Sato [ in *New Diamond Science and Technology* Eds. R. Messier, J.T. Glass, J.E. Butler, R. Roy (MRS Int. Conf. Proc., Washington DC, September 1990) p.637-41 ]

[11] Ch. Wild, N. Herres, P. Koidl [ *J. Appl. Phys. (USA)* vol.68 no.3 (1990) p.973-8 ]

[12] E. Wörner, C. Wild, W. Müller-Sebert, R. Locher, P. Koidl [ *Diam. Relat. Mater. (Switzerland)* vol.5 (1996) p.688-92 ]

[13] E. Boettger, A. Bluhm, X. Jiang, L. Schäfer, C.-P. Klages [ *J. Appl. Phys. (USA)* vol.77 no.12 (1995) p.6332 ]

[14] N. Fujimori. A. Ikegaya, T. Imai, N. Ota, T. Shibata [ Eur. Patent Application 0 469 626 A2 ]

[15] S. Jin, T.D. Moustakas [ *Appl. Phys. Lett. (USA)* vol.65 no.4 (1994) p.403 ]

[16] S. Bohr, R. Haubner, B. Lux [ *Appl. Phys. Lett. (USA)* vol.68 no.8 (1996) p.1075-7 ]

[17] N. Fujimori, Y. Nishibayashi [ *Diam. Relat. Mater. (Netherlands)* vol.1 (1992) p.665-8 ]

[18] O. Matsumoto, H. Toshima, T. Katagiri, Y. Kanzaki [ in *ISPC-7,* Eindhoven 1985, Paper A-1-3 ]

[19] A. Aleksov et al [ *Diam. Relat. Mater. (Switzerland)* vol.8 (1999) p.941-5 ]

[20] X.H. Wang et al [ *Diam. Relat. Mater. (Netherlands)* vol.1 (1992) p.828-35 ]

[21] X.Z. Liao, R.J. Zhang, C.S. Lee, S. Tong Lee, Y.W. Lam [ *Diam. Relat. Mater. (Switzerland)* vol.6 (1997) p.521-5 ]

[22] J.H. Won et al [ *Appl. Phys. Lett. (USA)* vol.68 no.20 (1996) p.2822-4 ]

[23] P. Wurzinger, P. Pongratz, P. Hartmann, R. Haubner, B. Lux [ *Diam. Relat. Mater. (Switzerland)* vol.6 (1997) p.763-8 ]

[24] P. Hartmann et al [ *Refract. Metals Hard Mater.* vol.16 (1998) p.223-32 ]

[25] T. Kolber, K. Piblits, R. Haubner, H. Hutter [ *Fresenius J. Anal. Chem. (Germany)* vol.365 (1999) p.636-41 ]

[26] P. Wurzinger, M. Joksch, P. Pongratz [ *Inst. Phys. Conf. Ser. (UK)* no.146 (1995) p.523-8 ]

[27] Z.L. Wang, J. Bentley, R.E. Clausing, L. Heatherly, L.L. Horton [ *J. Mater. Res. (USA)* vol.9 (1994) p.1552 ]

[28] P. Hartmann, R. Haubner, B. Lux [ *Diam. Relat. Mater. (Switzerland)* vol.6 (1997) p.456-62 ]

[29] H. Spicka et al [ *Diam. Relat. Mater. (Switzerland)* vol.5 (1996) p.383-7 ]
[30] G. Dollinger, A. Bergmaier, T. Faestermann, C.M. Frey [ *Fresenius. J. Anal. Chem. (Germany)* vol.353 (1995) p.311 ]
[31] K. Miyata, K. Nishimura, K. Kobashi, Y. Kawate, J.T. Glass [ in *Science and Technology of New Diamond* Eds. S. Saito, O. Fukunaga, M. Yoshikawa (1990) p.145-8 ]
[32] E.N. Farabaugh, L. Robins, A. Feldman, C.E. Johnson [ *J. Mater. Res. (USA)* vol.10 no.6 (1995) p.1448-54 ]
[33] R. Haubner, B. Lux [ in *Surface Modifications Technologies III* Eds. T.S. Sudarshan, D.G. Bhat (TMS, Warrendale, USA, 1990) p.835-9 ]
[34] P. Koidl, C. Wild, R. Locher, W. Müller-Sebert [ presented at *Diamond Films 95* Barcelona, Spain, 10-15 Sept. 1995, pres.5.1 ]
[35] J. Mort, M.A. Machonkin, K. Okumura [ *Appl. Phys. Lett. (USA)* vol.59 no.24 (1991) p.3148 ]
[36] A. Badzian, T. Badzian [ *Appl. Phys. Lett. (USA)* vol.62 no.26 (1993) p.3432 ]
[37] R. Locher, C. Wild, N. Herres, D. Behr, P. Koidl [ *Appl. Phys. Lett. (USA)* vol.65 no.1 (1994) p.34-6 ]
[38] B.V. Spitsyn, A.E. Alexenko [ *Proc. Electrochem. Soc. (USA)* vol.91-8 (1991) p.597-604 ]
[39] A.E. Alexenko, B.V. Spitsyn [ *Diam. Relat. Mater. (Netherlands)* vol.1 (1992) p.705-9 ]
[40] R. Ramesham, C. Ellis, T. Roppe [ in *Materials Science Monographs, 73, Applications of Diamond Films and Related Materials* Eds. Y. Tzeng et al (Elsevier, Amsterdam, 1991) p.411-6 ]
[41] W.A. Yarbrough, A.R. Badzian, D. Pickrell, Y. Liou, A. Inspector [ *J. Cryst. Growth (Netherlands)* vol.99 (1990) p.1177-82 ]
[42] R. Haubner, S. Bohr, B. Lux [ *Diam. Relat. Mater. (Switzerland)* vol.8 (1999) p.171-8 ]

# B1.6 Diamond growth by microwave plasma chemical vapour deposition (MPCVD)

A.J.S. Fernandes, Qi H. Fan, R.F. Silva and F.M. Costa

May 2000

## A INTRODUCTION

Since it was conclusively demonstrated that industrial synthesis was also possible from the gas phase [1], diamond film growth by chemical vapour deposition has been widely investigated during the last two decades in order to achieve the technological solutions to introduce it as a competitive, scaled-up and reliable material.

Some recent work proved that the presence of hydrogen is no longer understood as an essential condition for CVD diamond growth: a multi-laser based method [2] and an argon/carbon process, the former having been developed at Argonne National Laboratory, are clear examples of that. Specifically, MPCVD nano-crystalline diamond films were successfully grown in plasmas containing argon instead of hydrogen [3]. However, until now the conventional CVD methods using hydrogen and hydrocarbon gases demonstrated the highest deposition rates.

The chemistry involved in conventional CVD diamond growth is mainly coupled to the atomic hydrogen production ability. In fact, atomic hydrogen concentration turns out to be one of the most important factors for both the growth rate and quality of diamond deposits [4]. Moreover, the cost of CVD diamond production is in direct proportion to that of producing and capturing atomic hydrogen [5]. Roughly, in CVD diamond technology atomic hydrogen is produced in two ways, each one related to its group of CVD methods. The first one is by thermal decomposition of hydrogen gas (hot filament, hot graphite disk, etc.), in which the H concentration is highly heterogeneous, varying very much with the distance to the hot zone. The other method is plasma activated generation of H (microwave, RF, DC, Arc-jet, etc.), in which the supplied energy couples directly to the free electrons, providing homogeneous atomic hydrogen profiles [4]. Microwave plasmas hold the advantage of being sustained without electrodes, avoiding both contamination and degradation of the reactors. Moreover, they provide a highly reliable gas activation source, capable of long deposition periods without any noticeable changes in the plasma characteristics. A very large portion of the research work on diamond growth has been carried out using microwave systems, which have attained high quality material with acceptable growth rates [6].

## B THE MICROWAVE APPROACH FOR CVD SYNTHESIS OF DIAMOND

The first approach to coupling microwave technology to CVD of diamond was reported in 1982, when the Japanese NIRIM group presented an assembly of a 2.45 GHz microwave source normally directed to a quartz tube inside which the plasma was formed, a vacuum pump and a controlled gas feed-through system. The plasma was ignited at low pressure (~1 mbar) in a mixture of 99 vol.% $H_2$ and 1 vol.% $CH_4$, resulting in a quasi-neutral combination of highly reactive hydrogen free atoms and hydrocarbon species. Under these conditions, microwave radiation accelerates free electrons to energies corresponding to several thousand degrees, while the neutral species and heavy ions are kept at much lower temperatures—the so-called non-isothermal plasma. The pressure was then raised to >10 mbar for the deposition, these plasmas now allowing the electrons to heat up the gas species—the so-called isothermal or just thermal plasmas. Thermal plasmas permit better gas activation efficiency, thus leading to higher growth rates, typically up to 3 μm/hr for a 1.5 kW NIRIM type reactor. Despite its simple and inexpensive set-up, the scaling up to industrial level is hampered by the small deposition area and easy contamination with silica, when the plasma stays too close to the walls.

In order to overcome both the contamination and area problems, a different type of reactor was later presented (Bachmann et al, D. Smith and ASTeX collaborators, 1987), the 'bell jar' apparatus. Here, a silica dome takes the place of a quartz tube, the 2.45 GHz microwaves are conducted by a rectangular wave-guide and coupled via a centro-symmetrically emitting antenna located at the top, and a ball-shaped stable plasma forms away from the reactor walls. This system presented the novelty of controlling the substrate temperature independently of the plasma. Growth rates of 0.2–0.3 μm/hr (with a 1.0 kW source) together with the difficulty of the reactor load/unload procedure are the main limitations of this system.

By 1990, ASTeX presented what can be called an up-grade of the 'bell jar' system, the HPMS reactor. HPMS stands for high pressure microwave source, and it differs from the 'bell jar' mainly in its metallic structure with a top mounted quartz microwave window. A double walled water cooled cylinder, fitted with inspection windows and a loading door, is connected by its top to an antenna mount that emits the microwave radiation through the reactor's quartz window. In this case, the main principle of rectangular-to-circular coupling of microwave radiation into a cylindrical reactor is retained, resulting in a plasma-ball formation above the substrate. When the 1.0 kW or 1.5 kW sources were used with this system, an external RF induction heater and graphite substrate stage was available, keeping the substrate at the correct deposition temperature. In 1992, a new 5 kW magnetron microwave source was introduced, resulting in an immediate growth rate increase up to 14 μm/hr. Also, the high microwave power led to a direct plasma heating of the substrate, leaving the external heater often obsolete. However, using the HPMS reactor, it is only possible to operate inside a well defined zone of the pressure/power diagram, as otherwise unstable plasmas may be generated that may touch the steel walls or jump to the microwave window, destroying it. Another limitation is the radius dependent nature of the deposits, caused by the spherical symmetry of the plasma, even if this can be minimised using well-engineered substrate stages. In order to improve both the uniformity and the growth rate, ASTeX offered another solution, still based on centro-symmetrical microwave coupling, but this time delivering the raised 8 kW radiation power through the bottom of the reactor. The substrate stage acts as the microwave coupling electrode, which flattens the plasma and improves the homogeneity of the diamond film throughout the substrate area. This also provides an easy way to load/unload the reactor using a top cover door. In addition, the plasma being closer to the substrate confers higher growth rates, up to ~20 μm/hr.

A quite different approach is that made by Wavemat, Inc., a company dealing with plasma processing materials. Mainly, Wavemat's MPDR (microwave plasma disk reactor) systems differ from those by ASTeX in the microwave coupling inside the reactor. The plasma is formed and kept inside a silica glass cup by correctly positioning the sliding cavity end plate (on the top) and the antenna (normal to the reactor, by its perimeter). Once the tuning is found, a disk shaped plasma forms inside the silica glass cup. The substrate can be immersed into the plasma or it can be placed at a desired distance from it. Furthermore, another deposition mode uses a biased metal grid to extract and accelerate ionised species to the substrate surface. This system, once fitted with permanent magnets, allows ECR (electron cyclotron resonance) operation. The possibility of selecting the substrate distance from the plasma is a key advantage in operating at high power levels and with thermal plasmas because it is a way to control the deposition temperature. However, high density thermal plasma can etch the walls and deform them (as in the NIRIM reactor), contaminating the plasma with the glass cup material.

At Fraunhoffer Institute, another interesting microwave system was developed, this time based on an ellipsoidal reactor chamber where the microwave antenna emits at the first focus and the plasma is formed at the region around the other one. Based on this approach, AIXTRON has recently presented the AIX P6/P60 models for diamond deposition with reported growth rates up to 12 μm/hr and white diamond quality large wafers [7,8].

## C OTHER MICROWAVE APPROACHES

### C1 ECR (Electron Cyclotron Resonance)

ECR operation was addressed by many groups to overcome the non-uniformity problem and also in the search for low temperature deposition of diamond. The main advantage of ECR based plasmas is that the activated area is usually much wider and uniform, allowing a very large number of objects or objects with larger dimensions to be deposited. However, ECR requires low operation pressures (<0.1 torr) at which the deposition rate is rather low. Along with this, these diamond deposits are quite defective, containing important amounts of non-diamond phase. Although remarkable achievements were attained with this method—temperature sensitive substrates like polymers were successfully diamond coated at temperatures as low as 200°C [9]—only very specific applications demanding limited diamond quality can be explored by ECR diamond deposition.

### C2 Microwave 'Plasma Jet'

Still another way to allow increased deposition areas with uniformly distributed plasmas involves a pump forced gas stream that passes through a high density microwave induced plasma, resulting in a 'plasma jet' directed to a furnace where the samples to be coated are placed. This describes the ASTeX system LADS (large area deposition system). As in ECR systems, high quality growth in this case is problematic, not to mention that it requires huge amounts of gas, implying a closed-loop recycling operation with gas pumping, filtering and chilling.

### C3 Microwave 'Plasma Torch'

Increased pressure in microwave plasmas seems to favour growth rate at the cost of deposition area, pointing out the advantage of thermal plasmas. This becomes evident with the early experiment (1989) of Y. Mitsuda with his atmospheric pressure 2.45 GHz microwave plasma torch: at 5 kW, growth rates of 30 μm/hr were achieved for only 5 $cm^2$ of deposited area.

## D THE MICROWAVE FREQUENCY QUESTION

A very important matter is the frequency used to produce microwave plasmas. Most of the systems are based on 2.45 GHz—the typical domestic microwave oven frequency—mainly for three reasons: i) industrial attribution of operating frequencies; ii) widespread technology for 2.45 GHz wave-guides, tuners, antennas, etc., which availability leads to affordable prices; iii) the microwave components, with size often proportional to wavelength, are small enough to give the systems fairly compact dimensions, resulting in advantages in weight and logistics. However, since the trend seems to be the increase of microwave power input in order to reach industrial scale efficiency growth, the 2.45 GHz frequency shows its limitations: the maximum power for available commercial sources is 8 kW. Another difficulty is the easy disturbance of the electrical field by edges or pointy zones of substrates with complex shapes, which is minimised with longer wavelengths. Plasmas from 915 MHz sources have already proved to be efficient for diamond MPCVD and the wavelength involved is longer (32.8 cm against 12.2 cm), allowing wider deposition areas. Actually, a 915 MHz system can be seen as a 'magnified' 2.45 GHz system. Moreover, there is no power limitation as with the 2.45 GHz radiation sources, it being normal to find power supplies delivering up to many tens of kW. Thus, the scale-up necessary for the industrial production of CVD diamond cutting tools or heat sinks requires better technological resources. This explains why the leading companies turned to the 915

MHz technology: ASTeX proposed a 75 kW system; Wavemat is also offering 30 kW CVD machines [10]; more recently, AIXTRON presented the already mentioned novel ellipsoidal reactor featuring a 60 kW power source. All these systems are supposed to be capable of large area deposition with high diamond quality and growth rate.

## E RESIDUAL STRESS DEVELOPMENT AND ADHESION OF MPCVD DIAMOND FILMS

As shown above, microwave CVD synthesis has turned out to be one of the most suitable ways of mass producing diamond film. Since MPCVD systems generally depend on a plasma zone above the substrate, the geometry of the plasma ball reflects on the uniformity of the resulting films along the radius of the sample's stage. A consequence of such a specific feature is the development of inhomogeneous thermal and intrinsic residual stresses, with obvious implications for the adhesion especially at the sample's edges, where usually the electric fields are more concentrated. This leads to thermal profiles that show hotter zones with thicker deposits at the centre and at the border of the samples.

Residual stresses in CVD diamond consist of thermal and intrinsic components. Polycrystalline diamond MPCVD is usually conducted at elevated temperatures (600≅1000°C). The thermal stress is generated during cooling by the difference in thermal expansion coefficient of the diamond film and substrate. The simplest equation commonly used to estimate the thermal stress $\sigma_{th}$ is the thin film approximation given by

$$\sigma_{th} = \int_{T_0}^{T} \frac{E}{1-\tau}(\alpha_f - \alpha_s)dT \qquad (1)$$

where E and $\tau$ are the Young's modulus and the Poisson's ratio of the diamond, $\alpha_f$ and $\alpha_s$ are coefficients of thermal expansion for the film and substrate, T is the deposition temperature and $T_0$ is room temperature. A few factors need to be considered when estimating the thermal stress. Firstly, plastic deformation in relatively soft substrates, like Cu, steel, Si etc., usually leads to a lower stress value in the film as compared with the calculated result from EQN (1) [11]. Second, the sample bending causes significant stress relaxation throughout the film depth [12,13]. This effect becomes more pronounced with increasing film thickness or with decreasing substrate thickness. Third, temperature gradients in the sample radial and transverse directions are inevitable for an MPCVD system due to inhomogeneous heating from the plasma ball and due to the substrate cooling. Actually, the trend in CVD diamond mass production is to use hotter thermal plasmas at higher pressures, which implies considerable thermal flow across the samples. As an example, the authors observed that the thickness of a diamond film grown by MPCVD on a Co substrate 5 × 5 mm$^2$ in size changed from 240 μm at the sample centre to 280 μm at the edge! Intrinsic stress is associated with the growth mechanism. A general opinion is that the intrinsic stress arises from impurities and defects such as twins, dislocations, sp$^2$-bonded carbon and voids incorporated during diamond growth [14]. Although the effect of these defects on internal stress in the diamond is hardly quantified or modelled, some microstructure-related stress sources, such as coherency strains, surface energy effects and disclinations, have been studied intensively and quantitatively or semi-quantitatively [13,15]. FIGURE 1 shows X-ray patterns of two free-standing diamond films grown under identical conditions but with different growth time. Variation in the film morphology with the growth evolution is obvious. Therefore, significant change in the growth strain is expected.

Residual stresses can be evaluated by a number of techniques, including Raman spectroscopy, substrate curvature, X-ray diffraction, vibrating membrane etc. A detailed review has been given by

Windischmann and Gray [17]. Of all these methods, Raman spectroscopy is the most commonly utilised. When using the substrate curvature method, one should be careful about plastic deformation in the substrate. With the X-ray method, a curved sample surface sometimes makes the determination of the incident angle complicated. An interesting Raman result was noticed by the authors when diamond films grown on silicon nitride evidenced slight tensile stress instead of compressive stress, despite the lower thermal expansion coefficient of diamond.

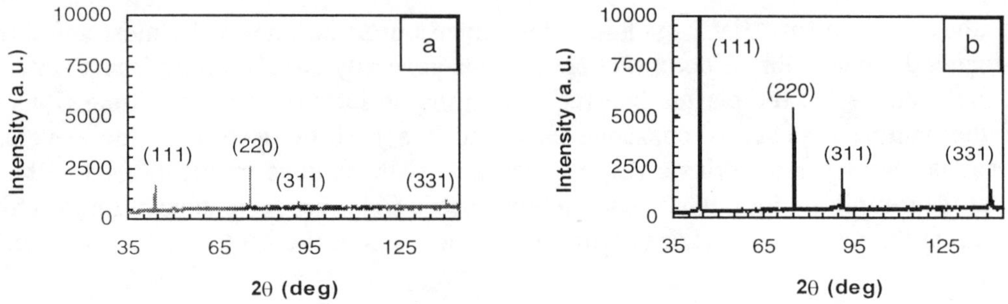


FIGURE 1 X-ray patterns of two diamond films of (a) 4 μm in thickness and (b) 30 μm in thickness.

Adhesion of MPCVD diamond on a substrate results from two main mechanisms: chemical bonding and mechanical interlocking. The strength of the chemical bonds depends on the carbon affinity of the substrates. The substrates can be classified into three groups: strong carbide forming materials, such as Si and Ti; non-carbon affinity or little carbon affinity materials, like Cu and Au; strong carbon dissolving materials, such as steel and Co. Diamond films grown on strong carbide forming materials show excellent adhesion, while diamond films deposited on the second type of material exhibit nearly no adhesion at all. Diamond growth on the carbon dissolving materials is found to be interesting. Usually a thick polycrystalline graphite layer mixed with amorphous carbon forms at the initial stage. The diamond then starts to nucleate from this layer and to grow under the control of the synthesis conditions. The resultant film exhibits no adhesion and is ready to peel off from the substrate.

To achieve reasonable adhesion on little carbon affinity and strong carbon dissolving materials, a single or multi-intermediate layers of strong carbide forming materials that block the carbon/substrate mutual diffusion are common solutions. The interlayers can be deposited by sputtering, evaporating, electroplating and hot-wall CVD. An interesting example of the mechanical interlock effect is that some authors obtained impressive film adhesion by using a Ni interlayer electroplated with diluted diamond powders [17,18]. One could deduce that better adhesion might be achieved if the transition metal, Ni, is replaced by a strong carbide forming metal, like Ti.

Evaluation of the adhesion of diamond coatings is still a challenge. Conventional scratch tests encounter some problems with possible damage to the scratch tip as the coating hardness is close to the value for a natural diamond [19]. Pull-off tests are shown in many cases to be limited by the strength of the bonding glue [20]. Static indentation tests seem to provide more reliable information [21]. However, the contribution of both elastic and plastic deformation of the substrate must be taken into account, which renders this method complex to interpret.

The effect of residual stress on the diamond film adhesion seems not so straightforward. Adhesion strength is obviously reduced when it has to accommodate residual stresses. However, significant differences in the residual stresses are usually found in diamond films grown on different substrate materials, while for the same kind of substrate material the residual stresses are usually close. Therefore, it is difficult to quantitatively correlate the film adhesion to residual stresses, as many factors, such as the diamond nucleation density, the substrate surface state and the carbide bond strength, influence the film adhesion. Interestingly, we found that an as-grown diamond film deposited on a copper substrate with a titanium interlayer exhibited higher residual stress and better adhesion compared with the same sample annealed in vacuum [22].

## F CONCLUSION

Diamond growth by MPCVD has been discussed, along with other microwave approaches. The frequency used to produce the microwave plasmas has generally been 2.45 GHz, but the leading companies are turning to 915 MHz technology. The development of residual stresses and the adhesion of the diamond films on different substrates have been described.

## REFERENCES

[1] B.V. Spitsyn, L.L. Bouilov, B.V. Deryagin [ *J. Cryst. Growth (Netherlands)* vol.52 (1981) p.219-26 ]
[2] A. Badzian et al [ *Diam. Relat. Mater. (Switzerland)* vol.7 no.1 (1998) p.64-9 ]
[3] D.M. Gruen [ *MRS Bull. (USA)* vol.23 no.9 (1998) p.32-5 ]
[4] D.G. Goodwin, J.E. Butler [ in *Handbook of Industrial Diamonds and Diamond Films* Eds. M.A. Prelas, G. Popovici, L.K. Bigelow (Marcel Dekker, New York, 1998) p.527-81 ]
[5] J.E. Butler, H. Windischmann [ *MRS Bull. (USA)* vol.23 no.9 (1998) p.22-7 ]
[6] P.K. Bachmann [ in *Handbook of Industrial Diamonds and Diamond Films* Eds. M.A. Prelas, G. Popovici, L.K. Bigelow (Marcel Dekker, New York, 1988) p.821-50 ]
[7] M. Füner, C. Wild, P. Koidl [ *Surf. Coat. Technol. (Switzerland)* vol.74-75 (1995) p.221-6 ]
[8] http://www.aixtron.com/products/aixp6-60.htm
[9] A. Hatta, A. Hiraki [ in *Handbook of Industrial Diamonds and Diamond Film* Eds. M.A. Prelas, G. Popovici, L.K. Bigelow (Marcel Dekker, New York, 1998) p.887-99 ]
[10] http://www.wavemat.com/
[11] V.G. Ralchenko, E.D. Obraztsova, K.G. Korotoushenko, A.A. Smolin, S.M. Pimenov, V.G. Pereverzev [ *Mater. Res. Soc. Symp. Proc. (USA)* vol.383 (1995) p.153 ]
[12] J. Michler, M. Mermoux, Y. von Kaenel, A. Haouni, G. Lucazeau, E. Blank [ *Thin Solid Films (Switzerland)* vol.357 (1999) p.189 ]
[13] Qi H. Fan, J. Grácio, E. Pereira [ *J. Appl. Phys. (USA)* vol.87 (2000) p.2880 ]
[14] P.R. Chalker, A.M. Jones, C. Johnston, I.M. Buckley-Golder [ *Surf. Coat. Technol. (Switzerland)* vol.47 (1991) p.365-74 ]
[15] I.J. Vlasov, V.G. Ralchenko, E.D. Obraztsova, A.A. Smolin, V.I. Konov [ *Thin Solid Films (Switzerland)* vol.308-309 (1997) p.168-72 ]
[16] S. Nijhawan, S.M. Jankovsky, B. Sheldon [ *J. Mater. Res. (USA)* vol.14 (1999) p.1046 ]
[17] H. Windischmann, K.J. Gray [ *Diam. Relat. Mater. (Switzerland)* vol.4 (1995) p.837 ]
[18] C.R. Lin, C.T. Kuo [ *Diam. Relat. Mater. (Switzerland)* vol.7 (1998) p.903 ]
[19] M. Alam, D.E. Peebles, J.A. Ohlhausen [ *J. Adhes. Sci. Technol. (Netherlands)* vol.7 no.12 (1993) p.1309-22 ]
[20] Qi H. Fan, E. Pereira, J. Grácio [ *J. Mater. Sci. (UK)* vol.34 (1999) p.1353-65 ]
[21] J.W. Ager, M.D. Drory [ *Phys. Rev. B (USA)* vol.48 (1993) p.2601-7 ]
[22] Qi H. Fan, J. Grácio, E. Pereira [ *J. Appl. Phys. (USA)* vol.86 (1999) p.5509 ]

CHAPTER B2

# NANOCRYSTALLINE DIAMOND FILMS

# B2.1 Growth of nanocrystalline diamond films

D.M. Gruen

January 2000

## A INTRODUCTION

Diamond's many unique properties make it a candidate material for a variety of important applications. Each of the potential uses requires that diamond properties be optimized to fulfil a different function: hardness, surface smoothness, optical transparency, electrical conductivity, field emission, etc. The optimization can only be achieved by operating on the microstructure, because one obviously cannot change diamond's molecular structure or its composition.

Two methodologies have evolved that can reproducibly lead to the synthesis of nanocrystalline diamond films. The first, exemplified by the work of Michler et al [1], has shown that if in a $CH_4/H_2$ mixture, the $CH_4$ content is continuously increased while the substrate temperature is kept constant, the morphology changes from faceted microcrystals to ball-shaped clusters of nanocrystals ('ballas-like') to graphitic feather-like morphologies with nanometre (nm) sized diamond inclusions. In the rather narrow ballas-like regime (e.g. 800°C substrate temperature, 9% $CH_4$ concentration) electron diffraction reveals no other phase but diamond. Bright field and annular dark field images from diamond {111} show that the films are nanocrystalline. The nanocrystals exhibit a high density of twins and stacking faults that are responsible for streaks in the corresponding diffraction pattern. The transition to nanocrystallinity appears to be initiated by two-dimensional nucleation on {111} facets leading to a twinned and faulted microstructure with domain sizes ranging between 30 and 50 nm.

The second methodology also involves changes in gas phase chemistry but maintains a constant carbon content (e.g. 1% $CH_4$ or $C_{60}$) while substituting a noble gas (e.g. Ar) for hydrogen. It was shown by Zhou et al [2] that going from 1% $CH_4$/99% $H_2$ to 1% $CH_4$/99% Ar in a microwave discharge, the morphology changes in a continuous fashion from faceted microcrystals to equiaxed nanocrystals. The randomly oriented nanocrystals have a narrow size distribution maximizing between 3 and 5 nm. Extensive characterization work [3] has shown that the phase-pure diamond films synthesized by this methodology have many highly unusual tribological, electrochemical, and field emission properties, due to the fact that approximately 10% of the carbon is situated at grain boundaries. Because of the order of magnitude larger number of grain boundaries in the $CH_4$/Ar synthesized films compared to the ballas-like films, the former are referred to as ultrananocrystalline films to distinguish them from the latter. The subsequent discussion in Section B will be entirely concerned with ultrananocrystalline diamond films (UNCD) synthesized from carbon-containing noble gas microwave plasmas, since the mechanical, electrical, and optical properties of this unique material depend profoundly on the $\pi$-bonded grain boundary carbon.

## B GROWTH OF ULTRANANOCRYSTALLINE DIAMOND FILMS (UNCD)

Conventional diamond film CVD uses hydrocarbon-hydrogen gas mixtures, and it generally has been believed that hydrogen plays a central role in the various diamond CVD processes developed over the years. In particular, it has been thought that atomic hydrogen is an absolutely essential ingredient of the vapour from which the films are grown [4–6]. The growth of phase-pure UNCD films from carbon containing noble gas microwave plasmas has required a re-examination and extension of current ideas concerning diamond CVD.

Absorption spectroscopy of the UNCD plasma shows that the methyl radical is virtually absent and that the principal active carbon species, carbon dimer, $C_2$, reaches concentrations of $10^{12}$–$10^{13}$ $cm^{-3}$ [7]. The growth rate of UNCD films has been correlated with $C_2$ concentration at constant total carbon content by varying the noble gas pressures. In other work, the $C_2$ concentration has been shown to decrease rapidly as the plasma hydrogen content is increased, as expected from the law of mass action. Rotational analysis of the $C_2$ Swan-band spectrum gives a neutral gas temperature of 1600 K for the Ar gas at 100 torr. Production mechanisms for $C_2$ involve:

a) $2CH_4 \xrightarrow{1600\,K} C_2H_2 + 3H_2$

$$C_2H_2 \xrightarrow{Ar^+,\,Ar^*} C_2H_2^{\,+} + Ar$$

$$C_2H_2^{\,+} \xrightarrow{e-} C_2 + H_2$$

b) $C_{60} \xrightarrow{Ar^+,\,Ar^*} C_{60}^{\,+} + Ar$

$$C_{60}^{\,+} \xrightarrow{e-} C_{58} + C_2 \xrightarrow{1600\,K} C_{56} \ldots C_{30} + nC_2$$

Charge exchange cross sections for $Ar^+$ with $CH_4$ and $C_{60}$ are very high, as are cross sections for dissociative electron recombination. Interestingly, $C_2$ production cross sections for $C_2$ are an order of magnitude higher for $C_{60}$ than for $CH_4$, thus requiring an order of magnitude lower total carbon content when $C_{60}$ is used as a precursor [7].

Typical growth conditions for UNCD films are: 1% $CH_4$ or 0.1% $C_{60}$ in 99% Ar at 100 torr; 100 sccm total flow rate; 1 kW microwave power; 800°C substrate temperature; 0.5 μm/hr growth rate. Because $C_2$ is a very highly active chemical species, it can insert directly both into C–C and C–H bonds [8]. The experimental results summarized above suggest that $C_2$ rather than $CH_3$• (or $C_2H_2$) is the growth species involved in UNCD film growth.

In order to place this supposition on a firm theoretical basis, density functional calculations have been carried out for growth on both the (110) face and the (100) face of diamond [8]. The results show that $C_2$ addition to diamond (110) is highly exothermic with small activation barriers. A schematic depiction of the $C_2$ growth mechanism on diamond (110) is shown in FIGURE 1.

## C CONCLUSION

Two methodologies exist to synthesise nanocrystalline diamond. This Datareview has summarised the situation in the microwave plasma technique used to produce ultrananocrystalline diamond. Typical growth conditions and the production mechanisms have been summarised.

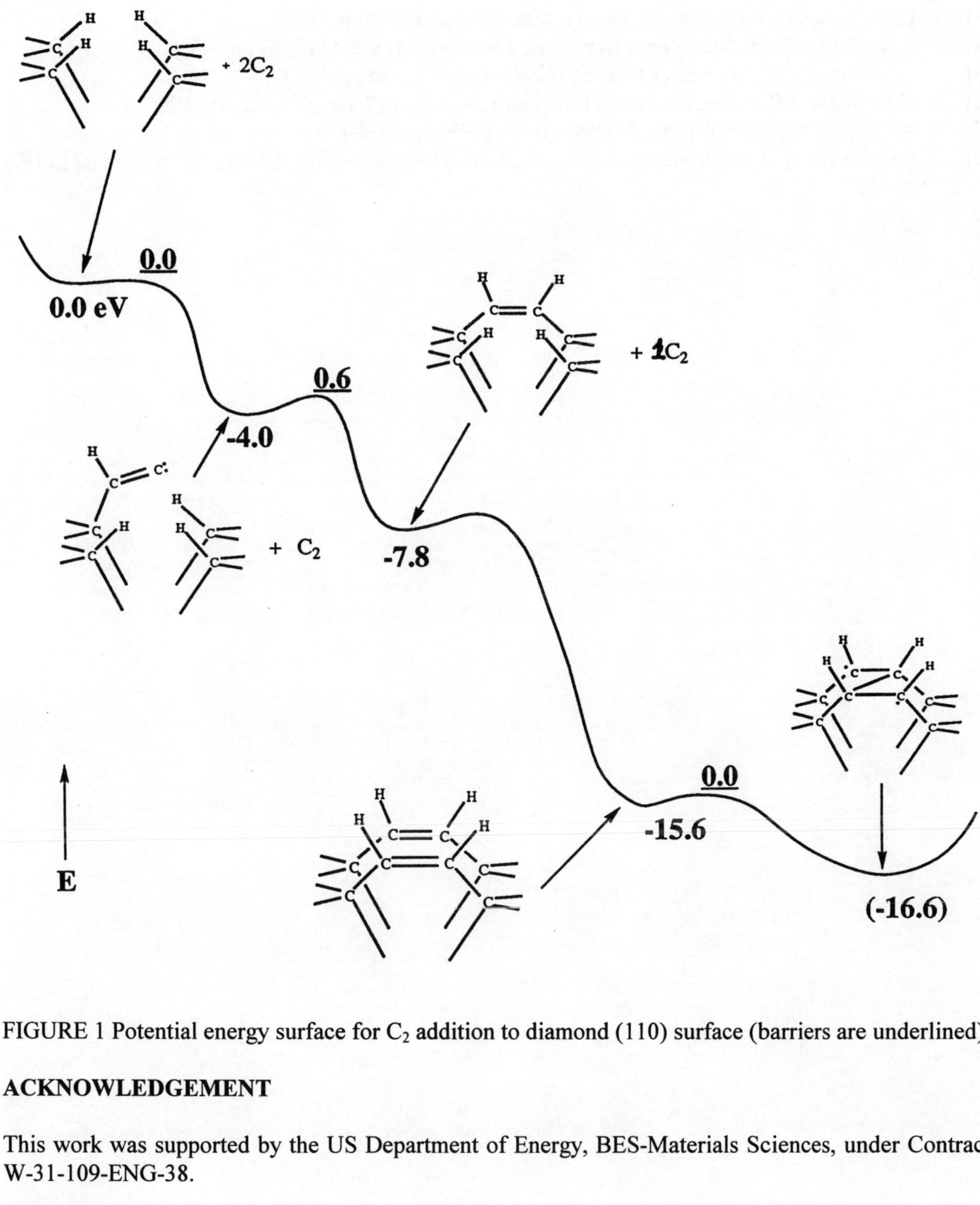


FIGURE 1 Potential energy surface for $C_2$ addition to diamond (110) surface (barriers are underlined).

## ACKNOWLEDGEMENT

This work was supported by the US Department of Energy, BES-Materials Sciences, under Contract W-31-109-ENG-38.

## REFERENCES

[1] J. Michler, S. Laufer, H. Seehofer, E. Blank, R. Haubner, B. Lux [ *Proc. 10th Int. Conf. on Diamond and Diamond-like Materials* Prague, Czech Republic, 12-17 Sept. 1999, paper 5.231 ]

[2] D. Zhou, D.M. Gruen, L.C. Qin, T.G. McCauley, A.R. Krauss [ *J. Appl. Phys. (USA)* vol.84 (1998) p.1981 ]
[3] D.M. Gruen [ *Annu. Rev. Mater. Sci. (USA)* vol.29 (1999) p.211 ]
[4] F.G. Celii, J.E. Butler [ *Annu. Rev. Phys. Chem. (USA)* vol.42 (1991) p.643 ]
[5] J.C. Angus, C.C. Hayman [ *Science (USA)* vol.241 (1988) p.913 ]
[6] K.E. Spear, J.P. Dismukes (Eds.) [ *Synthetic Diamond* (Wiley, New York, 1994) ]
[7] A.N. Goyette et al [ *J. Phys. D (UK)* vol.31 (1998) p.1975 ]
[8] P.C. Redfern, D.A. Horner, L.A. Curtiss, D.M. Gruen [ *J. Phys. Chem. (USA)* vol.100 (1996) p. 11654 ]

# B2.2 Nucleation of ultrananocrystalline diamond films

D.M. Gruen

January 2000

Growth of ultrananocrystalline (UNCD) films on diamond (110) and (100) has been delineated in Datareview B2.1. According to classical nucleation theory, for spherical or cubic crystallites, the following relationship holds:

$$\text{Nucleation rate } (\text{cm}^{-2}\ \text{s}^{-1}) = \frac{\text{Growth rate }\left(\text{cm s}^{-1}\right)}{\text{Crystallite size }\left(\text{cm}^{3}\right)}$$

This relationship is shown graphically in FIGURE 1. For conventional microcrystalline diamond CVD from hydrocarbon-hydrogen mixtures, the microstructure is typically composed of micrometre-sized crystallites. For UNCD film growth, the microstructure consists of 3–5 nm crystallites. An examination of FIGURE 1 shows that in order for this situation to obtain, the nucleation rate has to increase from $10^4$ $\text{cm}^{-2}$ $\text{s}^{-1}$ to on the order of $10^{10}$ $\text{cm}^{-2}$ $\text{s}^{-1}$. The $10^6$ order of magnitude increase in (secondary) nucleation rates suggests that the mechanism of CVD diamond nucleation needs to be closely examined so that a fundamental understanding of diamond crystallite size as a function of plasma chemistry can be achieved.

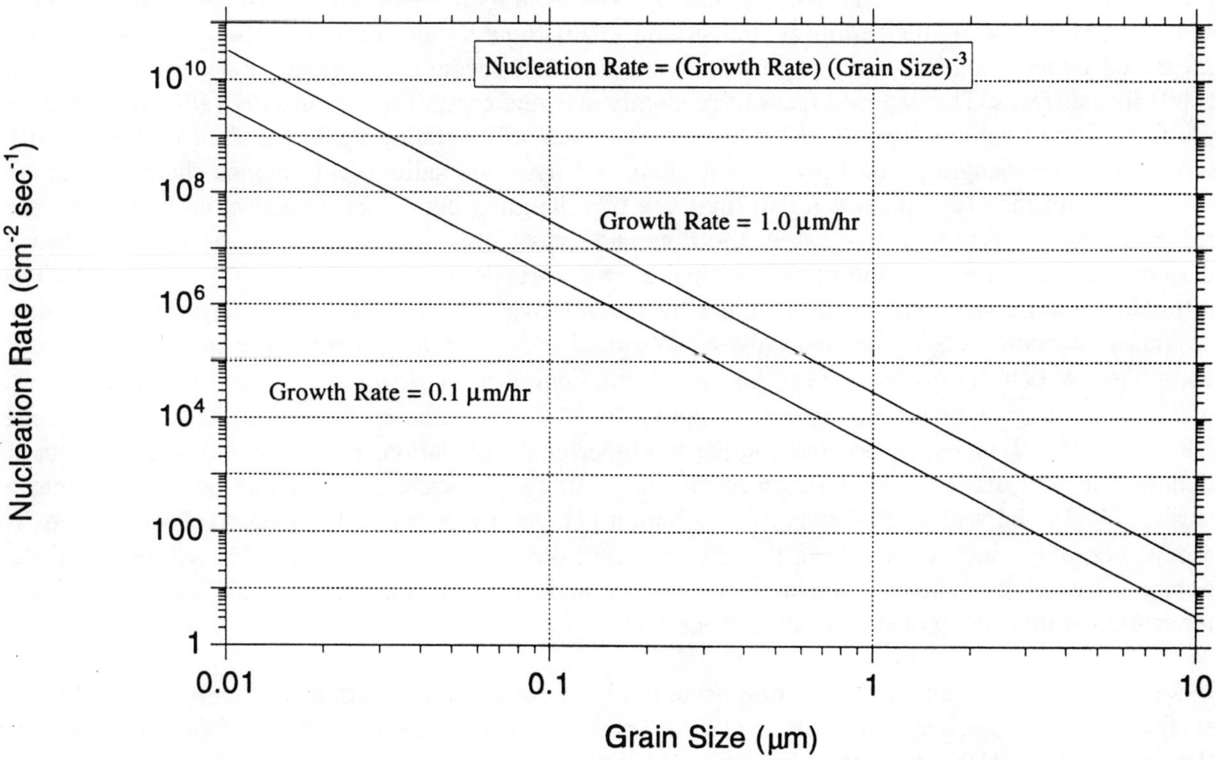


FIGURE 1 Functional relationship between nucleation rate and grain size displayed graphically.

It is important to point out that considerations relating to secondary or re-nucleation have not formed an important part of theoretical considerations of conventional diamond CVD, presumably because of the relatively low re-nucleation rates. On the other hand, a great deal of attention has been paid to the initial nucleation site density, which depends strongly on substrate preparation prior to deposition.

However, the development of diamond film growth from hydrogen-poor plasmas and the attendant nanocrystallinity of the films make secondary nucleation a subject of key importance. Because $C_2$ is the observationally important species in the hydrogen-poor microwave plasmas [1], theories of diamond re-nucleation have been elaborated using $C_2$ as the nucleating species on the (100) surface [2]. Density functional theory and ab initio molecular orbital theory have been used to calculate the energetics of $C_2$ insertion into $C_9H_{12}$ and $C_9H_{14}$ clusters that model unhydrided and monohydrided (100) diamond surfaces respectively. The reaction of $C_2$ with either the $C_9H_{12}$ or $C_9H_{14}$ cluster is exothermic by more than 100 kcal/mol, but the lowest energy product is different for the two clusters. The reaction of singlet $C_2$ with the C=C double bond of the $C_9H_{12}$ cluster leads to either carbene structures or a cyclobutyne-like structure, with the former having the lower energy at both the HF/6-31G* and B3LYP/6-31G* levels of theory. No barrier for insertion into the C=C double bond of the $C_9H_{12}$ cluster was found at the HF/6-31G* and B3LYP/6-31G* levels of theory. The reaction of singlet $C_2$ with the HC–CH single bond or C–H bonds of the $C_9H_{14}$ cluster leads to a structure having a cyclobutene-like geometry. The disparate nucleation rates of diamond crystallites grown in hydrogen-rich versus hydrogen-poor $C_{60}$/Ar microwave plasmas are accounted for qualitatively by these results on the basis of the following considerations.

The nucleation begins with the insertion of one end of the $C_2$ molecule into the $\pi$ bonds of unhydrided portions of the reconstructed (100) surface of diamond after which the unattached carbon atom can react with other $C_2$ molecules from the gas phase to nucleate a new diamond crystallite. Since most of the diamond (100) surface dimers are hydrided under even hydrogen-poor experimental conditions, it is necessary to see if nucleation at unhydrided (100) dimers can account for the nucleation rates measured experimentally. It is reasonable to assume that secondary nucleation occurs primarily at (100)-like surfaces. The exposed faces of randomly oriented crystallites are usually (100)-like because the face-centred cubic diamond lattice has only two sets of crystallographic planes, (111) and (110), having just one dangling bond per carbon atom. All other crystallographic planes share a common character with the (100) planes in that there are two dangling bonds per carbon atom before surface reconstruction. Also, in many cases, the diamond crystallites are observed to be cubic. When a diamond crystallite grows, the three principal growth directions are <110>, <111>, and <100>. The ultimate crystal shape will be determined by the relative growth rates along the three directions, assuming secondary nucleation and growth takes place only after the growth of the parent crystallite is completed. When the growth rate ratios satisfy the conditions $\nu_{100}/\nu_{111} \leq 1/\sqrt{3}$ and $\nu_{100}/\nu_{110} \leq 1/\sqrt{2}$, where $\nu$ is the growth rate or growth velocity, the ultimate shape of the crystallite will be a cube enclosed by (100) faces. As pointed out, density functional calculations have shown that the growth of diamond nanocrystallites can take place on the (110) faces, where a dimer can be inserted into a surface C–H bond with a low energy of activation [3]. As the growth rate along <110> directions is almost certainly much greater than that along <100> directions ($\nu_{110} >> \nu_{100}$), this will result in the enlargement of the (100) surface area, which would in turn enhance the probability of secondary nucleation at this newly created (100) surface.

If we assume that secondary nucleation occurs only at the carbon dimers of reconstructed (100)-like surfaces, then the nucleation density is at most about 8 sites/nm$^2$, i.e. the density of carbon dimers on the reconstructed (100) surface. A 15 nm crystallite requires about 50 s (at 1 µm/hr growth rate) to complete its growth while developing about 300 nm$^2$ of (100) surface for each grain. This grain size would give rise to a grain density of about $4 \times 10^{11}$ grains/cm$^2$. Combining this with the nucleation site density of 8 sites/nm$^2$ (or equivalently about $2 \times 10^3$ sites/grain), one finds that the maximum rate of creation of nucleation sites would be $2 \times 10^{13}$ sites cm$^{-2}$ s$^{-1}$, which is about 3 orders of magnitude higher than the experimentally determined nucleation rate of about $10^{10}$/cm$^2$ s. Thus, in order for nucleation at unhydrided carbon dimers to account for experimental nucleation rates, only 0.05% of surface carbon dimers need be unhydrided [2].

Possible ways for secondary nucleation to proceed from further addition of $C_2$ to the energetically most favourable carbene structure would be to grow either a new three-dimensional diamond crystallite or two-dimensional graphene sheets. The latter possibility is unlikely on energetic grounds. The graphene sheets, with the diamond (100) face acting as a template, are forced to maintain the distance of 2.53 Å, just as atoms in the diamond (100) plane. Resultant π-orbital repulsion energies make growth of graphite-like layers normal to the surface plane, as proposed in some previous studies, highly unlikely. To clarify this point, the calculated total energy per atom of hexagonal graphite in AA stacking sequence as a function of interlayer distance is shown in FIGURE 2. The difference in binding energy per carbon atom between uncompressed and uniaxially compressed (to 2.53 Å) graphite is 0.41 eV. It is likely that such a compression of graphite sheets due to the templating effect of the diamond lattice will result in a phase transition into a more stable phase, which is diamond under these growth conditions. It must be remembered that diamond is only 0.02 eV less stable than graphite.

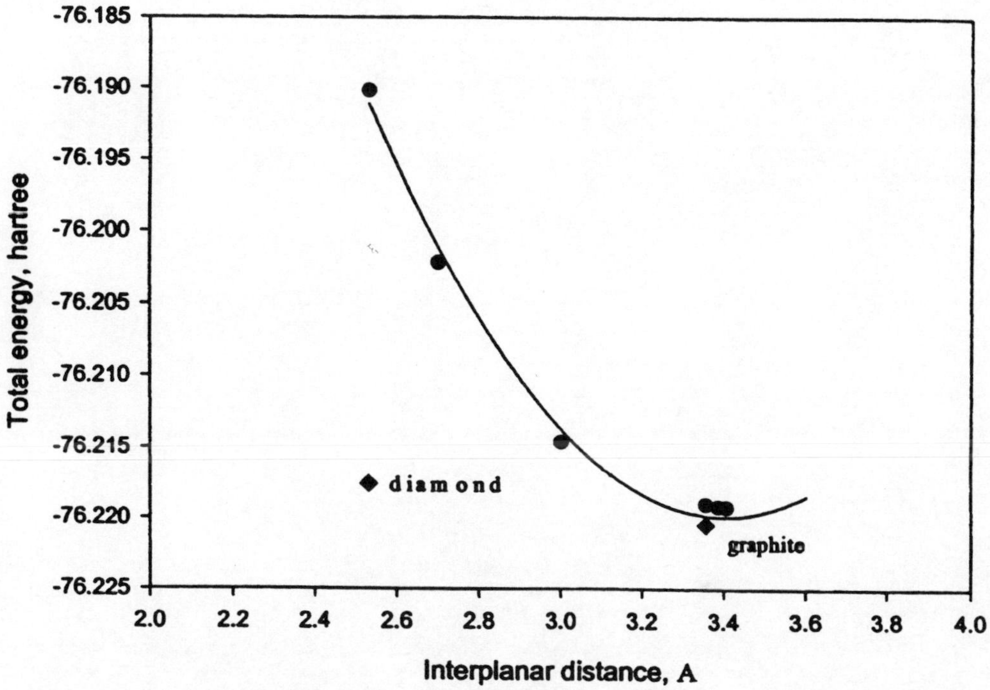


FIGURE 2 Total energy per unit cell versus interlayer distance for hexagonal graphite in AA stacking. Energies of diamond and graphite are given for comparison.

## ACKNOWLEDGEMENT

This work was supported by the US Department of Energy, BES-Materials Sciences, under Contract W-31-109-ENG-38.

**REFERENCES**

[1] A.N. Goyette et al. [ *J. Phys. D (UK)* vol.31 (1998) p.1975 ]
[2] D.M. Gruen, P.C. Redfern, D.A. Horner, P. Zapol, L.A. Curtiss [ *J. Phys. Chem. B (USA)* vol.103 (1999) p.5459 ]
[3] P.C. Redfern, D.A. Horner, L.A. Curtiss, D.M. Gruen [ *J. Phys. Chem. (USA)* vol.100 (1996) p.11654 ]

# B2.3 Microstructure and grain boundaries of ultrananocrystalline diamond films

D.M. Gruen

January 2000

Nanostructured materials can have profoundly different mechanical, electrical, and optical properties from microstructured materials because complex and collective phenomena such as phonon and quantum confinement effects manifest themselves on nanometre length scales. Diamond, in particular, undergoes fundamental changes on the nanoscale because of carbon's unique ability to rehybridize, with the result that grain boundary carbon chemistry differs drastically from the chemistry of the carbon atom in bulk diamond. The reason grain boundaries become of particular significance in nanocrystalline diamond, and most especially in ultrananocrystalline diamond with 3–5 nm grain size (UNCD), is that ~10% of all of the carbon atoms in UNCD are situated at the grain boundaries. A detailed molecular dynamics study has been made of the nature of twist grain boundaries, in both nanocrystalline diamond and nanocrystalline silicon [1,2]. Diamond and silicon have the same cubic structure, and their surface reconstructions bear many similarities. Differences in bonding across grain boundaries in these two materials reflect the fundamental differences in the bonding of carbon and silicon. Nanocrystalline diamond twist grain boundaries are largely π-bonded. On the other hand, silicon atoms in nanocrystalline silicon grain boundaries are largely $sp^3$-bonded. Typical grain boundaries in UNCD and nanocrystalline silicon are shown in FIGURE 1. These are high-energy, high-angle twist grain boundaries characteristic of randomly oriented crystallites. Calculations of the work of adhesion of such high-angle twist grain boundaries in diamond compared to the perfect crystal cleavage energy suggests that the UNCD grain boundaries might be mechanically stronger than single-crystal diamond. By implication, the dominant brittle fracture mode in UNCD might be intra- rather than intergranular.

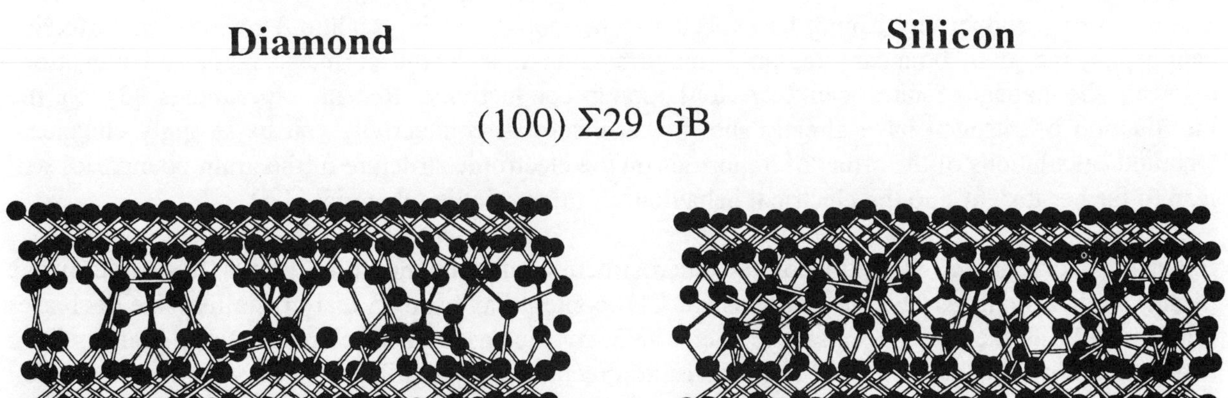


FIGURE 1 Projected structures of the high-temperature relaxed (100) Σ29 twist boundaries in diamond and Si. All the nearest-neighbour bonds between grain boundary atoms are shown.

The grain boundaries appear to be responsible for the electrical conductivity of UNCD films. FIGURE 2 compares the local electronic density of states (DOS) averaged over the atoms in the grain boundary region (solid curve) with that in the perfect, bulk crystal (dashed curve) [2]. The perfect crystal exhibits a well-defined bandgap (similar to the experimental value of 5.45 eV) separating the σ (bonding) states in the upper region of the valence band and the σ* (antibonding) states in the lower region of the conduction band. The presence of $sp^2$-bonded dimers and chain segments and $sp^3$-hybridised dangling bonds in the grain boundary introduces states in the perfect-crystal bandgap [2].

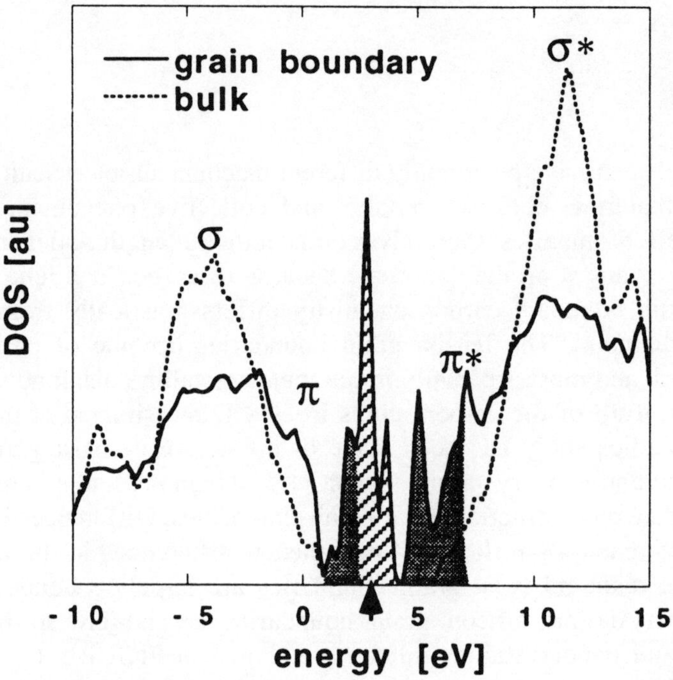


FIGURE 2 Comparison of the local electronic density of states in the diamond (100) Σ29 grain boundary averaged over the atoms in the grain boundary region (solid curve) with that in the perfect bulk crystal (dashed curve). The shaded region indicates states localised at $sp^2$ dimers; the striped region indicates states localised at $sp^3$ sites. The Fermi level at 3.05 eV is indicated by an arrow.

The electrical conductivity cannot be expected to be metallic as in graphite. However, the effective bandgap in the grain boundary region is much smaller than in the perfect crystal, and transitions between the bandgap states can lead to hopping conductivity. Recent experiments [3] on the introduction of nitrogen have already shown that electrical conductivity can be strongly enhanced. Detailed calculations of the effect of impurities on the electronic structure of the grain boundaries will lead to further insight into the electrical behaviour of this intriguing material [4,5].

Extensive work on the evolution of the microstructure of diamond films grown from microwave plasma with varying $Ar/H_2$ ratios at constant $CH_4$ content have shown that crystallite size decreases reproducibily as the Ar concentration is continuously increased. The microstructural changes have been studied by a wide variety of characterisation techniques, including plan view and cross section scanning electron microscopy (SEM), high resolution transmission electron microscopy (HRTEM), electron diffraction, X-ray diffraction (XRD), electron energy loss spectroscopy (EELS), visible and UV Raman spectroscopy and C1s photoabsorption [6].

Illustrative of the results are SEM cross section measurements (FIGURE 3) showing the transition from a columnar, large grain size structure (1% $CH_4$/99% $H_2$) to a UNCD microstructure (1% $CH_4$/99% Ar). An electron diffraction pattern taken from a typical UNCD film is shown in FIGURE 4. It reflects the 3–5 nm randomly oriented crystallites revealed by an HRTEM lattice image shown in FIGURE 5. Detailed analysis of the electron diffraction pattern reveals that the films contain less than

5% of a non-diamond phase. One has at hand a new kind of diamond film whose properties are largely determined by the bonding of carbon at the grain boundaries.

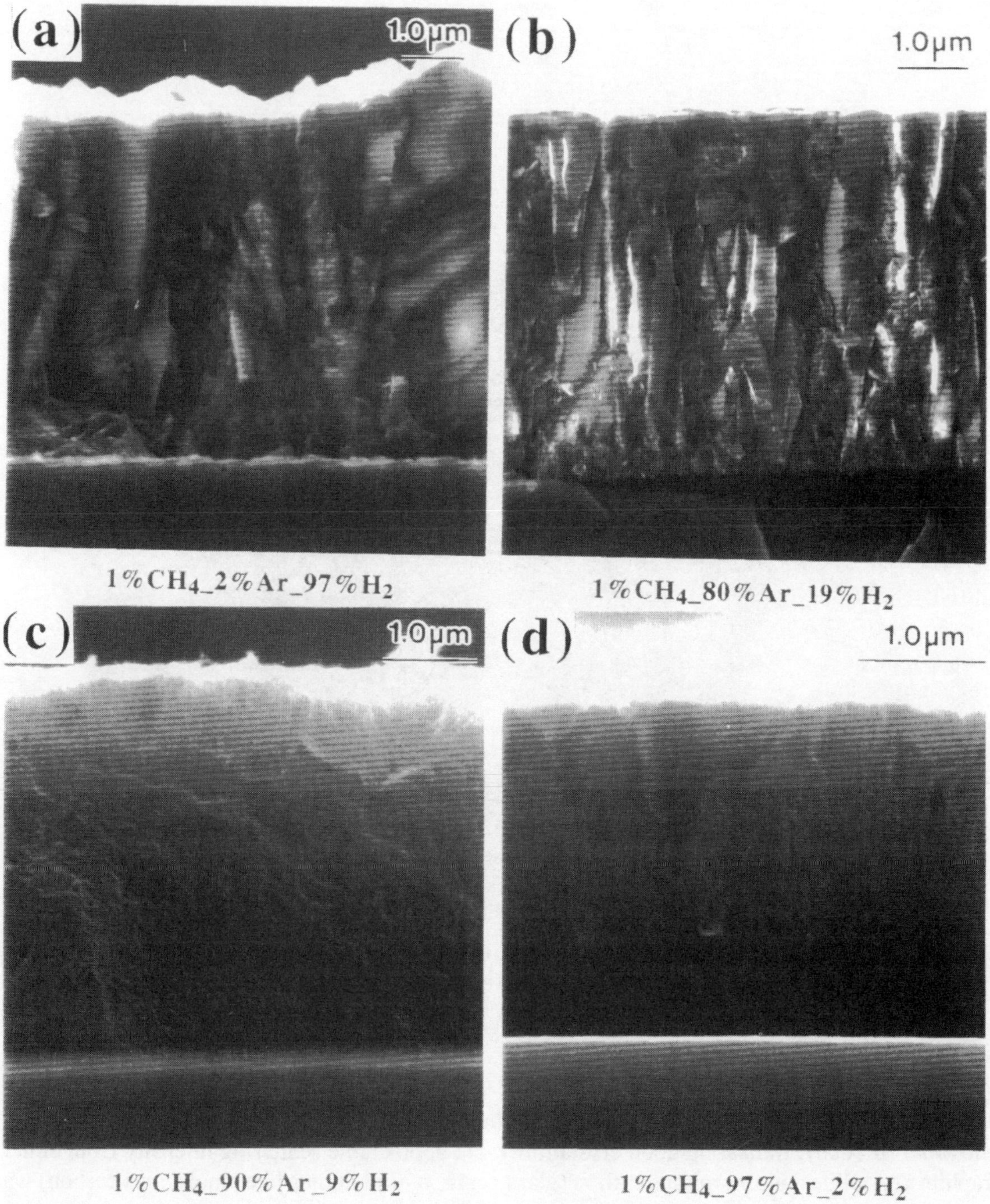


FIGURE 3 Cross-section SEM images of the as-grown films prepared from microwave plasmas with volume percentages of gases listed in the order Ar and $H_2$. The $CH_4$ percentage is always 1%. The changes in growth morphology are dramatically illustrated: (a) 2, 97; (b) 80, 19; (c) 90, 9; (d) 97, 2.

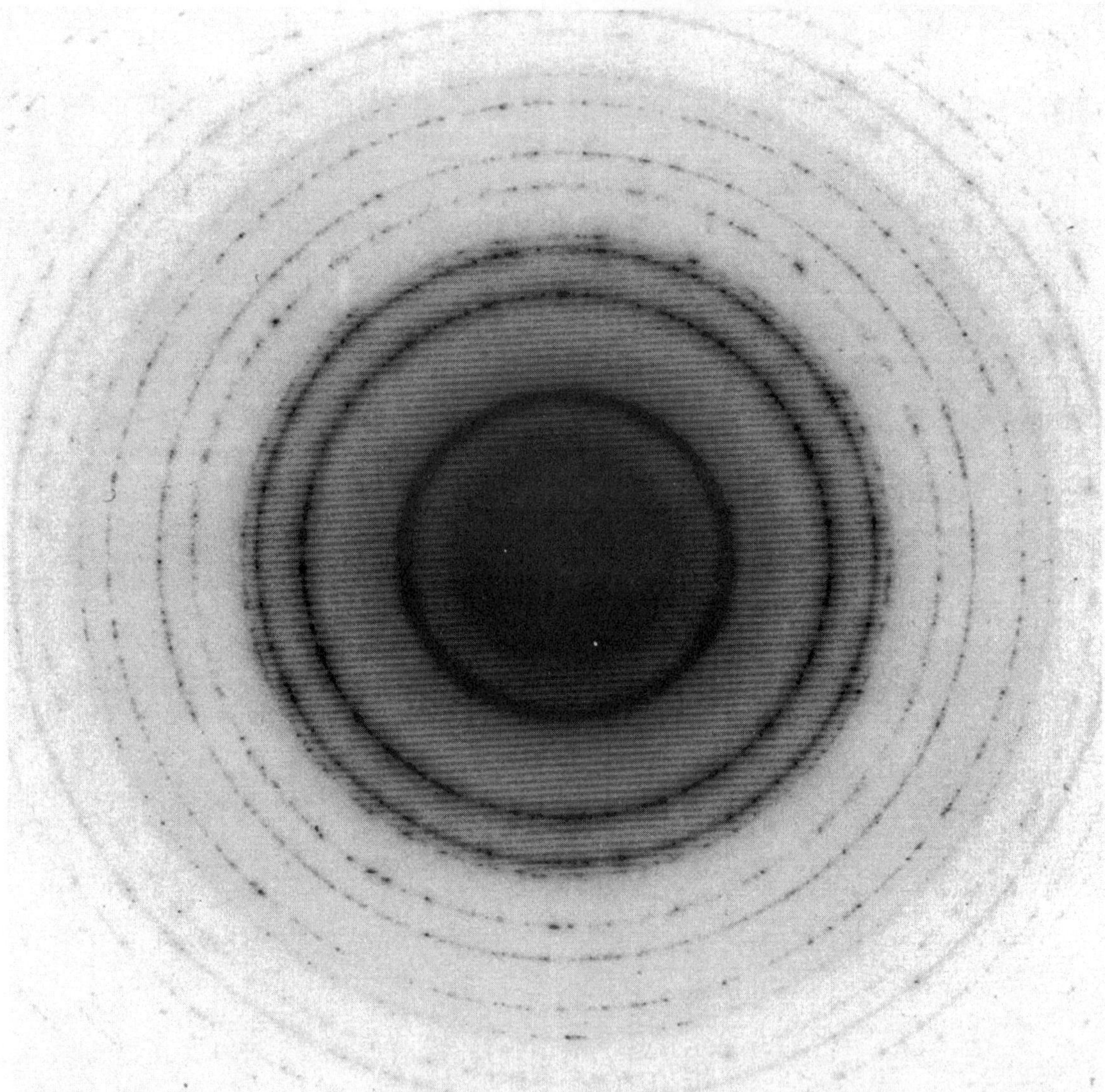

FIGURE 4 SAED pattern from the nanodiamond thin film. Sharp Bragg reflections are visible up to the order of {800}, indicating good crystallinity. No appreciable scattering intensity from either graphite crystallites or amorphous carbon (glassy carbon or diamond-like amorphous carbon) was observed, suggesting that the film is phase-pure diamond.

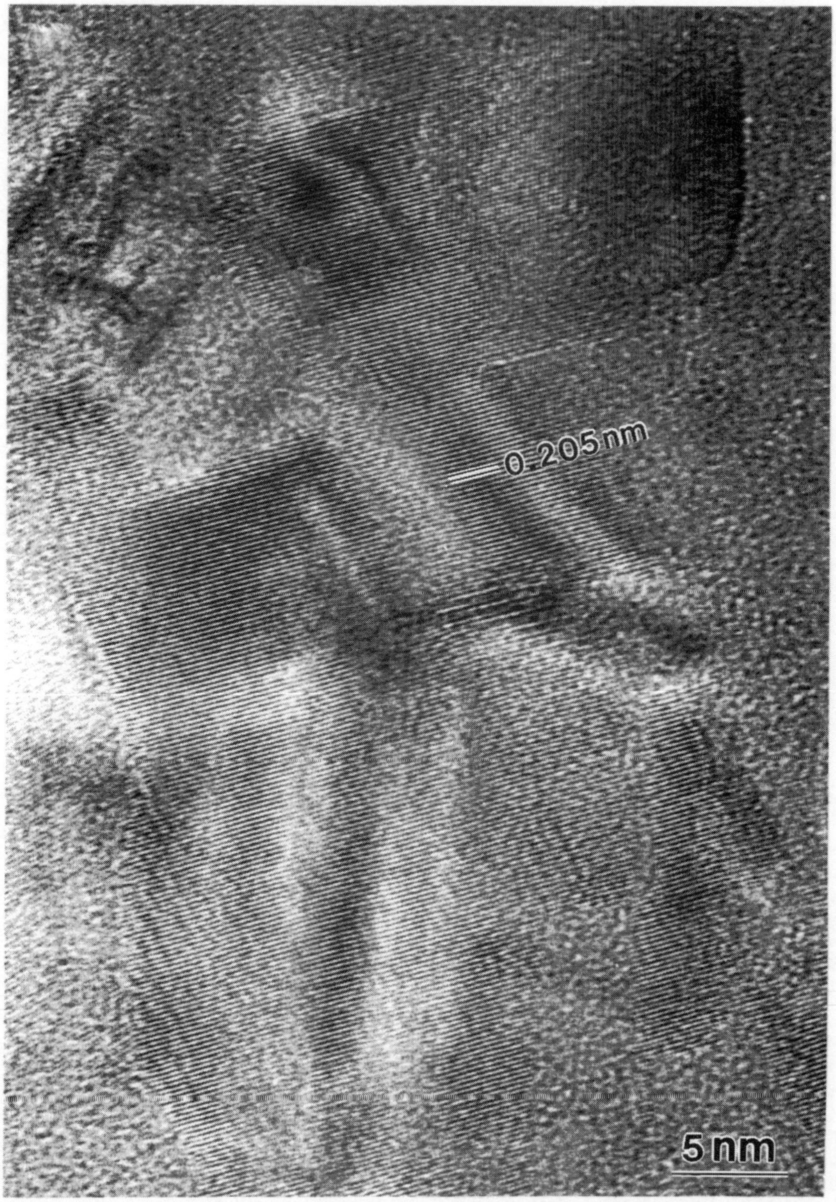


FIGURE 5 HREM image showing selectively the diamond crystallites whose {111} lattice planes are oriented parallel to the incident electron beam. Crystallite size was measured by the largest dimension where {111} lattice fringes are clearly visible.

**ACKNOWLEDGEMENT**

This work was supported by the US Department of Energy, BES-Materials Sciences, under Contract W-31-109-ENG-38.

## REFERENCES

[1] P. Keblinski, D. Wolf, S.R. Phillpot, H. Gleiter [ *J. Mater. Res. (USA)* vol.13 (1998) p.2077 ]

[2] P. Keblinski, D. Wolf, F. Cleri, S.R. Phillpot, H. Gleiter [ *MRS Bull. (USA)* vol.23 (1998) p.36 ]

[3] T.D. Corrigan, A.R. Krauss, D.M. Gruen, D. Zhou, M.Q. Ding, R.P.H. Chang [ manuscript in preparation ]

[4] P. Zapol, L.A. Curtiss, D.M. Gruen, M. Sternberg, T. Frauenheim [ *Proc. Amorphous and Nanostructured Carbon Symp.* MRS Fall 1999 Mtg., Boston, MA (in press) ]

[5] M. Sternberg, T. Frauenheim, P. Zapol, L.A. Curtiss, D.M. Gruen [ *Proc. Multiscale Phenomena in Materials Symp.* MRS Fall 1999 Mtg., Boston, MA (in press) ]

[6] D.M. Gruen [ *Annu. Rev. Mater. Sci. (USA)* vol.29 (1999) p.211 ]

# B2.4 Applications of ultrananocrystalline diamond films

D.M. Gruen

January 2000

## A INTRODUCTION

The special properties of ultrananocrystalline diamond (UNCD) films described in Datareviews B.2.1–B2.3 make possible a variety of new applications. This Datareview outlines the situation in several of these applications.

## B MICROELECTROMECHANICAL SYSTEMS (MEMS)

Using conventional photolithographic and etching techniques, UNCD films have been microfabricated to obtain a variety of two- and three-dimensional shapes with ~ 100 nm feature resolution (FIGURES 1 and 2) [1]. The work on UNCD films compares with similar work done on microcrystalline diamond films (MCD), where the feature resolution was limited to several micrometres because of the grain size. The much higher feature resolution achievable with UNCD versus MCD films is illustrative of a remark made by Richard Feynman in his classic 1959 lecture "There's Plenty of Room at the Bottom":

> "On the other hand, the materials that we use have grain structure, and this would be very annoying at small scale because the material is not homogeneous."

Because of its much smaller grain size, UNCD can achieve feature resolutions at least an order of magnitude higher than MCD.

The very high chemical reactivity of the UNCD growth species, $C_2$, allows growth on oxide materials such as $SiO_2$. By etching an $SiO_2$ sacrificial layer, free-standing two- and three-dimensional UNCD structures are obtained, which retain their shape when freed from the substrate. The combination of low residual stress, low friction and wear coefficients, high mechanical strength, outstanding thermal stability, micromachinability, and submicrometre feature resolution make UNCD film a candidate material to replace silicon, for example in microelectromechanical moving mechanical assemblies (MEMS-MMAs). UNCD films are likely to find uses in a variety of biotechnology applications, particularly where biocompatibility and corrosion resistance are required.

## C TRIBOLOGY

Previous research has demonstrated that the frictional behaviour of diamond and diamond-like carbon films is governed by a combination of physical, chemical and mechanical interactions at their sliding interfaces [2–11]. The amount of friction resulting from a sliding contact can be divided into two major components: mechanical ploughing and physical/chemical attraction or adhesion [12]. Rough diamond films were shown to cause ploughing, hence higher friction [2,9,13]. Apparently, sharp asperity tips of rough diamond films can penetrate the softer counterfaces, giving rise to higher friction and wear. However, for smooth and polished diamond films or cleaved natural diamond, the plough component of friction is thought to be minimal, and the amount of friction is then largely controlled by the extent of adhesive interaction across the sliding contact interfaces [3,7].

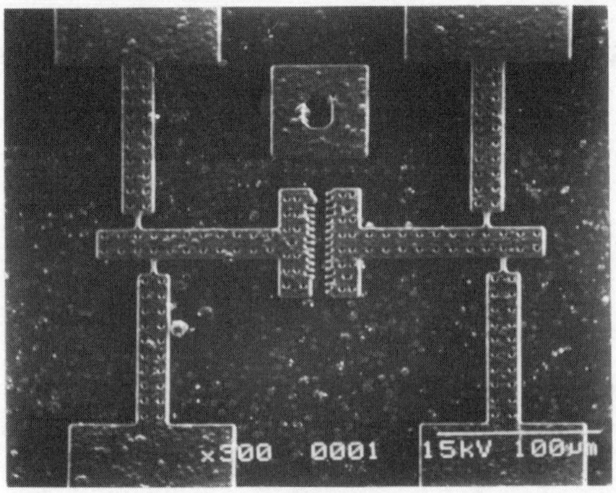


(a)

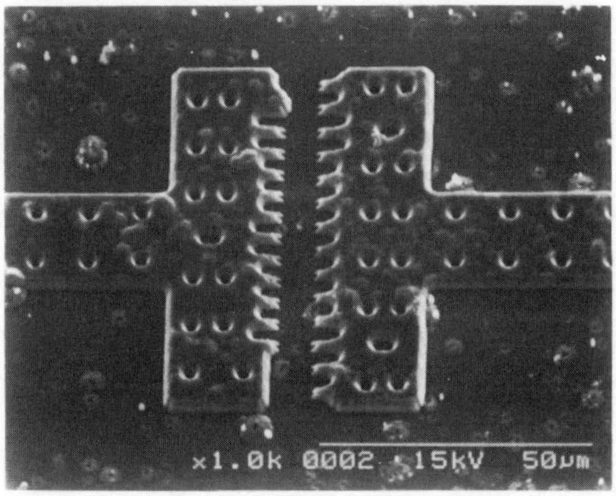


(b)

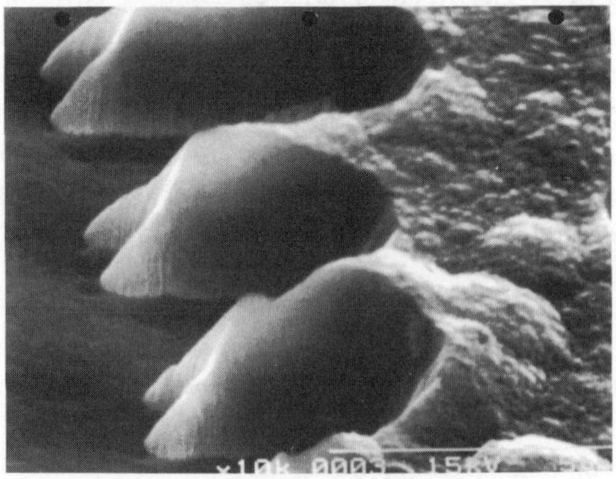

(c)

FIGURE 1 SEM showing the morphological characteristics of NCD-Si coated cantilever-type free structures: (a)–(c) show views of the same structure at increasingly higher magnifications before releasing the cantilever structure.

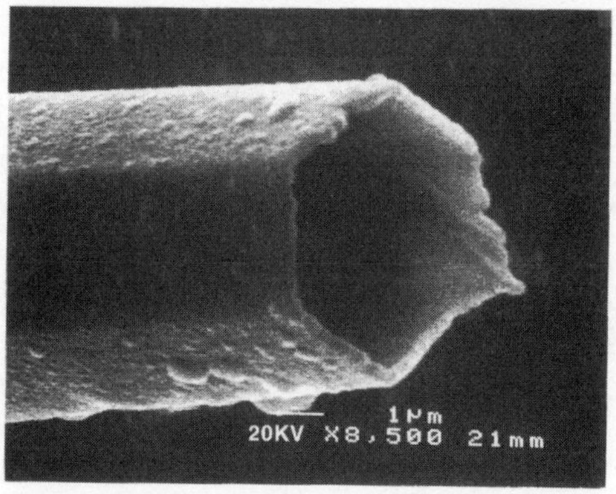


(a)

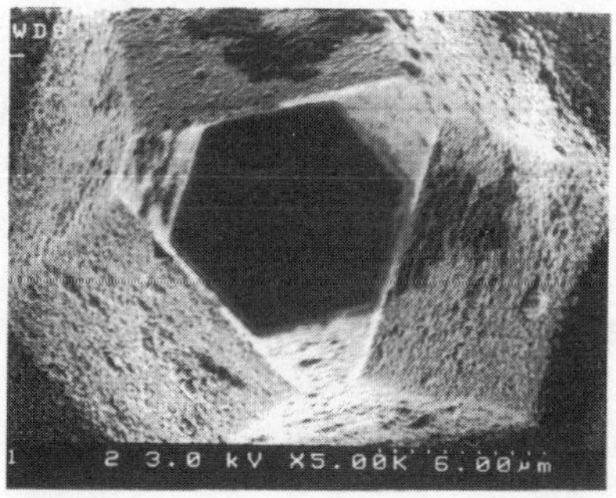


(b)

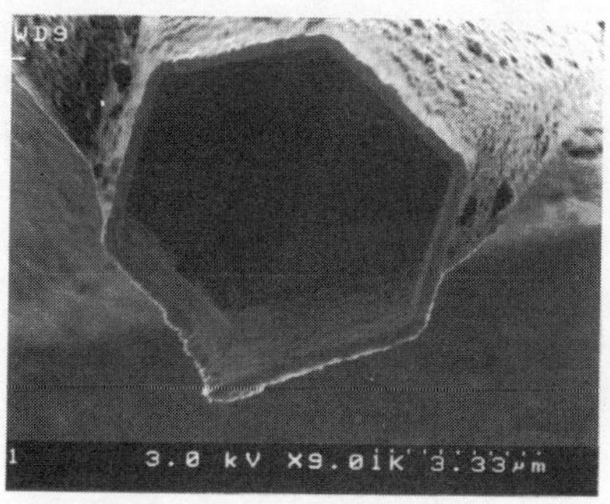


(c)

FIGURE 2 Free-standing diamond hexagonal cylinder produced by conformal coating of a Si microwhisker with subsequent removal of the Si template ((a), (b) and (c)).

Results of sliding friction studies on UNCD films have been encouraging [14–16]. In air, long duration tests (> $5 \times 10^4$ s) against $Si_3N_4$ balls give friction coefficients in the range 0.1–0.15 and wear rates of the diamond films of $3 \times 10^{-7}$/nm. These wear rates appear to reflect initial wear asperities. After wear track smoothing, wear rates become essentially immeasurable. In recent work, UNCD coated SiC seal rings rotating at 3600 rpm have been tested at 100 psi water pressure. Because of a substantial reduction in torque and an increase in wear life, such coated seals could be a significant advance over the current state of the art in liquid pump seals.

## D FIELD EMISSION

It has been empirically observed that various forms of diamond and related materials act as cold cathode electron emitting materials with extremely low effective work functions. The mechanism, once thought to be related to the negative electron affinity of various hydrogen-terminated diamond faces, is complex and not well understood. UNCD is an excellent electron-emitting material with a threshold field of ~ 2 V/μm, one of the lowest reported values [17,18]. Most of the materials exhibiting low field emission are mixed phase materials, and it is hard to characterize the morphological and electronic properties of the electron-emitting regions. UNCD has a much better-defined morphology, and it is possible to relate the morphology to possible electron emission mechanisms. A wide variety of both experimental and computer simulation studies indicate that the grain boundary provides: (a) a path for electrical conductivity from the substrate to the diamond/vacuum interface where emission occurs; (b) an enhanced density of interband electronic states that significantly reduces the width of the Schottky barrier; (c) a large enhancement in the local electric field at the emission sites as a result of the extreme narrowness (3.5 Å) of the conducting grain boundary. Mapping of the electron emission sites using scanning tunnelling microscopy has shown that field-emitted electrons originate at minima or inflection points in the surface topography and not at surface asperities, supporting the grain boundary emission model.

## E ELECTROCHEMICAL ELECTRODES

The unique bulk and surface properties of diamond make it extremely attractive for applications in electroanalysis and electrochemical synthesis that could include sensor electrodes for in vitro and in vivo biomedical applications and sensors for long-term environmental studies, or use in molten salts or in highly radioactive environments. Diamond electrodes may find uses in chemically aggressive environments, for instance electrosynthesis of fluorinated compounds, anodic destruction of organic wastes, ozone synthesis, or molten-salt processes. Reactions at unusually high cathodic potentials, for example reduction of active metals, may be performed on diamond. However, much still needs to be learned about electrochemistry on diamond electrodes before their potential can be fully realized [19]. Diamond is an insulator with a bandgap of 5.5 eV. Doping with boron to levels of > $10^{18}$ $cm^{-3}$ causes overlap of wave functions on neighbouring acceptor centres. An impurity band develops imparting semi-conductivity and even metallic conductivity to diamond. Recent [20] experiments on UNCD electrodes gave results that are quite similar to those obtained with conventionally grown microcrystalline but boron-doped electrodes. The advantages of nanocrystalline films in this context would appear to be that boron doping is not required and that corrosion protection of the underlying substrate can be achieved at considerably lower diamond film thicknesses than is the case for microcrystalline material.

## F CONCLUSION

Diamond films can be used in microelectromechanical systems by etching sacrificial layers of $SiO_2$ away to leave free-standing structures, in coated seals in pumps, in field emission cathodes and in electrochemical applications, where operation in aggressive and/or biomedical environments is required.

## ACKNOWLEDGEMENT

This work was supported by the US Department of Energy, BES-Materials Sciences, under Contract W-31-109-ENG-38.

## REFERENCES

[1] O. Auciello et al [ *J. Microelectromech. Syst. (USA)* (in press) ]
[2] I.P. Hayward [ *Surf. Coat. Technol. (Switzerland)* vol.49 (1991) p.554 ]
[3] M.N. Gardos, K.V. Ravi [ *Diam. Films Technol. (Japan)* vol.4 (1994) p.139 ]
[4] K. Miyoshi, R.L.C. Wu, A. Garscadden, N. Barnes, L. Jackson [ *J. Appl. Phys. (USA)* vol.74 (1993) p. 4446 ]
[5] S. Chandrasekar, B. Bhushan [ *Wear (Switzerland)* vol.153 (1992) p.79 ]
[6] K. Miyoshi, R.L.C. Wu, A. Garscadden [ *Surf. Coat. Technol. (Switzerland)* vol.54/55 (1992) p.428 ]
[7] M.T. Dugger, D.E. Peebles, L.E. Pope [ *Surface Science Investigations in Tribology, Experimental Approaches* Eds. Y.W. Chung, A.M. Homolo, G.B. Street (American Chemical Society, Washington, DC, 1992) vol.485 p.72 ]
[8] A. Erdemir, C. Bindal, G.R. Fenske, P. Wilbur [ *Tribol. Trans. (USA)* vol.39 (1996) p.735 ]
[9] M. Kohzaki, K. Higuchi, S. Noda, K. Uchida [ *J. Mater. Res. (USA)* vol.7 (1992) p.1769 ]
[10] A. Erdemir, C. Bindal, J. Pagan, P. Wilbur [ *Surf. Coat. Technol. (Switzerland)* vol.76/77 (1995) p.559 ]
[11] Y. Liu, A. Erdemir, E.I. Meletis [ *Surf. Coat. Technol. (Switzerland)* vol.82 (1996) p. 48 ]
[12] F.P. Bowden, D. Tabor [ *Friction and Lubrication of Solids, Part 1* (Clarendon, Oxford, UK, 1994) ]
[13] I.P. Hayward, I.L. Singer, L.E. Seitzman [ *Wear (Switzerland)* vol.157 (1992) p.215 ]
[14] A. Erdemir, C. Bindal, G.R. Fenske, C. Zuiker, A. Krauss, D. Gruen [ *Diam. Relat. Mater. (Switzerland)* vol.5 (1996) p.923 ]
[15] C. Zuiker et al [ *Thin Solid Films (Switzerland)* vol.270 (1995) p.154 ]
[16] A. Erdemir et al [ *Tribol. Trans. (USA)* vol.40 (1997) p. 667 ]
[17] T.G. McCauley et al [ *Mater. Res. Soc. Symp. Proc. (USA)* vol.498 (1998) p.227]
[18] A.R. Krauss et al [ *Mater. Res. Soc. Symp. Proc. (USA)* vol.495 (1998) p.299 ]
[19] G.M. Swain, A.B. Anderson, J.C. Angus [ *MRS Bull. (USA)* (Sept. 1998) p.56 ]
[20] B. Fausett, M.C. Granger, M.L. Hupert, J. Wang, G.M. Swain, D.M. Gruen [ *Electroanalysis (Germany)* (in press) ]

CHAPTER B3

# DOPING

# B3.1 Doping of diamond using ion implantation

R. Kalish and C. Uzan-Saguy

January 2000

## A INTRODUCTION

The doping of diamond is a crucial issue if diamond is to be rendered electrically conductive so that its unique electrical properties can be exploited. This doping is, however, a particularly difficult task for the case of diamond due to its unusual properties: the tightness of the diamond crystal makes it difficult for impurities to occupy the required (usually substitutional) sites in the lattice where they are to act as donors or acceptors; for the same reason, diffusion of impurities in diamond is extremely slow [1], and is hence, as yet, impractical for doping purposes. The energy levels for electrons (or holes) in the diamond energy gap are rather deep, too deep to yield reasonable conductivities at room temperature for low or medium doping concentrations. The formation of ohmic contacts to this wide bandgap semiconductor is also difficult.

Ion-implantation is a technique widely used to dope common semiconductors. In this method atoms are forced into the crystal, regardless of solubility or diffusivity considerations; hence it is one of the attractive methods for the introduction of the desired impurities into any crystal, including diamond [2]. Unfortunately, ion-implantation is always accompanied by the creation of many defects in the material. These have to be removed by proper annealing. For the case of diamond, the annealing of defects is more complex than in other group IV semiconductors due to the metastability of the $sp^3$ (diamond) bonding with respect to the $sp^2$ (graphitic) bonding. Broken $sp^3$ bonds and other implantation-related defects or graphitic clusters can give rise to electrical effects which may be mistaken for real chemical doping due to the presence of dopants.

The search for suitable dopants in diamond is not only confined to implantation doping, but is rather general. The well-known dopants in diamond, to date, are boron, a p-type impurity with an activation energy of 0.37 eV, and nitrogen, a deep donor, with an energy level 1.7 eV above the conduction band [3]. These two impurities, which are adjacent to C in the periodic table, readily enter the crystal. They can be introduced into diamond during geological synthesis, during HPHT growth, during CVD diamond film growth, and, as will be shown below, also by ion-implantation.

Of particular importance is the quest for suitable n-type doping of diamond. Many attempts were made on this topic, mostly without success. Only recently was it demonstrated (by CVD methods) that phosphorus has a donor state in the diamond gap, with reported activation energies ($E_a$) ranging from 0.46 to ~0.6 eV [4]. The doping efficiency and the electrical properties of this dopant are still poor, and much effort is devoted to their improvement. Sulphur has most recently been reported to act as a donor ($E_a$ = 0.37 eV) in CVD diamond [5]; however this finding still needs to be confirmed. Hence the search for ion-implantation doping of diamond (in particular n-type) is most intensive, however so far with only limited success. Below we review, among others, these attempts, which were not always successful, and hence remained unpublished.

Several reviews have been written on implantation doping of diamond and on the role of defects on the electrical properties of diamond [2,6,7].

Below we will concentrate on the fundamental issues and on the more recent results on doping of diamond by ion-implantation and on the related electrical effects that ion-implantation has on diamond.

The discussion below includes a section dealing with the electrical effects of implantation related defects, and on their annealing, a section on p-type doping of diamond by boron ion-implantation and a section reviewing attempts to achieve n-type doping of diamond by suitable ion-implantations.

## B DAMAGE

Ion-implantation, being a violent process, is always accompanied by bond breakage and the creation of defects in the material. Unless removed or avoided by proper annealing or implantation schemes, these defects may overshadow the real chemical effects of the doping. The diamond $sp^3$ bonds, when broken, can rearrange to form the more stable and electrically conductive $sp^2$ bonds or graphitic domains. The process by which this occurs and the nature of the transition between diamond and graphite have been extensively studied and several reviews on this have been published [8–10]. Those aspects by which the non-annealed damage influences the electrical properties of the diamond are of relevance to implantation doping. These can be manifested in the following processes: (a) creation of defect-related energy levels within the bandgap which may either give rise to donor or acceptor states, or compensate doping levels; (b) creation of conduction pathways (by hopping or percolation), in particular via $sp^2$ bonded states or clusters; (c) creation of scattering centres which limit the carrier mobilities; (d) creation of dopant-defect complexes which passivate the dopants.

### B1 Ion Beam Induced Damage

Damage introduced into diamond by non-dopant (C, noble gases) ion-implantations induce conduction which depends on the volume density of the defects and on the implantation temperature [2,8]. The resistivity versus dose curves $\rho(D)$ in diamond exhibit three distinct regions, regardless of ion: for doses lower than a critical dose $D_c$, $\rho$ remains almost constant, for $D \approx D_c$ a sharp drop in $\rho$ occurs and for $D >> D_c$ $\rho$ saturates at a value comparable to that of polycrystalline graphite. Prawer et al [9] have demonstrated that the critical dose $D_c$ is ion and temperature dependent and that the different regimes can be simulated by percolation. The conduction in this transition region is described by variable range hopping conductivity ($\rho(T) \propto T^{1/4}$) with carriers hopping between highly conductive small regions. The very high conductivity observed in the saturation regime is that of graphite (metallic, $\rho(T)$ = constant) and occurs when the graphitic islands percolate.

With regard to the type of conductivity induced by the implantation damage Fang et al [11] and Uzan-Saguy et al [12] found that low temperature Xe and Kr implantations yield p-type conductivity with poor mobilities (~10 $cm^2/V$ s). However implantation at higher temperature or post implantation annealing of the damage [12] as well as electron irradiation [13] were found to yield n-type conductivity. Since dopant ion implantation always requires annealing, one can expect the residual damage following incomplete annealing to act as donor centres in the diamond. These will either give rise to n-type conduction, or compensate acceptor states. Indeed this damage related n-type conduction was observed [11,14] and even used [15].

To assure that real chemical doping is achieved by ion-implantation, control experiments are essential. In these, non dopant ions have to be implanted into identical samples under conditions which create damage profiles similar to those due to the implanted dopants, and subjected to identical annealing protocols. Only if the non-dopant implanted sample returns to its pristine high resistivity following annealing, whereas the chemically doped one shows electrical effects, can these be attributed to the presence of the implanted dopant. Not all published data do include such important control experiments.

### B2 Defect-Dopant Interaction

The presence of defects related to the implantation process may influence the electrical properties of the doped layer either by interacting with the dopants, thus forming inactive complexes, or by creating defect levels in the gap which may compensate the desired electrical activity of the dopants. Prins [15] realized, by using electrical and optical characterisation of B implanted diamonds, that non-annealed implantation induced damage forms a compensating donor level located about 1.2 eV above the top of the valence band. This is in accord with the finding of Reznik et al [16] who showed that C implantations into type IIa diamond, to a dose below the graphitization limit, exhibit electrical conduction which is characterised by an activation energy of 1.2 eV.

The effect that point defects have on the electrical properties of doped (both n- and p-type) diamonds has been studied by Reznik et al [16]. Hydrogen irradiations through B and P doped homoepitaxial layers were used to introduce well-known numbers of point defects into the doped layers. The changes in resistivity, mobility and carrier concentration were measured as a function of increasing H dose. It was found that, for B doped layers, the introduction of point defects increases the resistivity of the implanted diamond to that of a pristine diamond without changing the activation energy of the p-type carriers (0.37 eV). It does, however, increase the compensation ratio $N_d/N_a$ linearly with increasing defect density and it considerably reduces the mobility of the p-type carriers. The temperature dependence of the mobility also changes from the expected dependence for a doped semiconductor to only a weak dependence governed by defect scattering [16]. It was also found that the defect density at which loss of the conductivity for n-type P doped homoepitaxial layers sets in for H irradiations occurs at a dose 3 orders of magnitude higher than that needed for the p-doped sample.

These findings clearly support the hypothesis that the ion implantation induced defects are responsible for a donor level in the bandgap. For p-type doped diamond, these defects compensate the acceptors (B) and reduce the hole mobility. For n-doped layers these defects may passivate the dopants or form neutral defects, thus reducing strongly the mobility of n-carriers, as observed in P doped diamond layers [4].

## C p-TYPE DOPING

p-Type, boron containing, diamonds exist in nature (type IIb natural diamond), and their optical and electrical properties have been extensively studied and documented [3]. These have shown that the activation energy of the B acceptor is 0.37 eV above the valence band. Very high hole mobilities in the range 1000–2000 $cm^2\ V^{-1}\ s^{-1}$ were reported. More recently B doped p-type diamonds were realized by both HPHT and CVD techniques. The best p-type artificial diamonds are obtained nowadays by very careful and slow CVD homo-epitaxial growth on <100> natural diamond substrates. These have yielded a record mobility of 1800 $cm^2\ V^{-1}\ s^{-1}$ and a carrier concentration of $2.3 \times 10^{14}\ cm^{-3}$ at room temperature. The reason for the outstanding properties of B containing p-type diamond is, most probably, the small size of the B atoms which enables them to easily enter the diamond lattice and occupy substitutional sites where they act as donors.

The optimal properties of p-type diamond, to serve as a guideline for realizing p-type diamond also by ion implantation, are thus well known, as are the methods of obtaining good ohmic contacts to such conductive layers. Because of the well-known and excellent properties of B containing p-type diamond, no further extensive research on attempts at implantation doping of other potential group III acceptors in diamond has been undertaken, to the best of our knowledge. Very old results on p-type doping of diamond with Al ion-implantations were not promising enough and have not been pursued.

We present here a chronological review of only the main stages in the progress of ion-implantation p-type doping (with B) of diamond.

The first attempt to dope IIa diamond by ion implantation was reported by Vavilov et al in 1970 and in 1974 [17]. These pioneering studies, however, suffered from the lack of control experiments; hence effects due to the presence of incompletely annealed implantation related defects could not be properly eliminated. Almost ten years later in 1983 Braunstein et al [18] reported on the fabrication of highly conducting thin p-type diamond layers obtained by high dose B implantation (to a dose that exceeded the critical dose for graphitization), which, following high temperature annealing (1450°C) and graphite removal in boiling acids, resulted in a very thin B doped layer (20 nm) reaching all the way to the surface. This method, of high dose ion-implantation followed by HT annealing and graphite removal, has later been further perfected, and it is now commonly used as a method to obtain a $p^{++}$ ohmic contact to p-type diamonds [19].

Prins [15] has considered the role of the implantation temperature on the nature of the defects and hence on the resulting electrical properties of the implanted layer. Three distinct implantation-temperature regions were identified: (1) a low temperature region (77 K) where both vacancies and interstitials (intrinsic point defects as well as implanted dopant atoms) are immobile; (2) an intermediate region (300 K<T<800 K) where only interstitials are mobile: in this regime some interstitials recombine with vacancies and some diffuse out of the implanted layer leaving behind a vacancy-rich region; and (3) a high temperature region where both interstitials and vacancies are mobile during the hot-implantation, resulting in a dislocated diamond layer. Following this theory, Prins devised a new implantation/annealing procedure, the so called cold-implantation-rapid-annealing (CIRA) technique. In this method the diamond is implanted at low temperature, rapidly heated to 800 K, to induce interstitial-vacancy recombination before the interstitial can diffuse away from the implanted layer, and then annealed at high temperature (1700 K) to further reduce the radiation damage and increase the electrical activation of the dopants.

The main idea of this method is that the B-interstitials should successfully compete against the self-interstitials to fill in vacancies which is most likely if the B-interstitials are located in a vacancy-rich volume. Two ways of increasing the dopant-vacancy spatial overlap are suggested [20]: one is to broaden the implanted layer by implanting the dopants at different energies, the other is to use carbon co-implantation to create damaged volume into which the dopant atoms are implanted. These methods, though interesting, did not lead to any significant improvement in the electrical properties of the B implanted sample over those achieved by the CIRA method.

In the last few years, many attempts to optimize the implantation/annealing procedure have appeared in the literature, mainly based on the CIRA method, or on the high-dose graphite removal methods. The various implantation and annealing parameters which influence the quality (mainly mobility) of the p-type diamond obtained by B ion-implantation were studied [21–23]. The outcome is that ion implantation into cold (77 K) diamond is helpful in avoiding the formation of extended defects (which are very difficult to anneal out) or the out-diffusion of the interstitials from the implanted region. Multiple energy implantations to create a 'box profile' of the B concentration were preferred over C co-implantations, since adding non-dopant related defects by the C implantations considerably reduced the hole mobility [24]. In-situ rapid annealing of the implanted diamond to medium temperatures (up to ~1300 K) was also not conclusively proven to improve the doping [22] even though others report that rapid annealing following the cold implantation stage does increase the carrier mobility [23]. Post-implantation, ex-situ annealing to the highest possible temperatures (1700

K under vacuum, with no residual oxygen) was found to be essential to further increase the carrier concentration and mobility and to reduce the compensation ratio [24]. Typical results of Hall effect measurements (as a function of temperature) obtained for diamonds B implanted according to the CIRA procedure followed by a further high temperature anneal at 1450°C are: $p = 6 \times 10^{13}$ $cm^{-3}$, $\mu =$ 400 $cm^2$ $V^{-1}$ $s^{-1}$ and $N_d/N_a = 0.005$ where $N_a$ and $N_d$ are the acceptor and donor concentrations respectively [25]. Other higher values of mobility are quoted in the literature but are not supported by any detailed data.

Prawer et al [26] have shown that by creating a deeply buried B rich layer in diamond by mega-volt (MeV) ion implantation improved mobilities can be obtained. The reason for this is possibly the improved annealing due to the internal pressure exerted on the damaged layer by the overlying, only lightly damaged, diamond layer. A novel method for making ohmic contacts to the buried layer based on local graphitization by laser pulses was devised. The electrical properties of these B doped, buried layers, reported by Uzan-Saguy et al [27], are characterised by high mobilities (600 $cm^2$ $V^{-1}$ $s^{-1}$) and a low compensation ratio (5%). Furthermore, it was shown that the cap layer, overlying the buried, doped layer, is highly insulating, opening the possibility of using such technology for realizing field effect transistor (FET) devices in diamond.

The optical signature of B in implanted and annealed diamond has been investigated [28,29] by cathodoluminescence measurements at 77 K. New luminescence lines were found, one related to the exciton bound to B in substitutional sites at 5.22 eV (observable only after annealing at 1000°C), and one at 4.5 eV. Prins [29,30] showed that the 4.5 eV band is quenched when introducing defect related donor centres into the implantation doped layer. Moreover, Prins showed [30] that the two bands related to B are observable in the cathodoluminescence spectrum of diamond samples even though their electrical measurements indicate that $N_a < N_d$, in contradiction of previous conclusions that claim that these lines appear only if $N_a > N_d$.

Very high p-type doping ($p = 2 \times 10^{19}$ $cm^{-3}$) suitable for contacting p-type diamond was achieved by Kalish et al [31,32] by B implanting polycrystalline CVD diamond layers by two different methods: (I) high dose B implantations (above the critical dose for graphitization upon annealing), followed by annealing (1300 K) and chemical removal of the graphitized surface layer to expose a thin, highly conductive over-doped layer [31]; (II) high dose high current (current densities up to 500 $A/cm^2$) B implantations into poorly heat-sunk samples [32], thus heating the sample during implantation to temperatures of the order of 1100°C. This hot implantation leads, without any further need for annealing, to a highly doped relatively undamaged layer which exhibits high conductivity.

To summarize, the following main points describe the current status of obtaining p-type diamond by ion-implantation:

1. Boron is the best (and only thoroughly studied) p-type dopant in diamond. B ion implantation is a well controlled process which has been optimized in the last 5 years resulting in doping efficiencies of the order of 2–5%, hole mobilities of up to 600 $cm^2/V$ s and compensation ratios below 5%.
2. The optimal implantation/annealing scheme is comprised of a multiple energy cold implantation followed by medium temperature and a high temperature annealing (T>1400°C) stage.
3. The fabrication of over-doped highly conductive layers can be achieved either by hot implantations, or by high dose implantations followed by annealing and graphite removal. These are widely used for realizing ohmic contacts to p-type diamond.
4. Optical characterisation of the B acceptors in substitutional sites is well understood and is a useful tool in the study of defect-dopant interaction.

## D n-TYPE DOPING

Among the group V elements, which are potential substitutional donors in group IV semiconductors, only N and P seem to enter the diamond crystal and contribute to its electrical properties. Nitrogen is abundant in natural diamonds, and can be synthesized under HPHT laboratory conditions. P and N can be introduced into the diamond also during CVD growth. Group I elements occupying interstitial sites (Na and Li) were theoretically predicted by Kajihara et al [33] to act as donors with activation energies of 0.1 and 0.3 eV, respectively. Most recently attempts to dope diamond n-type with group VI elements (O and S) have been reported, though the electrical effects of these still need to be verified [5,34].

The formation energy for bonding with C of N is predicted to be negative (–3.4 eV), in contrast with the high positive formation energies predicted for P (10.4 eV), Li (5.5 eV) and Na (15.3 eV). This suggests that, with the exception of N, the solubility of all these impurities in diamond is very low. Therefore, ion implantation, which enables the introduction of impurities under non-equilibrium conditions, should, in principle, be very appropriate for the search for n-type doping of diamond. Indeed ion-implantation doping of diamond with the above elements has been the subject of intensive studies, though with only limited success.

### D1 Nitrogen

Nitrogen is the dominant impurity in natural diamond. Both N atoms located in substitutional sites in type Ib diamonds and those accumulated in aggregates in type IaA are donors in diamond with activation energies of 1.7 and 4 eV respectively. Due to these very high ionization energies, N containing diamonds are electrically insulating at room temperature and it is impossible to study the N donors by conventional electrical measurements. Photoconductivity, cathodoluminescence (CL) and electron paramagnetic resonance (EPR) techniques are more suitable techniques to study deep levels, like N, in diamond. Indeed the various defect complexes which N forms in the crystal, the amount of N incorporated in substitutional sites and the N bonding configuration have been extensively studied in natural diamonds [3]. The question of whether N implantation into diamond can reproduce the well known properties of natural N containing diamonds was addressed by Kalish et al [35]. Implantation/annealing schemes that were found to yield good p-type doping for B implantations were applied. The N implanted and annealed diamonds were checked by resistivity $\rho(T)$, EPR and CL measurements after implantation and after different annealing stages. It was found that following annealing, almost 50% of the implanted N is located in substitutional sites and that defects found in type Ib diamond exist also in the N implantation-doped diamond. Two important conclusions can be drawn from these results: (i) the CIRA procedure (developed for p-type implantation-doping of diamond) is suitable for N; (ii) substitutional N containing diamonds can be obtained by selective area ion-implantation thus enabling us to use the improved field emission properties of N containing CVD diamond also for N implantation doped diamond.

Nevertheless, N cannot find application as a standard n-type dopant in diamond in room temperature devices due to its deep level.

### D2 Phosphorus

Phosphorus on a substitutional site in diamond was predicted to act as a shallow donor ( 0.1 eV) [33]. Recently P containing homoepitaxial diamond layers were successfully grown by CVD [4] and were shown, by detailed Hall effect measurements, to produce a donor level with an ionization energy of about 0.5 eV below the bottom of the conduction band. The mobility of the P doped layers varies between 30 and 180 $cm^2\ V^{-1}\ s^{-1}$ and typical room temperature carrier concentrations are of the order of $10^{13}$–$10^{14}$ $cm^{-3}$. This finding provides a guideline for P doping of diamond by ion-implantation.

In 1994, Prins [36] proposed n-type doping by using low dose implantations followed by low temperature annealing (LDLT). He applies his method in parallel to B, C and P implantations. Doses as low as $10^9$ to $10^{11}$ $cm^{-2}$ were used and the samples were annealed at 500°C and 1200°C after each implantation step.

The P implanted diamond was reported to be n-type, however only by hot probe measurements, with an activation energy of 0.1 eV. It should be noted that no Hall measurements were reported in that work, and that the resistance versus T measurements suffered from hysteresis, perhaps indicating non-ohmicity and curing of the contacts.

Ran [37] has attempted, in a very systematic way, to obtain n-type doping of type IIa diamonds by implanting P ions following various implantation/annealing schemes: (i) by using the CIRA procedure followed by high temperature annealing; (ii) by using implantation into a hot diamond followed by various annealing steps; (iii) by repeating the LDLT procedure utilized by Prins. In all these experiments, controls were included by subjecting one half of the diamond to P implantation and the second half to identical Ar (a non-dopant ion producing the same damage profile as P) implantation. The two implanted parts were therefore obviously subjected to exactly the same annealing conditions. The resistivities of the two parts of each sample were measured by using the van der Pauw configuration, so that possible non-ohmic contacts did not influence the results. In all the above described procedures, no significant difference could be found between the resistivity of the P and Ar implanted samples. Hence, it seems as if, to date, no conclusive n-type conduction due to real chemical doping of diamond by P ion-implantation has been demonstrated.

The location of ion-implanted P in diamond has been determined both by PIXE [38] and electron-emission [39] channelling techniques. The results of [37] and [38] show that about 70% of the implanted P occupies substitutional sites following annealing at 1200°C. Very recently, EPR measurements were performed by Deneuville et al [40] on type IIa diamond implanted with P at increasing doses by using the CIRA procedure. A peak in the EPR spectra, associated with P, was observed after an implantation to a dose of $6 \times 10^{12}$ $cm^{-2}$ following the CIRA procedure.

Even though both EPR and channelling measurements show that a substantial fraction of the implanted P occupies the substitutional site, it seems that, so far, no useful way of obtaining a donor level due to ion-implanted P in diamond has been found.

It can be speculated that the introduction of the large P atom into the tight diamond crystal induces strain which is likely to attract defects to form complexes with the P which are electrically inactive. The fact that the implantation process in its very nature involves the formation of defects in the vicinity of the implanted ion may enhance the formation of such complexes thus reducing the n-type doping efficiency of P implanted diamond.

## D3 Lithium (and Na)

Even before the prediction of Kajihara that interstitial Li should exhibit a shallow donor level in diamond [33] Li implantation doping attempts were made by the Russian group [41], however with no proper control experiments to eliminate radiation damage effects. The encouraging results reported in these early works ($E_a$~0.1 eV, mobility ~ $10^3$ $cm^2/V\ s$) could unfortunately not be verified in later, more controlled, experiments. Prawer et al [42] have implanted, at low temperatures, Li and C (as a control non-dopant ion) into type IIa diamonds to doses below and above the threshold for graphitization. For the low dose implanted samples, no significant electrical signature could be detected. The high dose implanted sample, following 1400°C annealing for 10 min, and graphite removal, did show electrical conductivity. From the temperature dependence of the resistivity an activation energy of 0.22 eV over the temperature range between 400 K and 700 K could be determined. However, variable range hopping conduction could not be eliminated. Unfortunately, no Hall effect measurements were possible on these samples.

Job et al [43] have performed high dose Li implantations into diamonds heated to 850°C and subsequently annealed at 900°C. As expected, no radiation induced graphitization was observed, as instantaneous annealing occurs for such hot implantations. n-Type conductivity could be verified by hot probe measurements only. An activation energy of 0.23 eV was deduced from R(T) measurements.

Buckley-Golder et al [44] have indirectly deduced n-type indications for hot Li implanted p-type diamond by observing rectifying p-n features. However, it is impossible to associate these with a definite donor level related to the implanted Li ions. A possible activation energy of 0.22 eV was deduced from R(T) data.

High current density implantations [31] into CVD diamonds of Li (C as a control) and P (Ar as a control) ions to high doses were performed in a fashion similar to that successfully applied to the high dose p-type B implantation-doping. No clear indications of doping due to the presence of Li or P could be identified.

The location of Li implanted into diamond at low doses was reported by Restle et al [45] using the emission channelling technique and by Braunstein et al [38] using nuclear reaction channelling analysis. A fraction of 40% of the implanted Li was found in [45] to occupy interstitial lattice sites and 17% to occupy substitutional sites. No change in the fractions of implanted Li on these two lattice sites with increasing implantation temperature from 100 K to 900 K could be measured, indicating that no diffusion of Li takes place in this temperature range. No clear electrical n-type signature could be identified. (It should be noted that if Li on a substitutional site is expected to act as an acceptor and on an interstitial site as a donor, these could compensate each other.)

Data for diamond ion-implantation doped with Na ions is scarce, and, once more, no significant indication for the existence of a donor level associated with Na in the crystal could be found [42].

Very early reports on electrical activities due to heavy group V elements implanted into diamond (As and Sb) seem all to have suffered from non annealed defects which were misinterpreted as being due to the chemical doping of these potential (very large!) donor atoms.

Once more, lattice location studies have verified that the heavy implants do occupy predominantly substitutional sites in the diamond; however, this did not warrant their expected electrical activity.

## E CONCLUSION

Doping of diamond using the ion implantation technique has been discussed. The ways in which non-annealed damage influences the electrical properties of the diamond were described. Boron is the best p-type dopant in diamond and boron ion implantation has been extensively studied. Potential n-type dopants include nitrogen, phosphorus, lithium and sodium, but there has only been limited success in ion implantation doping of diamond with these elements.

## REFERENCES


[1] R. Kalish [ in *Properties and Growth of Diamond* Ed. G. Davies (INSPEC, IEE, London, UK, 1994) ch.6 p.189-215 ]

[2] M.S. Dresselhaus, R. Kalish [ *Ion Implantation in Diamond, Graphite and Related Materials* (Springer-Verlag, Berlin Heidelberg,1992) ]

[3] A.T. Collins [ in *Properties and Growth of Diamond* Ed. G. Davies (INSPEC, IEE, London, UK, 1994) ch.9 p.261-88 ]

[4] S. Koizumi et al [ *Diam. Relat. Mater. (Switzerland)* vol.7 (1998) p.540-5 ]

[5] I. Sakaguci et al [ *Phys. Rev. B (USA)* vol.60 (1999) p.R2139-41 ]

[6] J.F. Prins [ *The Physics of Diamond* (IOS Press, 1997) p.411-75 ]

[7] R. Kalish [ *The Physics of Diamond* (IOS Press, 1997) p.373-407 ]

[8] S. Prawer, R. Kalish [ *Phys. Rev. B (USA)* vol.51 (1995) p.15711-22 ]

[9] S. Prawer, A. Hoffman, R. Kalish [ *Appl. Phys. Lett. (USA)* vol.57 (1990) p.2187-9 ]

[10] R. Kalish [ *Diam. Relat. Mater. (Switzerland)* vol.2 (1993) p.621-33 ]

[11] F. Fang, C.A. Hewett, M.G. Fernandes, S.S. Lau [ *IEEE Trans. Electron Devices (USA)* vol.36 (1989) p.1783 ]

[12] C. Uzan-Saguy, V. Richter, S. Prawer, Y. Lifshitz, E. Grossman, R. Kalish [ *Diam. Relat. Mater. (Switzerland)* vol.4 (1995) p.569-74 ]

[13] C.D. Clark, P.J. Kemmey, E.W.J. Mitchell [ *Discuss. Faraday. Soc. (UK)* vol.31 (1961) p.96 ]

[14] J.F. Prins [ *Appl. Phys. Lett. (USA)* vol.197 (1982) p.950-2 ]

[15] J.F. Prins [ *Phys. Rev. B (USA)* vol.38 (1988) p.5576-84 ]

[16] A. Reznik, V. Richter, R. Kalish [ *Phys. Rev. B (USA)* vol.56 (1997) p.7930-5 ]

[17] V.S. Vavilov, M.A. Gukasyan, M.I. Guseva, T.A. Karatygina, E.A. Konorova [ *Sov. Phys.-Semicond. (USA)* vol.8 (1974) p.471-3 ]

[18] G. Braunstein, R. Kalish [ *J. Appl. Phys. (USA)* vol.54 (1983) p.2106-8 ]

[19] J.F. Prins [ *J. Phys. D (UK)* vol.22 (1989) p.1562 ]

[20] J.F. Prins [ *Phys. Rev. B (USA)* vol.39 (1989) p.3764-70 ]

[21] C. Uzan-Saguy [ private communication ]

[22] P.R. de la Houssaye, C.M. Pechina, J.R. Zeidler, R.G. Wilson [ *J. Appl. Phys. (USA)* vol.71 (1992) p.3220-4 ]

[23] C.A. Hewett, R. Nguyen, J.R. Zeidler, R.G. Wilson [ *Diam. Relat. Mater. (Switzerland)* vol.2 (1993) p.17 ]

[24] J.R. Zeidler, C.A. Hewett, R.G. Wilson [ *Phys. Rev. B (USA)* vol.47 (1993) p.2065-71 ]

[25] F. Fontaine, C. Uzan-Saguy, B. Philosoph, R. Kalish [ *Appl. Phys. Lett. (USA)* vol.68 (1996) p.2264-6 ]

[26] S. Prawer, K.W. Nugent, D.N. Jamieson [ *Diam. Relat. Mater. (Switzerland)* vol.7 (1997) p.106 ]

[27] C. Uzan-Saguy, R. Kalish, R. Walker, D.N. Jamieson, S. Prawer [ *Diam. Relat. Mater. (Switzerland)* vol.7 (1998) p.1429-32 ]

[28] H. Sternshulte, K. Tonke, R. Sauer [ *Phys. Status Solidi (Germany)* vol.172 (1999) p.37 ]

[29] J.F. Prins [ *Advances in New Diam. Sci. Technol.* (1994) p.443-8 ]

[30] J.F. Prins [ *Appl. Phys. Lett. (USA)* vol.73 (1998) p.2308-10 ]

[31] R. Kalish, C. Uzan-Saguy, A. Samoiloff, R. Locher, P. Koidl [ *Appl. Phys. Lett. (USA)* vol.64 (1994) p.2532-4 ]

[32] R. Kalish, C. Uzan-Saguy, B. Ran, H. Ferber, H. Guettler, R. Zachai [ *Diam. Relat. Mater. (Switzerland)* vol.7 (1999) p.877-81 ]

[33] S.A. Kajihara, A. Antonelli, J. Bernholc, R. Car [ *Phys. Rev. Lett. (USA)* vol.66 (1991) p.2010-3 ]

[34] R. Kalish, A. Reznik, C. Uzan-Saguy, C. Cytermann [ *Appl. Phys. Lett. (USA)* to be published in the issue of February 7 (2000) ]

[35] R. Kalish et al [ *Diam. Relat. Mater. (Switzerland)* vol.6 (1997) p.516-20 ]

[36] J.F. Prins [ *Diam. Relat. Mater. (Switzerland)* vol.4 (1995) p.585-90 ]

[37] B. Ran [ MSc Thesis, Technion, Israel, 1996 ]

[38] G. Braunstein, R. Kalish [ *Appl. Phys. Lett. (USA)* vol.38 (1981) p.416-8 ]

[39] H. Hoffsass, M. Dalmer, M. Restle, C. Ronning [ *J. Appl. Phys. (USA)* vol.81 (1997) p.2566-9 ]

[40] N. Casanova, A. Deneuville, E. Gheeraert, C. Uzan-Saguy, R. Kalish [ to be published in the proceedings of the E-MRS (Strasbourg, France) (1999) ]

[41] V.S. Vavilov, M.A. Gukasyan, M.I. Guseva, T.A. Karatygina, E.A. Konorova [ *Sov. Phys.-Semicond. (USA)* vol.9 (1976) p.962 ]

[42] S. Prawer, C. Uzan-Saguy, G. Braunstein, R. Kalish [ *Appl. Phys. Lett. (USA)* vol.63 (1993) p.2502-4 ]

[43] R. Job, M. Werner, A. Denisenko, A. Zaitsev, W.R. Fahrner [ *Diam. Relat. Mater. (Switzerland)* vol.5 (1996) p.757-60 ]

[44] I.M. Buckley Golder et al [ *Diam. Relat. Mater. (Netherlands)* vol.1 (1991) p.43-7 ]

[45] M. Restle et al [ *Appl. Phys. Lett. (USA)* vol.66 (1995) p.2733-5 ]

# B3.2 Large dopants in diamond

J.F. Prins

January 2000

## A INTRODUCTION

Both boron (B) and nitrogen (N) atoms are similar in size to carbon (C) atoms and they can be incorporated to occupy single substitutional sites during diamond growth. They form acceptor and donor levels respectively with activation energies of ≈ 0.37 [1] and ≈ 1.7 [2] eV. These energies, especially the one for the donor level, are high, and alternative dopant atoms are thus coveted. Other possible substitutional dopant atoms are larger. Diamond has the highest atomic density of any known material [3]. One thus expects, and it is actually found [4], that the larger the possible dopant atom, the less likely it will be to take up a substitutional position. Most attempts, using larger dopant atoms, concentrated on the generation of shallower donor states. Phosphorus is the obvious next choice after nitrogen, and therefore a large number of studies have concentrated on the possibility of incorporating and activating phosphorus atoms.

## B ACTIVATION DURING GROWTH

After confirmation of the claims by Russian scientists [5] that diamonds can be grown by plasma-assisted-chemical-vapour-deposition (PACVD) of hydrogen-diluted, hydro-carbon gases like methane [6], attempts were reported to dope diamond n-type by mixing gases containing P-atoms into the plasma (see for example [7]). Koizumi et al [8] finally reported conclusive evidence that this is possible in 1997, using phosphine ($PH_3$) as the dopant gas. The reported activation energies ranged from ≈ 0.42 eV to more than 0.5 eV [8–10]. Cathodoluminescence measurements on diamonds which had been very lightly doped with phosphorus showed luminescence peaks which are typical for donor-acceptor pair recombination, and from these measurements the ionisation energy for the P-donors emerged as ≈ 0.65 eV [11].

Compared to the B-acceptors, these donors are still deep. The mobilities found for the electrons by means of Hall effect measurements are, to date, well below 200 $cm^2/V$ s, compared to values of over 2000 $cm^2/V$ s determined for electron injection into undoped, insulating diamonds [12]. Furthermore, the control over the density of activated dopants is poor. Although the density increases with increasing $PH_3$ concentration in the plasma, a maximum is reached beyond which the quality of the diamond layers deteriorates with a concomitant increase in the resistivity.

Oxygen can be a beneficial element to mix into the plasma gas during PACVD growth. Very few, if any, oxygen atoms actually become incorporated into diamond during the growth process, but, in many cases, the presence of oxygen improves the quality of the diamond by turning it from a less transparent substrate to a more transparent one (see for example [13]). During experiments where $H_2S$ was mixed into the plasma to determine whether sulphur will play a similar role, it was discovered that the diamond layers became conducting [14]. Hall measurements at first confirmed n-type conduction with an electron mobility of ≈ 600 $cm^2/V$ s and an activation energy of ≈ 0.38 eV. The latter value compares favourably with the B-acceptors. However, subsequent measurements [15] showed p-type conduction that could be ascribed to boron contamination during growth. Eliminating the boron contamination also eliminated the observed conduction. It would thus seem that sulphur, like its group VI counterpart, oxygen, does not easily become incorporated into the diamond lattice during PACVD growth.

Nishimori et al reported growing n-type layers by means of molecular beam epitaxy using methane and tri-n-butylphosphine [16]. They measured an activation energy for conduction of only 0.12 eV. The conductivity did not approach a saturated value at temperatures below 400°C, as one would expect for such a low activation energy. This may indicate a very high compensation ratio or some type of hopping conduction.

## C ION IMPLANTATION STUDIES

Since the early 1960s many attempts have been made to dope diamond by means of ion implantation (for reviews, see [17–19]). During ion implantation, radiation damage is caused by the impinging dopant ions. This has to be removed by annealing. Owing to its metastability, diamond cannot be melted or even heated to temperatures (> 1900°C) where diffusion of thermally activated point defects can assist in the annealing process. At these temperatures it will convert to graphite. Annealing temperatures have to be restricted to much lower values. Even in the best vacuum possible it becomes risky to exceed 1700°C. Some residual defects are thus expected to remain, and they will lower the mobility of carriers in the conduction or valence bands.

Another limitation is that the implantation dose has to remain below a critical value (the graphitisation threshold [20]). Above this dose, the density of defects is so high that the diamond will revert to graphite, even for temperatures at which it will normally not do so. Any implantation-annealing process has to be designed to achieve the activation of the dopant atoms, as well as the required reduction in radiation damage, by controlling the interactions between the very defects which are created during ion irradiation. A knowledge of the types of defects and their interactions during annealing is, thus, a prerequisite to designing and developing implantation-doping routines which can lead to high quality p- and n-type conductive layers.

During ion implantation atoms are kicked out of their lattice sites, and these atoms can again collide with other atoms to displace them, in this way forming a cascade of collisions. There are two schools of thought about the nature of the radiation damage created in these cascades. In one model it is assumed that each cascade creates a well-defined region (sometimes called a displacement spike) which contains a very high density of defects modified by the thermal heating, or spike, that accompanies its creation. Assuming such regions to be spherical, it has been postulated that 'the passage of each ion leaves in its wake a 'trail' of damaged spheres of average radius r', and that these spheres consist of 'conducting graphitic material' [21]. From this it follows that the critical dose relating to the graphitisation threshold is reached when these displacement spikes overlap sufficiently. If this model is correct, most of the implanted dopant atoms, especially the larger, more massive ones, will end up within the resultant graphitic regions. It will then be near impossible to activate them into diamond lattice sites. Using emission channelling, it was found that, even for an ion as massive as indium (In), 'a weak channelling effect is observed' that 'indicates that the diamond lattice is not amorphised and a small fraction of the In atoms occupy at least near substitutional sites' [4]. It is thus unlikely that displacement spikes play a role in diamond for lighter ions.

In the other model it is postulated that the defects formed in the collision cascades consist primarily of individual vacancies and displaced atoms [18]. The graphitisation threshold is reached when the vacancy density reaches a certain critical value [22]. This postulate is supported by the observation that electron irradiated diamond, where it is known that only interstitial atoms and vacancies form, and ion irradiated diamond, show similar annealing stages [4,18]. The first stage is above room temperature and below 600°C and is caused by interstitial atom diffusion [23], while the second stage is above 600°C, where vacancy diffusion dominates [24]. Annealing of the point defects and activation of the implanted dopant atoms relies on the diffusion of interstitial atoms. Self-interstitials that recombine with vacancies reduce the damage, and dopant interstitials doing the same become activated. Ideally, one would want the dopant atoms to all move into vacant lattice sites and the self-interstitials to mop up the rest of the vacancies. However, the implanted layer has a limited depth-width $\omega$. Interstitial atoms can thus diffuse out of and escape from the layer before combining with any vacancies. Vacancies can act both as compensating donors situated at $\approx$ 1.2 eV [25,26] above the valence band and as compensating acceptors by forming so-called ND1 [27] centres situated at $\approx$ 2.6 eV below the conduction band (> 3.15 eV at $\mathbf{k}$ = 0 in the bandgap). Even when annealing the vacancies, it is found that not many of them diffuse away, but that they rather form other defects that are still electronically active [18,28].

When implanting at a target temperature where interstitial atoms can diffuse, it is found that the residual damage, for the same ion dose, decreases for higher temperatures [18]. In terms of point defects, this implies that the number of interstitial atoms that escape from the layer decreases. This behaviour can be explained by postulating that the activation energy for interstitial atom diffusion $E_D$ is smaller than the activation energy $E_C$ that is involved when an interstitial atom combines with a vacancy [29]. If, on average, a self-interstitial needs to make $\mu_e$ jumps to escape from $\omega$, and $\mu_c$ jumps to find a vacancy and recombine with it when the average vacancy density within $\omega$ is $N_v$, then the probability P that an interstitial atom will combine with a vacancy before it can escape follows as $P = \mu_e/(\mu_e + \mu_c)$, and this can be expressed as [30]:

$$P = \frac{\mu_e}{\mu_e + \mu_c} = \frac{\mu_e N_v}{\mu_e N_v + Q\left(1 + q\exp\left[\frac{E_C - E_D}{k_B T}\right]\right)} \qquad (1)$$

where q and Q are constants for a given crystal structure. It is clear that P will increase with increasing temperature T.

The cold-implantation-rapid-annealing (CIRA) routine is based on the latter observation. The dopant atoms are implanted at a low temperature (typically liquid nitrogen) where neither interstitial atoms nor vacancies can diffuse. By increasing their densities in this way, the probability that interstitial atoms will find vacancies before diffusing away during subsequent annealing, is enhanced (see EQN (1)). After implantation, the diamond is subjected to rapid thermal annealing (RTA) by heating it extremely fast to a suitably high temperature which will further curtail interstitial atom escape. According to EQN (1), one would like T, $N_v$ and $\mu_e$ to be as large as possible. However, there are limitations. For temperatures above $\approx$ 1200°C, the vacancies agglomerate so fast that extended defects are formed before the interstitials can annihilate enough of them [31]. Increasing the initial vacancy density, $N_v$, unfortunately always causes the residual damage after annealing to also increase [18]. $\mu_e$ increases with the depth-width $\omega$ of the implanted layer. For this reason implantations are usually done over a range of energies. It should be noted that when $\mu_e$ increases to larger values, the significance of the temperature-related term in the denominator decreases. The RTA phase of the process becomes less important. In fact, for $\mu_e$ large enough $P \approx 1$ for any annealing temperature at which interstitial atom diffusion can take place.

Although the highest possible annealing temperature will result in the lowest residual vacancy-related defect density, it may not result in the best activation of the implanted dopant atoms. The self- and dopant-interstitial atoms compete to fill vacancies. For example, when doping diamond with boron, the higher the RTA temperature, the fewer boron acceptors become activated [32]. There is thus an optimum annealing temperature at which the highest fraction of acceptors is activated for the lowest residual damage. For keV (30–200 keV) implanted boron ions, with a depth-width $\omega \approx 0.25$ μm, this temperature is ≈ 1000°C [32]. Subsequent anneals at higher temperatures, up to 1600°C, lead to further improvements in the quality of such layers. Quite good boron-doped layers with mobilities up to 1000 $cm^2$/V s can be generated by means of CIRA when using high quality substrates, and the first field effect transistor (FET) on diamond which showed pinch-off was manufactured on a CIRA-doped layer [33]. If the number of activated dopants required is higher than the critical ion dose to reach the graphitisation threshold, multiple CIRA steps can be used [34].

Large, massive, dopant ions create more radiation damage per ion than lighter ions and their penetration into the substrate is less for the same ion energy. The initial vacancy density will be higher and $\mu_e$ lower, thus causing a drastic increase in residual radiation damage after annealing. Phosphorus CIRA implantation using the same doses and annealing steps which rendered the best results for boron did not produce measurable n-type conduction. The first indication that such layers can be produced by $P^+$-CIRA was obtained for very low dose implantations followed by RTA to 500°C, plus short anneals at 1200 and 1600°C [35]. Owing to the low dopant density the resistances measured were still very high. An attempt was made to rectify this by multiple CIRA steps. Although this at first helped, the resistance subsequently started to deteriorate. It was later realised [36] that anneals at temperatures where the vacancies can diffuse deactivated the P-donors. In retrospect this should have been expected because nitrogen donors also become deactivated when annealed with vacancies present [3].

By doing multiple $P^+$-CIRA steps with RTAs to 500°C and not subjecting the diamond substrates to anneals above 600°C, in order to inhibit any vacancy diffusion, successful phosphorus doping ensued [37]. Owing to the low temperature (in conjunction with a higher initial vacancy density and lower $\mu_e$) the residual damage in these layers is very high. Consequently the compensation of the donors is high and the electron mobility low. For very low dose steps, the activation energy for conduction was measured as ≈ 0.62 eV, which corresponds to the cathodoluminescence measurements [11]. With increasing steps the activation energy came down to ≈ 0.43 eV. These experiments were done using ion energies up to 170 keV. The quality of such doped layers is expected to improve when higher energies, up to and within the MeV range, are employed.

As already mentioned, oxygen is not incorporated in large quantities into diamond during growth. It was reasoned that if the CIRA process with a low RTA is used, it may be possible to quench oxygen atoms into substitutional sites to form high-energy, metastable states which may act as donors. This was, indeed, achieved and electron conduction with an activation energy of ≈ 0.32 eV measured [38]. Oxygen has two donor electrons, and by annealing the layer, which deactivated some of the oxygen, an activation energy of ≈ 0.52 eV was measured at higher temperatures [39]. Although oxygen is a less massive atom than phosphorus, the residual vacancy damage, and concomitant compensation, is still too high. Sulphur can also be activated using CIRA with low temperature annealing [40]. Activation of sulphur donors has subsequently been reported when implanting very high quality homo-epitaxial PACVD diamond at only 400°C [41]. Again a low temperature was needed to quench these atoms into substitutional sites. Owing to it being more massive than oxygen, which results in even more compensating vacancies, oxygen should be the preferred group VI element to dope diamond n-type.

In an attempt to reduce the residual vacancies further, especially when one is equipment-limited to small values of ω, the LODDI (low-damage-drive-in) routine was developed [42]. In the case of CIRA, the object is to prevent interstitial atoms from diffusing out of ω. The LODDI routine uses the dopant interstitials that do actually escape from the layer which is being implanted. Ion implantation, using isochemical carbon or lighter ions (for example $He^+$), is used to pre-damage a layer of depth-width $\omega_V$. This damage is kept very low in order to create just the required density of vacancies needed for the process. The dopant ions are then implanted to a very high dose into an adjacent layer, such that during annealing, which may occur simultaneously with the implantation, dopant-atom interstitials can escape and diffuse into the pre-damaged layer where they then meet up with vacancies to become activated.

Phosphorus and aluminium have been activated using LODDI implantation into a shallower layer than the pre-damaged one [35,40]. The shallower layers were etched off after the drive-in anneals. However, during these experiments some extra vacancies did succeed in entering the pre-damaged layer. Whether this happened during the dopant implantation-annealing step, or when the upper layer was etched off, is not yet clear. More experiments are needed to develop this approach further.

## D CONCLUSION

It is possible to activate P donor atoms during diamond growth. There exists an upper limit to the density of these atoms that can be activated during growth before the diamond quality deteriorates substantially. This is not the case when using ion implantation. The density can be increased continuously by using multiple CIRA steps. The claim that S-doped layers can be generated during PACVD growth seems to be incorrect. This goal can, however, be achieved by ion implantation when using low annealing temperatures. The ion-implanted P- and S-doped layers are, like the P-doped grown layers, highly compensated. To improve their quality, a drastic reduction in the compensating vacancies is needed. New implantation-doping routines like LODDI may lead to such a reduction. Oxygen atoms can be inserted into diamond to act as shallow donors. Owing to the fact that these atoms are smaller and less massive than P and S, this development may lead to good quality n-type doped layers with dopant densities that can be pre-selected.

## REFERENCES

[1] A.T. Collins, A.W.S. Williams [ *J. Phys. C (UK)* vol.4 (1971) p.1789-800 ]
[2] R.J. Farrer [ *Solid State Commun. (USA)* vol.7 (1969) p.685-7 ]
[3] J.E. Field [ *Properties of Natural and Synthetic Diamond* (Academic Press, 1992) ]
[4] C. Ronning, H. Hofsäss [ *Diam. Relat. Mater. (Switzerland)* vol.8 (1999) p.1623-30 ]
[5] B.V. Spitsyn, L.L. Bouilov, B.V. Deryaguin [ *J. Cryst. Growth (Netherlands)* vol.52 (1981) p.219-26 ]
[6] S. Matsumoto, Y. Sato, M. Kamo, N. Setaka [ *Jpn. J. Appl. Phys. (Japan)* vol.21 (1982) p.L183-5 ]
[7] K. Okano, H. Kiyota, T. Kuruso, M. Iida [ *Diam. Relat. Mater. (Switzerland)* vol.3 (1993) p.35-40 ]
[8] S. Koizumi, M. Kamo, Y. Sato, H. Ozaki, T. Inuzuka [ *Appl. Phys. Lett. (USA)* vol.71 (1997) p.1065-7 ]
[9] S. Koizumi et al [ *Diam. Relat. Mater. (Switzerland)* vol.7 (1998) p.540-4 ]
[10] R. Kalish, A. Reznik, C. Uzan-Saguy, S. Koizumi, M. Kamo [ *Advanced Materials '98* (NIRIM, 1998) p.105-8 ]
[11] H. Sternschulte, K. Thonke, R. Sauer, S. Koizumi [ *Phys. Rev. B (USA)* vol.59 (1999) p.12974-7 ]
[12] F. Nava, C. Canali, M. Artuso, E. Gatti, P.F. Manfredi [ *IEEE Trans. Nucl. Sci. (USA)* vol.NS-26 (1979) p.308-11 ]
[13] P.K. Bachmann [ *The Physics of Diamond* (IOS, 1997) p.63-70 ]
[14] I. Sakaguchi et al [ *Phys. Rev. B (USA)* vol.60 (1999) p.R2139-41 ]
[15] R. Kalish, A. Reznik, C. Uzan-Saguy, C. Cytermann [ *Appl. Phys. Lett. (USA)* vol.76 (2000) p.757-9 ]

[16] T. Nishimori, K. Nakano, H. Sakamoto, Y. Takakuwa, S. Kono [ *Appl. Phys. Lett. (USA)* vol.7 (1997) p.945-7 ]

[17] V.S. Vavilov [ *Radiat. Eff. (UK)* vol.37 (1978) p.229-36 ]

[18] J.F. Prins [ *Mater. Sci. Rep. (Netherlands)* vol.7 (1992) p.271-364 ]

[19] M.S. Dresselhaus, R. Kalish [ *Ion implantation in Diamond, Graphite and Related Materials* (Springer-Verlag, 1992) p.129-58 ]

[20] C. Uzan-Saguy, C. Cytermann, R. Brener, V. Richter, M. Shanaan, R. Kalish [ *Appl. Phys. Lett. (USA)* vol.67 (1995) p.1194-6 ]

[21] S. Prawer, R. Kalish [ *Phys. Rev. B (USA)* vol.51 (1995) p.15711-22 ]

[22] J.F. Prins [ *Phys. Rev. B (USA)* vol.31 (1985) p.2472-8 ]

[23] L. Allers, A.T. Collins, J. Hiscock [ *Diam. Relat. Mater. (Switzerland)* vol.7 (1998) p.228-32 ]

[24] G. Davies, S.C. Lawson, A.T. Collins, A. Mainwood, S.J. Sharp [ *Phys. Rev. B (USA)* vol.46 (1992) p.13157-70 ]

[25] H.B. Dyer, P. Ferdinando [ *Br. J. Appl. Phys. (UK)* vol.17 (1966) p.419-20 ]

[26] J.F. Prins [ *The Physics of Diamond* (IOS, 1997) p.427-8 ]

[27] G. Davies [ *Nature (UK)* vol.269 (1977) p.498-9 ]

[28] J.F. Prins [ *Phys. Rev. B (USA)* vol.39 (1989) p.3764-70 ]

[29] J.F. Prins [ *Phys. Rev. B (USA)* vol.44 (1991) p.2470-9 ]

[30] J.F. Prins [ *Phys. Rev. B (USA)* vol.61 (2000) p.7191-4 ]

[31] R.A. Spits, T.E. Derry, J.F. Prins [ *Nucl. Instrum. Methods Phys. Res. B (Netherlands)* vol.64 (1992) p.210-4 ]

[32] J.F. Prins [ *Nucl. Instrum. Methods Phys. Res. B (Netherlands)* vol.59/60 (1991) p.1387-90 ]

[33] C.A. Hewett, J.R. Zeidler [ *Diam. Relat. Mater. (Netherlands)* vol.1 (1992) p.688-91 ]

[34] J.F. Prins [ *Physica B (Netherlands)* vol.185 (1993) p.132-43 ]

[35] J.F. Prins [ *Diam. Relat. Mater. (Switzerland)* vol.4 (1995) p.580-5 ]

[36] J.F. Prins [ *Diam. Films Technol. (Japan)* vol.8 (1998) p.181-93 ]

[37] J.F. Prins [ *Diam. Relat. Mater. (Switzerland)* vol.8 (1999) p.1635-41 ]

[38] J.F. Prins [ *Phys. Rev. B (USA)* in press (2000) ]

[39] J.F. Prins [ *Diam. Relat. Mater. (Switzerland)* in press (2000) ]

[40] J.F. Prins [ *Proc. Electrochemical Soc. Meeting, Hawaii 1999* in press ]

[41] M. Hasegawa et al [ *Jpn. J. Appl. Phys. (Japan)* vol.38 (1999) p.L1519-22 ]

[42] J.F. Prins [ *Diam. Relat. Mater. (Switzerland)* vol.7 (1998) p.545-9 ]

# B3.3 Boron doping and characterisation of diamond

C. Johnston, A. Crossley, M. Werner and P.R. Chalker

February 2000

## A INTRODUCTION

One of the key process technologies for the use of diamond as an electronic semiconductor material is doping. It is known [1] from studies on natural diamond stones that boron acts as an acceptor with an acceptor level 0.368 eV above the valence band edge. Boron doping of diamond can be achieved in situ during growth [2] or ex situ by ion implantation [3] or high temperature diffusion [4].

Only in situ boron doping of CVD diamond will be discussed here.

## B BORON DOPING

The technologically important diamond growth methods [5] rely on chemical vapour deposition, either in microwave plasma assisted processes or thermally activated processes, such as thermal filament or DC plasma torch methods. Hence the introduction of boron material (solid, liquid or gas) into the growth media can result in the incorporation of boron in the deposit.

A convenient gaseous form of boron at room temperature is diborane [6]. This has been used extensively to dope CVD diamond films [7–9]. However, diborane is highly toxic and dangerous to handle. Safer and more easily handled solid and liquid sources of boron have been used successfully, e.g. boron powder [10], boron oxide [11], organo-boron complexes [2], and boron trichloride [12]. These alternative sources of boron have to be introduced into the reactant stream either by high temperature evaporation or by trace pickup through bubblers and other delivery systems.

Boron trichloride can be introduced as a dilute solute (< 1%) in argon or hydrogen and delivered to the growth chamber via a controlled gas stream. The organo-boron complexes must, however, be dissolved in a suitable solvent (hot alcohol). As the carrier gas, e.g. argon or hydrogen, is bubbled through the hot solution trace quantities of the boron complex are picked up and delivered to the growth chamber. Solid boron sources can be placed near the hot substrate and heated so that they can diffuse to the growth surface.

The probability for boron incorporation can be higher than $10^{-1}$ [5]. However, boron incorporation is dependent on the texture of the diamond film or the orientation of the single crystal diamond. Spytsin et al [13] observed enhanced boron incorporation in <111> single crystal diamond. This result was confirmed by Samlenski et al [14] applying nuclear reaction analysis in order to quantify the concentration of boron in doped homoepitaxial films. They found the boron incorporation probability in <111> oriented films to be up to one order of magnitude higher than in the <100> oriented films. Similar results [15] have also been obtained using secondary ion mass spectroscopy on boron doped polycrystalline CVD diamond films with different textures. The total amount of boron incorporated into CVD diamond films can be varied over several orders of magnitude up to concentrations of about $10^{21}$ cm$^{-3}$ without significant deterioration of the structural quality under appropriate deposition conditions.

The incorporation of dopants into growing diamond films can influence the morphology of the deposited material. By simulating the growth of single crystal diamonds [16] a growth parameter $\alpha$

can be found, which not only determines the crystal shape and the texture of polycrystalline diamond films, but can also be used to determine the stability against twinning. <100> faces are stable with respect to twinning for $2 \leq \alpha \leq 3$, whereas <111> faces are stable with respect to twins for $1 \leq \alpha \leq 1.5$. In the range $\alpha$ between 1.5 and 2, both faces are unstable against twinning and therefore the probability of growing fine grained films is enhanced. Nitrogen and boron can be used to shift $\alpha$ into regions of stable growth [17]. Applying laser interferometry in order to determine the growth rate on <100> and <111> oriented single crystals, respectively, nitrogen was found to increase the growth rate on <100>, thereby increasing $\alpha$. Boron, on the other hand, decreases $\alpha$ by increasing the growth rate on <111>. The strong dependence of the film structure on the admixture of small amounts of boron or/and nitrogen is shown in FIGURE 1.

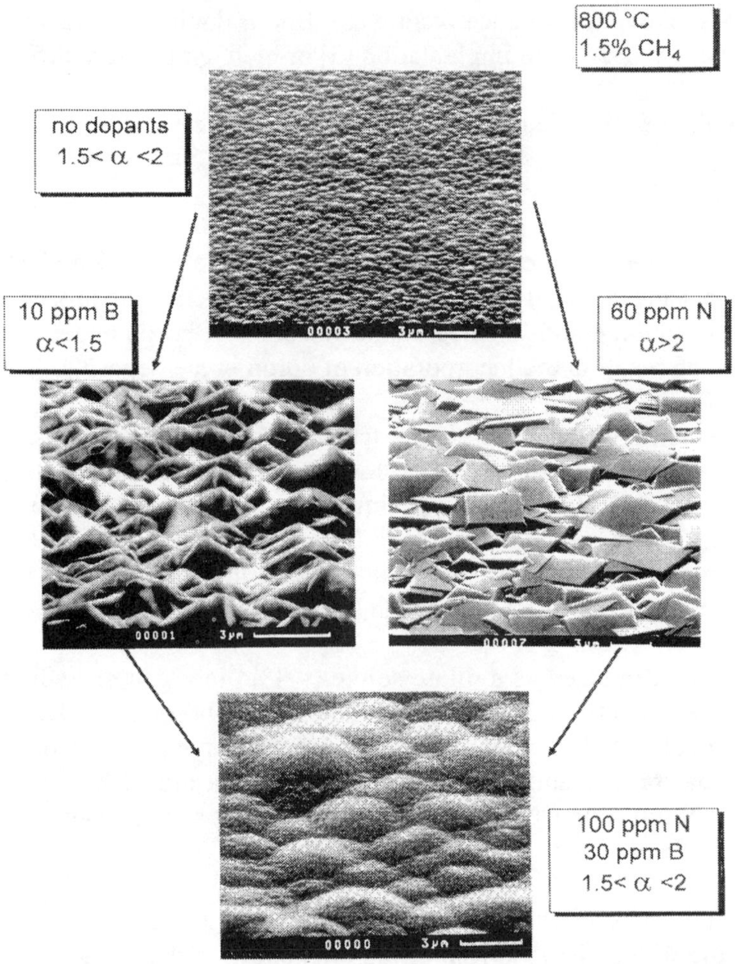


FIGURE 1 Effect of nitrogen and/or boron additions to the morphology of CVD diamond films as indicated by the $\alpha$ growth factor.

## C ELECTRICAL CHARACTERISTICS

Experimental results such as resistivity, Hall and thermoelectric measurements show that boron doped diamond behaves like a compensated p-type semiconductor.

FIGURE 2 shows a number of temperature dependent conductivity measurements on polycrystalline and homoepitaxial diamond films. Obviously the temperature dependence is clearly non-Arrhenius, leading to a pronounced tail at low temperatures. The conductivity increases with increasing temperature except for the most heavily doped films. This decrease of the activation energy can be explained in terms of impurity band conduction. With increasing doping concentration the acceptor band-width increases. Massarani et al [18] estimated values from 20 to 200 meV for diamond by increasing the doping concentration from $10^{17}$ to $10^{20}$ $cm^{-3}$ while assuming 50% compensation. At sufficiently high doping concentrations the impurity band merges with the valence band and metallic conduction is observed. A simple estimation [19] of the doping level concentration at which the metal-insulator transition occurs gives a doping concentration of $2 \times 10^{20}$ $cm^{-3}$. Experimental results [20] suggest a doping level concentration above $7 \times 10^{20}$ $cm^{-3}$ for the metal-insulator transition in B-doped polycrystalline diamond films and synthetic stones.

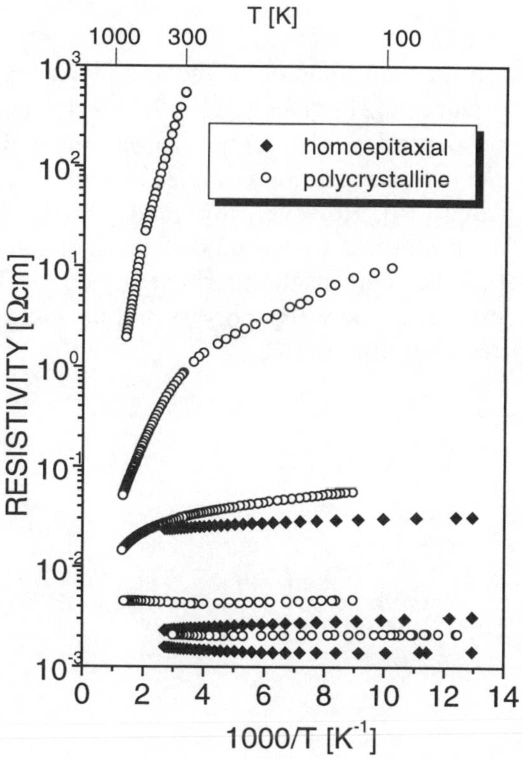


FIGURE 2 Temperature dependent conductivity measurements on polycrystalline and homoepitaxial diamond films.

For synthetic B-doped diamond stones, homoepitaxial diamond films and polycrystalline B-doped diamond it has been shown [21] that the conductivity at low temperatures can be expressed in terms of variable-range hopping. In general, it is difficult to distinguish between variable range hopping and nearest-neighbour hopping, but from systematic data collection [21] it is more successful to interpret the results in terms of nearest-neighbour hopping. Starting with the typical expression for an extrinsic, partially compensated semiconductor, the electrical conductivity σ can be approximated by

$$\sigma = \sigma_1 e^{-\frac{E_1}{kT}} + \sigma_2 e^{-\frac{E_2}{kT}} + \sigma_3 e^{-\frac{E_3}{kT}} \quad (1)$$

where k is the Boltzmann constant ($k = 8.62 \times 10^{-5}$ eV/K), T the absolute temperature, $\sigma_1$, $\sigma_2$ and $\sigma_3$ are constants and $E_1$, $E_2$ and $E_3$ describe energies associated with three different conduction mechanisms. The first term relates to valence-band conduction and the third term to nearest-neighbour hopping due to hopping of holes from occupied to unoccupied acceptor sites, made possible by the finite overlap of the wave functions. The second term is associated with conduction in an impurity band. This conduction mechanism [22] can only be observed in the intermediate concentration range for certain values of the compensation.

In order to simplify the analysis we can omit the second term in EQN (1). The first term is simply $q\mu p$ where q is the elementary charge, $\mu$ the band mobility and p the hole concentration with the activation energy $E_1 = 0.368$ eV. The pre-exponential factor of the remaining third term ($E_3$) can be written [23] in the form

$$\sigma_3 = \sigma_{03} e^{-\frac{\alpha N_A^{-1/3}}{a_H}} \tag{2}$$

where $\alpha$ is a constant, $a_H$ is the Bohr radius and $N_A$ is the acceptor concentration. The best fit for the experimental data from various authors [21] gives $\alpha/a_H = 2.16 \times 10^7\ cm^{-1}$ and $\sigma_{03} = 283.3\ \Omega^{-1}\ cm^{-1}$. The activation energy $E_3$ should depend on the compensation, but is difficult to determine because conventional theories predict erroneous activation energies at high doping levels. $E_3$ for low doping concentrations has been calculated [24]. However, this model for $E_3$ does not fit the measurement results at higher doping levels. In contrast to the model, the activation energy $E_3$ depends on the acceptor concentration, but not on the compensation ratio at doping levels between $10^{17}\ cm^{-3}$ and $10^{22}\ cm^{-3}$. Furthermore, no clear difference between polycrystalline and single crystalline samples is observed. Therefore, EQN (1) can be written as:

$$\frac{1}{\rho} = \sigma = q\mu p + \sigma_{03} e^{-\frac{\alpha N_A^{-1/3}}{a_H}}$$

$$p = \left[\sqrt{\frac{1}{4}\left(\frac{p_1}{2N_A} + K\right)^2 + \frac{p_1}{2N_A}(1-K)} - \frac{1}{2}\left(\frac{p_1}{2N_A} + K\right)\right] N_A \tag{3}$$

$$K = \frac{N_D}{N_A}; \qquad p_1 = N_V e^{-E_1/kT}$$

The last missing property is the band mobility, which is different for poly- and single crystal diamond. Again a fit-function for the description of the room temperature mobility versus effective doping concentration $N_A$–$N_D$ has to be used.

The conventional interpretation of Hall measurements provides information about the average value of the hole concentration and neglects the impact of a second or third current conduction mechanism working in parallel.

FIGURE 3 shows the average hole concentration from various authors versus the effective doping concentration at room temperature. The solid lines were calculated from $(q\mu\rho)^{-1}$ with the aid of EQN (3) and fit functions for different donor concentrations. The dashed lines were calculated from Fermi–Dirac statistics alone. The onset of hopping is clearly around $N_A$–$N_D = 10^{19}\ cm^{-3}$. The compensating donor concentrations are in the range $N_D = 10^{15}$ to $10^{17}\ cm^{-3}$.

At low doping levels Fermi–Dirac statistics are valid, whereas around $10^{19}\ cm^{-3}$ a drop in the resistivity due to hopping is apparent. The agreement between calculated and measured results from various groups is satisfactory. The resistivity of the polycrystalline samples is obviously slightly higher due to the reduced mobility.

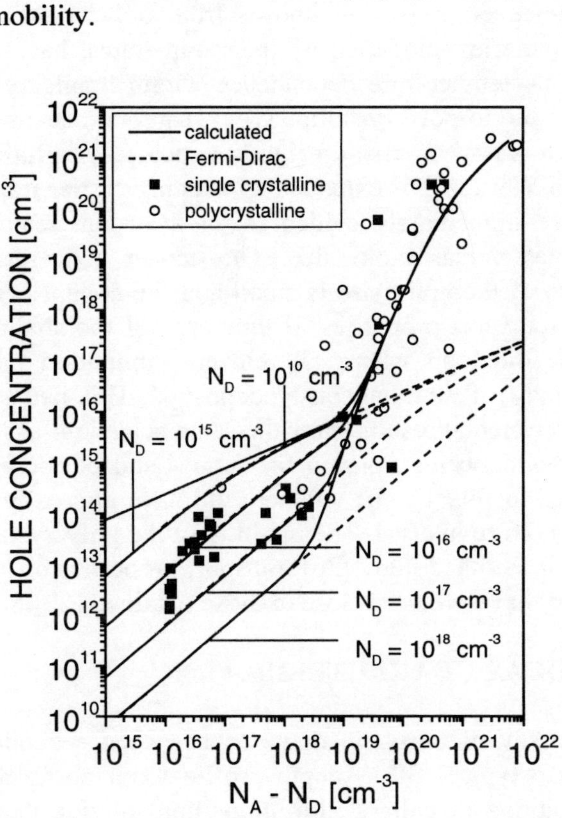


FIGURE 3 Average hole concentration as a function of effective dopant concentration at room temperature.

The temperature and doping level dependence of the activation energy of the conductivity can be calculated using a temperature dependent mobility by

$$E_A = -\frac{\partial \ln \sigma}{\partial\left(\frac{1}{kT}\right)} = -\frac{q}{\sigma}\left[\mu \frac{\partial p}{\partial\left(\frac{1}{kT}\right)} + \frac{\partial \mu}{\partial\left(\frac{1}{kT}\right)} p - \sigma_3 E_3 e^{-E_3/kT}\right] \tag{4}$$

The activation energy drops with increasing doping concentration $N_A$–$N_D$ and depends on the compensation level at low temperatures. In engineering terms, the resistive temperature detectors (RTDs) are described by their temperature coefficient of resistance (TCR). The TCR and the activation energy are related by

$$TCR = \frac{1}{R}\frac{\partial R}{\partial T} = -\frac{E_A}{k}\frac{1}{T^2} \tag{5}$$

Therefore the TCR depends on carrier concentration, the mobility, the density of states at the valence band edge and the compensation ratio. The temperature dependence of the conductivity is widely used for several sensing applications at the engineering prototype level.

Only very limited information on the impact of grain boundaries on the current conduction mechanism is available. However, it is well known from polycrystalline silicon that the current conduction across grain boundaries, modelled by their trap states, has a considerable effect on the electrical conductivity and its temperature dependence. Grain boundary models of polysilicon are generally not directly applicable to polycrystalline CVD diamond because all known models assume total dopant ionisation which is not the case for lightly doped polycrystalline diamond. To overcome this shortcoming Rutsch and Werner [25] extended an existing carrier trapping model for incomplete ionisation. Two things were immediately evident from the room temperature simulations of the electrical transport properties versus grain size. The carrier concentration is only significantly different from its bulk value, if the grain size is small and the dopant concentration is not too high. The traps only change the transport properties, if they are not too shallow. Few electrical data are available on polycrystalline diamond where all relevant simulation parameters are indicated. A notable exception is the study of simultaneously deposited, B-doped single- and poly-crystalline diamond [22] where the experimental results show the same behaviour as the simulations from Rutsch and Werner [25]. The carrier concentrations in the mono- and poly-crystalline diamond are fairly close, within a factor of two to four of one another, and have approximately the same temperature behaviour over a large temperature interval. The mobility of the polycrystalline material is one or two orders of magnitude lower. In another study [26] transport properties of large grained polycrystalline diamond films were reported that were very close to those obtained [22] for single crystalline values.

## D PHYSICO-CHEMICAL CHARACTERISATION

The primary tool for the assay of boron in diamond has been secondary ion mass spectrometry (SIMS). FIGURE 4 shows a typical SIMS depth profile for a boron doped polycrystalline CVD diamond film. The concentrations are calibrated from ion implantation standards.

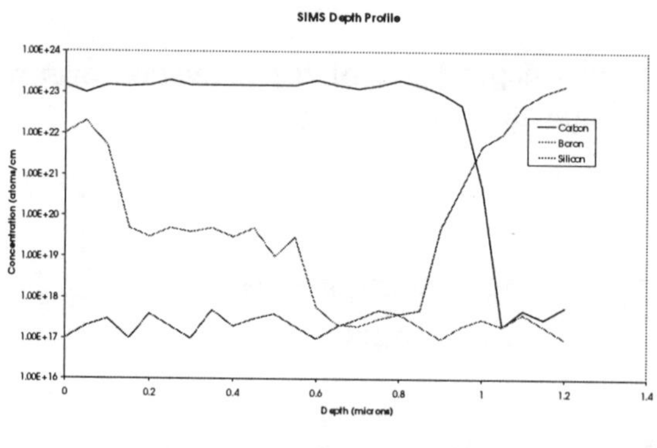


FIGURE 4 SIMS depth profile of B doped CVD diamond film showing a thin heavily doped surface region which facilitates low contact resistance formation. The active doped region is ~ 0.5 μm thick, $\sim 5 \times 10^{19}$ B atoms $cm^{-3}$.

It is clear that for polycrystalline films there is a linear relationship between the gas phase concentration of boron and the solid phase concentration of boron in CVD diamond over the electrically useful range, $10^{16}$–$10^{21}$ $cm^{-3}$. The actual efficiency of boron incorporation into the growing diamond film is different for each growth method and even individual deposition chambers.

With increasing boron concentration, from $10^{16}$–$10^{21}$ $cm^{-3}$, the average crystallite size drops by an order of magnitude, typically for a low power microwave plasma system from 10 μm to ~ 1 μm. Although all boron doped CVD diamond films exhibit a cubic structure there is significant lattice strain at the very high boron doping concentrations (> $10^{21}$ $cm^{-3}$). It is also evident from a number of TEM studies that the crystallite size reduction is accompanied by an increase in the incidence of twinning. The twins appear to originate at secondary nucleation sites on the growing <111> facets.

Raman spectroscopy has become a benchmark for assessing the quality of diamond films. The Raman spectra of boron doped diamond films (FIGURE 5) show a marked transition point around $10^{19}$ $cm^{-3}$. At boron concentrations below this level the Raman spectra exhibit the classic diamond Raman signature of a single sharp band at 1332 $cm^{-1}$. The width of the triplet degenerate $F_{2g}$ phonon increases with increasing boron concentration which can be attributed to an increase in lifetime commensurate with a decrease in defect concentration up to the transition point. Above ~ $10^{19}$ $cm^{-3}$ boron the Raman band becomes asymmetrical, widens and decreases in intensity. At concentrations above approximately $10^{20}$ $cm^{-3}$ the 1332 $cm^{-1}$ band disappears and the Raman spectrum resembles that from a metal surface, although some structure may be apparent from graphite (1500–1580 $cm^{-1}$) and disorder induced defect bands (~ 1200 $cm^{-1}$).

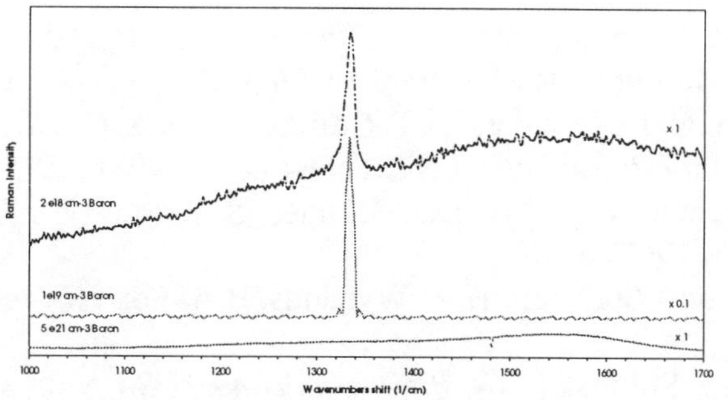


FIGURE 5 Raman spectra obtained from boron doped polycrystalline CVD diamond films. Exciting wavelength 514 nm, power ~ 80 mW. Micro-focused ~ 2 μm spot.

## E CONCLUSION

Boron doping of CVD diamond films is easily achieved by a number of means; however, the extent of material incorporated into the growing film can affect the film morphology and the degree of activation. The electrical characterisation of boron doped material is complex with competing conduction mechanisms operating depending on the amount of boron present within the film. Care must be exercised when assessing the quality of doped films since the Raman spectrum, the primary QA tool, is strongly influenced by the amount of boron doping.

## REFERENCES

[1] J.E. Field [ *The Properties of Diamond* (Academic Press, 1979) p.324 ]
[2] R. Locher et al [ *Mater. Sci. Eng. B (Switzerland)* vol.29 (1995) p.211 ]
[3] F. Fontaine, C. Uzan-Saguy, B. Philosoph, R. Kalish [ *Appl. Phys. Lett (USA)* vol.68 no.16 (1996) p.2264 ]
[4] W. Tsai et al [ *IEEE Electron Device Lett. (USA)* vol.12 no.4 (1991) p.157 ]
[5] P.K. Bachmann [ in *Properties and Growth of Diamond* Ed. G. Davies (INSPEC, IEE, London, UK, 1994) p.349 ]

[6] R.E. Harper, C. Johnston, N.G. Blamires, P.R. Chalker, I.M. Buckley-Golder [ *Surf. Coat. Technol. (Switzerland)* vol.47 (1991) p.344 ]

[7] J. Mort et al [ *Appl. Phys. Lett. (USA)* vol.55 no.11 (1989) p.1121 ]

[8] N. Fujimori, H. Nakahata, T. Imai [ *Jpn. J. Appl. Phys. (Japan)* vol.29 no.5 (1990) p.824 ]

[9] K. Miyati, K. Kumagai, K. Nishimura, K. Kobashi [ *J. Mater. Res. (USA)* vol.8 no.11 (1993) p.2845 ]

[10] S.A. Grot, C.W. Hatfield, G.S. Gildenblat, A.R. Badzian, T. Badzian [ *Appl. Phys. Lett. (USA)* vol.58 (1991) p.1542 ]

[11] X.K. Zhang et al [ *Appl. Phys. A (Germany)* vol.56 (1993) p.425 ]

[12] P.R. Chalker, C. Johnston [ *Phys. Status Solid A (Germany)* vol.154 (1996) p.455 ]

[13] B.V. Spytsin, L.L. Bouilov, B.V. Derjaguin [ *J. Cryst. Growth (Netherlands)* vol.52 (1981) p.219 ]

[14] R. Samlenski, C. Haug, R. Brenn, C. Wild, R. Locher, P. Koidl [ *Diam. Relat. Mater. (Switzerland)* vol.5 (1996) p.947 ]

[15] R. Locher, J. Wagner, F. Fuchs, M. Maier, P. Gonon, P. Koidl [ *Diam. Relat. Mater. (Switzerland)* vol.4 (1995) p.678 ]

[16] C. Wild, R. Kohl, N. Herres, W. Müller-Sebert, P. Koidl [ *Diam. Relat. Mater. (Switzerland)* vol.3 (1994) p.373 ]

[17] C. Wild, R. Locher, P. Koidl [ *Mater. Res. Soc. Symp. Proc. (USA)* vol.416 (1996) p.75 ]

[18] B. Massarani, J.C. Bourgoin, R.M. Chrenko [ *Phys. Rev. B (USA)* vol.17 no.4 (1978) p.1758 ]

[19] A.W.S. Williams, E.C. Lightowlers, A.T. Collins [ *J. Phys. C (UK)* vol.3 (1970) p.1727 ]

[20] M. Werner et al [ *Appl. Phys. Lett. (USA)* vol.64 no.5 (1993) p.595 ]

[21] M. Werner, R. Locher, W. Kohly, D.S. Holmes, S. Klose, H.J. Fecht [ *Diam. Relat. Mater. (Switzerland)* vol.6 (1997) p.308 ]

[22] D.M. Malta, J.A. von Windheim, H.A. Wynands, B.A. Fox [ *J. Appl. Phys. (USA)* vol.77 no.4 (1995) p.1536 ]

[23] B.I. Shoklovskii, I.S. Shlimak [ *Sov. Phys.-Semicond. (USA)* vol.6 no.1 (1972) p.104 ]

[24] A. Miller, E. Abrahams [ *Phys. Rev. (USA)* vol.120 no.3 (1960) p.745 ]

[25] G. Rutsch, M. Werner [ *Proc. Micro Materials Conference 1997* (Druckhaus Dresden GmbH, 1997) p.602 ]

[26] T. Sugino, K. Karasutani, H. Kataoka, K. Kobashi, J. Shirafuji [ *Advances in New Diamond Science and Technology* (MYU, Tokyo, 1994) p.705 ]

# B3.4 Nitrogen and phosphorus doping in CVD diamond

G.Z. Cao

December 1999

## A INTRODUCTION

Diamond promises high-power, high-frequency, and high-temperature electronic applications due to its unique physical properties such as high carrier mobility (0.16 $m^2/V$ s), high thermal conductivity (up to 1.5 × $10^4$ W/m K), and wide bandgap energy (5.5 eV). p-Type conduction is known in both natural and high-temperature high-pressure (HTHP) synthetic boron containing diamond, and such diamond films can be easily made by chemical vapour deposition (CVD) with boron doping. For fabrication of many electronic devices, n-type diamond is required. Nitrogen and phosphorus are the possible donor elements in diamond [1,2].

## B NITROGEN DOPING

Nitrogen is known as the most prevalent impurity in both natural and HTHP synthetic diamond and can be readily incorporated into CVD diamond films using either $N_2$ or $NH_3$ as doping precursors. A concentration as high as 6 × $10^{19}$ atoms/$cm^3$, determined by secondary ion mass spectrometry (SIMS), could be readily achieved with hot-filament CVD using $N_2$ as a doping precursor [3]. The (110) facets always incorporate the highest concentration of nitrogen, followed by (111) facets, whereas (100) facets incorporate the least amount of nitrogen [3]. However, nitrogen in diamond always acts as a deep donor; therefore, diamonds containing nitrogen are electrical insulators [4].

The addition of either $N_2$ or $NH_3$ during deposition was found to enhance the growth rate of diamond films [5,6]; the highest growth rate would be achieved with an $N_2/CH_4$ ratio of 0.5% [6]. However, the enhancement of growth rate is orientation dependent [7]. The addition of nitrogen resulted in the greatest enhancement of the growth rate of (100) facets, followed by (111) facets, with the least enhancement in (110) facets. As a result, with the addition of nitrogen, (100) facets have the highest growth rate in comparison with (110) and (111) facets. {100} textured diamond films could be grown with the addition of nitrogen [7,8], which is partly attributed to the highest growth rate of (100) facets [7]. In addition, growth of {100} textured films may be at least partly due to the fact that the preferential incorporation of nitrogen on the (111) facets resulted in twinning and eventually destruction of the continued growth [7]. The incorporation of nitrogen was also found to have significant influence on the morphology and crystal quality of the resulting diamond films [7].

## C PHOSPHORUS DOPING

Phosphorus has been considered as one of the most promising donor candidates for making n-type semiconducting diamond films. Modelling, using the plane wave pseudopotential method, shows that phosphorus would be a shallow donor in diamond, at a level of 0.2 eV relative to the bottom of the conduction band [2]. However, phosphorus has a large positive formation energy (10.4 eV) and, thus, a very small equilibrium solubility in diamond due, at least partly, to the large radius of the phosphorus atom (1.10 Å), as compared to the radius of the carbon atom (0.77 Å), and to the rigidity of the diamond lattice.

Early work demonstrated that doping of phosphorus into diamond was very difficult and typically only a very low doping concentration could be achieved. A reasonable doping concentration could be achieved only when other impurities, such as boron, were present [9]. Due to the donor-acceptor compensation, no n-type conduction was observed. It was also reported that for phosphorus doping using phosphine as a precursor, a large amount of hydrogen was also incorporated into the diamond [10,11]. It is considered that the hydrogen atoms may passivate the donor P atoms in the diamond lattice and, thus, no n-type electrical conduction was found.

A breakthrough in the growth of n-type conducting diamond film was made only recently and Hall-effect measurements confirmed the n-type conduction in P-doped diamond thin films grown epitaxially using triethylphosphine, $P(C_2H_5)_3$, as a dopant source [12–15]. A lower methane concentration (~0.1%), a lower P/C ratio of $5 \times 10^{-3}$, and a higher substrate temperature (~950°C) than usual were required to obtain higher crystalline perfection for the P-doped diamond thin films. The hydrogen incorporation stimulated by P doping has not been observed. A lower RT electrical conductivity of 0.33 S $cm^{-1}$ was found with an activation energy of 0.12 eV [12]. The Hall mobility and concentration of the activated carriers at about 500 K were 38 $cm^2/V$ s and $3.8 \times 10^{16}$ atoms/$cm^3$, respectively, whereas the phosphorus doping concentration was $1 \times 10^{20}$ atoms/$cm^3$ [13].

It was also found that the incorporation of phosphorus exhibited an appreciable influence on the growth behaviour of the diamond films. Similar to nitrogen, the doping of phosphorus would result in an increase in the growth rate which would reach a maximum at a $PH_3/CH_4$ ratio of approximately 300 ppm [6]. A high concentration would reduce the growth rate and may even prevent the growth of epitaxial film. For example, with an atomic ratio of P:C = 1:10 in the gas phase, only amorphous carbon layers were grown on the (100) facets [16]. Introduction of phosphine into the CVD system would also result in significant changes in the morphology and the quality of the resulting diamond films [6,17,18].

## D NITROGEN AND PHOSPHORUS CO-DOPING

Nitrogen and phosphorus co-doping was proposed to achieve a sufficient doping level of phosphorus in epitaxial diamond films [3,16]. It was speculated that nitrogen atoms have a covalent radius of 0.70 Å, smaller than carbon (0.77 Å), and therefore nitrogen would be a viable co-doping element to promote the incorporation of relatively large phosphorus (1.10 Å) into the diamond lattice. Both solid compound, $NH_4H_2PO_4$, and gaseous $PH_3$, $NH_3$ and $N_2$ precursors were used. Epitaxial diamond films were grown by hot filament CVD.

High doping concentrations of nitrogen and phosphorus were achieved with both solid and gaseous precursors [16]. It was found that the concentration of nitrogen is always approximately one order of magnitude higher than that of phosphorus, while the doping level of phosphorus has been appreciably enhanced. The highest concentration of phosphorus incorporated into diamond films was $3 \times 10^{19}$ atoms/$cm^3$, whereas the nitrogen concentration was $4 \times 10^{19}$ atoms/$cm^3$. With co-doping, it was found that the phosphorus would be preferentially incorporated into the (111) facets as compared to both (110) and (100) facets, which is different from the case for nitrogen. However, no n-type conduction was found in these samples with high doping concentrations of nitrogen and phosphorus. The enhanced incorporation of phosphorus by the co-doping of nitrogen into the diamond lattice was well explained by cluster models and the atom superposition and electron delocalization molecular orbital theory calculations [19]. The theory shows that substitutional N compensates substitutional P, creating stable $P^+$–$N^-$ pairs. These ionized systems are calculated to be deep donors, which explains the absence of n-type conduction.

Co-doping of nitrogen was also found to reduce the influence of phosphorus on the growth rate of diamond films [16]. In addition, the incorporation of nitrogen and phosphorus reduced the tensile stress in CVD epitaxial diamond films, which is commonly observed. This reduction in tensile stress may be attributable to the fact that a phosphorus atom occupies a much larger space than a carbon atom in the diamond lattice.

## E CONCLUSION

Although doping of phosphorus and nitrogen was initially aimed at the development of n-type diamond films for potential applications in electronic devices, the research resulted in the discovery of the influences of nitrogen and phosphorus doping on the growth behaviour and morphologies of CVD diamond films. In particular, the doping of nitrogen was found to enhance the growth rate and resulted in the formation of (100) textured diamond films. The recent confirmation of n-type conduction in phosphorus doped diamond films may reactivate the research efforts in the development of high temperature semiconducting diamond devices.

## REFERENCES

[1] S.A. Kajihara, A. Antonelli, J. Bernholc, R. Car [ *Phys. Rev. Lett. (USA)* vol.66 (1991) p.2010-3 ]

[2] M.E. Zvanut, W.E. Carlos, J.A. Freitas Jr., K.D. Jamison, R.P. Hellmer [ *Appl. Phys. Lett. (USA)* vol.65 (1994) p.2287-9 ]

[3] G.Z. Cao, L.J. Giling, P.F.A. Alkemade [ *Diam. Relat. Mater. (Switzerland)* vol.4 (1995) p.775-9 ]

[4] R.G. Farrer [ *Solid State Commun. (USA)* vol.7 (1969) p.685 ]

[5] G.Z. Cao, W.J.P. van Enckevort, L.J. Giling, R.C.M. de Kruif [ *Appl. Phys. Lett. (USA)* vol.66 (1995) p.688-90 ]

[6] R. Haubner, S. Bohr, B. Lux [ *Diam. Relat. Mater. (Switzerland)* vol.8 (1999) p.171-8 ]

[7] G.Z. Cao, J.J. Schermer, W.J.P. van Enckevort, W.A.L.M. Elst, L.J. Giling [ *J. Appl. Phys. (USA)* vol.79 (1996) p.1357-64 ]

[8] S. Jin, T.D. Moustakas [ *Appl. Phys. Lett. (USA)* vol.65 (1994) p.403-5 ]

[9] S.N. Schauer, J.R. Flemish, R. Wittstruck, M.I. Landstrass, M.A. Plano [ *Appl. Phys. Lett. (USA)* vol.64 (1994) p.1094-6 ]

[10] S. Koizumi et al [ *Diam. Relat. Mater. (Switzerland)* vol.7 (1998) p.540-4 ]

[11] S. Koizumi, M. Kamo, Y. Sato, H. Ozaki, T. Inuzuka [ *Appl. Phys. Lett. (USA)* vol.71 (1997) p.1065-7 ]

[12] T. Nishimori, J. Utasumi, H. Sakamoto, Y. Takakuwa, S. Kono [ *Diam. Films Technol. (Japan)* vol.8 (1998) p.323-30 ]

[13] T. Saito et al [ *J. Cryst. Growth (Netherlands)* vol.191 (1998) p.723-33 ]

[14] T. Saito et al [ *Jpn. J. Appl. Phys. (Japan)* vol.37 (1998) p.L543-6 ]

[15] T. Saito, M. Kameta, K. Kusakabe, S. Morooka, H. Maeda, T. Asano [ *Diam. Relat. Mater. (Switzerland)* vol.7 (1998) p.560-4 ]

[16] G.Z. Cao, F.A.J.M. Driessen, G.J. Bauhuis, L.J. Giling, P.F.A. Alkemade [ *J. Appl. Phys. (USA)* vol.78 (1995) p.3125-31 ]

[17] R.S. Tsang, P.W. May, M.N.R. Ashfold, K.N. Rosser [ *Diam. Relat. Mater. (Switzerland)* vol.7 (1998) p.1651-6 ]

[18] M. Shimomura, T. Nishimori, T. Aukawa, Y. Takakuwa, H. Sakamoto, S. Kono [ *J. Appl. Phys. (USA)* vol.85 (1999) p.3931-3 ]

[19] A.B. Anderson, L.N. Kostadinov [ *J. Appl. Phys. (USA)* vol.81 (1997) p.264-7 ]

CHAPTER C1

# OPTICAL AND ELECTRONIC APPLICATIONS

# C1.1 Optical properties of diamond and applications as radiation windows

C. Wild and P. Koidl

January 2000

## A INTRODUCTION

Diamond is an ideal radiation window material in particular for applications involving high power levels and mechanical, thermal or chemical load. Due to its large bandgap (5.5 eV) and the lack of infrared active fundamental vibrational modes, diamond is optically transparent over a large wavelength range. Even at elevated temperatures, diamond remains transparent, since the large bandgap does not allow the formation of free carriers. This Datareview summarizes the properties of chemical vapour deposition (CVD)-diamond relevant to radiation windows applications. Emphasis is put on windows for high-power $CO_2$ lasers and microwave (Gyrotron) tubes. For a more detailed discussion of optical properties and applications the reader is referred to a recent review [1].

## B INFRARED ABSORPTION

Diamond is known for its broadband optical transparency covering the UV, visible, near and far IR. In this range the optical transmission exhibits only minor intrinsic absorption bands arising from two-phonon (1332–2664 $cm^{-1}$) and three-phonon (2665–3994 $cm^{-1}$) transitions. The maximum absorption coefficient amounts to 14 $cm^{-1}$ at 2158 $cm^{-1}$ [2]. This holds true for optical grade CVD-diamond as shown in FIGURE 1.

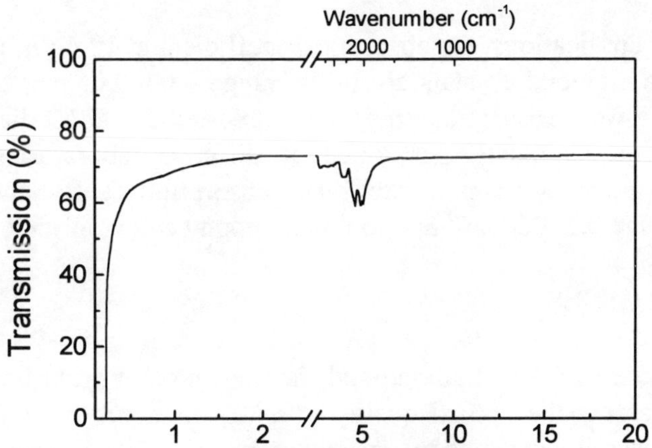


FIGURE 1 Transmission spectrum of a high-quality CVD-diamond window (thickness: 150 µm).

The absorption around 10 µm is of particular interest for $CO_2$-laser components and because many IR sensors operate within the 8–12 µm atmospheric window. A detailed analysis of diamond long-wave IR absorption has been performed by Thomas et al [3]. According to their conclusions, there is an intrinsic absorption floor in the 10 µm spectral range caused by acoustic–acoustic two-phonon interactions. This two-phonon red wing can be approximated by [3–5]:

$$\alpha\left(\text{cm}^{-1}\right)=\frac{146\omega^5}{\omega_{max}^6} \quad \left(\text{for } \omega < \omega_{max}, \quad \omega_{max} = 1143\ \text{cm}^{-1}\right)$$

FIGURE 2 shows the absorption coefficient around 10 μm for two CVD-diamond samples and a type IIa natural diamond crystal. The smooth curve represents the theoretical absorption coefficient. Laser calorimetry data are shown as open symbols.

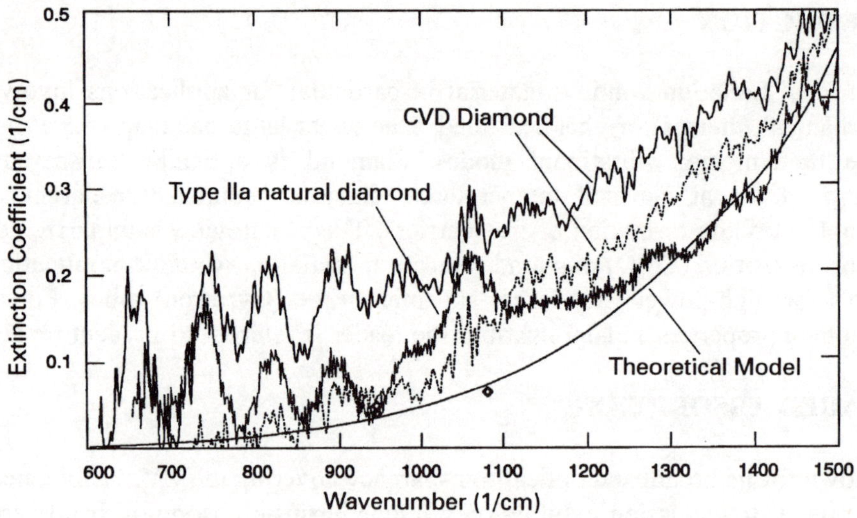


FIGURE 2 Infrared absorption of natural type IIa diamond and two high-quality CVD-diamond samples [3,4].

For high-power $CO_2$-laser applications, the absorption coefficient at 10.6 μm is of central importance. Typical values of type IIa diamond crystals are in the range 0.03–0.05 $cm^{-1}$, as determined by laser calorimetry [2,6–8]. The lowest absorption coefficient reported for CVD-diamond is comparable to these values [8]. In [9], measurements performed by different laboratories on three high-quality samples from various suppliers are summarized. The absorption coefficients at room temperature were found to be in the range 0.1–0.3 $cm^{-1}$ and to double upon heating to ~500°C.

## C SCATTERING LOSSES

The polycrystalline structure of CVD-diamond and the concomitant grain boundaries raise the issue of optical scattering. To assess the optical quality of CVD-diamond, the bidirectional transmission distribution function has been investigated by various groups [1,2,10]. For transparent polished CVD-diamond windows, integrated scattering intensities (integrated from 2° to 80°) ranging from 0.79% to above 50% were observed [1].

Scattering in the infrared is usually an order of magnitude lower [11]. Values reported in the literature are 1% [1,12] and 0.15% [13] (10.6 μm, CVD-diamond, 2.5°–70°). No change in the optical scattering was found upon heating to 500°C [2,9,10]. Type IIa single crystals are roughly an order of magnitude lower in scatter than the best CVD-diamond samples [10,12].

## D THERMO-OPTICAL AND THERMO-MECHANICAL PROPERTIES

The performance of CVD-diamond IR windows at high temperatures or high optical power densities can be modelled provided that basic material parameters and the temperature dependence are known. In FIGURE 3 the thermal conductivity, the thermal expansion coefficient, the thermal coefficient of the refractive index, and the variation of the optical thickness are plotted as functions of temperature [1]. The data can be approximated by the following equations:

$$\kappa(T) = 5.62 + 16.295 \cdot \exp(-T/204.9) \qquad (T = 0, ..., 500°C)$$
$$\alpha(T) = 8.19E{-}7 + 1.107E{-}8 \cdot T - 1.48E{-}11 \cdot T^2 + 1.08E{-}14 \cdot T^3 \qquad (T = 20, ...,500°C)$$
$$dn/dT = 7.80E{-}6 + 9.80E{-}8 \cdot T - 9.84E{-}11 \cdot T^2 + 4.74E{-}14 \cdot T^3 \qquad (T = 20, ...,700°C)$$
$$(nd(T) - nd(25°C))/nd = -8.1E{-}5 + 3.02E{-}6 \cdot T + 2.91E{-}8 \cdot T^2 - 2.29E{-}11 \cdot T^3 + 9.52E{-}15 \cdot T^4 \qquad (T = 25, ...,800°C)$$

where the temperature T is in °C, the thermal conductivity $\kappa$ is in W/cm K, the expansion coefficient $\alpha$ is in $K^{-1}$, dn/dT is in $K^{-1}$ and $\Delta nd/nd$ is dimensionless.

## E LASER-INDUCED DAMAGE THRESHOLDS

Laser-induced damage threshold measurements on diamond under various conditions have been reviewed by Klein [14] and Wood [15]. They pointed out that dielectric breakdown is the dominant damage mechanism for highly transparent diamond samples, and demonstrated the applicability of the Bettis–House–Guenther (BHG) scaling law. After scaling to a spot size of 100 μm and a pulse duration of 1 ns, the data for natural and CVD-diamond are in remarkable agreement (10–20 GW/$cm^2$ at 10.6 μm, 6–10 GW/$cm^2$ at 1.06 μm and around 2 GW/$cm^2$ at 532 nm) [1,14].

## F DIELECTRIC PROPERTIES

The application of CVD-diamond as vacuum windows for high-power microwave (Gyrotron) tubes has found considerable interest. These Gyrotron tubes are used for the electron cyclotron heating of fusion plasmas [21]. Power levels exceeding 1 MW at frequencies of around 100 GHz have been demonstrated. Until recently the output window of these devices has been the most critical component limiting the maximum output power or the pulse duration. In this context CVD-diamond windows with water edge-cooling were found to be very promising [22–24]. The extremely high power level requires very low dielectric losses. CVD-diamond exhibits loss tangent values as low as $10^{-5}$ (at 140 GHz) [25–27]. Below 350–400°C there is practically no temperature dependence [26,27]. In the 10–145 GHz range the loss tangent decreases with frequency as 1/f [27] or as $1/f^{0.5}$ [26]. Performance tests of CVD-diamond windows for high-power microwave applications are reported in [28–30].

## G CONCLUSION

Diamond is optically transparent over a large wavelength range, even at high temperatures, and this Datareview covers those properties which are relevant to its application in radiation windows. Transmission and absorption data are presented and scattering losses in CVD-diamond are discussed. Modelling of performance can be done using known temperature dependences of optical and mechanical properties.

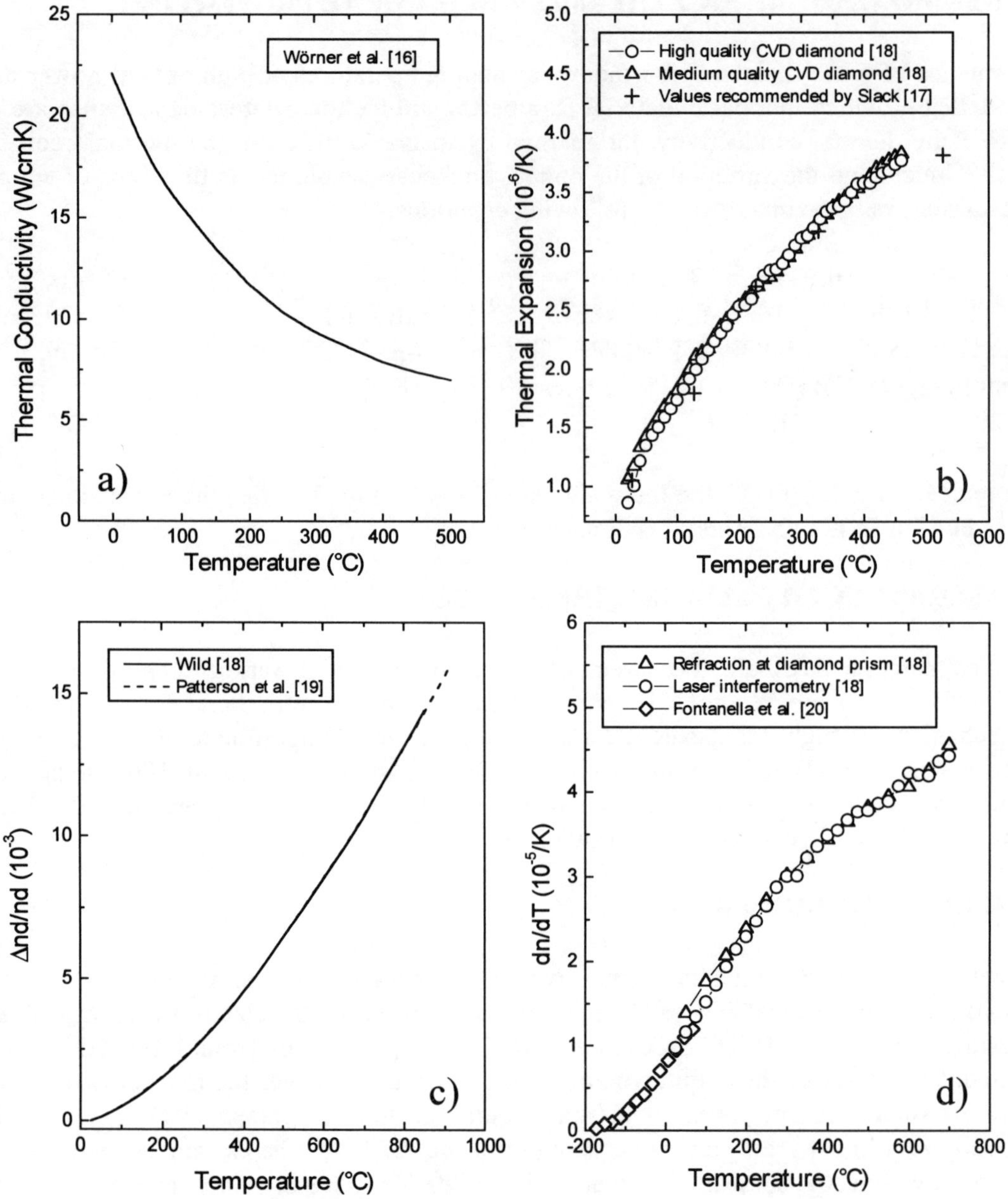


FIGURE 3 The thermo-optical and thermo-mechanical properties of diamond [1]. (a) The thermal conductivity of a CVD-diamond sample. (b) The thermal expansion of two CVD-diamond samples. (c) The variation of the optical thickness nd for natural diamond. (d) The thermal coefficient of the refractive index for natural diamond.

## REFERENCES

[1] C. Wild [ in *Low-Pressure Synthetic Diamond: Manufacturing and Applications* Eds. B. Dischler, C. Wild (Springer, Heidelberg, 1998) p.189-205 ]

[2] D.C. Harris [ *Proc. SPIE (USA)* vol.2286 (1994) p.218 ]

[3] M.E. Thomas, W.J. Tropf, A. Szpak [ *Diam. Films Technol. (Japan)* vol.5 (1995) p.159 ]; M.E. Thomas [ *Proc. SPIE (USA)* vol.2286 (1994) p.152 ]

[4] M.E. Thomas, W.J. Tropf [ *Proc. SPIE (USA)* vol.2286 (1994) p.144 ]

[5] M.E. Thomas, W.J. Tropf [ *Johns Hopkins APL Tech. Dig. (USA)* vol.14 (1993) p.16 ]

[6] S.P. McGeoch, D.R. Gibson, J.A. Savage [ *Proc. SPIE (USA)* vol.1760 (1992) p.122 ]

[7] K. Harris, G.L. Herrit, C.J. Johnson, S.P. Rummel, D.J. Scatena [ *Appl. Opt. (USA)* vol.30 (1991) p.5015 ]

[8] M. Massart, P. Union, G.A. Scarsbrook, R.S. Sussmann, P.F. Muys [ *Proc. SPIE (USA)* vol.2714 (1996) p.177 ]

[9] D.C. Harris [ Naval Air Warfare Ctr. Weapons Division (USA) TP 8210 (1994) ]

[10] D. C. Harris [ in *Applications of Diamond Films and Related Materials: Third International Conference* Eds. A. Feldmann et al, NIST special publication (USA) vol.885 (1995) p.539 ]

[11] C.F. Hickey, J. DeRosa, K.A. Snail [ *Proc. SPIE (USA)* vol.1760 (1992) p.154 ]

[12] K.V. Ravi [ *Proc. SPIE (USA)* vol.2286 (1994) p.174 ]

[13] M.B. McIntosh, J.R. McNeely, R.E. Clausing, W.B. Snyder [ *Proc. SPIE (USA)* vol.1995 (1993) p.246 ]

[14] C.A. Klein [ *Proc. SPIE (USA)* vol.2428 (1995) p.517 ]

[15] R.M. Wood [ *Proc. SPIE (USA)* vol.2428 (1995) p.594 ]

[16] E. Wörner, C. Wild, W. Müller-Sebert, R. Locher, P. Koidl [ *Diam. Relat. Mater. (Switzerland)* vol.5 (1996) p.688 ]

[17] G.A. Slack, S.F. Bartram [ *J. Appl. Phys. (USA)* vol.46 (1975) p.89 ]

[18] C. Wild [ to be published ]

[19] M.J. Patterson, J.L. Margrave, R.H. Hauge, Z. Ball, R. Sauerbrey [ *Electrochem. Soc. Proc. (USA)* vol.95-4 (1995) p.503 ]

[20] J. Fontanella, R.L. Johnston, J.H. Colwell, C. Andeen [ *Appl. Opt. (USA)* vol.16 (1977) p.2949 ]

[21] M.Q. Tran [ *Plasma Phys. Control. Fusion (UK)* vol.40 no.8A (1998) p.53 ]

[22] J.P. Calame, D.K. Abe [ *Proc. IEEE (USA)* vol.87 (1999) p.840 ]

[23] V.V. Alikaev, D.V. Khmara, V.A. Khmara, V.E. Myasnikov, V.V. Parshin [ *ITG-Fachber. (Germany)* vol.150 (1998) p.357 ]

[24] M. Thumm [ *Int. J. Infrared Millim. Waves (USA)* vol.19 (1998) p.3 ]

[25] V. Ralchenko et al [ *Diam. Relat. Mater. (Switzerland)* vol.8 (1999) p.189 ]

[26] R. Spörl, R. Schwab, R. Heidinger, V.V. Parshin [ *ITG-Fachber. (Germany)* vol.150 (1998) p.369 ]

[27] B.M. Garin et al [ *Tech. Phys. Lett. (USA)* vol.25 (1999) p.288 ]

[28] O. Braz et al [ *Int. J. Infrared Millim. Waves (USA)* vol.18 (1997) p.1495 ]

[29] K. Takahashi et al [ *17th IEEE/NPSS Symposium Fusion Engineering* New York, USA (IEEE, 1998) vol.1 of 2 p.441 ]

[30] K. Sakamoto et al [ *Rev. Sci. Instrum. (USA)* vol.70 (1999) p.208 ]

# C1.2 CVD diamond for X-ray lithography masks

H. Windischmann

January 2000

## A INTRODUCTION

The resolution limit of optical lithography, perceived to be about 100 nm, is defined by diffraction and scattering as the feature size approaches the exposing wavelength. X-ray lithography, which uses significantly shorter radiation (0.8–1.5 nm versus 300–400 nm) offers a technical path to achieving the higher resolution. However, several factors (technical and non-technical) have delayed the implementation of X-ray lithography on the production line for IC fabrication.

The major non-technical factor is related to the huge built-in optical technological infrastructure which has continued to make significant improvements in the past two decades to the existing technology, by

- using step and repeat exposure tools,
- incorporating multilevel resist,
- employing contrast enhancers,
- using shorter wavelength radiation,
- designing higher numerical aperture optics,

which has effectively delayed the implementation of X-ray lithography. In 1980 the predicted lower limit for optical lithography was perceived to be 1000 nm, at which point X-ray lithography would dovetail into the production line. Today, feature sizes printed by optical techniques have achieved 350 nm and prospects exist for printing features down to 250 nm.

The technical barriers to X-ray lithography implementation include:

- the absence of a reliable, high volume, low defect density X-ray mask technology,
- a high speed X-ray resist,
- a high speed, low cost exposure/alignment tool.

This Datareview will deal with the X-ray mask technology issue from a materials point of view, concentrating on diamond film properties pertinent to the application.

## B MATERIAL SELECTION CRITERIA

Over the past 30 years a wide variety of materials [1–3] have been evaluated for use as the X-ray mask support membrane. The earliest demonstration that chemical vapour deposition (CVD) diamond membranes several micrometres thick could be deposited in tension over a 75 mm diameter was reported in 1990 [4]. Three key considerations in selecting the optimum material are: optical transparency, X-ray transmission and mechanical stiffness. From an operational point of view, the first two properties affect product throughput while the mechanical stiffness bears on the dimensional stability of the mask which impacts the critical dimension alignment budget. TABLE 1 presents [5] candidate materials based on these three factors, although there are certainly other factors to consider such as surface roughness, flatness, metallization suitability and secondary electron emission, which are processing specific characteristics. Generally, the best mask material has low atomic number since the X-ray transparency improves with decreasing atomic number.

TABLE 1 Comparison of materials based on X-ray transparency $t_x$, optical transparency $t_o$, and mechanical stiffness S.

| Material | $t_x$ (μm) | $t_o$ (μm) | $S = Et_o/(1-\nu)$ (GPa μm) |
|---|---|---|---|
| Polyimide | 8.7 | >5 | 43 |
| Mylar | 8.5 | >5 | 42 |
| Ti | 1.09 | opaque | – |
| Be | 42.0 | opaque | – |
| Si | 9.6 | 2 | 362 |
| $Al_2O_3$ | 0.8 | 1 | 448 |
| $SiO_2$ | 4.99 | 5 | 444 |
| $Si_3N_4$ | 7.74 | 6 | 728 |
| BN | 7.94 | 4 | 420 |
| SiC | 6.81 | 3 | 1590 |
| Diamond | 5.9 | 2 | 2336 |

The second column presents the calculated thickness $t_x$, to achieve 50% transmission at an X-ray wavelength of 0.693 nm. The third column gives the membrane thickness $t_o$, required to achieve 50% transmission in the visible, at the He-Ne laser line. The last column presents the mechanical stiffness factor, S, which is the product of the biaxial Young's modulus, $E/(1-\nu)$, and the membrane thickness (using the smaller of $t_x$ or $t_o$). The stiffness factor S represents the spring constant of the membrane and is a measure of the mechanical rigidity of the membrane. This quantity defines dimensional stability and should be as high as possible, within the transparency constraints. The argument here is that the absorber-supporting membrane should be mechanically stiff in relation to the absorber to minimise in-plane and out-of-plane distortion and, therefore, minimise overlay error.

TABLE 1 reveals the weakness of polymers as membrane material candidates. They are not only hygroscopic but are mechanically soft and therefore easily distorted. The metals Ti and Be are moderately stiff; however, their opacity is troublesome, but not pathological, since alignment windows can be etched in the membrane after overcoating with polyimide to support the alignment pattern. Beryllium would be excellent were it transparent, discounting its toxicity. Si and its nitride and oxide are good from an X-ray and optical transparency viewpoint but lack the mechanical stiffness of the refractories like SiC, BN and diamond. Si has the significant advantage of a large installed technology base and capital equipment availability.

As can be seen, diamond has the highest stiffness factor S of any material. Of course, there are other factors to consider in selecting a material such as: scale-up of fabrication process, X-ray-induced degradation, surface smoothness, flatness, secondary electron emission induced by the X-rays, adhesion of metallization. Diamond's low mass absorption coefficient and low density make it compatible with a variety of X-ray sources. TABLE 2 presents the source material and its corresponding wavelength, the mass absorption coefficient (MAC) and the calculated X-ray transmission for a 2 μm thick diamond membrane. As shown, the X-ray transmission is greater than 60% for an Al radiation source and increases to 99% for copper radiation.

The rotating anodes, commonly used in the early days of mask development as the X-ray 'light bulb', have been displaced by the synchrotron X-ray source with capability for multiple exposure stations that provide an economy of scale. The investment in a synchrotron source, however, can be substantial and may limit the number of participants in this technology.

TABLE 2 Transmission of a 2 μm thick CVD diamond membrane for various rotating X-ray sources.

| Source | Wavelength (nm) | MAC ($cm^2/g$) | X-ray transmission (%) |
|---|---|---|---|
| Cu | 0.154 | 4.24 | 99.5 |
| Ag | 0.415 | 89.9 | 93.9 |
| Pd | 0.437 | 109.0 | 92.6 |
| Mo | 0.541 | 213.0 | 86.1 |
| W | 0.698 | 337.0 | 79.0 |
| Si | 0.712 | 457.0 | 72.6 |
| Al | 0.834 | 719.0 | 60.4 |

The diamond film is usually deposited 1–3 μm thick on a silicon wafer, which is then back-etched [1] over a selected area to create a free-standing membrane supported at the periphery by a silicon frame. The X-ray absorber film is deposited by sputtering either before or after back-etching and patterned using an electron beam lithography to print features below 100 nm. CVD diamond films are deposited primarily by two techniques [4]: hot filament and microwave plasma CVD. The critical process variables which affect the morphological, mechanical and optical properties of the diamond film are the hydrocarbon (usually methane) fraction in the hydrogen gas supply and deposition temperature.

## C PROPERTIES

To align the pattern on an X-ray mask with that of an existing pattern on the coated silicon wafer requires a minimum transparency. The theoretical transparency of diamond in the visible part of the spectrum is 71%, due to 29% reflective losses from both surfaces. Anti-reflective coating may be required to increase the transparency. FIGURE 1 presents the near IR-visible-UV transmission spectrum of nominally 1 μm thick diamond membranes prepared by microwave plasma CVD [5] as a function of methane fraction, from 0.3% to 3%.

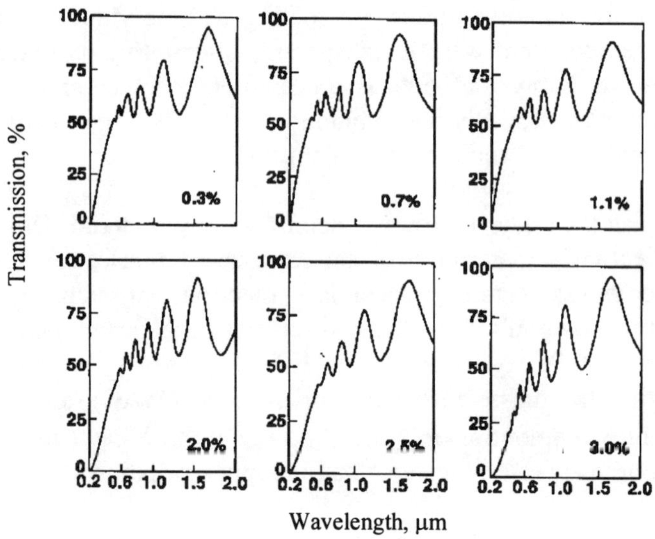


FIGURE 1 Transmission of 1 μm thick CVD diamond film deposited at various methane fractions.

The arithmetic mean transmission between 500 and 1000 nm is 55%, which is below the theoretical value due to absorption by the non-diamond phase, which increases with methane fraction, and grain-boundary scattering, which decreases with methane fraction. The loss due to scattering is evidenced by the monotonic, smooth decrease in the transmission and amplitude of the interference pattern as one approaches the blue part of the spectrum. At least 50% transmission is desirable to use optical alignment techniques based on phase shift technology or interferometry. The loss in transparency with increasing methane is accompanied by:

- increased deposition rate,
- reduced surface roughness,
- loss in mechanical stiffness.

To achieve critical dimensional control at the 100 nm scale requires surface roughness of the order of 5–10 nm. The natural tendency of CVD diamond to form a faceted, columnar grain structure, which has a variety of crystallographic orientations at the free surface, is problematic and possibly a limiting factor to its utility, unless a planarization process is used. The roughness scales with film thickness and deposition conditions. FIGURE 2 presents the variation of the free surface RMS roughness, as a function of methane fraction for a 2 μm thick film prepared by MWCVD. While the roughness decreases with increased methane fraction in the process gas, this comes at the expense of the optical transparency. Reducing the diamond film thickness is an option but with the undesirable consequence of increased fragility and in-plane distortion, which leads to overlay error.

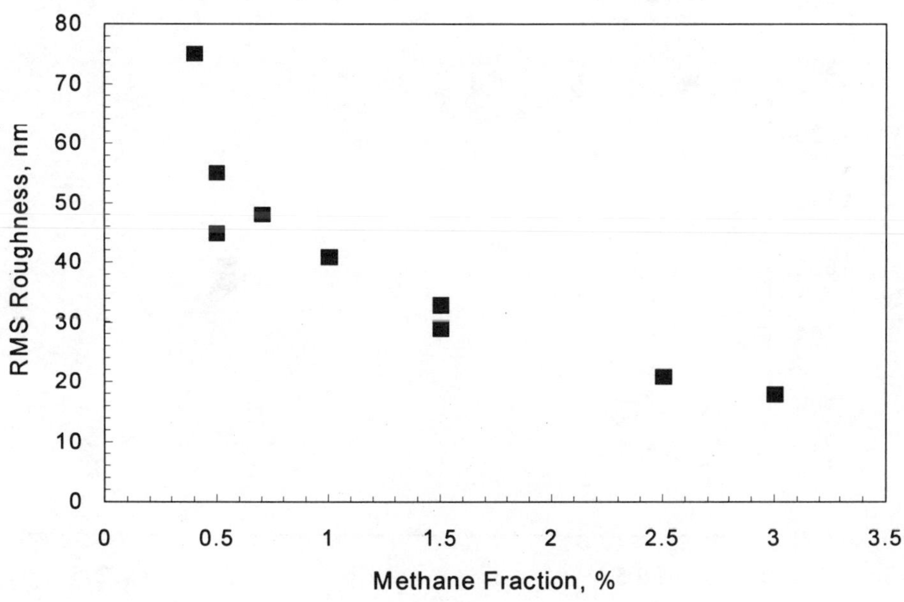


FIGURE 2 RMS roughness variation with methane fraction for a 2 μm thick film.

Several approaches have been investigated as smoothing agents such as reactive plasma etching in an oxygen:fluorine plasma or planarizing with a polymer over-coating. The difficulty is that the plasma may preferentially attack the grain boundaries and possibly increase the rejection rate due to defect generation. Alternatively, the nucleation surface, rather than the free growth surface, may be a more suitable choice for patterning since it replicates the 5–10 nm RMS roughness of the silicon substrate.

Distortion is an important [7] issue as it effects overlay error. The printed pattern distortion consists of in-plane and out-of-plane components which result from intrinsic and/or thermal stresses generated during sputtering, evaporation or the electroplating of the absorber on top of the diamond membrane. In addition, there are residual stresses in the diamond membrane, which are necessary to maintain flatness, but if excessive can distort the silicon support wafer frame or, in the extreme, exceed the fracture strength of the diamond which is typically 1–3 GPa for films 1–2 μm thick [8]. Comprehensive stress studies of CVD diamond films [9] show the stress can be modulated in type and magnitude by two primary process variables: methane fraction and temperature. Distortion can be greatly reduced by using a membrane whose elastic modulus is much larger than that of the absorber material. Diamond meets this condition even for the stiffest refractory metal, tungsten. Measurement of distortions [7] on a patterned diamond membrane versus silicon membrane confirms model calculations, which predict an inverse relation between distortion and elastic modulus.

The elastic modulus of CVD diamond is sensitive to deposition conditions, and can approach the value of the metal at high methane fraction in the supply gas. The variation of the biaxial elastic modulus with methane fraction is shown in FIGURE 3. At low methane fraction, film with the highest purity is deposited. As a result the biaxial modulus is nearly identical to the bulk value. The modulus decreases monotonically with methane fraction as the theoretical carbon coordination number is reduced below 4 with the incorporation of 3-fold coordinated $sp^2$ carbon and atomic hydrogen in the lattice.

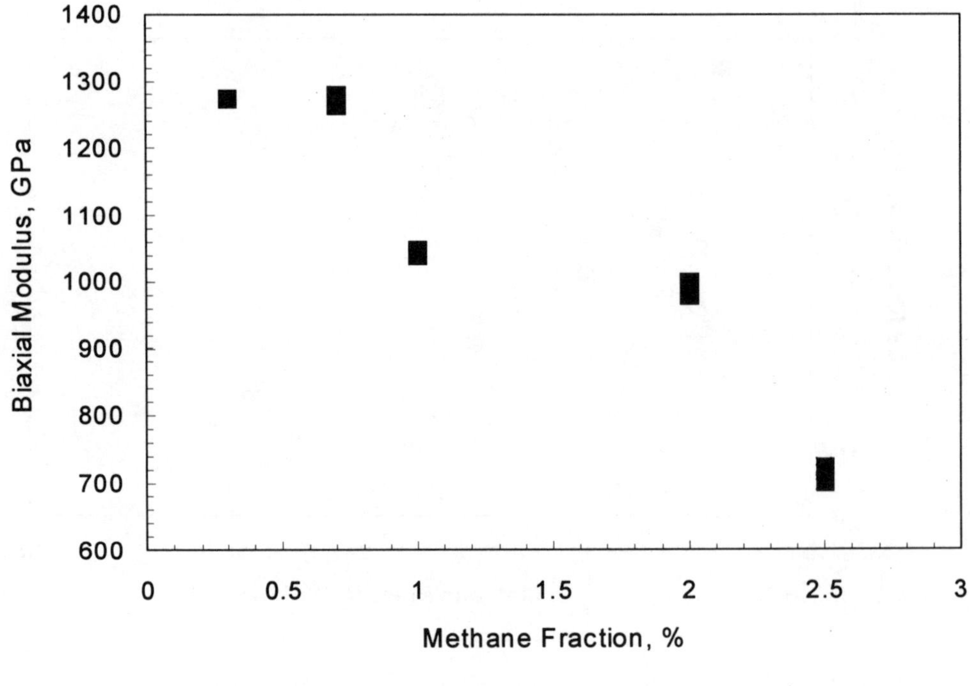


FIGURE 3 Decrease in biaxial modulus variation with methane fraction.

The fracture strength of brittle materials is not a unique quantity but depends on a variety of factors such as: flaw size and its distribution, film thickness, whether the nucleation or growth surface is placed in tension, the sample size, the fracture test method, and deposition conditions. The ultimate strength is limited by the largest flaw. The theoretical fracture strength of diamond is 200 GPa, as calculated from the Orowan relationship [10]. However, the measured strength is significantly below that value due to structural defects, such as voids at grain boundaries and impurities, incorporated in the film. The fracture strength of CVD diamond films 1–3 μm thick, prepared as a function of methane fraction, has been measured [6]. Values as high as 2–4000 MPa were observed for 1–2 μm thick films prepared at 0.5% methane fraction. This provides a sufficient margin of safety in operation, since the membrane flatness requirement for patterning requires 50–100 MPa tension as a practical minimum.

## D CONCLUSION

CVD diamond has the required mechanical, optical and X-ray properties, in addition to mechanical stiffness and radiation hardness, to meet the stringent requirements of X-ray mask technology better than any other material. However, the outstanding issue remains volume production using specialized equipment and defect control to achieve high yield and the ancillary technology of sensitive X-ray resist, alignment tool and economical X-ray source. The outstanding non-technical issue which has delayed its adoption on the production line is optical lithography's installed base and the existence of complementary technology, e-beam, which may relegate X-ray lithography to high value added, niche markets.

## REFERENCES


[1] M.F. Ravet, F. Rousseaux [ *Diam. Relat. Mater. (Switzerland)* vol.5 (1995) p.812-8 ]

[2] A.R. Shikunas [ *Solid State Technol. (USA)* (1984) p.192-9 ]

[3] W.D. Buckley, J.F. Nester, H. Windischmann [ *J. Electrochem. Soc. (USA)* vol.128 (1981) p.1117-20 ]

[4] H. Windischmann [ *Proc. SPIE (USA)* vol.1263 (1990) p.241-50 ]

[5] H. Windischmann, G.F. Epps [ *J. Appl. Phys. (USA)* vol.68 (1990) p.5665-73 ]

[6] M.A. Prelas, G. Popovici, L.K. Bigelow [ *Handbook of Industrial Diamonds and Diamond Films* (Marcel-Dekker, New York, 1998) ]

[7] N.I. Maluf, R.F. Pease, H. Windischmann [ *J. Vac. Sci. Technol. B (USA)* vol.8 no.6 (1990) p.1584-8 ]

[8] H. Windischmann, G.F. Epps [ *Diam. Relat. Mater. (Netherlands)* vol.1 (1992) p.656-64 ]

[9] H. Windischmann, G.F. Epps [ *J. Appl. Phys. (USA)* vol.69 (1991) p.2231-7 ]

[10] J.E. Field [ *The Properties of Natural and Synthetic Diamond* (Academic Press, London, 1992) p.478 ]

# C1.3 Applications of diamond in electron beam devices

V.V. Zhirnov and J.J. Hren

March 2000

## A INTRODUCTION AND DEFINITIONS

### A1 Diamond and Diamond Materials

There is a diversity of carbon-based materials that are referred to as 'diamond' or 'diamond material'. As a very general classification, for purposes of application to electron emission, all carbon can be subdivided into two categories: non-conductive ('transparent carbon or wide bandgap carbon materials', such as natural diamond, diamond-like carbon, etc.), and conductive ('black carbon materials', such as graphite, nanotubes, etc.). In this Datareview we reserve the term 'diamond' to describe the entire class of non-conductive (or poorly conducting) materials. Categories of 'diamond materials' used in this review are:

- Macroscopic single crystals (natural and high pressure synthetic or HPHT), which are further categorized as Ia, Ib, IIa, and IIb [1].
- Chemical vapour deposited (CVD) carbon films [1] comprised of: polycrystalline diamond (pc-D), amorphous diamond (a-D), diamond-like carbon (DLC), etc.
- Diamond powder: HPHT micrometre or submicrometre powders 0.05–5 μm in size and explosion-type nanodiamond or ultradispersed diamond (UDD) 1–10 nm in size [2].

### A2 Electron Affinity of Diamond

Electron affinity is defined as the energy difference between the conduction band minimum and the vacuum level. Diamond has a low electron affinity (LEA) that is typical for wide bandgap materials. A diamond surface terminated by monolayer coatings of hydrogen [3–5], as well as some metals (e.g. Cs, Ti, Ni, Co, Zr, and Cu [4,5]), have a negative electron affinity or NEA. (Also see TABLE 1). The NEA plays an important role for high photoelectron and secondary electron yields from diamond surfaces [3–5].

TABLE 1 Measured electron affinity of various diamond surfaces.

| Surface | Electron affinity (eV) | Ref |
|---|---|---|
| B-doped C(111):H | –0.9 | [4] |
| B-doped C(100):H | –0.5 | [4] |
| N-doped C(100):H | –0.2 (at 400°C) | [4] |
| C(100) (clean) | 0.65 | [5] |
| C(100):O | 1.50 | [5] |
| UDD:H | –0.2 | [2] |

### A3 Diamond in Electron Beam Devices (EBD)

In this Datareview we use the term EBD to describe all devices using electron beams that contain functional parts made of diamond. These encompass:

- Diamond source electron beam devices (FIGURE 1)

- Diamond target electron beam devices (FIGURE 1)
- Diamond-based passive components of electron beam devices.

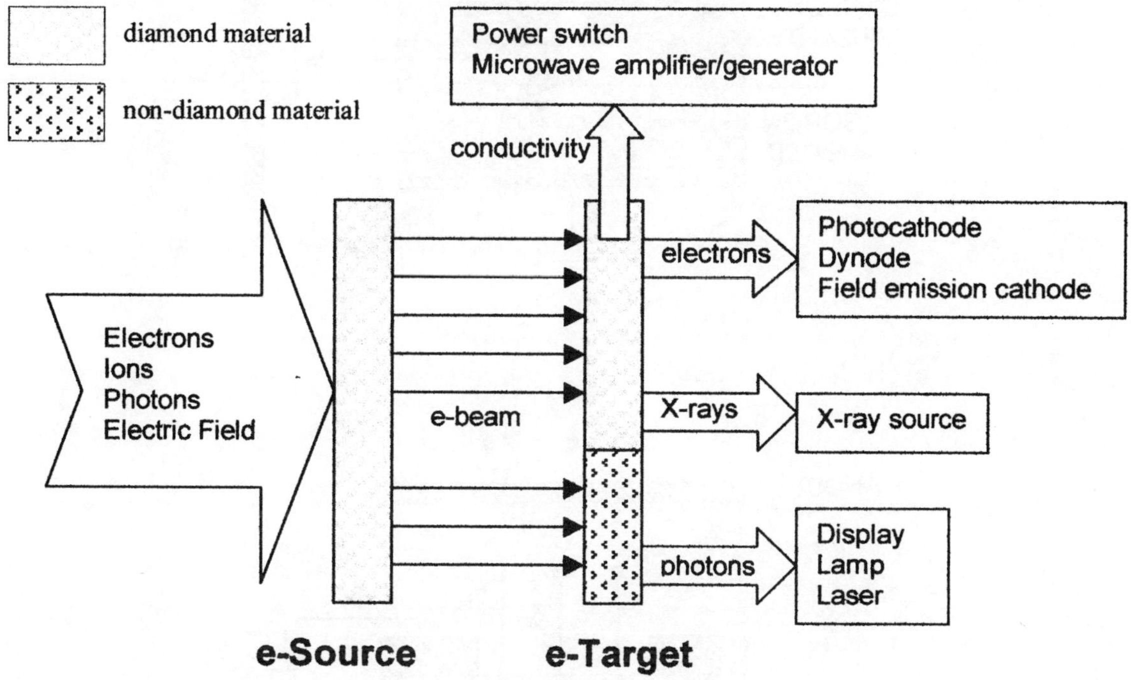


FIGURE 1 Diamond electron beam devices.

## B DIAMOND ELECTRON SOURCES

### B1 Diamond Junction Cathodes

In junction cathodes, a biased Schottky barrier, p-n junction or semiconductor heterojunction is used to create a high population of electrons in the conduction band of diamond (FIGURE 2). If the diamond surface has a low, or negative, electron affinity, the injected electrons escape into vacuum with a minimal applied field. The cathode efficiency $\gamma$ is defined as the ratio of the emission current $I_e$ to the total current passing through the junction $I_D$; that is: $\gamma = I_e/I_D$. A summary of some emission characteristics of several diamond junction cathodes is given in TABLE 2.

TABLE 2 Diamond junction cathodes.

| Ref | Year | $I_e$ max | $\gamma$ (at $I_e$ max) | Material |
|---|---|---|---|---|
| [6] | 1991 | 200 nA | 0.002% | IIb, C-doped |
| [7] | 1995 | 0.01 pA | $10^{-6}$% | Ib, P, B-doped |
| [8] | 1998 | 20 pA | 0.1% | CVD, polycryst. |

### B2 Diamond Field Emission Cathodes

Three approaches to fabricating diamond field emitters have been employed. The first approach assumed that only the material properties (e.g. NEA) of the diamond are the key to controlling the emissivity. This approach has resulted in the attempt to fabricate 'planar' cold cathodes, some of whose properties are summarized in FIGURE 3(a), and TABLE 3.

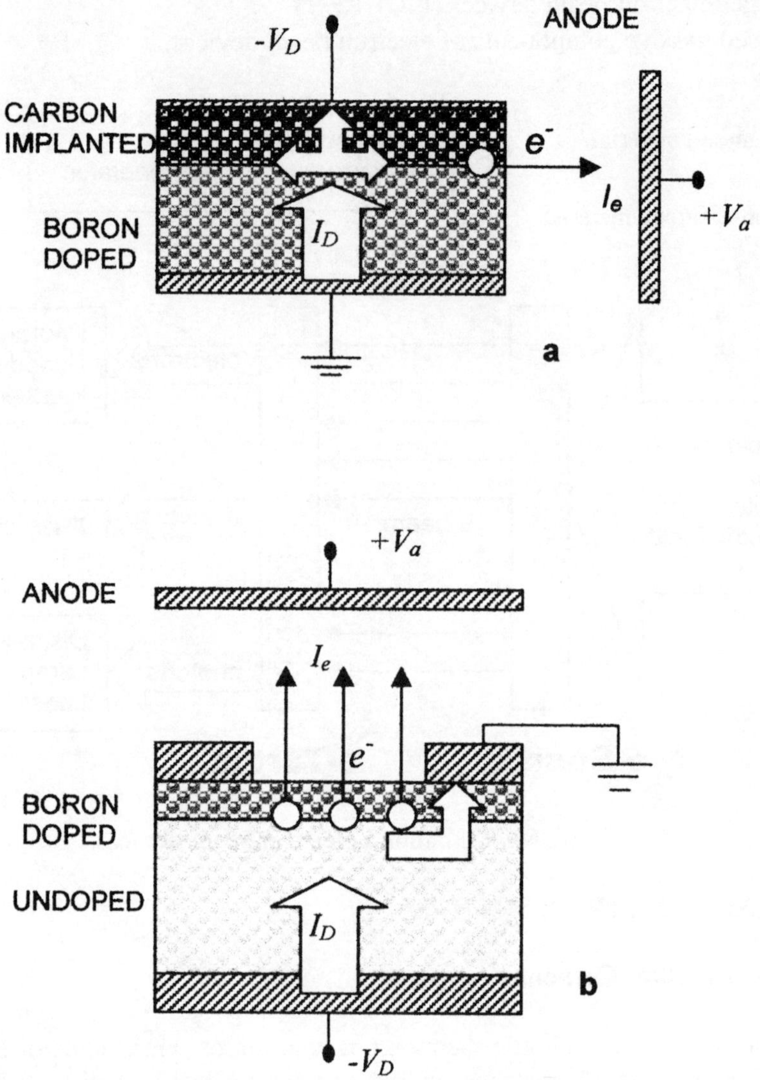


FIGURE 2 Diamond junction cathodes: (a) p-n junction, junction parallel emission from the junction region [6]; (b) MIS structure, junction normal emission through diamond film [8].

TABLE 3 Planar field emission diamond cathodes.

| Ref | Year | Material | Threshold point | | Maximum point | |
|---|---|---|---|---|---|---|
| | | | $J_{th}$, mA/cm$^2$ ($I_{th}$, mA) | $F_{th}$, V/μm ($V_{th}$, V) | $J_{max}$, mA/cm$^2$ ($I_{max}$, mA) | $F_{max}$, V/μm ($V_{max}$, V) |
| [9] | 1993 | Laser-ablated a-D | $10^{-1}$ | 20 | 100 | >40 |
| [10] | 1995 | CVD pc-D | $10^{-2}$ | 15 | 10 | 58 |
| [11] | 1996 | HPHT | $10^{-4}$ | 7 | 1000 | 35 |
| [12] | 1996 | CVD pc-D:N | $10^{-4}$ | 0.5 | $2 \times 10^{-2}$ | 2 |
| [13] | 1998 | CVD | $10^{-1}$ | 6 | 100 pulse | 10 |
| [14] | 1999 | CVD pc-D | ($10^{-2}$ mA) | (0.25 V) | (3.58 mA) | (7 V) |
| [15] | 2000 | CVD | – | – | 500 (500 mA) pulse | 10 (1000 V) |

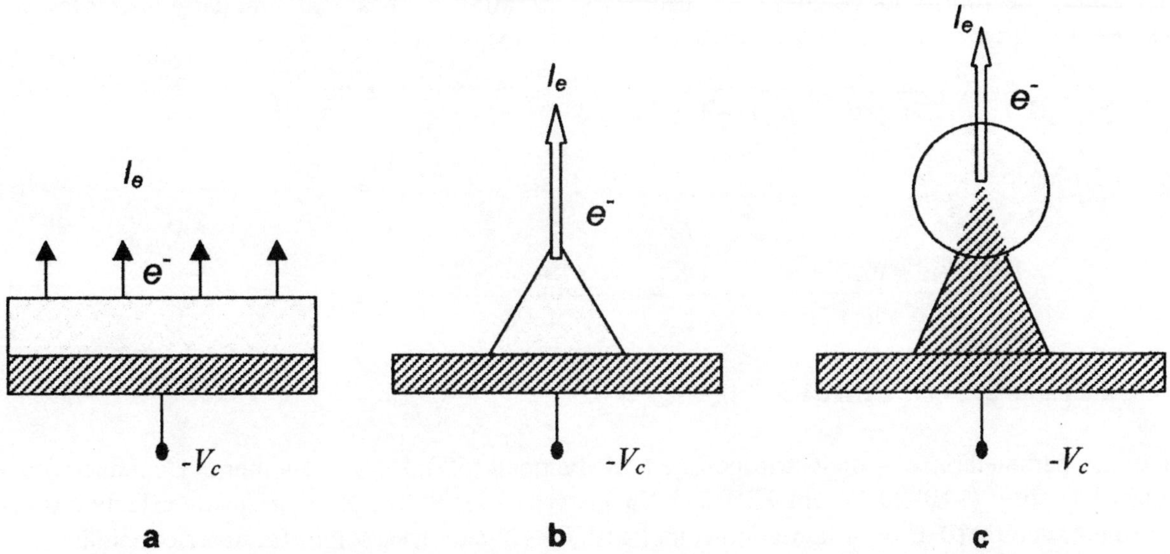


FIGURE 3 Three basic designs of diamond field emitter: (a) planar emitter; (b) pointed 'full-diamond' emitter; (c) pointed composite emitter.

The second approach to fabricating diamond field emitters is based on shaping pure diamond cathodes into cones and pyramids (see FIGURE 3(b)) by various methods, including the transfer mould technique, CVD growth, ion milling by laser ablation, etc. This approach assumes that the well known concepts of field emission from needles of conducting materials also applies directly to diamond materials. Indeed emission at very low applied voltages is observed, but only very small currents have been achieved so far from such emitters. Recently, Chubun [16] and Kang et al [17] have reported low-voltage operation of gated diamond field emission arrays with good emission stability as well (TABLE 4).

TABLE 4 Gated diamond field emission arrays.

| Ref | Year | Material | Design | Total current (current density) | Voltage (field) |
|---|---|---|---|---|---|
| [16] | 1998 | DLC cones, $10^3$–$10^6$ Ω/sq | 961 (31 × 31) tips (0.001 $cm^2$) | 3 μA (3 mA/$cm^2$) DC, 1 mA (1A/$cm^2$) pulse | 50–150 V (25–75 V/μm) |
| [17] | 1999 | B-doped ($10^{19}$ $cm^{-3}$) diamond pyramids | 2 × 2 tips (4 × $10^{-6}$ $cm^2$) | 4 μA | 1–5 V (1–5 V/μm) |

The third approach employs pointed composite cathodes. This approach attempts to combine the most attractive features of conductive microtips (e.g. high electron supply and maximum field enhancement) with the most attractive properties of diamond (e.g. high thermal conductivity, physical robustness, chemical inertness, and NEA). The resulting emitters consist of a conductive pointed core (or substrate) with a thin coating of diamond (FIGURE 3(c)). Examples of such pointed composite emitters with diamond coatings are shown in FIGURE 4. The effect of coatings on emission parameters of such microtips can be characterized by several improvement factors, such as the maximum current, threshold voltage and emission stability [18]. An example of the I–V characteristics of a Si tip before and after such diamond coating is shown in FIGURE 5. Note that the effects of the diamond coating on the emission properties also depend upon coating thickness. For example, the maximum current ($I_{max}$) linearly increases with thickness, as shown in FIGURE 6. On the other hand, the threshold voltage ($V_{th}$) decreases for thin coatings, but gradually increases with thickness for thicker coating. These dependences can be expressed as

$$I_{max}(D) = C_1\sqrt{\lambda_{Me}}\, r(C_2\sqrt{\lambda_D}D + 1)$$

and

$$V_{th} = \frac{B\beta^{-1}\varepsilon(m^*/m_0)^{-1/2}\varphi_B^{-3/2} + K_1}{K_2 + \ln(I_{th}^{-1}A\beta^2 r^2)} + \frac{(I/r^2 + ev_s N_t)D^2}{2v_s\varepsilon\varepsilon_0}$$

where the parameters are defined as follows:

Materials parameters: $\varepsilon$ – dielectric constant of diamond (5.7); $\lambda_D$, $\lambda_{Me}$, – thermal conductivity of diamond coating (~10–20 W cm $K^{-1}$) and tip material (~1 W cm $K^{-1}$); $m^*/m_0$ – effective-to-free electron mass ratio (0.2); $v_s$ – saturation velocity (2.7 × $10^7$ cm/s); $\varphi_B$ – interface barrier height.

Cathode design parameters: D – thickness of diamond coating, cm; r – tip radius, cm; $\beta$ – field enhancement factor ($\beta$~1/5r for many practical cases).

Fowler–Nordheim constants: $A = 1.54 \times 10^{-6}$ A eV $V^{-2}$, $B = 6.83 \times 10^7$ $eV^{-3/2}$ V $cm^{-1}$.

Working or approximation constants: $C_1$, $C_2$ – constants, dependent on material and geometry of the tips, for Si emitters the experimentally determined values are $C_1 = 2$, $C_2 = 2 \times 10^5$; $K_1 = 300\ln10 \approx 691$; $K_2 = 6\ln10 \approx 13.82$.

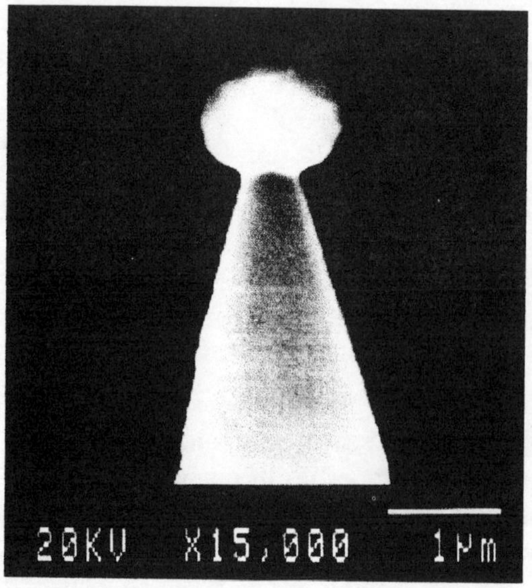


FIGURE 4 Pointed composite emitters: Si tips with diamond coatings [45].

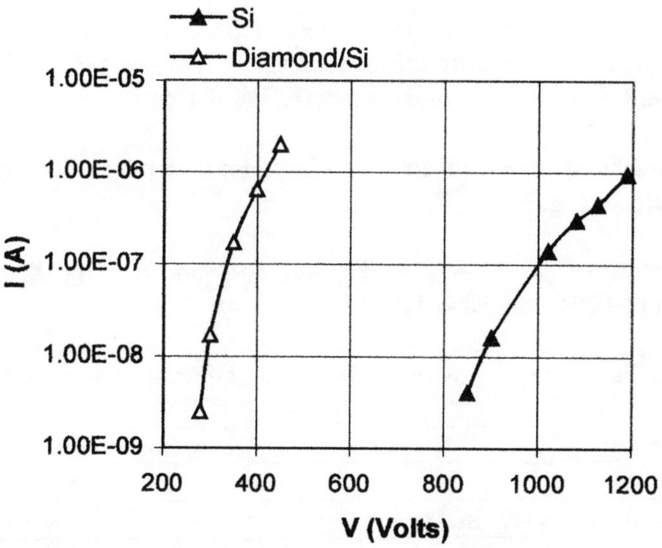


FIGURE 5 Emission characteristics of a Si tip before and after diamond coating. A considerable decrease of emission threshold occurred for the diamond-coated emitter [18].

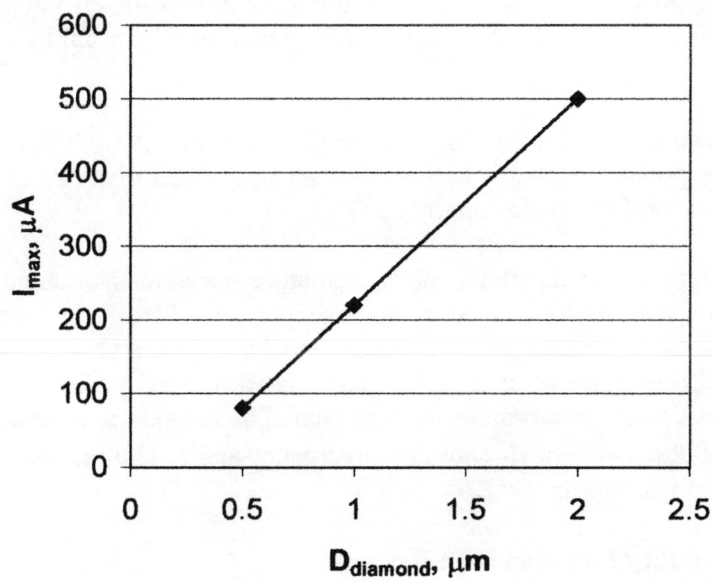


FIGURE 6 The relationship between coating thickness of CVD diamond and maximum current.

Correspondingly the improvement factors for current and voltage are given by:

$$f_I = \frac{I_{max}^{D/Me}}{I_{max}^{Me}} = C_2\sqrt{\lambda_2}D + 1$$

and

$$f_V = \frac{V_{th}^{Me}}{V_{th}^{D/Me}} = \frac{BK_1^{-1}\beta^{-1}\phi^{-3/2} + 1}{BK_1^{-1}\beta^{-1}\varepsilon(m^*/m_0)^{1/2}\,\varphi_B^{-3/2} + 1}(D \to 0)$$

The parameter of merit for device applications of field emitters is the integral current density, i.e. the total current emitted divided by the entire cathode area. A realistic practical challenge would be to demonstrate a total current of 1–10 A from an integral area of 1 $cm^2$.

Applications of diamond field emission cathodes include: flat panel displays, lamps, RF devices, sensors, and miniature electron guns.

The fabrication and testing of prototypes of displays and lamps using diamond coated composite emitters was reported in [19,20]. (See also TABLE 5.)

TABLE 5 Parameters of diamond based field emission display and lamp.

| Parameter | Display | Lamp |
|---|---|---|
| Active area | 1" × 1" | 1" × 1" |
| Emission site density, $cm^{-2}$ | $10^6$ | $4 \times 10^5$ |
| Anode voltage, V | 280 (DC) | 3000 (pulse) |
| Luminance, $cd/m^2$ | 100 | 2000 |
| Continuous operation measurements, hr | – | 2000 |

FEPET Inc. fabricated an efficient large-area diamond cathode with an emission site density of $2 \times 10^5$ $cm^{-2}$, and a projected lifetime of 20 000 hr (7000 measured) for applications in displays and lamps [13].

RF applications of diamond emitters are discussed in [14] and [21]. A high frequency field emission device has also been proposed [21] that is based on an ungated planar diamond emitter with decreased capacitance, for operation at frequencies up to 100 GHz.

A general tendency today is the development of miniature portable instruments, such as displays, plasma reactors, mass-spectrometers, electron microscopes, etc. Miniature e-beam devices require cold electron sources that are much more compact and have lower power consumption than hot filament sources. In [22] the application of a diamond-coated Si tip (by Containerless Research Inc.) for a microscale ion trap mass spectrometer is discussed. The cathode requirements for miniature e-guns are: current 20–100 μA, and small emission divergence angle. The reported divergence angle of a diamond-coated pointed emitter is ~3° [23].

## B3 Diamond Secondary Emission (SE) Devices

Required properties of diamond: NEA, electrical conductivity (B-doped).
Characteristic parameter: secondary yield $\delta = I_{emitted}/I_{incident}$.

Mearini et al [24] demonstrated very high secondary yields (as high as 48) for hydrogenated diamond surfaces. They built and tested prototypes of secondary multipliers with Pd and Mo dynodes coated with CVD B-doped diamond (FIGURE 7(a)). The multipliers demonstrated gains as high as $9 \times 10^5$; however degradation of the secondary yield occurred due to hydrogen desorption from the diamond surfaces. The decreased yield followed an exponential decay function of incident electron energy E and fluence F, given by: $\delta(E,F) = \delta_0 e^{-kF}$ [25]. The decay constant k = 2–6 × $10^9$ $cm^2$ $mA^{-1}$ $s^{-1}$ for E = 1 keV [25]. Extensive reports on the properties of secondary electron emission from diamond can be found in [26].

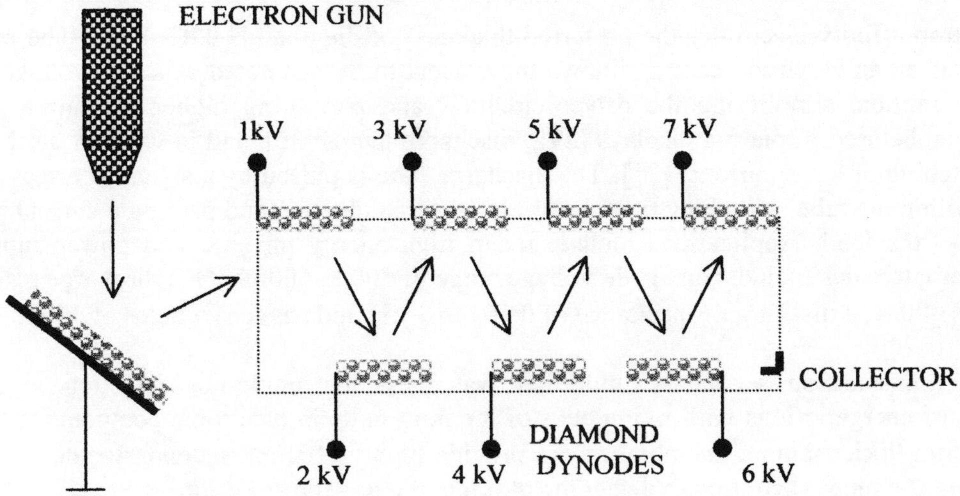


**a**

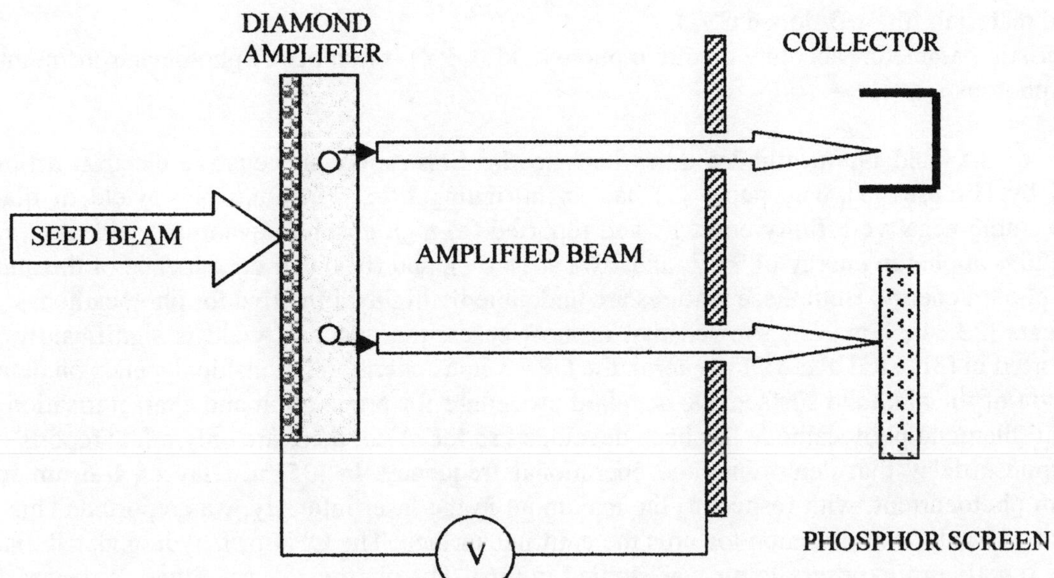


**b**

FIGURE 7 Reflection (a) and transmission (b) type diamond-based electron multipliers.

Other applications of SE from diamond include: crossed-field amplifiers; photo-electron multipliers [27]; image intensifiers [27]; flat panel displays [28,29]; replacements for thermionic cathodes, cathode ray tubes, and discharge tubes. The materials requirements for each differ slightly; for example, B-doped, porous diamond films for transmission type devices; CVD films or nanodiamond particles for discharge tube cathodes. Both reflection and transmission type electron multipliers were proposed in [27,28] (see FIGURE 7). For a 60 keV electron beam the recommended diamond thickness is 15 μm. The theoretical current gain produced by a 10 keV electron is about 600. In an experimental diode assembly current densities exceeding 20 A/cm$^2$ were generated.

In low-electron affinity electrodes the preferred thickness of diamond is 0.02–1 μm. The presence of LEA material on an electrode coating allows the plasma to be generated at lower voltage, reducing power consumption, simplifying the driver circuitry, and permitting higher resolution. The LEA electrodes can be used in plasma displays [29] , discharge lamps [30], and in vacuum discharge tubes for fast switching of large currents [31]. The discharge tube is pulsed by a signal to effect an electric discharge within the tube, switching the tube to electrical conduction and passing a current pulse from the source to the load. Applications include radar, high energy physics, and power supplies. The required characteristics include: an anode voltage range of 500 to 6000 V DC; discharge peak currents of 300 to 18 000 A; a discharge capacitance of 0.165 to 4 μF; and trigger voltages of 400 to 2000 V.

A diamond transmission secondary multiplier was used for single-ion detection in studies of interactions of energetic ions with living cells or memory units in electronic components [32]. CVD B-doped, 2 μm thick diamond membranes can provide highly efficient secondary-electron yields for many months at a time when used to detect the passage of energetic MeV ions.

### B4 Diamond Photocathodes

Required properties of diamond: NEA, electrical conductivity (B-doped).
Preferred material: IIb or B-doped CVD.
Characteristic parameter: absolute quantum photoyield (QPY) = number of photoelectrons/number of incident photons.

The idea of diamond photocathodes dates back to the first report on negative electron affinity of diamond by Himpsel [3]. The paper [3] had an intriguing title, 'Quantum photoyield of diamond (111): A stable negative-affinity emitter', and reported (a) high absolute quantum yield of C(111):H surface (20% at photon energy of 9 eV, and 70% at 18 eV), and (b) linear dependence of the quantum yield on photon energy. Both these features are undoubtedly highly attractive for photocathodes. Later experiments [33,34], however, showed that in most cases, the quantum yield is significantly lower than reported in [3] (FIGURE 8). In general, the QPY versus energy relationship depends on details of preparation of the diamond surface. No standard procedure for preparation and characterisation of an 'optimum' diamond photocathode has been developed so far. Also, there are only a few reports on the photoresponse delay that determines the operational frequency. In [35] a delay of 4–6 nm for the maximum photocurrent, with respect to the maximum in the laser intensity, was reported. This delay was ascribed to electron diffusion towards the emitting surface. The QPY of a hydrogenated diamond film as a function of exposure to air was studied in [36]. The photoyield was found to degrade from 12% to 6% (at 140 nm) after 1000 hr of exposure and to saturate at about 5%. However, the original, high QPY can be completely restored by rehydrogenation (e.g. by hydrogen plasma treatment).

The general requirements for photocathodes include: (1) a high QPY enabling high current densities and high brightness, (2) a linearly increasing, or constant, QPY versus photon energy, (3) a fast response, and (4) reliability. Photocurrent densities up to 10 A/cm$^2$ were reported [36]. No data on the reliability of diamond photocathodes have so far been reported. It should be noted that the energy range for operation of diamond photocathodes is >5.5 eV. In general this is not very practical, since there are a limited number of light sources available in this energy range.

## C DIAMOND TARGET ELECTRON BEAM DEVICES

Diamond target electron beam devices (EBDs) include a diamond layer in the path of an electron beam and are based on the effects within the diamond under electron irradiation (FIGURE 1).

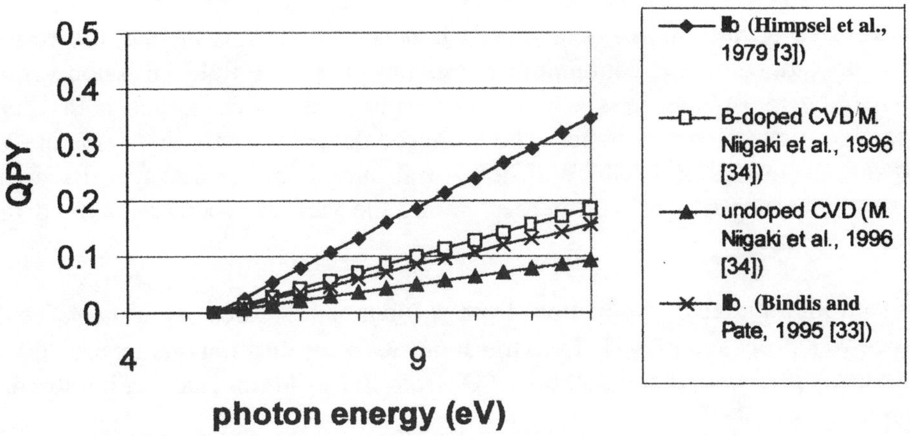


FIGURE 8 Linear approximations of QPY of diamond reported by different authors.

## C1 X-ray Sources

Required properties of diamond: electrical conductivity, thermal conductivity and mechanical strength.
Preferred diamond material: natural, IIb.

When diamond is bombarded with electrons, bremsstrahlung and X-rays of the carbon K characteristic wavelength (4.47 nm or 0.277 eV) are emitted. These soft X-rays have considerable potential for X-ray lithography. In order to maximise the intensity of the emitted X-rays, efficient cooling of the target should be provided. Because of diamond's outstanding tensile strength, diamond targets can be made extremely small and thin and such a target dissipates heat very efficiently. Type IIb diamond is the preferred target material because it is less susceptible to charging effects (a resistivity of 10–1000 Ω cm for type IIb). The amount of power that can be dissipated in an X-ray target is a function of its thermal conductivity (14 W cm $K^{-1}$ for type IIb), its maximum operating temperature, and its tensile strength (>1.2 GPa). For example, a target made from IIb diamond with diameter 2.5 mm and thickness of 10 μm can dissipate 63.5 kW of power [37].

## C2 Fast High Power Switches and Amplifiers

Useful or required properties of diamond: large bandgap (5.45 eV); dielectric properties: high resistivity ($10^{16}$ Ω cm) and dielectric strength ($10^7$ V/cm); high thermal conductivity (20 W cm $K^{-1}$ for type IIa).

A semiconductor electron beam device operates on the principle that it will normally have a low electrical conductivity. Electrical carriers, however, can be generated in the semiconductor (e.g. diamond) in response to electron bombardment allowing electrical power to flow from the power source through the diamond layer to the load [38]. Power switches and amplifiers can be built on this principle.

Diamond target electron beam switches in which a diamond target conducts electrical current in an electrical circuit when the diamond target is illuminated by electrons are described in [38,39]. The thickness of the diamond films ranges from 1 to 10 μm. In this case, electron energies of the order of a few tens of kilovolts are sufficient to generate a uniformly conducting electron–hole plasma in the diamond film. This enables a miniature high power, impulse generator to be made. The switch is expected to produce 10–100 kW pulses at a 10–100 MHz repetition rate. In [40] a compact EBD for amplifying microwave or millimetre wave signals is described. In this device, the diamond target is mounted within an output circuit containing a resonant cavity. A field emission array produces a density-modulated beam of electrons which illuminates a diamond target in a diamond switch connecting a load to high voltage switches. The proposed device is expected to operate at frequencies of 1–100 GHz and at powers of 1–10 kW. The natural pure single crystal type IIa diamond has the best and the most consistently stable properties among the various types of diamond for high speed and high power applications.

The use of electron-stimulated desorption from a diamond surface as a negative hydrogen or deuterium source was proposed in [41]. The largest negative currents obtained were 700 pA ($4.3 \times 10^9$ ions/s) of $H^-$ from CVD diamond and 200 pA of $D^-$ from Ib C(100), at electron bombardment currents of 40 μA with energies of 9–12 keV.

Diamond detectors for the dosimetry of high-energy electron beams are a traditional application of diamond. Such detectors have small size, good tissue equivalence, high resistance to radiation damage (0.05% $kGy^{-1}$), high sensitivity (0.05 μC $Gy^{-1}$), high stability of detector response (relative standard deviation over 13-week period of 0.67%), low leakage current (<5 pA), good temperature stability (better than 1%/K), and good time resolution [42].

## D DIAMOND PASSIVE COMPONENTS OF EBD

### D1 Field Emission Display with Spacers made of Diamond [43]

Properties used: dielectric properties: high resistivity ($10^{16}$ Ω cm) and dielectric strength ($10^7$ V/cm); thermal conductivity (10–20 W cm $K^{-1}$).

### D2 DLC Resistor [44]

Properties used: controllable resistivity ($10^3$ Ω cm – $10^{11}$ Ω cm), good adhesion, chemical-durability, and a simple low temperature fabrication process.

The resistive layer is discontinuously and locally formed on the cathodes by the CVD method. The required resistivity is achieved by doping of DLC with $PH_3$. A field emission device in which resistors are used comprises a substrate, cathodes formed on the substrate, a resistor layer continuously formed on the cathodes, and microtips formed on the resistor layer.

## E CONCLUSION

Significant progress in the application of diamond in electron beam devices has been achieved during the last several years. Nevertheless, these applications are not in a commercial phase yet. For most applications, higher integral current densities than achieved at this point are required. For applications in displays, a uniform emission site density is a critical parameter. While for many applications CVD diamond/carbon films are quite satisfactory, single crystal bulk diamond is preferable for high power switches and for amplifiers and X-ray sources. The values of the parameters obtained for diamond electron beam devices as reported by a number of different authors are summarized in this review.

**REFERENCES**


[1] M.H. Nazare [ see *Editor's Note* in the preliminary pages of this book ]
[2] E. Maillard-Schaller, O.M. Küttel, L. Dietrich, L. Schlapbach, V.V. Zhirnov, P.I. Belobrov [ *Diam. Relat. Mater. (Switzerland)* vol.8 (1999) p.805-8 ]
[3] F.J. Himpsel, J.A. Knapp, J.A. Van Vechten, D.E. Eastman [ *Phys. Rev. B (USA)* vol.20 (1979) p.624-7 ]
[4] L. Diederich, O.M. Küttel, P. Aebi, L. Schlapbach [ *Surf. Sci. (Netherlands)* vol.418 (1998) p.219-39 ]
[5] P.K. Baumann, R.J. Nemanich [ in *Low-Pressure Synthetic Diamond* Eds. B. Dischler, C. Wild (Springer-Verlag, 1998) ch.15 p.281-303 ]
[6] M.W. Geis et al [ *IEEE Electron Device Lett. (USA)* vol.12 (1991) p.456-9 ]
[7] G.R. Brandes, C.P. Beetz, C.A. Feger, R.L. Wright [ *Diam. Relat. Mater. (Switzerland)* vol.4 (1995) p.586-90 ]
[8] A. Hatta, T. Ito, A. Hiraki [ *Isr. J. Chem. (Israel)* vol.38 (1998) p.113-20 ]
[9] C. Xie, N. Kumar, C.B. Collins, T.J. Lee, H.K. Schmidt, S. Wagal [ *Tech. Digest of 6th Int. Vacuum Microelectronics Conf.* Newport, RI, USA, 1993 ]
[10] W. Zhu et al [ *Appl. Phys. Lett. (USA)* vol.67 (1995) p.1157-9 ]
[11] M.W. Geis, J.C. Twichell, T.M. Lyszczarz [ *J. Vac. Sci. Technol. B (USA)* vol.14 (1996) p.2060-3 ]
[12] K. Okano, S. Koizumi, S. Ravi, P. Silva, G.A.J. Amaratunga [ *Nature (UK)* vol.389 (1996) p.140 ]
[13] R.L. Fink, Z.L. Tolt, Z. Yaniv [ *Surf. Coat. Technol. (Switzerland)* vol.109 (1998) p.570-6 ]
[14] W. Mueller, H.S. Dewan, H. Chen, G.T. Mearini, I.L. Krainsky [ *Appl. Surf. Sci. (Netherlands)* vol.146 (1999) p.328-33 ]
[15] A.T. Rakhimov et al [ *J. Vac. Sci. Technol. B (USA)* vol.18 (2000) p.76-81 ]
[16] N. Chubun [ *Technical Digest of 11th Int. Vacuum Microelectronics Conf.* Asheville, NC, USA, 1998 ]
[17] W.P. Kang et al [ *J. Vac. Sci. Technol. B (USA)* vol.17 (1999) p.740-3 ]
[18] V.V. Zhirnov, J.J. Hren [ *MRS Bull. (USA)* vol.23 (1998) p.42-8 ]
[19] E.I. Givargizov, V.V. Zhirnov, N.N. Chubun, A.N. Stepanova [ *J. Vac. Sci. Technol. B (USA)* vol.15 (1997) p.450-3 ]
[20] A.N. Alimova, N.N. Chubun, P.I. Belobrov, V.V. Zhirnov [ *J. Vac. Sci. Technol. B (USA)* vol.17 (1999) p.715-8 ]
[21] S.V. Johnson, J.E. Jaskie [ US Patent 5801486 (1998) ]
[22] O. Kornienko, P.T.A. Reilly, W.B. Whitten, J.M. Ramsey [ *Anal. Chem. (USA)* vol.72 (2000) p.559-62 ]
[23] E.I. Givargizov, V.V. Zhirnov, N.N. Chubun, A.B. Voronin [ *J. Vac. Sci. Technol. B (USA)* vol.15 (1997) p.442-5 ]
[24] G.T. Mearini, I.L. Krainsky, J.A. Dayton [ *Surf. Interface Anal. (UK)* vol.21 (1994) p.138-43 ]
[25] J.B. Miller, G.R. Brandes [ *J. Appl. Phys. (USA)* vol.82 (1997) p.4538-45 ]
[26] A. Shih, J. Yater [ Datareviews in this book: *A3.3 and A3.4 Properties of secondary electron emission from diamond* ]
[27] M. Niigaki, T. Hirohata, H. Kan [ US Patent 5986387 (1999) ]
[28] L.H. Sverdrup, S.-H. Lin, E.J. Korevaar, B.A. Spivey, K.Y. Tang [ US Patent 6005351 (1999) ]
[29] S. Jin, G.P. Kochansky, W. Zhu [ US Patent 5982095 (1999) ]
[30] N.A. Fox, P.M. Tyler [ US Patent 5880559 (1999) ]
[31] G.E. Boettcher [ US Patent 5739637 (1998) ]
[32] T. Kamiya et al [ *Appl. Phys. Lett. (USA)* vol.71 (1997) p.1875-7 ]

[33] C. Bandis, B.B. Pate [ *Phys. Rev. B (USA)* vol.52 (1995) p.12056-71 ]
[34] M. Niigaki et al [ *Diam. Film Technol. (Japan)* vol.6 (1996) p.359 ]
[35] S.M. Pimenov, A.A. Beloglazov, V.I. Konov, A.V. Karabutov, V. Nassisi, V.I. Polyakov [ *Diam. Relat. Mater. (Switzerland)* vol.6 (1997) p.1650-7 ]
[36] A. Laikhtman et al [ Appl. *Phys. Lett. (USA)* vol.73 (1998) p.1433-5 ]
[37] D.A. Nelson, A.L. Ruoff [ US Patent 4266138 (1981) ]
[38] S-H. Lin [ US Patent 4993033 (1991) ]
[39] B.A. Fox, J.T. Glass, D.L. Dreifus, L. Lojk [ US Patent 5592053 (1997) ]
[40] T.A. Treado, S.-H. Lin [ US Patent 5355093 (1994) ]
[41] C. Goeden, G. Dollinger [ *Appl. Surf. Sci. (Netherlands)* vol.147 (1999) p.107-13 ]
[42] W.U. Laub, T.W. Kaulich, F. Nuesslin [ *Phys. Med. Biol. (UK)* vol.44 (1999) p.2183-92 ]
[43] Z. Yaniv [ US Patent 5959710 (1999) ]
[44] J. Kim [ US Patent 5998916 (1999) ]
[45] V.V. Zhirnov, E.I. Givargizov, P.S. Plekhanov [ *J. Vac. Sci. Technol. B (USA)* vol.13 (1995) p.418-21 ]

# C1.4 Diamond SAW devices

B.A. Fox and D.L. Dreifus

January 2000

## A INTRODUCTION

There has been increasing interest in large acoustic velocity materials for use in high frequency surface acoustic wave (SAW) devices. Diamond is an attractive material for SAW devices due to its high acoustic velocity, which exceeds that of any other known material. A schematic of a SAW device is shown in FIGURE 1 where the device operates in the following way:

- incoming electrical signals are converted to acoustic waves by interdigitated transducer (IDT) electrodes on a piezoelectric material,
- the acoustic waves travel in the material to an output IDT, where
- the acoustic waves are converted to output electrical signals by a piezoelectric material.

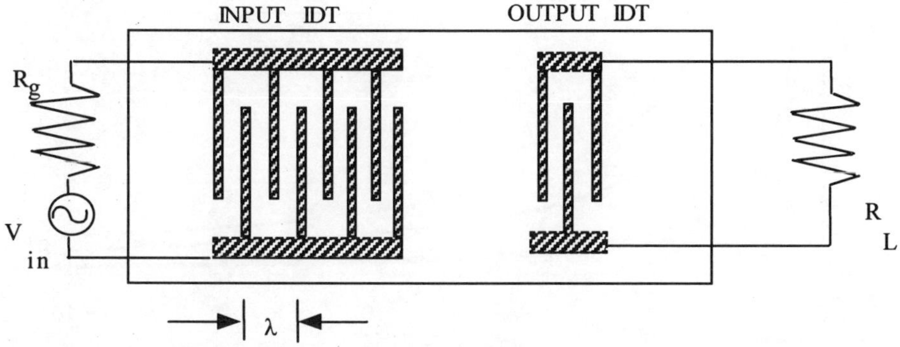


FIGURE 1 A surface acoustic wave device typically consists of an input interdigitated electrode (IDT) and output IDT on top of a piezoelectric medium.

The wavelength of the device is determined by the spacing of the electrodes and the frequency of the device is simply

$$f = \frac{v}{\lambda}$$

where v is the velocity and λ is the wavelength. Typical device designs employ IDT electrodes with a feature size of λ/4 or λ/8. New requirements for signal processing at higher frequencies continue to place increasing demands on SAW devices [1]. To achieve operational frequencies above 2–3 GHz using standard SAW materials, feature sizes approaching 0.25 μm are required. The tolerances on metal thickness and lateral variation for the interdigitated transducer (IDT) electrodes used in SAW designs are many more times more stringent than those used in traditional semiconductor devices. At operating frequencies above a few hundred megahertz, devices employing λ/8 designs are difficult and often impractical. Higher order harmonics may be used to attain high frequency operation; however, such devices are increasingly sensitive to manufacturing tolerances and resonant filter coupling degrades with the overtone squared. Therefore, higher velocity materials, such as diamond, are being developed to attain high frequency operation with viable feature sizes.

Surface acoustic waves propagate on the surface at phase velocities near the shear wave velocity of a material. Pseudo-SAW modes, which exhibit a bulk-wave character, propagate at velocities that approach the longitudinal phase velocity. For (001) cut, [110] propagating acoustic waves on diamond, the shear velocity is 12 788 m/s and the longitudinal velocity is 18 277 m/s [2]. These values approximate the velocity of SAW and pseudo-SAW waves, respectively. The high acoustic velocity of diamond, three to four times higher than traditional SAW materials, allows high frequencies to be reached with correspondingly wider IDT spacings. This is attractive for high frequency applications that are 2 GHz and above where control of submicrometre IDT line-widths across several hundreds of micrometres of IDT electrode length is increasingly difficult to attain using conventional photolithographic techniques. The IDT spacing versus frequency of operation for various SAW materials is shown in FIGURE 2.

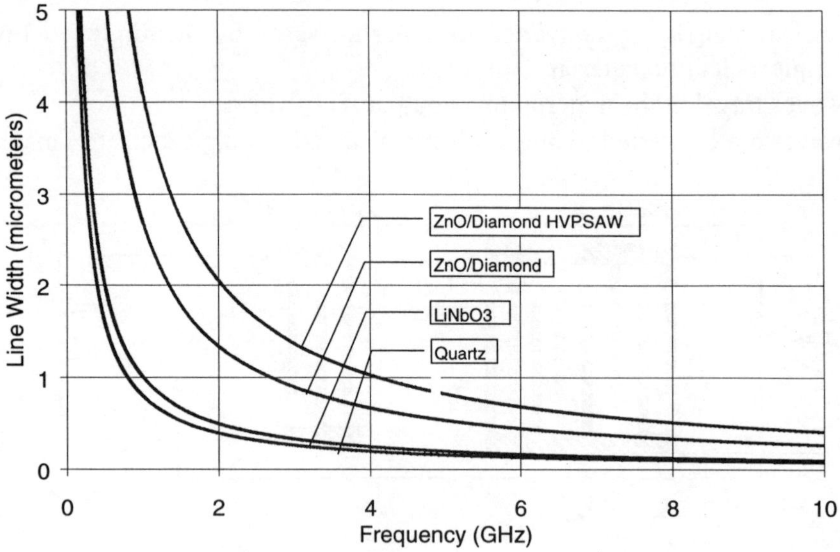


FIGURE 2 The interdigitated transducer electrode spacing varies with material and with centre operating frequency. The spacings are for λ/4 device designs and depend on the SAW velocity of the particular material.

Surface acoustic waves are often generated by employing a piezoelectric material and IDT electrodes. However, diamond is not piezoelectric. As a result, diamond must be part of a multi-layer structure for SAW devices. This multi-layer structure typically consists of a substrate, a diamond layer, a piezoelectric layer and metal IDTs as shown in FIGURE 3. Sometimes a metal shorting plane is used to modify the wave velocity and coupling. Metal shorting planes or IDTs can be located at any of the interfaces as well as at the surface of the multi-layer. In the multi-layer structure, the surface acoustic wave travels in both the diamond and the piezoelectric, and therefore, the resulting device performance depends on the material properties of both diamond and the piezoelectric, and on the device design.

## B MATERIAL PARAMETERS

The primary material parameters that determine the functionality of SAW devices are velocity, electromechanical coupling, propagation loss and frequency temperature coefficient. For multi-layer structures, these parameters are difficult to summarise because values differ for each device structure, for each orientation and for each piezoelectric thickness. To provide a basic understanding of the interaction between material properties and material parameters for SAW devices, calculation of monocrystalline material parameters will be discussed. Pertinent material properties are shown in TABLE 1. For multi-layer structures, one model used will be discussed and graphical results from the model will be presented.

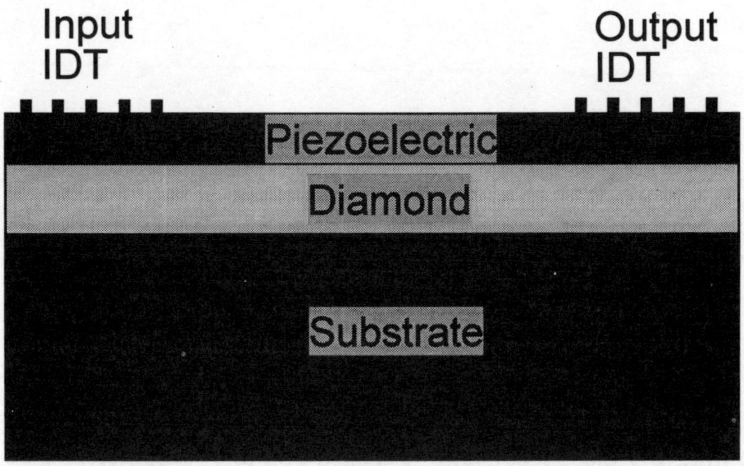


FIGURE 3 A typical diamond surface acoustic wave device consists of interdigitated electrodes (IDT), a piezoelectric layer, a diamond layer and a substrate.

TABLE 1 The elastic constants $C_{ij}$, density ρ, piezoelectric constants $e_{ij}$, and dielectric constants $\varepsilon_{ij}$ are shown for materials of interest for SAW devices.

| Material | $C_{11}$ (× $10^{11}$ $N/m^2$) | $C_{13}$ (× $10^{11}$ $N/m^2$) | $C_{33}$ (× $10^{11}$ $N/m^2$) | $C_{44}$ (× $10^{11}$ $N/m^2$) | ρ ($g/cm^3$) | $e_{11}$ ($C/m^2$) | $e_{12}$ ($C/m^2$) | $e_{14}$ ($C/m^2$) | $\varepsilon_{11}$ | $\varepsilon_{33}$ | Ref |
|---|---|---|---|---|---|---|---|---|---|---|---|
| Quartz | 0.867 | 0.119 | 1.07 | 0.579 | 2.65 | 0.171 | −0.171 | −0.0406 | 3.92 | 4.1 | [3] |

| Material | $C_{11}$ (× $10^{11}$ $N/m^2$) | $C_{13}$ (× $10^{11}$ $N/m^2$) | $C_{33}$ (× $10^{11}$ $N/m^2$) | $C_{44}$ (× $10^{11}$ $N/m^2$) | ρ ($g/cm^3$) | $e_{15}$ ($C/m^2$) | $e_{31}$ ($C/m^2$) | $e_{33}$ ($C/m^2$) | $\varepsilon_{11}$ | $\varepsilon_{33}$ | Ref |
|---|---|---|---|---|---|---|---|---|---|---|---|
| Fused Quartz | 0.785 | 0.161 | 0.785 | 0.312 | 2.21 | --- | | | 3.75 | 3.75 | [4] |
| $LiNbO_3$ | 2.03 | 0.752 | 2.24 | 0.595 | 4.64 | 3.76 | 0.23 | 1.33 | 44.3 | 27.9 | [4] |
| $LiTaO_3$ | 2.298 | 0.812 | 2.798 | 0.968 | 7.45 | 2.72 | −0.38 | 1.09 | 42.6 | 42.8 | [4] |
| ZnO | 2.10 | 1.05 | 2.11 | 0.423 | 5.67 | −0.480 | −0.573 | 1.321 | 8.55 | 10.2 | [4] |
| Diamond | 10.8 | 1.25 | 10.8 | 5.76 | 3.51 | --- | --- | --- | 5.70 | 5.70 | [4] |
| Sapphire | 4.97 | 1.64 | 4.98 | 1.47 | 3.98 | --- | --- | --- | 8.28 | 10.2 | [3] |

## B1 Monocrystalline Materials

The velocity, propagation loss, coupling and temperature coefficient are shown in TABLE 2 for some of the common piezoelectric materials. For monocrystalline piezoelectric materials, the velocity, electromechanical coupling, propagation loss and frequency temperature coefficient may be understood from some basic equations. The SAW acoustic velocity is the speed at which the acoustic wave traverses the crystal. Exact expressions for the SAW velocity have been derived [8]. Surface acoustic waves are primarily comprised of shear components whose velocity is given by the expression

$$v_s = \sqrt{\frac{C_{11} - C_{12}}{2\rho C_{44}}}$$

where $C_{ij}$ are the elastic constants and ρ is the density. The change in the elastic constants with an applied voltage determines the electromechanical coupling $k^2$, which is the per cent of electrical energy that is transferred into acoustic energy. For a piezoelectric material, the elastic constant is increased or stiffened with the presence of an applied voltage. The stiffened elastic constant $C^D{}_{ij}$ is given by the expression

$$C_{ij}^{D} = C_{ij}\left(1 + \frac{e^2}{C_{ij}\varepsilon}\right) = C_{ij}\left(1 + k^2\right)$$

where e is the magnitude of the piezoelectric constant, ε is the magnitude of the dielectric constant and $k^2$ is the coupling coefficient given by the expression

$$k^2 = \frac{e^2}{C_{ij}\varepsilon}$$

In addition to stiffening, the elastic constant is also a complex parameter. This is due to the time-dependent elastic deformation and results in a loss of energy known as the propagation loss. The out of phase component of the elastic modulus is related to the propagation loss and may be understood by considering the time-dependent stress equation

$$\sigma_{ij} = C_{ij} \bullet \gamma_{ij} + \eta\frac{\partial\gamma_{ij}}{\partial t}$$

where $\gamma_{ij}$ is the strain and η is the viscosity. The viscosity represents the magnitude of the time dependent response to the strain. The propagation loss $\alpha_s$ is given by

$$\alpha_s = \frac{2\pi^2\eta f^2}{\rho v_s}$$

TABLE 2 The velocity, propagation loss, coupling $k^2$ and frequency temperature coefficient FTC are shown for a variety of structures and orientations.

| Material | Orientation | Velocity (m/s) | Propagation loss | $k^2$ (%) | FTC (ppm/°C) | Ref |
|---|---|---|---|---|---|---|
| Quartz | ST | 3158 | | 0.16 | 0 | [4] |
| Quartz | Yx | 3159 | 1.3 mdB/um @ 1 GHz | 0.22 | –24 | [5] |
| $LiNbO_3$ | 128Y cut, X prop | 3980 | | 5.5 | 75 | [4] |
| $LiNbO_3$ (pseudo-SAW) | 64Y cut , X prop | 4450 | 18 mdB/um @ 1GHz | 10.9 | 70 | [6] |
| $LiTaO_3$ | X cut 112 Y prop | 3290 | | 0.75 | 18 | [4] |
| $LiTaO_3$ (pseudo-SAW) | X cut 112 Y prop | 4100 | 18 mdB/um @ 1GHz | 6.5 | 35 | [4] |
| ZnO/glass | Z-axis | 2600 | | 1.9 | 25 | [4] |
| ZnO/diamond/Si | Z-axis | 10 710 | | --- | 30 | [4] |
| ZnO/diamond/Si pseudoSAW | Z-axis | 16 400 | 3.7 mdB/um @ 547 MHz | --- | --- | [7] |
| $SiO_2$/ZnO/diamond/Si | Z-axis | 11 710 | | 1.5 | 0 | [4] |
| $LiNbO_3$/diamond/Si | Z-axis | 11 890 | | 9 | --- | [4] |
| $LiTaO_3$/diamond/Si | Z-axis | 10 610 | | 3.6 | --- | [4] |
| ZnO/sapphire | $(11\bar{2}0)/(01\bar{1}2)$ | 5200–5500 | 9 mdB/um @ 2 GHz | 4.0–4.7 | 43 | [6] |

The propagation loss depends on the material and its quality. Scattering from grain boundaries, inhomogeneities, and surface roughness contribute to the propagation loss. The propagation loss also increases as the square of the operating frequency and plays an increasingly important role in any SAW device operating beyond 3 GHz.

The materials properties also vary with temperature and can dramatically affect device performance. This temperature performance can be quantified as either the frequency temperature coefficient (FTC) or temperature coefficient of delay (TCD), which represent the change in frequency or delay with a change in temperature. It is determined by a combination of the changes in material properties as a function of temperature, as well as the thermal expansion, which impacts the line spacing of the interdigitated electrodes.

**B2 Multi-Layer Materials**

As previously discussed, diamond SAW devices require a piezoelectric material to form a multi-layer SAW device structure. The surface acoustic wave travels in both the diamond and the piezoelectric material and the performance is determined by the properties of both materials. In general, to model the behaviour of acoustic waves in a multi-layer system, one would have to solve simultaneous second order wave equations for four variables for each layer, matching boundary conditions at each interface. This can be computationally intensive. An alternative approach is to employ a propagator matrix method which enables one to develop a first-order matrix differential equation in eight variables [9]. From this analysis, one can determine propagating solutions in the multi-layer, the associated acoustic velocities, particle displacements, and electromechanical coupling.

The material parameters for multi-layer structures are significantly more complex than those of monocrystalline materials due to the combined contribution of each material. Since surface waves are confined to the surface within about a wavelength, propagating modes of different wavelengths are contained within different component volumes of the multilayer. Thus, the velocity and electromechanical coupling become dispersive with frequency and component layer thickness. Additionally, the multi-material system allows for additional modes in which there are multiple phase transitions of particle displacement along the propagating wave. In this discussion, these modes are denoted by type and order. For example, the first order Rayleigh mode (a surface wave polarised in the sagittal plane) is denoted as the 'R1' mode, and it exhibits one phase transition. The second order Rayleigh mode with two phase transitions is referred to as the R2 mode, etc.

The dispersive and multi-mode behaviour is best understood in graphical form. The modelled dispersive behaviour for acoustic phase velocity and coupling are depicted in FIGURE 4 for interdigitated electrodes on the surface (see FIGURE 1) for the following structure: a 1.0 μm-thick ZnO layer on a 22 μm-thick diamond layer, located on a semi-infinite Si substrate. Both the velocity and electromechanical coupling are shown as a function of frequency for the first five Rayleigh modes. Device behaviour can be predicted by employing a load line for specific wavelengths. The intersection of velocity mode with the load line determines the centre frequency of operation. This can be seen for 20 μm and 5 μm wavelengths in FIGURE 4. Note that even small thickness variations result in a shifting of the dispersion relationships to lower (thicker ZnO) or higher (thinner ZnO) frequencies. Rapidly changing dispersive relationships thus result in devices with widely varying performance. Thus, minimally dispersive designs are desirable since they are less sensitive to variations in layer thickness encountered during multilayer fabrication.

The materials parameters for multi-layer SAW structures are dispersive and cannot be generalised. The materials parameters for specific multi-layer structures are given in TABLE 2. Currently, AlN and ZnO are the only piezoelectrics that have been experimentally evaluated on diamond. Other materials, such as $LiNbO_3$ and $LiTaO_3$, offer excellent promise due to their higher electromechanical coupling coefficients. Unfortunately, they have not yet been deposited as a thin film where they retain their excellent piezoelectric properties.

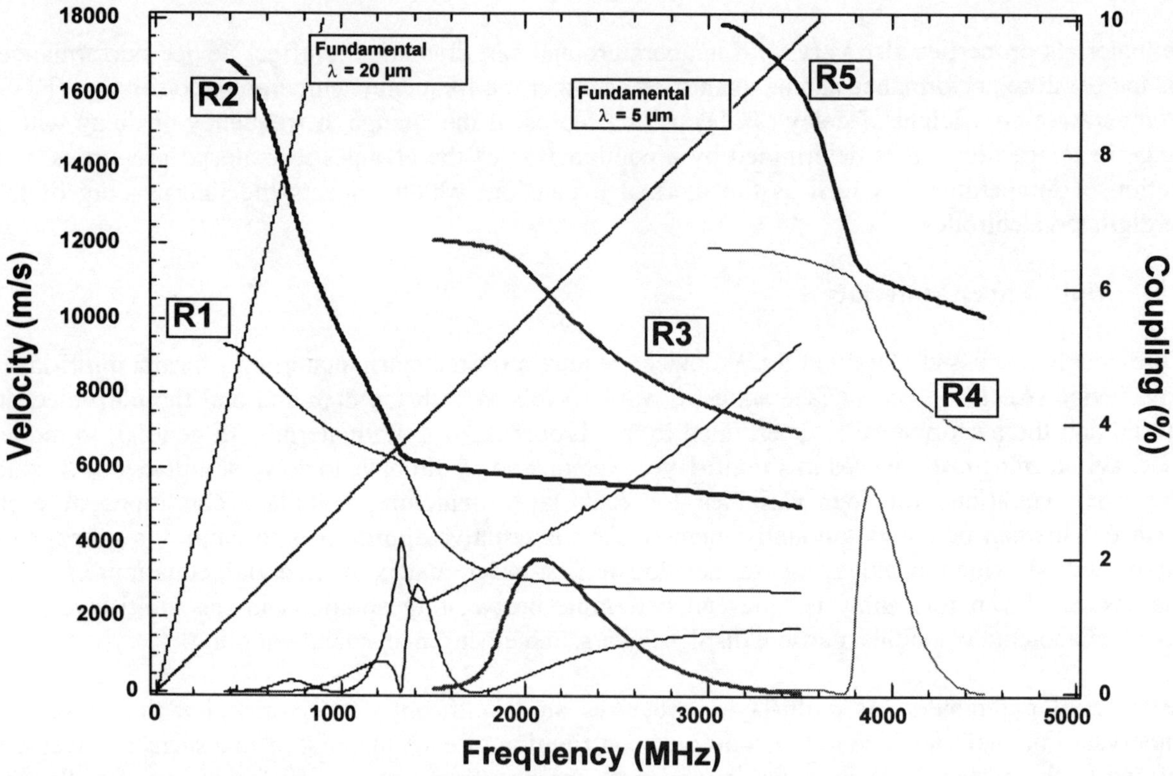


FIGURE 4 Multi-layer materials exhibit multiple Raleigh modes where the velocity and coupling are dispersive. The load line for a specific wavelength is determined by IDT spacing. Intersection of the load line and velocity defines the operating frequency of the device.

Device performance may be engineered by altering the structure for each multi-layer device. This is controlled by the material selection, layer thickness, IDT spacing and electrode placement. Each choice, however, is accompanied by corresponding trade-offs including velocity, temperature dependent behaviour, ease of fabrication, and degree of dispersiveness. Examples of specific device structures and performance characteristics will be discussed in the next section.

## C DEVICE DESIGNS

Surface acoustic wave device design strongly impacts device performance. The primary device performance characteristics are the frequency, bandwidth, insertion loss and FTC or TCD. There are two main device types: resonator and transverse frequency selective devices. In SAW resonators, standing waves are formed in a fixed cavity bounded by distributed reflectors that are constructed from either shorted periodic metal electrode stripes or etched ridges. In transverse SAW devices, waves propagate from one interdigitated transducer to the other. Resonators are typically employed as lower loss, narrower bandwidth devices, while transversal devices have linear phase, sharp bandwidth edge and higher insertion loss. Transversal devices require smaller feature sizes. For many resonator filter applications, the device performance parameters of interest are the insertion loss frequency, bandwidth, power handling, and temperature dependent performance. Simple examples of SAW filter two-port transversal and resonator structures are shown in FIGURES 1 and 5, respectively. Optimised device performance involves proper mode and structure selection, improved materials properties, and appropriate device design. A summary of diamond-based SAW device performance is given in TABLE 3.

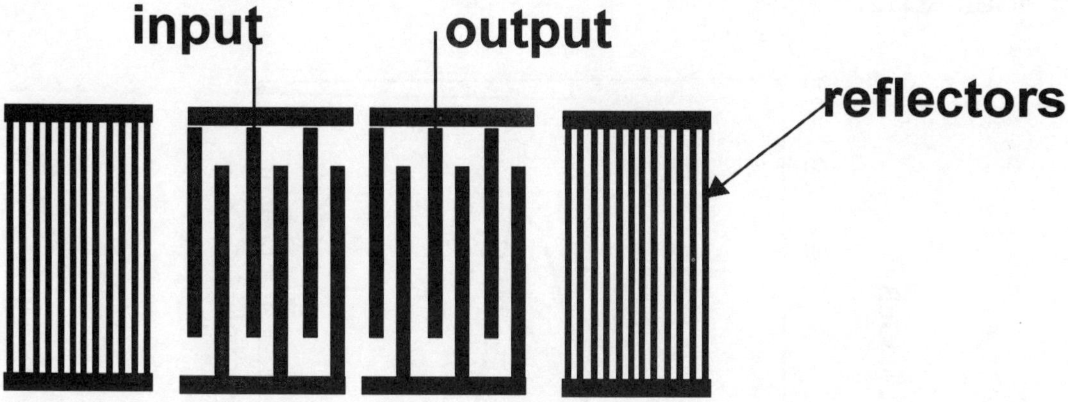


FIGURE 5 A two-port surface acoustic wave resonator device structure includes reflectors.

TABLE 3 Diamond-based surface acoustic wave devices have exhibited high frequency and low insertion loss.

| Material | IDT spacing (μm) | Frequency (GHz) | Bandwidth | IL (dB) | Ref |
|---|---|---|---|---|---|
| ZnO/diamond/Si | 1.5 | 1.45 | | | [10] |
| ZnO/IDT/diamond/Si | 0.6 | 4.7 | --- | --- | [4] |
| ZnO/IDT/diamond/Si | 1 | 2.55 | 0.8% | 15.5 | [4] |
| ZnO/diamond/Si (pseudoSAW) | 22 | 0.547 | 4.9% | 46 | [7] |
| ZnO/diamond/Si (high power) | 1 | 2.896 | 0.5% | 19 | [11] |
| ZnO/diamond/Si | ~ 1 | 2.42 | 1.1% | 3.5 | [12] |

## C1 High Frequency/High Velocity

A 1 μm IDS spacing was used on a ZnO/IDT/diamond/Si structure with 310 nm of ZnO [4]. The design excited the R2 mode with a velocity of 10 200 m/s. The device exhibited a 15.5 dB insertion loss at 2.55 GHz. Using 0.75 and 0.6 μm spacings, 3.5 GHz and 4.7 GHz devices were also fabricated. Alternatively, very high velocity devices can be fabricated using the R2 pseudo-SAW mode [7]. This high velocity allows high frequencies to be attained with wide IDT spacings. Using a 30 μm spacing in a ZnO/diamond/Si structure and 0.9 μm of ZnO, the phase velocity was 16 400 m/s. The minimum unmatched insertion loss of 46 dB occurs at 547 MHz as shown in FIGURE 6. The electromagnetic coupling for this device structure is much less than 1% and is responsible for the large insertion loss.

## C2 Low Insertion Loss

Using a composite longitudinal mode resonator device design with reflectors outside the IDTs and a 1.5 μm IDT spacing, the minimum insertion loss was 8.7 dB at 1.45 GHz [10]. The insertion loss, although much improved from earlier devices, still limits the applicability of devices fabricated from this multi-layer system, especially in RF transceivers where the insertion loss must be less than 3 dB.

A lower insertion loss was obtained using the third order Rayleigh (R3) mode on a ZnO/diamond structure with resonator filter elements arranged in lattice and ladder designs [12]. Example topologies for ladder and lattice designs are shown in FIGURE 7. In order to maximise coupling, a lattice device topology was selected. While less common than ladder-type designs, lattice structures can exhibit twice the bandwidth. This is especially important for dispersive media in which the coupling is reduced as compared to non-dispersive materials. The IDT spacing in the filter elements was ~ 1 μm. A matched insertion loss of 3.5 dB at 2.4 GHz with a 1.1% fractional bandwidth is shown in FIGURE 8 [12].

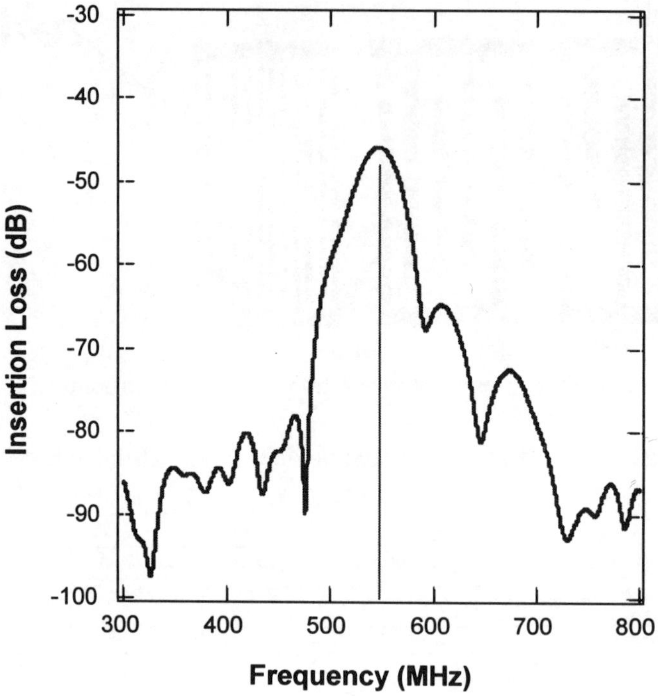


FIGURE 6 The filter characteristics for a ZnO/diamond pseudo-SAW device exhibited an unmatched insertion loss of 46 dB at 547 MHz. The acoustic velocity was 16 400 m/s (from [7]).

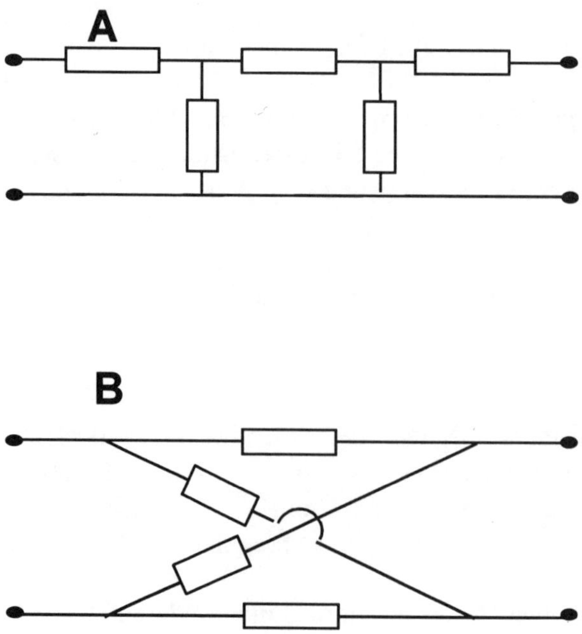


FIGURE 7 Surface acoustic wave filter elements may be combined into ladder (A) and lattice (B) topologies.

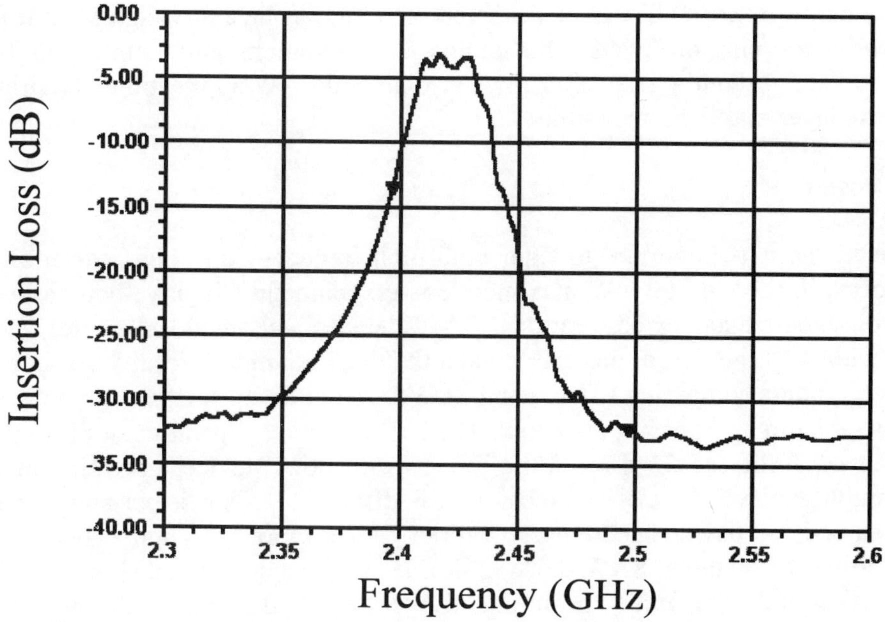


FIGURE 8 The filter characteristics for a lattice design with ZnO/diamond filter elements exhibited a 3.5 dB insertion loss at 2.42 GHz (from [12]).

## C3 Wide Bandwidth

The fractional bandwidth of resonator-based ZnO/diamond SAW devices is typically limited to 0.5–2% for modes exhibiting large SAW phase velocities. This is due to the relatively low electromechanical coupling coefficient of the ZnO/diamond system for thicknesses and modes of interest. The electromechanical coupling coefficient can be as large as 4–8% (see FIGURE 3 for the R2 mode) for this system; however, this occurs for modes with phase velocities of less than 6 km/s and thus does not provide differentially greater performance enhancement as compared to traditional monocrystalline-based SAW materials. Higher coupling materials, such as $LiNbO_3$ and $LiTaO_3$, have not yet been deposited in oriented form for multi-layer applications. These materials may increase the coupling up to 9% while the velocity remains above 10 500 m/s [4]. This large coupling will allow the bandwidth to be up to 6%. Other materials such as lead zinc titanate (PZT) may potentially allow even wider bandwidths to be attained. Use of a higher coupling piezoelectric would be a tremendous improvement over ZnO and would allow diamond to be used for wide band applications where the required bandwidth is 3% or more. Development of thin film piezoelectric materials is required to attain wide bandwidth multi-layer devices.

## C4 Zero Frequency Temperature Coefficient

For many applications thermal stability is critical. Traditionally, ST-quartz is used because it has a frequency–temperature coefficient of zero. The frequency–temperature coefficient of the Rayleigh 2 mode for ZnO/diamond is –20 to –35 ppm/°C [4]. To minimize the sensitivity of the device frequency to temperature, a material with positive temperature coefficients ($SiO_2$) was used to cap the multi-layer structure. Models were developed to calculate the impact of varying $SiO_2$ thickness on a ZnO/diamond structure. Increasing the $SiO_2$ thickness increases the TCD (less negative). By depositing the appropriate thickness of $SiO_2$, the negative FTC for ZnO/diamond may be compensated to achieve a zero FTC. Depending on the multilayer structure and SAW mode of interest, the required

$SiO_2$ layer thickness may be prohibitively thick. Additionally, the limitation to adding $SiO_2$ is a potential decrease in the acoustic velocity due to the introduction of a layer with a lower SAW velocity. For example, a zero TCD can be achieved with $SiO_2$ using a Rayleigh 2 mode with a velocity of 8045 m/s and a coupling of 3.98%. This compares to a velocity and coupling of 10 524 m/s and 3.12% for a structure without a $SiO_2$ layer [4]. For diamond SAW applications requiring a zero FTC, use of a $SiO_2$ cap layer makes this possible.

## C5 High Power

In some applications, it is important to filter both high frequency and high power signals. Existing materials are often limited to 200 mW maximum power. Diamond's high velocity allows the IDTs to be spaced further apart than for conventional SAW devices and should allow for additional power handling. In addition, the large thermal conductivity of diamond results in improved thermal diffusivity. Experiments comparing $LiTaO_3$ and ZnO/diamond/Si multi-layer-based devices with IDTs identically spaced 1 μm apart were performed. The $LiTaO_3$ device, operating at 822 MHz, exhibited a power handling capability of 23 dBm while the ZnO/diamond device, operating at 2.9 GHz, was capable of handling power levels in excess of 36 dBm [11]. This experiment demonstrates the promise of diamond for power handling capabilities. For a more thorough validation of the power handling capabilities of diamond SAW devices, devices from each material should be optimised for a specific application and compared. The initial results suggest that diamond's power handling may complement its high velocity for high frequency, high power applications.

## D CONCLUSION

The materials properties for diamond-based SAW devices make them promising for high frequency and high power device applications. Diamond is not piezoelectric and a multi-layer structure including a piezoelectric is required. Currently, only a few piezoelectric materials, primarily ZnO and AlN, can be deposited as a thin film. Higher coupling piezoelectrics, such as $LiNbO_3$, $LiTaO_3$, and PZT, would considerably increase the bandwidth of diamond SAW devices and increase the number of application areas for diamond SAW devices. The multi-layer structure produces a dispersive velocity and coupling. The dispersive properties make diamond SAW design more complicated than monocrystalline SAW design. Overall, diamond-based SAW technology is still in its infancy. Recent developments have increased the likelihood of commercially viable devices fabricated from multi-layers with diamond. Three-inch diameter diamond substrates have become available [13]. In addition, significant progress has been made in modelling the effects of dispersive multilayers in order to synthesize devices [14]. Near theoretical materials properties have been demonstrated [15], and high power handling has been observed [11]. Additionally, work continues on depositing thin film piezoelectrics with larger coupling coefficients. There is still much work to be done on developing diamond for SAW applications; however, higher frequency SAW devices will become more important as communications systems continue to increase their operating frequency to achieve higher information transfer rates.

## REFERENCES


[1] C.S. Lam, D. Chen, B. Potter, V. Narayanan, A. Vishwanathan [ *International Workshop on Ultrasonics Applications* Nanjing, China, 1996 ]

[2] K. Yamanouchi, N. Sakurai, T. Satoh [ *Ultrasonics Symposium* (IEEE, 1989) p.351-3 ]

[3] A.J. Slobodnik Jr., E.D. Conway, R.T. Delmonico [ *Microwave Acoustics Handbook* (US Department of Commerce, NTIS AD780172, 1973) ]

[4] H. Nakahata et al [ *Ultrasonics Symposium* Seattle, Washington, USA, Eds. M. Levy, S.C. Schneider, B.R. McAvoy (IEEE, 1995) p.361-70 ]

[5] C. Campbell [ *Surface Acoustic Wave Devices and their Signal Processing Applications* (Academic Press, Inc., 1989) ]

[6] H. Ieki, H. Tanaka, J. Koike, T. Nishikawa [ *IEEE MTT-S (USA)* San Francisco, CA, USA (1996) p.409-12 ]

[7] D.L. Dreifus, R.J. Higgins, R.B. Henard, R. Almar, L.P. Solie [ *Ultrasonics Symposium* Toronto, Canada, Eds. M. Levy, S.C. Schneider, B.R. McAvoy (IEEE, 1997) p.191-4 ]

[8] B.A. Auld [ *Acoustic Fields and Waves in Solids* $2^{nd}$ edn (Robert E. Kreiger Publishing Company, 1990) ]

[9] E.L. Adler, J.K. Slaboszewicz, G.W. Farnell, C.K. Jen [ *IEEE Trans. Ultrason. Ferroelectr. Freq. Control (USA)* vol.3 (1990) p.215-23 ]

[10] S. Shikata et al [ *Applications of Diamond Films and Related Materials: Third International Conference* Eds. A. Feldman, Y. Tzeng, W.A. Yarborough, M. Yoshikawa, M. Murakawa (1995) p.29-36 ]

[11] K. Higaki et al [ *IEEE MTT-S (USA)* (IEEE, 1997) p.829-32 ]

[12] D.L. Dreifus et al [ *Frequency Control Symposium* Pasadena, California, USA, 1998, unpublished presentation ]

[13] S. Fujii et al [ *Ultrasonics Symposium* Toronto, Canada, Eds. M. Levy, S.C. Schneider, B.R. McAvoy (IEEE, 1997) p.183-6 ]

[14] A. Hachigo, D.C. Malocha [ *Ultrasonics Symposium* Toronto, Canada, Eds. M. Levy, S.C. Schneider, B.R. McAvoy (IEEE, 1997) p.205-9 ]

[15] B.R. Stoner, R.B. Henard, B.A. Fox, T. Onishi, I. Petrov [ *Frequency Control Symposium* Pasadena, California, USA, 1998, unpublished presentation ]

# C1.5 Diamond X-ray dosimeters

E. Vittone and C. Manfredotti

January 2000

## A INTRODUCTION

The interest in the use of diamond as a radiation dosimeter stems principally from its tissue equivalence. The atomic number of diamond ($Z = 6$) matches that of the biological tissue being exposed to high energy photon radiation (the effective number of human tissue is 7.42 for muscle and 5.92 for fat), and therefore the energy deposited by any beam of radiation for unit of mass will be virtually the same for tissue and for diamond. This fact makes diamond detectors more suitable than any other solid state dosimeters, e.g. the commonly used silicon diodes, for photon energies below 200 keV where the absorbed energy depends at least on the fourth power of the atomic number.

Other attractive properties of diamond as a dosimeter are that it is chemically stable, non-toxic, mechanically robust and relatively insensitive to radiation damage. Diamond detectors can be fabricated with small volumes (of the order of 1 $mm^3$) and then can be considered for in vivo dose measurements. As a consequence, diamond is particularly suitable for accurate dose measurements in small radiation fields such as in stereotactic radiosurgical beams where ion chambers are inappropriate because of their poor spatial resolution and their sensitivity to lateral electronic disequilibrium [1].

Furthermore, diamond is an extremely versatile material and its dosimetric properties can be exploited as a thermoluminescence (TL) or thermoconductive (TSC) dosimeter, as a solid state ionisation chamber (IC) and as a radiophotoluminescence detector [2,3].

So far, commercial diamond detectors are fabricated from natural gems. The main drawbacks of these dosimeters are the high cost and long waiting time [4], due to the severe selection of gems with suitable quality, and the poor reproducibility overall in the case of TL dosimeters. However, the recent availability of synthetic samples, grown under controlled conditions by both the high pressure-high temperature (HPHT) and chemical vapour deposition (CVD) techniques, makes possible, in principle, a widespread application of diamond in medical radiation dosimetry.

## B AS AN IONISATION CHAMBER

Almost all the works published up to now describe the dosimetric properties of pure IIa natural diamond solid state ionisation chambers [1,3–15]. Kozlov et al [5] have summarised the severe criteria for selecting gems suitable as radiation detectors: gems should have high resistivity ($\geq 10^{12}$ Ω cm) and low impurity content (nitrogen should be less than $10^{19}$ atoms/$cm^3$). The latter requirement is particularly important to guarantee a recombination rate almost independent of the electron–hole generation, i.e. an almost linear increase of photocurrent with dose rate and an almost complete charge collection. Furthermore, charge carriers in traps form a space charge which gives rise to a counter electric field which opposes the applied bias voltage (polarisation effect). Kozlov et al [5] first demonstrated that the use of special injecting contacts (e.g. boron-implanted contacts) greatly contributes to neutralising such space charge effects by a significant increase of dark current. Using contacts with different injecting properties, the linearity of the photoresponse can be extended down to a few mR $s^{-1}$ (about $10^{-5}$ Gy $s^{-1}$), even if a significant increase of dose rate threshold due to the increase of dark current has to be taken into account. Another commonly used method to maximise the detector response and to stabilise the current consists of a strong pre-irradiation (or priming) of the diamond detector with doses ranging from 1 to 10 Gy. This strongly recommended pre-treatment is essential to fill charge traps and to reach an equilibrium trap population which allows a stable response of the detector to be reached and settled for long times [10].

Diamond detectors show a response that in general slightly decreases with dose rate. The empirical expression of the dependence of the photocurrent $I_{ph}$ on dose rate $\dot{D}$ is

$$I_{ph} = I_0 + S\left(\dot{D}\right)^{\Delta} \tag{1}$$

where $I_0$ is the dark current; Δ and S are the phenomenological parameters indicating the slight sublinearity of the response and the sensitivity, respectively. Some values of Δ obtained by various authors are listed in TABLE 1. Δ ranges between 0.9 and 1, is to be considered almost constant, is very slightly dependent on photon beam energy [4], and is considered to be peculiar to the material.

The sensitivity S is calculated by the photocurrent/(dose rate) ratio where the denominator is evaluated by means of the exposure rate $\dot{X}$ measured by an ionisation chamber (IC) taking into account the mass energy absorption coefficient of carbon and air shown in FIGURE 1.

$$\dot{D} = \frac{W}{e}\frac{(\mu_{en}/\rho)_C}{(\mu_{en}/\rho)_{air}}\dot{X} \tag{2}$$

where $\dot{D}$ and $\dot{X}$ are expressed in terms of SI units (Gy/s and C/(kg s), respectively) and (W/e) is the mean energy expended in air per ion pair formed and per electron charge (W/e = (33.97 ±0.06) J/C [16]).

Carbon is nearly water equivalent and, as a consequence, the diamond response is nearly directly proportional to the absorbed dose rate of water. The mass energy absorption coefficients ($\mu_{en}/\rho$) of carbon, air, silicon and LiF relative to water as a function of photon energy are shown in FIGURE 2. The assumption of energy-independent correction factors to convert diamond response to dose, involves a systematic error of less than 7% to be compared with the changes of more than 40% relevant to the silicon to water ratio [15].

The mean carrier lifetime τ can then be evaluated [7] by assuming a simple photoconductive mechanism, fully depleted detectors, ohmic contacts and 13 eV as the mean energy to generate electron–hole pairs [6]. Apart from the result obtained by Burgemeister [7], all the values of τ relevant to natural diamond lie between 1 and 10 ns, as stated by Kozlov et al [5]. Burgemeister [7] obtained exceptionally high results for one small, very accurately selected IIa diamond sample which showed a photoconductive gain larger than 100.

Very little information is currently available on synthetic diamonds used as solid state ionisation chambers. Keddy et al [11] showed that diamond samples with nitrogen content around 100 ppm synthesised by the HPHT method have responses both to alpha particle and to gamma-ray radiation comparable with natural gems.

High quality CVD diamond detectors have been recently fabricated [17,18]. These preliminary results demonstrate that CVD diamond detectors show dosimetric properties close to those measured in natural diamond but with a much higher reproducibility and lower costs. Moreover, the fabrication of microdosimeters realised by the deposition of thin diamond films on metallic tips makes realistic the possibility of making 'pin-point' detectors [3] with sensitivity/mass ratio close to those of ionising chambers, for special treatment, as in brachytherapy [13] and stereotactic radiosurgery [14], where high spatial resolution is required.

TABLE 1 Dosimetric properties of diamond detectors working as ionisation chambers. T = thickness, V = active volume, S = sensitivity [radiation energy], S/M = sensitivity to mass ratio, $I_0$ = dark current [at operating bias voltage]. Lifetime τ is evaluated considering a mean carrier mobility of 2000 V $s^{-1}$ $cm^{-2}$ and a mean energy to create an electron/hole pair of 13 eV [31], Δ = sublinear exponent.

| Detector | T (mm) | V ($mm^3$) | S (C $Gy^{-1}$) | S/M (C $J^{-1}$) | τ (ns) | $I_0$ (pA) | Δ |
|---|---|---|---|---|---|---|---|
| Natural diamond [5] D-10 | 0.4 | 7.6 | $2.6 \times 10^{-6}$ [1250 keV] | 0.01 | 3.5 | 0.04 [300] | |
| Natural diamond [6] C1 | 1 | 3 | $0.06 \times 10^{-6}$ [661 keV] | 0.006 | 2.6 | 1 [150] | 0.91 |
| Natural diamond [7] | 1 | 0.8 | $31 \times 10^{-6}$ [≈1000 keV] | 11 | 14700 | ≤0.6 [50] | |
| Natural diamond [8] | 0.33 | 1.4 | $0.175 \times 10^{-6}$ [1250 keV] | 0.037 | 2.6 | ≤1 [100] | 0.98 |
| Natural diamond [9] | (0.2–0.3) | ≈3 | ≈$0.5 \times 10^{-6}$ [1250 keV] | ≈0.05 | 1.9 | ≤1 [100] | 1 |
| Natural diamond [4,5] | 0.32 | 3 | $0.41 \times 10^{-6}$ [4–20 MV photon beam] | ≈0.04 | 2.6 | ≤3 [100] | 0.963 |
| Natural diamond [15] | 0.26 | 1.9 | $0.22 \times 10^{-6}$ | 0.03 | 1.4 | | |
| HPHT diamond [11] | | | $0.01 \times 10^{-6}$ [1250 keV] | | | ≤2.5 [60] | |
| CVD diamond [17] | 0.15 | 210 | $8.1 \times 10^{-6}$ [1250 keV] | 0.01 | 0.1 | 10 [150] | |
| CVD diamond [18] | 0.4 | 24.8 | $2.7 \times 10^{-6}$ [6 MV photon beam] | ≈0.03 | 3.2 | 50 [100] | |
| CVD diamond tips [18] | ≈0.02 | $<3 \times 10^{-2}$ | $0.007 \times 10^{-6}$ [6 MV photon beam] | 0.07 | | 1 [100] | |
| Silicon diode [32] | 0.06 | 0.4 | $0.21 \times 10^{-6}$ [20 MeV electron beam] | ≈0.15 | | | |
| IC [33] | | 144 | 0.005* [1250 keV] | | | | |

*Absorbed dose to water

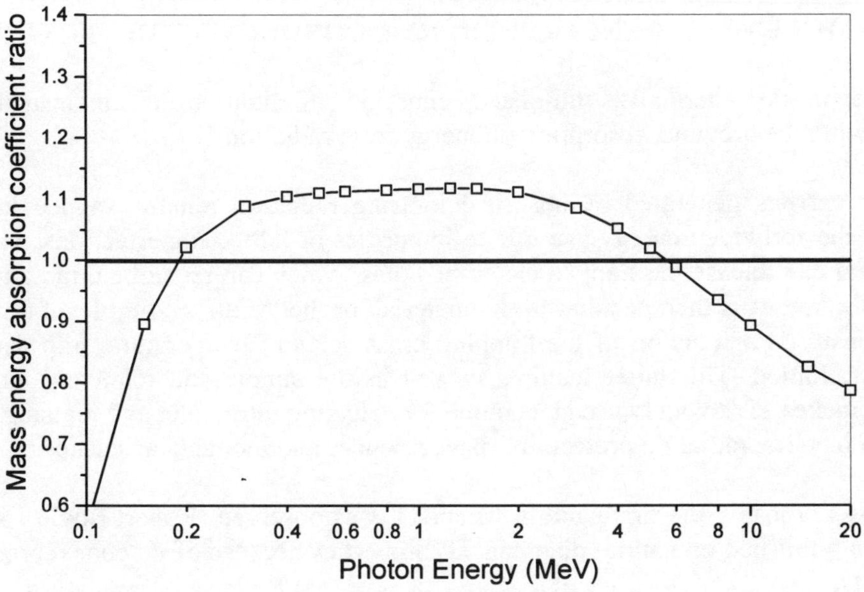


FIGURE 1 Mass energy attenuation coefficients $\dfrac{\left(\mu_{en}/\rho\right)_C}{\left(\mu_{en}/\rho\right)_{air}}$ of carbon relative to air as a function of photon energy [30].

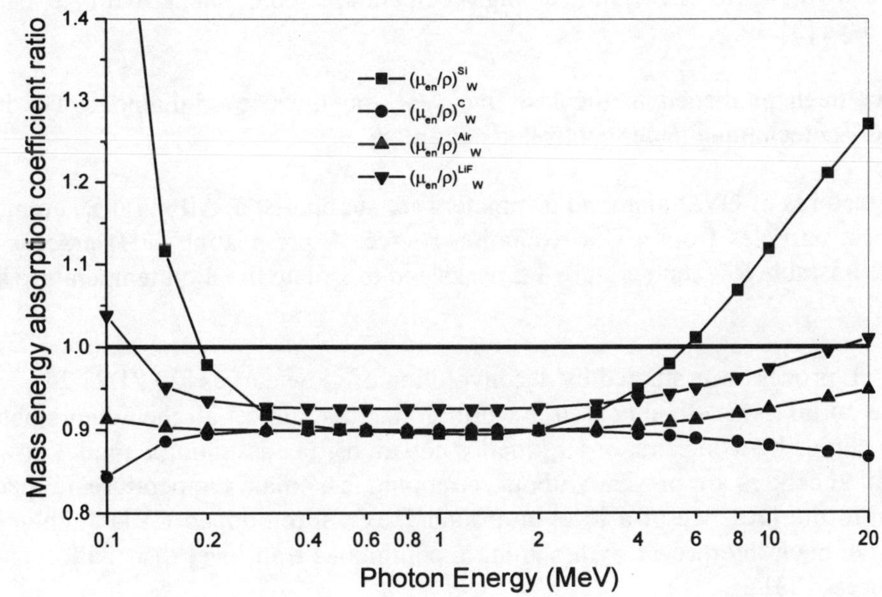


FIGURE 2 Mass energy attenuation coefficients $\left(\mu_{en}/\rho\right)^{xxxx}_{w} = \dfrac{\left(\mu_{en}/\rho\right)_{xxxx}}{\left(\mu_{en}/\rho\right)_{water}}$ (xxxx = C,air,Si,LiF) relative to water as a function of photon energy [30].

## C AS A THERMOLUMINESCENCE OR THERMOCONDUCTIVE DETECTOR

Thermoluminescence is the thermally stimulated emission of light from an insulator or a semiconductor following the previous absorption of energy from radiation [19].

Some of the charge carriers generated by absorbed ionising radiation remain trapped in discrete energy levels within the forbidden energy gap due to impurities or lattice imperfections. Subsequent heating of the material can release the trapped electrons/holes, which can recombine radiatively. The energy of the emitted photons is then peculiar to the material, or, better, to the trap level distribution, whereas the TL intensity is a function of the trapping cross section, temperature, trap density and absorbed energy of radiation. This latter feature, as well as the simple and relatively inexpensive experimental set-up, makes TL a widely used technique for detecting ultraviolet and ionising radiation in the areas of health physics, radiation protection, space sciences and biomedical sciences.

Although diamond was probably the first solid in which TL was observed (Robert Boyle 1663), very few papers have been published on natural diamond TL properties because of its poor reproducibility and low signal [20,21].

Nam and Keddy [22] first studied the performance of HPHT diamonds with controlled amounts of impurities. They showed that the TL sensitivity of diamonds with low nitrogen content (less than 10 ppm) is over three orders of magnitude higher than in gems with 100 ppm of nitrogen. Boron content of the order of 1 ppm increases the linearity from less than two decades to over three decades, whereas higher concentrations introduce supralinearity into the TL response. They showed that a controlled introduction of boron which partially compensates the nitrogen content allows diamond thermoluminescent dosimeters to be fabricated with a linearity of response extending from 0.01 to 10 Gy, sensitivities equivalent to LiF and a high-temperature component with a half-life of approximately 200 days [3].

Several papers have been published in the last five years on the use of diamond TL dosimeters synthesised by the CVD technique under controlled conditions.

In TABLE 2 some features of CVD diamond dosimeters are summarised. All the measurements were carried out using beta particles from a $^{90}$Sr excitation source. A pre-heating (PH) process was also used to suppress the unstable low temperature TL peaks and to isolate the high temperature structures (PT).

The kinetics of the TL process was studied by deconvolution of glow curves [20,21,23,24], isothermal decay [21,24] or the initial rise method [20]. It is worth noting that almost all the authors obtained the best fit of the glow curves by using first order kinetics equations, i.e. assuming a model in which one trap-one recombination centres are present without retrapping. The main temperature TL peak can be reasonably ascribed to the presence of a level at around 1 eV. Boron-doped CVD diamonds exhibit glow curves that have been interpreted by assuming a continuous trap level distribution and a more complex kinetic process [23].

Furetta et al [25] observed no fading for a period of one month for one sample stored in a Pb box at a constant (21°C) temperature. Optical bleaching has been reported by Vittone et al [24] if the sample is illuminated with light in the spectral range 450–550 nm. A linearity range extending up to 20 Gy and a flat energy response for X-ray energies between 0.1 and 1.25 MeV is reported by Furetta et al [25]. Comparative studies of the TL response to beta irradiation of diamond and LiF dosimeters have demonstrated comparable sensitivities of the two materials [20,26]. The spectrum of light emission shows a broad band centred at 498 nm [20]. If such a spectral feature is peculiar to the TL mechanism and valid for all the samples, a large margin of improvement in TL sensitivity can be foreseen by using photocathodes with more suitable quantum efficiency spectral response.

TABLE 2 Dosimetric properties of diamond samples working as thermoluminescence or thermoconductive dosimeters. PH = pre-heating temperature, β = heating rate, PT = peak temperature, E = activation energy, s = frequency factor, b = kinetic order.

| Detector | Excitation source | PH (K) | β (K $s^{-1}$) | PT (K) | E (eV) | s (Hz) | b |
|---|---|---|---|---|---|---|---|
| CVD diamond [20]-TL | $^{90}$Sr | 500 | 0.5 | 540 | 0.86 ±0.07 | $4.2 \times 10^6$ | 1 |
| CVD diamond [21]-TL | $^{90}$Sr | 573 (1 hr) | 2 | 550-560 | 1.275<br>0.646 | $4.79 \times 10^9$<br>$2.85 \times 10^4$ | 1 |
| CVD diamond [24]-TL | $^{90}$Sr | 473 (30 s) | 8 | 513 | 1.1 | $4.6 \times 10^{10}$ | 1 |
| CVD diamond [26]-TL | $^{90}$Sr | 493 (3 min) | 2 | 571 | 0.92 | $2.92 \times 10^7$ | 1 |
| Boron doped CVD diamond [23]-TL | $^{90}$Sr | | 2 | 377<br>442<br>492<br>525 | 0.62<br>0.75<br>1.13<br>1.64 | $7.1 \times 10^7$<br><br><br>$3.3 \times 10^{12}$ | 1.6<br>1.4<br>1.1<br>1.2 |
| CVD diamond [27]-TSC | Deuterium lamp | 543 (10 min) | ≈0.2 | 555 | 1.86 ±0.05 | $3 \times 10^{14}$ | 1 |
| CVD diamond [28]-TSC | Deuterium lamp | | ≈0.2 | 520 | 1.42<br>0.87 | | 1 |
| IIa diamond [28]-TSC | Deuterium lamp | | ≈0.2 | 330 | 0.68<br>0.78 | | 1 |

Other investigations have been carried out recently to gain better understanding of bandgap states. In particular, the thermally stimulated current technique was recently used for trap level spectroscopy and for UV dosimetry [27,28].

## D AS A RADIOPHOTOLUMINESCENCE DETECTOR

Araikum et al [29] studied the properties of diamond as a radiophotoluminescence dosimeter. The sample was first exposed to ionising radiation and then subjected to de-excitation illumination (usually UV). The intensity of the photoluminescence peak (526 nm) was directly proportional to the absorbed dose in the range from 0.1 to 10 kGy. Possible applications of such dosimeters can be found in sterilisation by radiation fields [3].

## E CONCLUSION

Diamond is an ideal candidate as a radiation detector since it is chemically stable, inert and essentially soft tissue equivalent. Moreover, diamond is an extremely versatile material and its application as a detector for ionising radiation can be considered in various modalities. Some key parameters of the performances of diamond dosimeters as measured and reported by different authors are summarised in this Datareview.

**REFERENCES**


[1] M. Heydarian, P.W. Hoban, A.H. Beddoe [ *Phys. Med. Biol. (UK)* vol.41 (1996) p.93-110 ]

[2] C. Furetta, G. Kitis, A. Brambilla, C. Jany, P. Bergonzo, F. Foulon [ *Radiat. Prot. Dosim. (UK)* vol.84 no.1-4 (1999) p.201-5 ]

[3] R.J. Keddy, T.L. Nam [ *Radiat. Phys. Chem. (UK)* vol.41 no.4-5 (1993) p.767-73 ]

[4] W.U. Laub, T.W. Kaulich, F. Nüsslin [ *Med. Phys. (USA)* vol.24 (1997) p.535-6 ]

[5] S.F. Kozlov et al [ *IEEE Trans. Nucl. Sci. (USA)* vol.NS-24 no.1 (1977) p.235-7 ]

[6] B. Planskoy [ *Phys. Med. Biol. (UK)* vol.25 no.3 (1980) p.519-32 ]

[7] E.A. Burgemeister [ *Phys. Med. Biol. (UK)* vol.26 no.2 (1981) p.269-75 ]

[8] M. Heydarian, P.W. Hoban, W.A. Beckam, I.M. Borchardt, A.H. Beddoe [ *Phys. Med. Biol. (UK)* vol.38 (1993) p.1035-42 ]

[9] S. Vatnisky, H. Järvinen [ *Phys. Med. Biol. (UK)* vol.38 (1993) p.173-84 ]

[10] W.U. Laub, T.W. Kaulich, F. Nüsslin [ *Phys. Med. Biol. (UK)* vol.44 (1999) p.2183-92 ]

[11] R.J. Keddy, T.L. Nam, R.C. Burns [ *Phys. Med. Biol. (UK)* vol.32 (1987) p.751-9 ]

[12] P.W. Hoban, M. Heydarian, W.A. Beckam, A.H. Beddoe [ *Phys. Med. Biol. (UK)* vol.39 (1994) p.1219-29 ]

[13] S.N. Rustgi [ *Phys. Med. Biol. (UK)* vol.43 (1998) p.2085-94 ]

[14] S.N. Rustgi, A.K. Rustgi, S.B. Jiang, K.M. Ayyangar [ *Phys. Med. Biol. (UK)* vol.43 (1998) p.3509-18 ]

[15] S.N. Rustgi [ *Med. Phys. (USA)* vol.22 no.5 (1995) p.567-70 ]

[16] P. Andreo, J.R. Cunningham, K. Hohlfeld, H. Svenson [ *Absorbed dose determination in photon and electron beams* Technical reports series no.277, International Atomic Energy Agency, Vienna, 1987 ]

[17] A. Brambilla, D. Tromson, N. Aboud, C. Mer, P. Bergonzo, F. Foulon [ *Nucl. Instrum. Methods Phys. Res. B (Netherlands)* to be published ]

[18] C. Manfredotti et al [ *Nucl. Instrum. Methods Phys. Res. B (Netherlands)* to be published ]

[19] S.W.S. McKeever [ *Thermoluminescence of Solids* Cambridge Solid State Science Series (Cambridge University Press, 1985) ]

[20] M. Benabdesselam, P. Iacconi, D. Briand, D. Lapraz, J.E. Butler [ *Radiat. Prot. Dosim. (UK)* vol.84 no.1-4 (1999) p.257-60 ]

[21] E. Borchi, C. Furetta, G. Kitis, C. Leroy, R.S. Sussmann, A.J. Whitehead [ *Radiat. Prot. Dosim. (UK)* vol.65 no.1-4 (1996) p.291-5 ]

[22] T.L. Nam, R.J. Keddy [ *Med. Phys. (USA)* vol.14 no.4 (1987) p.596-601 ]

[23] C. Furetta, G. Kitis, C.-H. Kuo [ *Nucl. Instrum. Methods Phys. Res. B (Netherlands)* vol.160 (2000) p.65-72 ]

[24] E. Vittone, C. Manfredotti, F. Fizzotti, A. Lo Giudice, P. Polesello, V. Ralchenko [ *Diam. Relat. Mater. (Switzerland)* vol.8 (1999) p.1234-9 ]

[25] C. Furetta, G. Kitis, A. Brambilla, C. Jany, P. Bergonzo, F. Foulon [ *Radiat. Prot. Dosim. (UK)* vol.44 no.1-4 (1999) p.201-5 ]

[26] F. Bogani, E. Borchi, M. Bruzzi, C. Leroy, S. Sciortino [ *Nucl. Instrum. Methods Phys. Res. A (Netherlands)* vol.388 (1997) p.427-30 ]

[27] P. Gonon, S. Prawer, D. Jamieson [ *Appl. Phys. Lett. (USA)* vol.70 no.22 (1997) p.2996-8 ]

[28] D. Tromson, P. Bergonzo, A. Brambilla, C. Mer, F. Foulon, V.N. Amosov [ *Phys. Status Solidi A (Germany)* vol.174 (1999) p.155-64 ]

[29] S. Araikum, T.L. Nam, R.J. Keddy [ *Radiat. Phys. Chem. (UK)* vol.44 no.4 (1994) p.359-70 ]

[30] NIST [ Physical Reference Data http://physics.nist.gov/PhysRefData/ ]

[31] C. Canali et al [ *Nucl. Instrum. Methods Phys. Res. (Netherlands)* vol.160 (1969) p.73-7 ]

[32] [ SCANDITRONICS Therados silicon semiconductor GR-p detector ]

[33] [ Wellhofer Dosimetrie IC10 Ionisation Chamber ]

# C1.6 Diamond optoelectronic devices

R.B. Jackman

May 2000

## A INTRODUCTION

Conventional solid state photodetectors, such as those made using silicon, are typically diode structures operated with a reverse bias placed across them. Low dark currents result; carriers photo-generated in the depletion region form a drift current which is the basis for light detection. Light with energy greater than the bandgap of the material can be seen. Silicon has a 1.1 eV bandgap and devices fabricated from this material therefore react to both ultraviolet (UV) and visible wavelengths. The high resistivity of diamond suggests that diode structures may not be needed to achieve low dark currents if this material were used for the fabrication of photodetectors. Devices can be considered which simply rely upon photoconductivity for their operation; such a device could possess high gain (also known as external quantum efficiency) since many carriers may be able to flow around the detector circuit during the lifetime of a photo-generated electron-hole pair. The wide bandgap (5.5 eV, 225 nm) of diamond implies that this form of photodetector will be capable of detecting deep UV light whilst being essentially 'blind' to visible wavelengths. This property is highly desirable; filtering conventional devices to make them visible blind significantly reduces their sensitivity to UV light. Moreover, filters often degrade in the presence of deep UV radiation leading to limited device lifetime. The physical and chemical robustness of diamond also suggests that such devices may be suitable for operation in hostile environments. Many industrial, military and environmental applications can thus be envisaged for diamond UV sensors. In this Datareview the design, fabrication and performance of diamond UV photodetectors is discussed, along with some of the uses for these devices. For some applications it may be desirable to fabricate diamond photodiodes and optically activated field effect transistors; the realisation of this type of device from p-type material is also addressed.

## B INTRINSIC AND EXTRINSIC PHOTOCONDUCTIVITY IN DIAMOND

In a real material defect states will enable light with energy lower than the bandgap energy to be absorbed. Some of these processes may lead to enhanced electrical conductivity and it is therefore important to understand the extrinsic, as well as the intrinsic, photoconductivity that is possible in diamond. The photoconductivity of natural diamond was investigated by Nahum and Halperin some years ago [1]. A strong photoconductive response with a threshold of around 225 nm originating from band to band transitions was recorded for a stone with properties between types I and II. Peaks in photoconductivity at longer wavelengths were strongly influenced by temperature; features present in a 300 K spectrum were often absent when investigated at 80 K. Indirect transitions to the conduction band by phonon interaction and thermal ionisation of carriers from states in the bandgap (as opposed to luminescence) were suggested to be the cause of this difference. Denham and co-workers investigated the influence of the nitrogen content on photoconductivity [2]. For a stone with little nitrogen (IIb), strong intrinsic absorption with a threshold of ~225 nm is apparent, but nitrogen containing stones (Ia) have a threshold around 300 nm; this was attributed to a nitrogen based donor level. Photoconductivity associated with the population of states by longer wavelength light (3.5 μm – 350 nm) has also been observed [3,4,9]. Sharp peaks within the spectral range 440 – 364 nm are thought to be caused by the presence of vacancies and these features have been widely observed following high energy electron irradiation [5,7,8].

The photoconductive properties of CVD diamond have also been investigated by a number of groups [11–15]. Allers and Collins [14] found that the dominant photoconductive response in the near infra-red (IR) and visible spectral regions was a broad band with a threshold of ~825 nm (1.5 eV). Whilst it was not clear whether this was a donor or acceptor state, the authors concluded that this defect was exclusive to diamond grown by CVD methods. Gonon and co-workers [13,15] identified three distinct peaks in this spectral region when they studied thin (10 μm) films supported on silicon, at 1127 (1.1 eV), 886 (1.4 eV) and 653 nm (1.9 eV). The latter were believed to be acceptor states associated with bulk diamond. It was suggested that the 1127 nm feature was associated with ionisation from acceptor states responsible for dark conductivity within the film, thought to be associated with disordered regions at the grain boundaries within the polycrystalline film.

## C PHOTOCONDUCTIVE UV DETECTORS

An ideal UV photodetector will be highly responsive to deep UV light, blind to visible wavelengths, support only a low dark current and rapidly reset after operation. It is apparent from the discussion above that significant extrinsic photoconductivity can be measured in most forms of diamond; the control, or eradication, of this property is therefore necessary for the realisation of high performance device structures from this material. Early attempts to fabricate UV photoconductive detectors from both single crystal and thin film diamond were only moderately successful [16]. Aluminium electrodes, separated by 300 μm, were evaporated onto type IIa natural, type Ib synthetic and CVD grown polycrystalline diamond samples. Relatively low sensitivity to 200 nm light was reported for the single crystal structures (gain <0.39); thin film diamond structures had sensitivities some $10^3$ times lower. Extrinsic photoconductivity prevented the device from being blind to visible light, with only 100:1 discrimination between 200 nm and visible wavelengths being achieved. Moreover, even the type IIa single crystal devices took >$10^4$ seconds to reset once they had been exposed to UV light. Vaitkus and co-workers [17] also measured a considerable photocurrent when exposing thin film polycrystalline diamond structures to visible light, once they had been exposed to deep UV wavelengths. Other reports have also shown that dark current levels can be strongly influenced by UV irradiation of thin film diamond structures [18].

More recent work has led to the realisation of device designs more appropriate for use with polycrystalline material and has succeeded in producing high performance diamond UV detectors [19–28]. CVD diamond films can comprise relatively large grains on the upper surface, whilst the substrate side of the film is fine grain material with significant graphitic inclusions. This region can therefore be expected to exhibit high dark conductivity through states that are likely to cause photoconduction from visible light. Furthermore, carriers may be trapped leading to recombination and reduction in carrier lifetimes. To reduce the influence of this region on the photoconductive response a structure where both electrodes are placed on the large grains of the top surface can be used. Carrier mean free paths in CVD diamond have been found to be in the range 1 – 10 μm [29]; optimal electrode spacing will therefore be of similar dimensions. An interdigitated electrode structure can be used to provide a large collection area for the device. The crystal structure within CVD diamond films can be controlled by variation of the growth conditions. An idealised arrangement would thus comprise planar electrodes whose spacing is well matched to the grain size within the film. The regions between the electrodes are then effectively single crystallites: hence carrier generation and recombination characteristics should be dominated by the CVD diamond as opposed to the grain boundary regions. In practice grains vary in size and orientation and the influence of grain boundaries cannot be completely ignored. In addition to careful device design it is necessary to treat the diamond in well chosen gaseous environments to achieve a low photoconductive response to visible light and high sensitivity to UV [19]. The photoconductive response of such a device to visible light is dramatically reduced by the application of a methane-air gas treatment. A sharp cut-off in responsivity at the bandgap edge (225 nm) is apparent. In addition to becoming visible blind ($10^6$:1 discrimination between 200 nm and visible wavelengths), the device also has a low dark current (<1 nA), improved turn-off characteristics and low operational bias level (1–10 V) and is highly sensitive [19–22]. An external quantum efficiency of 6 was demonstrated (i.e. 600% indicating gain), suggesting that material properties had been considerably improved, even when compared to single crystal structures [16]. The changes are believed to be associated with passivation of photoconductive defects during the gas treatments used [23].

Further work [24] led to the demonstration of devices with gains as high as $10^6$. Photoconductive devices can show gain since the number of electrons that will travel round the external circuit following the photogeneration of an electron-hole pair will depend upon the lifetime of the carriers. High gain requires high mobility, long carrier lifetime and closely spaced electrodes. The mobility-lifetime product, denoted $\mu\tau$, is a useful indicator of the potential for high performance operation of a diamond structure. The values reported [24] following gas treatments for diamond films with differing grain sizes (at $10^{-2}$–$10^{-5}$ $cm^2$ $V^{-1}$) are very high compared to those previously reported. For example, Salvatori et al [30] measured a value of $1 \times 10^{-6}$ $cm^2$ $V^{-1}$ for silicon supported polycrystalline diamond films; this low value was attributed to the presence of a high concentration of defects and impurities within the diamond film. These researchers fabricated planar photoconductive devices on Si supported thin film diamond and recorded gains up to 400, although a sharp cut-off in responsivity at the band edge was not measured. There are applications, such as in the fabrication of fast high voltage optoelectronic switches, where a short carrier lifetime is actually desirable. Indeed, Yoneda and co-workers [31] have cited the short carrier lifetimes measured in most polycrystalline CVD diamond as a reason that this material may be preferable to natural diamond for the fabrication of such devices.

The devices described above reset after exposure comparatively quickly, compared to all other examples, enabling these devices to be used within a number of applications. The application of several gas treatments actually enhances the speed at which the devices turn off [27]. The transient responses of one and five times treated devices to 193 nm excimer laser pulses at ~1.5 mJ $cm^{-2}$ $pulse^{-1}$ were compared. Both devices were subjected to a DC bias of 9 V and all other conditions were similar. The five times treated device is clearly faster than the device which received only a single treatment (FWHM 25 ns compared to 40 ns) and is comparable with the response of a vacuum photodiode, indicating that the true pulse duration of the 193 nm light is being recorded. The slower device, with a peak output response of 8 V, is around 30% more sensitive than the fast device. Excimer lasers can also output light with a wavelength of 157 nm; these devices have also proved to be fast and highly sensitive to this form of deeper UV light [28].

For deep UV beam imaging applications it is important to understand the homogeneity of the photoconductive signal from these devices. There is some evidence that the signal from devices fabricated on polycrystalline CVD material is stronger in certain crystalline regions compared to others [32]. However, provided all regions are sufficiently active that they 'see' light of the intensity used within a given application, and that the inhomogeneous nature of the signal intensity is stable, this does not prohibit the use of these devices; read-out electronics can simply be calibrated to account for this effect.

## D DIODES AND TRANSISTORS AS UV DETECTORS

In normal operation photodiodes do not display gain, although avalanche breakdown can be used to promote current amplification. However, since the photo-generated carriers need only traverse the depletion region to form a photocurrent, photodiodes can exhibit high speed characteristics. The absence of a reliable n-type dopant for diamond means that Schottky barrier diodes must be considered based on p-type material. Marchywka and co-workers [33] studied the UV response of the

Al Schottky diode on p-type (boron containing) single crystal material; at reverse bias levels greater than 5 V the leakage current exceeded the photocurrent detected. At lower voltages 100:1 discrimination between UV and visible light was reported.

The first purposefully designed thin film diamond photodiode structures utilised a two level Au Schottky contact on boron doped (~$10^{17}$ $cm^{-3}$) silicon supported thin film polycrystalline CVD diamond (~6 μm thick) [34]. A thin Au region enabled penetration of light through the contact into the depletion layer within the device whilst a thicker Au region was used as the contact pad. Ti-Ag-Au ohmic contacts were formed in the same plane. Low leakage currents (<2 pA) were measured for reverse bias levels of up to 100 V. With a reverse bias level of 50 V the spectral response of the device showed a sharp cut-off at the band edge (225 nm) and a discrimination of greater than $10^5$ between deep UV and visible wavelengths. A photoconductive device fabricated on similar material (without boron doping) showed a softer cut-off at the band edge and considerably higher dark currents, indicating the advantages offered by diode structures when using this type of fine grain thin film diamond. Salvatori and co-workers [30] have also fabricated Schottky barrier photodiodes; a field dependent photocurrent was measured at 185 nm, but the spectral response of the devices was not reported.

It has been demonstrated that near surface hydrogen can promote the formation of shallow acceptor states in diamond leading to p-type conductivity [35–40]. It has recently been shown [41] that Al contacts on hydrogenated CVD diamond films can be used for the formation of highly rectifying (Schottky) junctions, enabling high performance field effect transistors (MESFETs) to be realised. If these devices are designed such that the Al metallisation layer is very thin in the active area of the device, sufficient light penetration can occur such that an optically activated FET (Op-FET) can be formed [42]. Such devices enable high speed operation to be realised, since they rely on light absorption only in the depletion region for their activation, along with transistor gain. Enhancement mode MESFET structures, with an Al gate and Au source and drain contacts, show clear modulation of channel current as the gate bias is applied; for all gate bias values $I_{DS}$ rapidly saturates as $V_{DS}$ is increased, indicating channel pinch-off. The drain current of a similar device was plotted against the drain-source voltage, but with an electrically isolated (floating) gate which was illuminated with 200 nm light with varying intensity. The effect of the light is clearly to enhance the channel current level, with increasing optical powers giving higher saturated $I_{DS}$ values. A three-dimensional plot of the spectral responsivity of the device (the enhancement in the measured drain current compared to the incident light power) was made for differing $V_{DS}$ values as a function of illumination wavelength. It is immediately apparent that the response at the longer wavelengths is small compared to that for UV light; a pronounced increase in responsivity can be seen at around 320 nm, with a sharp rise at 220 nm which corresponds to the bandgap energy of diamond. This particular device displays a gain of 4 at 200 nm. These devices are not 'visible-blind' in the way that photoconductive structures can be, but they do offer the potential of high switching speed allied to high sensitivity.

Metal-semiconductor-metal (MSM) devices utilise adjacent metal contacts, which are Schottky in nature, as opposed to ohmic and Schottky metallisations found in conventional photodiode or Op-FET structures. In the case of deep UV photodetectors, the principle advantage of the MSM over an Op-FET lies in the removal of the need for light to penetrate the metal contacts themselves. For a planar device with a thin doped layer, and Schottky contacts with a high reverse bias breakdown potential, this structure can be operated such that the depletion layers extend laterally across the unmetallised region between the contacts. This allows light penetration directly into the diamond to activate the detector (similar to the case of a photoconductive device) enabling, for example, a 157 nm operation to be achieved. The photoresponse of an MSM device, fabricated from hydrogenated CVD diamond, to deep UV light (220 nm) of increasing optical power was investigated [43]. It is clear that the UV signal strongly influences the current from the device; at all incident powers the devices exhibit a low threshold voltage before the photocurrent begins to increase rapidly before reaching saturation at biases between 0.75 and 1 V. The spectral resistivity plot indicated a peak device responsivity at a wavelength of 200 nm and approximately 3 orders of discrimination between UV and optical wavelengths.

## E APPLICATIONS

The applications for deep UV photodetectors are diverse. Detectors for satellites and earth bound UV spectroscopy are required. Environmental monitoring is another application area. Water purifiers which operate with UV light require visible blind UV detectors as do flame detection systems within civilian and military sectors. Numerous manufacturing processes involve UV light and consequently require effective detectors. Within the semiconductor industry the move to 193 nm and 157 nm based lithography that will occur within the next few years offers yet another application area for diamond devices, since silicon detectors are simply not effective at this wavelength. This application area is particularly important, and devices from the author's laboratories are the first to be deployed within the semiconductor industry for this important function.

## F CONCLUSION

The design, fabrication and performance of diamond UV photodetectors has been discussed. Applications include environmental monitoring and many manufacturing processes, and diamond devices are being used in 193 nm and 157 nm based lithography in the semiconductor industry.

## REFERENCES


[1] J. Nahum, A. Halperin [ *J. Phys. Chem. Solids (UK)* vol.23 (1962) p.345 ]

[2] P. Denham, E.C. Lightowlers, P.J. Dean [ *Phys. Rev. (USA)* vol.161 (1967) p.762 ]

[3] A.T. Collins, E.C. Lightowlers [ *Phys. Rev. (USA)* vol.171 (1968) p.843 ]

[4] A.T. Collins, E.C. Lightowlers, P.J. Dean [ *Phys. Rev. (USA)* vol.183 (1969) p.725 ]

[5] R.G. Farrer, L.A. Vermeulen [ *J. Phys. C, Solid State Phys. (UK)* vol.5 (1972) p.2762 ]

[6] L.A. Vermeulen, R.G. Farrer [ *Diamond Res. (UK)* (1975) p.18 ]

[7] A. Halperin, L.A. Vermeulen [ *J. Phys. Chem. Solids (UK)* vol.43 (1982) p.691 ]

[8] L.A. Vermeulen, A. Halperin [ *J. Phys. Chem. Solids (UK)* vol.45 (1984) p.771 ]

[9] E. Pereira, L. Santos [ *Diam. Relat. Mater. (Switzerland)* vol.4 (1995) p.688 ]

[10] See, for example, P.K. Bachmann, I.M. Buckley-Golder, J.T. Glass, M. Kamo (Eds.) [ *Diamond Films '95* (Elsevier Science, 1996) ]

[11] L.S. Pan et al [ *Science (USA)* vol.255 (1992) p.830 ]

[12] R. Vaitkus, T. Inushima, S. Yamazaki [ *Appl. Phys. Lett. (USA)* vol.62 (1993) p.2384 ]

[13] P. Gonon, A. Deneuville, E. Gheeraert, F. Fontaine [ *Diam. Relat. Mater. (Switzerland)* vol.3 (1994) p.836 ]

[14] L. Allers, A.T. Collins [ *J. Appl. Phys. (USA)* vol.77 (1995) p.3879 ]

[15] P. Gonon, A. Deneuville, F. Fontaine, E. Gheeraert [ *J. Appl. Phys. (USA)* vol.78 (1995) p.6633 ]

[16] S.C. Binari, M. Marchywka, D.A. Koolbeck, H.B. Dietrich, D. Moses [ *Diam. Relat. Mater. (Switzerland)* vol.2 (1993) p.1020 ]

[17] R. Vaitkus, T. Inushima, S. Yamazaki [ *Appl. Phys. Lett. (USA)* vol.62 (1993) p.2384 ]

[18] P. Gonon, S. Prawer, Y. Boiko, D.N. Jamieson [ *Diam. Relat. Mater. (Switzerland)* vol.6 (1997) p.860 ]

[19] R.D. McKeag, S.S.M. Chan, R.B. Jackman [ *Appl. Phys. Lett. (USA)* vol.67 (1995) p.2117 ]

[20] S.S.M. Chan, R.D. McKeag, M.D. Whitfield, R.B. Jackman [ *Phys. Status Solidi A (Germany)* vol.154 (1996) p.445 ]

[21] M.D. Whitfield, R.D. McKeag, L.Y.S. Pang, S.S.M. Chan, R.B. Jackman [ *Diam. Relat. Mater. (Switzerland)* vol.5 (1996) p.829 ]
[22] R.D. McKeag, M.D. Whitfield, S.S.M. Chan, L.Y.S. Pang, R.B. Jackman [ *Mater. Res. Soc. Symp. Proc. (USA)* vol.416 (1996) p.419 ]
[23] S.S.M. Chan, R.B. Jackman [ *Diam. Relat. Mater. (Switzerland)* vol.6 (1997) p.374 ]
[24] R.D. McKeag, R.B. Jackman [ *Diam. Relat. Mater. (Switzerland)* vol.7 (1998) p.513 ]
[25] R.B. Jackman [ *SPIE (Laser in Syn., Charac. Proc. of Diamond)* vol.3483 (1999) p.283 ]
[26] O. Gaudin, S. Watson, S.P. Lansley, H.J. Looi, M.D. Whitfield, R.B. Jackman [ *Diam. Relat. Mater. (Switzerland)* vol.8 (1999) p.886 ]
[27] S.P. Lansley, M.D. Whitfield, R.D. McKeag, N. Rizvi, R.B. Jackman [ *Diam. Relat. Mater. (Switzerland)* vol.9 (2000) p.195 ]
[28] S.P. Lansley, M.D. Whitfield, R.D. McKeag, N. Rizvi, R.B. Jackman [ *Appl. Phys. Lett. (USA)* in press ]
[29] D.R. Kania, M.I. Landstrass, M.A. Plano, L.S. Pan, S. Han [ *Diam. Relat. Mater. (Switzerland)* vol.2 (1993) p.1012 ]
[30] S. Salvatori, E. Pace, M.C. Rossi, F. Galluzzi [ *Diam. Relat. Mater. (Switzerland)* vol.6 (1997) p.361 ]
[31] H. Yoneda, K. Ueda, Y. Aikawa, K. Baba, N. Shohata [ *Appl. Phys. Lett. (USA)* vol.66 (1995) p.460 ]
[32] J. Hiscock, A.T. Collins [ *Diam. Relat. Mater. (Switzerland)* vol.8 (1999) p.1753 ]
[33] M. Marchywka, J.F. Hochedez, M.W. Geis, D.G. Socher, D. Moses, R.T. Goldberg [ *Appl. Opt. (USA)* vol.30 (1991) p.5011 ]
[34] M.D. Whitfield, S.S.M. Chan, R.B. Jackman [ *Appl. Phys. Lett. (USA)* vol.68 (1996) p.290 ]
[35] M.I. Landstrass, K.V. Ravi [ *Appl. Phys. Lett. (USA)* vol.55 (1989) p.1391 ]
[36] H. Shiomi, Y. Nishibayashi, N. Fujimori [ *Jpn. J. Appl. Phys. (Japan)* vol.30 (1991) p.1363 ]
[37] T. Maki, S. Shikama, M. Komori, Y. Sakaguchi, K. Sakuta, T. Kobayashi [ *Jpn. J. Appl. Phys. (Japan)* vol.31 (1992) p.1363 ]
[38] K. Hayashi, S. Yamanaka, H. Okushi, K. Kajimura [ *Appl. Phys. Lett. (USA)* vol.68 (1996) p.376 ]
[39] H.J. Looi, J.S. Foord, R.B. Jackman [ *Appl. Phys. Lett. (USA)* vol.72 (1998) p.353 ]
[40] H.J. Looi, L.Y.S. Pang, A.B. Molley, F. Jones, J.S. Foord, R.B. Jackman [ *Diam. Relat. Mater. (Switzerland)* vol.7 (1998) p.550-5 ]
[41] H.J. Looi, L.Y.S. Pang, Y. Wang, M.D. Whitfield, R.B. Jackman [ *IEEE Electron Device Lett. (USA)* vol.19 (1998) p.112-4 ]
[42] S.P. Lansley, H.J. Looi, M.D. Whitfield, R.B. Jackman [ *Appl. Phys. Lett. (USA)* vol.74 (1999) p.615-7 ]
[43] H.J. Looi, S.P. Lansley, M.D. Whitfield, R.B. Jackman [ *Appl. Phys. Lett. (USA)* vol.74 (1999) p.3332-4 ]

CHAPTER C2

# MECHANICAL APPLICATIONS AND THERMAL MANAGEMENT

# C2.1 Sample requirements for diamond deposition

B. Lux and R. Haubner

April 2000

## A INTRODUCTION

The extreme hardness and wear resistance of low-pressure diamond predestines diamond as an ideal coating for certain wear part applications [1,2]. Major considerations for development and fabrication of high-quality tool coatings are:

(a) Appropriate surface pre-treatments of the tool substrate prior to diamond coating in order to obtain a perfect coating adhesion, high nucleation rates, etc.
(b) Adjustment of diamond deposition parameters for specific, different applications.
(c) Economic, large-volume or high growth rate synthesis methods with flexible diamond growth parameters permitting a high level of reproducibility.

In general, cutting tools require two equally important material properties: wear resistance and toughness. Ceramic materials with a high wear resistance generally offer low toughness. The high hardness of pure single-crystal diamond is not fully usable, due to its easy cleavage. One way to avoid cracking could be the use of thin polycrystalline 'Ballas' diamond [3].

Another way to increase toughness is a principle also well known from hardmetals (WC–Co), whereby the WC particles are consolidated by a metal binder phase. A similar type of composite material, called polycrystalline-diamond (PCD), is a product where diamond crystals are embedded in a Co binder phase. With increasing Co content, the ductility of the composite increases while the wear resistance decreases. Another successful approach to combine toughness and wear resistance for tools is to coat a hard, ductile and heat-conducting substrate with a thin surface layer, such as binder-free diamond. Its very high hardness and heat conductivity are useful for wear applications [4]. Last, but not least, also the economics, i.e. the final costs, play an important role.

## B COMPOSITE DIAMOND TOOLS FOR WEAR APPLICATIONS

CVD-diamond applied onto WC–Co can substantially increase lifetime during heavy-duty cutting or milling operations on materials such as graphite, fibre-reinforced plastics, laminated wood products, timber, concrete and various rock and stone products [5,6].

Two types of tools can be produced:

1. free-standing (thick) diamond layers brazed onto a carrier substrate,
2. thin diamond layers deposited directly onto an appropriate substrate.

### B1 Free-Standing Diamond Layers for Tools

Wear parts containing free-standing diamond layers are commercially available today. For tools with simple geometries the bonding of thick free-standing CVD-diamond sheets by brazing is very applicable. It competes with the classical PCD (polycrystalline-diamond) products which are produced by high-pressure sintering of diamond powder with a binder metal (mainly Co). PCD parts are limited in size—to about 70 mm in diameter—because of the high-pressure equipment.

Free-standing diamond layers can be produced in large sizes by various diamond-deposition methods which [7] in the case of the manufacturing of endmills and simple cutting tools are carried out in three steps.

First a thick, large-sized diamond layer is deposited on a dummy substrate. The diamond can be removed from the substrate or manufactured as is. The diamond sheet is cut by a laser into the pieces needed for the tools.

Second, the diamond sheets are bonded to the wear part.

Third, the tool and the layer are ground to their final geometry.

Low-pressure CVD-diamond layers are 100% diamond (it is a binder-free material); thus their wear resistance is even better than that of PCD [8]. However, during machining with thick CVD-diamond layers, edge deterioration may occur. This is due to the cleavage of diamond crystals, as the diamond CVD structure is composed of relatively coarse, elongated single crystals, perpendicularly oriented to the substrate layers. Polycrystalline 'Ballas' diamond [3,9] might provide a solution to such problems. Products using free-standing diamond layers are, for example, endmills and cutting tools.

### B2 In-situ Diamond-Coated Tools

For most industrial applications of diamond-coated tools hardmetal substrates are the obvious choice. A primary requirement for the industrial use of a diamond coating is its adhesion to the substrate [10]. Therefore, the use of an appropriate hardmetal grade, and the application of a surface pre-treatment are equally important. The rigorous control of all possible interfacial interactions occurring at the diamond/hardmetal interface is also necessary. For commercial realization the coating adhesion must be reproducible.

The diamond synthesis method applied and the correct choice of deposition parameters are also important in achieving a usable and economic product. Generally, the energy density during diamond synthesis—and thus the diamond growth rates—are lower for tool coating (growth rate around 1 μm/hr), in order to avoid thermal damage to the substrate properties, as compared to free-standing layer synthesis, in which growth rates may exceed 10 μm/hr.

The advantage of in-situ diamond deposition is that complex tool geometries can be coated.

## C DIAMOND DEPOSITION ON HARDMETAL SUBSTRATES

Hardmetals are the most valuable and important substrates for coated tools, due to their intrinsic properties and their wide range of mechanical properties. They basically consist of WC and Co with additions of TiC, (Ta,Nb)C, VC and so on, which mainly change their hardness and wear resistance [11]. The amount of Co binder is largely responsible for ductility or brittleness. Hardmetals have been used as wear parts and cutting tools for decades, with and without coating applications.

Successful diamond coatings on WC–Co substrates showing reasonable adhesion are reported in the literature on substrates that generally have no or a very low amount of cubic carbides (TiC) and also a relatively low Co content. Both Co and TiC additions increase the thermal expansion coefficient of the hardmetal and reduce the adhesion of the diamond coating [12,13].

## C1 The Co Binder and its Detrimental Influences

A high Co vapour pressure and its high mobility on the substrate surface influence diamond deposition. In the gas phase surrounding the substrate surface, Co catalyses the formation of non-diamond carbon phases, which can be deposited at the interface prior to the diamond formation. Thus, the adhesion between diamond and the substrate is jeopardised [14]. Such carbon deposits influence diamond nucleation and adhesion detrimentally. Special nucleation enhancement might become necessary to achieve reasonable nucleation densities and thus rapid layer formation [15,16].

Co droplets are formed by migration, on the substrate surface but also on top of the diamond layers [17,18]. How and why the Co drops reach the diamond coating surface is not yet fully understood. Surface forces might play an important role (FIGURE 1).

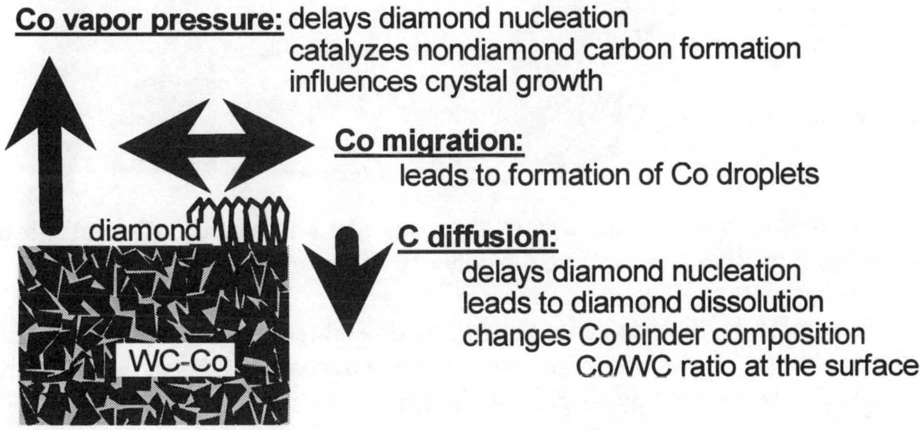


FIGURE 1 Co as part of the hardmetal binder phase and its influences on the diamond deposition [19].

## C2 The Carbon Balance within the WC–Co Substrate

During sintering, or heat treatments, of hardmetals the C and W contents in the Co binder phase can be changed by varying the H/O/C ratio in the gas phase. The C-balance in the Co binder phase becomes most important during diamond deposition, as carbon diffuses into the WC-Co substrate if the Co binder has a low C content.

Additionally, the adhesion of the diamond coating is influenced by carbon dissolution/diffusion processes. Industrially available hardmetal substrates are classified into two categories. Depending on the carbon saturation of the Co binder, different reactions occur during the diamond-coating process:

High carbon: Co is fully saturated with C while the W content is very low. Thus even formation of free C precipitates (= graphite) in the Co binder phase is possible during cooling.

Medium carbon: W is dissolved in the Co binder phase (up to 15%). The Co content in the binder phase becomes lower, but the amount of binder phase increases by forming a solid solution of W in Co. The carbon content in the binder is low, but is regulated by the W content (a high amount of W results in a low C content). This type is the most common hardmetal composition used for cutting tools.

During diamond nucleation and the early growth stage, C diffusion into the substrate occurs. Isolated diamond crystals, as well as the deposited diamond layer might dissolve at the diamond/hardmetal interface, and as long as there is a C gradient across the WC–Co substrate carbon diffuses into the substrate via the Co binder [19,20] (FIGURE 2).

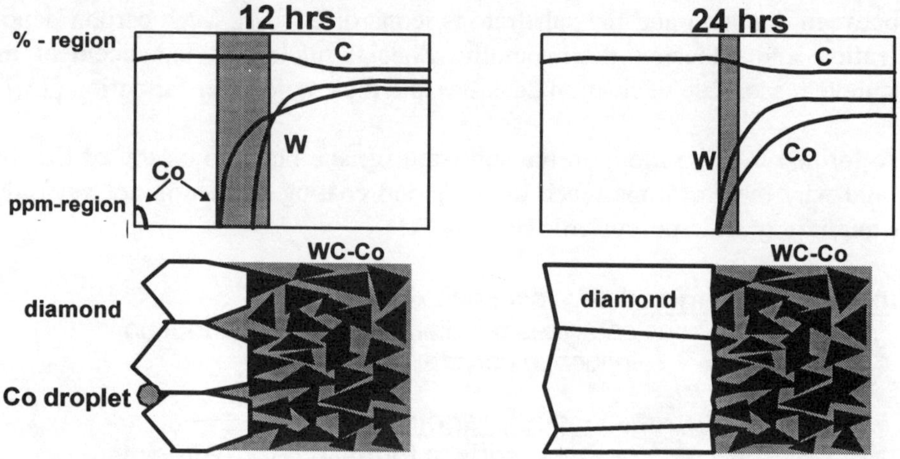


FIGURE 2 Illustration of diamond dissolution by the Co binder phase of a hardmetal substrate, based on the depth profiles of C, W and Co concentrations, measured by SIMS [19].

To obtain good adhesion of the diamond coating on a hardmetal tool, it is necessary to saturate the Co binder in WC–Co hardmetal grades with C prior to diamond deposition. Diamond dissolution into the Co binder phase can also be reduced by shortening the heating time (= deposition time) [19].

## C3 Surface Treatments of Hardmetal Substrates prior to Diamond Deposition to Improve the Adhesion of Diamond Coatings

Several approaches to reduce the Co vapour pressure and Co migration have been discussed, such as: local reduction of Co by surface pre-treatment via etching [21–23]; intermediate layers [24]; reducing Co mobility by forming stable Co compounds and so on [25] (FIGURE 3).

### C3.1 Removal of the Co binder by etching

A number of pre-treatment procedures aim at a depletion of the Co at the hardmetal substrate surface by selective chemical etching with acids such as $HNO_3$, HCl or $CH_3COOH$ [21–23,26]. Thus, Co migration is drastically reduced. A rough surface increases diamond nucleation, while a rough interface additionally improves the coating adhesion. The formation of excessively large pores, on the contrary, is detrimental. Thus for good adherence removal of Co from the surface must be carried out carefully [26].

A two-step pre-treatment by etching gives good diamond adhesion. First, the WC is removed by Murakami solution ($K_3[Fe(CN)_6]$ in KOH and $H_2O$), leaving an excess of Co on the hardmetal surface. Subsequently, Co is treated with an $H_2SO_4/H_2O_2$ solution. Systematic investigations have proved that only the combination of both etching steps can give improved adhesion results [23,27].

### C3.2 Co leaching and replacing by Cu

Treating a WC–Co substrate with a solution of a Cu salt leads to the substitution of Co by Cu at the substrate surface, due to a simple electrochemical reaction:

$$Co + Cu^+ + e^- \rightarrow Co^+ + Cu + e^-$$

After 1–2 min, the Co at the substrate surface is replaced by Cu. Adhesion of the diamond coating on Cu-treated substrates has been reported to be much better than on substrates from which Co was removed by chemical leaching only [28,29].

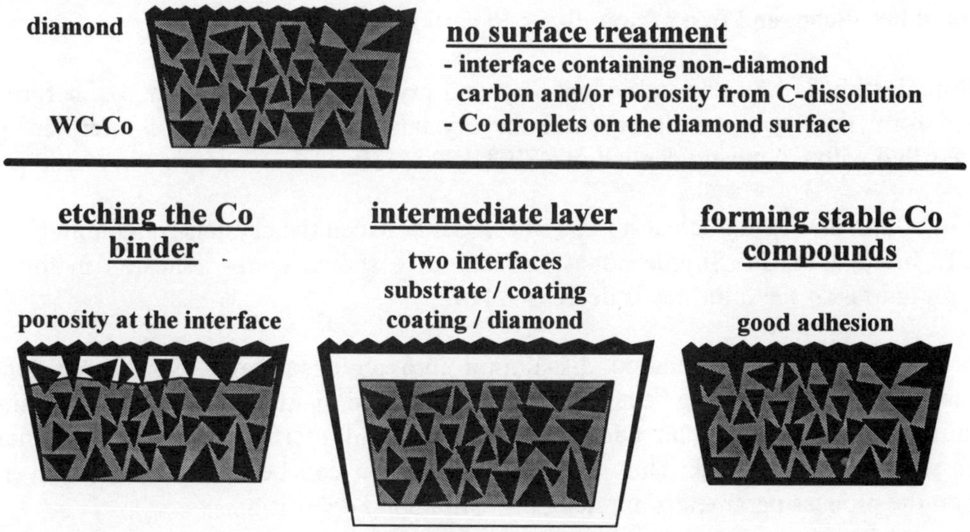


FIGURE 3 Different types of surface treatments of hardmetal substrates leading to improved adhesion of diamond coatings.

### C3.3 Intermediate interfacial layers

To separate the diamond coating from the hardmetal surface the application of intermediate layers has frequently been described as a promising technique for the deposition of adherent diamond films [24]. Intermediate layers which fully cover the substrate reduce the negative Co effects. The successful adhesion of a diamond coating deposited on such intermediate layers depends strongly on their chemical nature and composition. The problem of gaining acceptable adhesion of the diamond layer remains with most of the intermediate materials [30]. Only a few substrates seem to permit a better diamond adhesion than pure WC, which could thus be used as a successful intermediate layer itself.

### C3.4 Interfacial barriers formed by stable Co compounds

A successful way to improve the diamond-coating adhesion is based on the formation of Co compounds which are stable under diamond deposition conditions. Compound formation on the binder areas of the surface decrease the Co vapour pressure and hinder Co migration. Thus diamond layers formed on hardmetals treated in this way show well-faceted crystals, and no detrimental Co influences as described above [25].

A great number of elements (such as B, Si, Al, Ti, Hf, and Sc) can be used to form such stable Co compounds; some have proved to be of interest and applicable in industrial tools [31,32].

Hardmetal substrates pre-treated with B or Si vapour at elevated temperatures form Co borides or silicides on the binder surface [25]. Cobalt-aluminium-compounds such as aluminides or aluminates can be formed by substrate pre-treatment with solutions respectively suspensions of aluminium-salts in water. This method combines a soft etching process with the formation of stable compounds [33].

Also, thin sputtered layers of Ti or Cr resulted in increased diamond nucleation and diamond growth rates, and led to excellent diamond-coating adhesion.

## C4 Some Conclusions on the Adhesion of Diamond Coatings on Hardmetal Substrates

Many parameters, like substrate preparation and deposition conditions, influence the adhesion of diamond coatings on hardmetal substrates (FIGURE 1) [19]. Certainly, the diamond coating/substrate interface should be clean, and free of non-diamond carbon.

If diamond nucleation is low, the individual diamond crystals can increase in size before they grow together and form the layer. Voids between the crystals are formed, at the interface, which can, however, be filled with Co during deposition (FIGURE 2) [30].

In the case of a high diamond nucleation rate, the voids between the crystals remain much smaller and are generally filled with Co. Simple adhesion tests have shown better adhesion in this case, even though the roughness of the interface is decreased [19].

As already explained above, diamond dissolution into an unsaturated Co binder occurs at the diamond/substrate interface during heat treatment of deposited diamond. This process again leads to a reduction of the interfacial voids, but additionally the diamond interface roughness becomes smoother and further stresses are induced. Thus diamond dissolution can be detrimental to layer adhesion, depending on the process parameters applied during diamond deposition.

However, diamond dissolution does not occur with C-saturated hardmetal substrates, so that these problems could be avoided easily [19].

## D FERROUS METAL SUBSTRATES (HSS)

The ease of carbon diffusion into ferrous substrates is the main problem for the diamond deposition process. At normal diamond deposition temperatures, the austenitic phase is present, which has a high carbon solubility (up to 2.06 wt % C) [34]. Thus barrier layers have to be applied in order to inhibit carbon diffusion into the substrates [30], otherwise the deposition temperature has to be decreased below the ferrite–austenite transformation temperature [35].

Decreasing the deposition temperature below 723°C allows the deposition on ferrite which shows minimal solubility for carbon. 723°C is the ferrite–austenite transformation temperature in the pure Fe–C system. At such low deposition temperatures, the diamond growth rate decreases and almost reaches zero, when conventional deposition methods are used. However, diamond deposition on HSS is possible with high-power microwave equipment and substrate cooling. Only under such conditions can the carbon diffusion into the substrate, without intermediate layers, be avoided [35]. Although diamond coatings on HSS tools for industrial applications are of great interest, until now no commercial tools have become available.

Diffusion barrier layers such as TiN, TiC, Ti(CN), TiAlN and CrC have been applied on various steel substrates. The production of such interlayers is expensive, as a two-step PVD process is needed to reach a fully covered surface and to prevent carbon interaction with the substrate. Such barrier coatings have to fully cover the substrate, must be free of pores and cracks, and need to show excellent adhesion to the substrate and the diamond [30].

The combination of intermediate layers and low-temperature deposition seems to be a successful way to apply diamond coatings onto steel substrates [36]. Thereby, the steel substrate is coated in situ, prior to diamond deposition with cubic silicon carbide (3C-SiC), in order to prevent carbon diffusion and hydrogen brittleness [37].

In another approach, a pre-coating composite layer containing a Ni and diamond mixture is applied by electro-deposition. This layer serves as a nucleation layer for the diamond and provides good adhesion to the steel substrate. Diamond depositions with a duration of more than 24 hr have been made, whereby the hardness of a steel substrate remained unchanged [36].

## E CERAMIC SUBSTRATES

In the early days of diamond deposition, the fundamentals were studied using SiAlON substrates which are stable in the diamond deposition atmosphere at high temperature. The adhesion of the diamond coatings proved to be good and reproducible experiments were conducted [1].

On other ceramic materials diamond-deposition experiments (h-BN and $ZrO_2$) resulted in chemical reactions with atomic hydrogen and/or the carbon species. No chemical reactions between the gas phase and the substrates were observed during diamond-coating experiments on $Al_2O_3$, SiC and SiAlON substrates. However, thermodynamic calculations showed that some reactions are possible [38].

Also, various composite ceramics were tested and proposed as substrates for tool materials, such as ($Al_2O_3$ + TiC), ($Al_2O_3$ + SiC) and ($Al_2O_3$ + $ZrO_2$), but the diamond adhesion on SiAlON substrates always turned out to be higher [39].

Due to their relatively high shock sensitivity diamond-coated ceramic tools are rarely used in industrial applications today. The authors know only of an $Si_3N_4$ round tool which is produced for graphite machining [40].

## F CONCLUSION

Since Matsumoto [41] presented the first low-pressure diamond deposition, more than a decade has passed. He used the hot-filament method and gave the first technological approach to a simple diamond synthesis process allowing diamond deposition on various surfaces, larger and thinner than ever seen before. Although it was one of the first objectives, diamond-coated hardmetal tools have only recently started to become a serious industrial product. The main reasons for this delay were not due to problems with developing available methods of synthesis, but due to problems with adhesion and reliability of the diamond coatings on the hardmetal inserts. Last but not least, large-scale industrial production is also problematic. Overall, the problems have turned out to be far more complicated in comparison with classical CVD and PVD tool coatings, such as TiC, TiN, $Al_2O_3$ and so on, which were introduced industrially two to three decades ago. Currently, the advantages of the 'new diamond tools' in industrial applications have to be proved. A large domain of applications that call for precise shaping with chip forming is in prospect—from the machining of non-ferrous metals, carbon products and refractory materials over fibre-reinforced plastics to hard and wear-resistant parts, and other 'high-tech' structures. However, diamond surfaces for applications such as extruder dies, heavy-duty wear and friction parts, and composite tools combining extreme hardness, wear resistance and high heat conducting properties should also profit from these developments. Furthermore, new areas of research and development are open, to establish sophisticated composite structures and novel tools based on chipless modification of materials.

**ACKNOWLEDGEMENTS**

We thank the management of Sandvik, Stockholm and Balzers AG, Liechtenstein for sponsoring parts of our early works. Last, but not least, we wish to express our gratitude to the Austrian Science Foundation which currently sponsors our work, carried out under the auspices of the trinational cooperation 'D–A–CH' (Germany, Austria, Switzerland), on the 'Synthesis of Superhard Materials' initiated by K. Komarek (FWF; Projects S5907 and P12055).

**REFERENCES**

[1] B. Lux, R. Haubner [ in *Diamond Films and Coatings* Ed. R.F. Davis (NOYES Publications, 1993) p.184-243 ]

[2] B. Lux, R. Haubner [ in *Low Pressure Synthetic Diamond* Eds. B. Dischler, C. Wild (Springer-Verlag, 1998) ch.12 p.223-42 ]

[3] B. Lux, R. Haubner, H. Holzer, R.C. DeVries [ *Refract. Metals Hard Mater.* vol.15 (1997) p.263 ]

[4] J. Nedelik, W. Kremser, B. Lux [ *Diam. Films Technol. (Japan)* vol.8 no.1 (1998) p.37-50 ]

[5] H. Schachner, B. Lux, K. Stjernberg, A. Thelin, H. Tippmann [ European Patent 166708 (1986) ]

[6] M. Murakawa, S. Takeuchi [ *Surf. Coat. Technol. (Switzerland)* vol.49 (1991) p.359 ]

[7] T. Yashiki, T. Nakamura, N. Fujimori, T. Nakai [ in *18$^{th}$ Int. Conf. on Metallurgical Coatings and Thin Films (ICMCTF)* San Diego, CA (1991) ]

[8] [ NORTON Diamond Film, DIAMAPAK$^{TM}$ (prospectus) ]

[9] S. Bühlmann, E. Blank, R. Haubner, B. Lux [ *Diam. Relat. Mater. (Switzerland)* vol.8 (1999) p.194 ]

[10] K. Shibuki, M. Yagi, K. Saijo, S. Takatsu [ *Surf. Coat. Technol. (Switzerland)* vol.36 (1988) p.295 ]

[11] W. Schedler [ in *Hartmetall für den Praktiker* Ed. Plansee TIZIT GmbH (VDI-Verlag, Düsseldorf, 1988) p.43-6 ]

[12] R. Haubner, B. Lux [ *J. Phys. (France)* vol.C5 (1989) p.169 ]

[13] R. Haubner, A. Lindlbauer, B. Lux [ *Diam. Relat. Mater. (Switzerland)* vol.2 (1993) p.1505 ]

[14] H. Matsubara, J. Kihara [ in *Science and Technology of New Diamond* Eds. S. Saito, O. Fukunaga, M. Yoshikawa (KTK Scientific Publishers, Tokyo, 1990) p.89-93 ]

[15] K. Saijo, M. Yagi, K. Shibuki, S. Takatsu [ *Surf. Coat. Technol. (Switzerland)* vol.43/44 (1990) p.30 ]

[16] J.L. Chen, T.H. Huang, F.M. Pan, C.T. Kuo, C.S. Chang, T.S. Lin [ *Surf. Coat. Technol. (Switzerland)* vol.54/55 (1992) p.392 ]

[17] A.K. Mehlmann, A. Fayer, S.F. Dirnfeld, Y. Avigal, R. Porath, A. Kochman [ *Diam. Relat. Mater. (Switzerland)* vol.2 (1993) p.317 ]

[18] D. Satrapa [ Doctoral Thesis, Vienna University of Technology (1993) ]

[19] S. Kubelka, R. Haubner, B. Lux, R. Steiner, G. Stingeder, M. Grasserbauer [ *Diam. Films Technol. (Japan)* vol.5 (1995) p.105 ]

[20] R. Haubner, W.D. Schubert, B. Lux [ in *14th Int. Plansee Seminar* Reutte, Tirol, Austria, 12-16 May 1997, Eds. G. Kneringer et al, vol.3 p.16-30 ]

[21] T.H. Huang, C.T. Kuo, C.S. Chang, C.T. Kao, H.Y. Wen [ *Diam. Relat. Mater. (Netherlands)* vol.1 (1992) p.594 ]

[22] A.K. Mehlmann, S.F. Dirnfeld, Y. Avigal [ *Diam. Relat. Mater. (Netherlands)* vol.1 (1992) p.600 ]

[23] M.G. Peters, R.H. Cummings [ European Patent No. 0519 587 A1 (1992) ]

[24] T. Isozaki et al [ *Diam. Relat. Mater. (Switzerland)* vol.2 (1993) p.1146 ]

[25] S. Kubelka, R. Haubner, B. Lux, R. Steiner, G. Stingeder, M. Grasserbauer [ *Diam. Relat. Mater. (Switzerland)* vol.3 (1994) p.1360 ]
[26] X.X. Pan [ Doctoral Thesis, Vienna University of Technology (1989) ]
[27] R. Haubner, S. Kubelka, B. Lux, M. Griesser, M. Grasserbauer [ *J. Phys. IV (France)* vol.C5 (1995) p.753 ]
[28] M. Nesladek, K. Vandierendonck, C. Quaeyhaegens, M. Kerkhofs, L.M. Stals [ *Thin Solid Films (Switzerland)* vol.270 (1995) p.184 ]
[29] M. Sommer, R. Haubner, B. Lux [ *Diam. Relat. Mater. (Switzerland)* vol.9 (2000) in press ]
[30] S. Kubelka [ Doctoral Thesis, Vienna University of Technology (1994) ]
[31] W.G. Moffatt [ *The Handbook of Binary Phase Diagrams* (Genium, New York, 1976) ]
[32] J. Karner, E. Bergmann, M. Pedrazzini, I. Reineck, M. Sjöstrand [ Int. Patent WO 95/12009 (4 May 1995) ]
[33] A. Köpf [ Doctoral Thesis, Vienna University of Technology (2000) ]
[34] A. Lindlbauer, R. Haubner, B. Lux [ *Diam. Films Technol. (Japan)* vol.2 (1992) p.81 ]
[35] G. Rosenauer [ Diplom Thesis, Lehrstuhl Metalle (WTM), Universität Erlangen-Nürnberg (1997) ]
[36] C.-P. Klages, M. Fryda, T. Matthee, L. Schäfer, H. Dimingen [ in *14th Int. Plansee Seminar* Reutte, Tirol, Austria, 12-16 May 1997, Eds. G. Kneringer et al, vol.3 p.1-15 ]
[37] L. Schäfer, A. Blum, M. Sattler, R. Six, C.-P. Klages [ *in Applications of Diamond Films and Related Materials: Third Conference* Eds. A. Feldman et al (NIST Special Publication 885, Washington, 1995) p.399-402 ]
[38] A. Lindlbauer, R. Haubner, B. Lux [ *Wear (Switzerland)* vol.159 (1992) p.67-77 ]
[39] E. Cappelli, F. Pinzari, P. Ascarelli, G. Righini [ *Diam. Relat. Mater. (Switzerland)* vol.5 (1996) p.292 ]
[40] L.K. Bigelow [ NORTON, personal communication ]
[41] S. Matsumoto, Y. Sato, M. Tsutisimi, N. Setaka [ *J. Mater. Sci. (UK)* vol.17 (1982) p.3106 ]

# C2.2 CVD diamond for thermal management

H. Windischmann

January 2000

## A INTRODUCTION

Long-term package reliability is becoming an increasing concern as IC clock speeds and power densities increase while simultaneously the package footprint decreases, placing an ever-increasing demand on thermal management. Increasing the reliability and performance of IC packages requires reducing the package/junction temperature and improving the integrity of the interfaces between the constituents of the package. The heat generated is proportional to the computing power, which is a product of the number of transistors and the clock frequency. Over the last decade the computing power has increased by five orders of magnitude, yet the design rules have remained the same, i.e. the junction temperature must be maintained below 125°C, placing severe demand on the heat dissipation medium.

TABLE 1 lists materials that have been used and are being considered as potential candidates for managing the waste heat.

TABLE 1 Comparison of thermal management heat spreader materials

| Material | Thermal exp. coefficient[1] (ppm/°C) | Thermal conductivity[2] (W/m K) | Density (gm/cm$^3$) |
|---|---|---|---|
| Al | 23.6 | 171 | 2.7 |
| Cu | 17.6 | 391 | 8.9 |
| Kovar | 5.8 | 17 | 8.2 |
| Mo | 5.1 | 140 | 10.2 |
| Cu-75%W | 8.3 | 228 | 14.8 |
| Cu-85%W | 7.0 | 226 | 16.4 |
| BeO | 7.2 | 260 | 2.9 |
| $Al_2O_3$ | 6.3 | 29 | 3.9 |
| AlN | 3.8 | 180 | 3.3 |
| CVD diamond | 2.8 | 100–2500 | 3.5 |
| Si-50%Al | 11 | 140 | 2.5 |
| Si-30%Al | 6.8 | 120 | 2.4 |
| Be-30%BeO | 8.7 | 210 | 2.1 |
| Al-65%SiC | 7.9 | 158 | 3.0 |
| GaAs | 5.6 | 50 | 5 |
| Si | 3.5 | 130 | 2.3 |

[1] Data shown are room temperature values, which typically increase with temperature.
[2] Data shown are room temperature values, which typically decrease with temperature.

The materials presented are selected based on three criteria: thermal expansion coefficient, thermal conductivity and density, since these properties directly impact performance, reliability and package footprint/weight, respectively. For comparison, the data also include the characteristics of Si and GaAs. CVD diamond, with a thermal conductivity 3–5 times that of copper, a close match in expansion coefficient to silicon and GaAs, low density and high electrical resistivity, offers an attractive alternative to traditional metal heat spreader materials. Pure metals suffer from poor match in expansion coefficient with the IC die. Alloying sacrifices performance (by reducing the thermal conductivity) to achieve reliability (by improving the expansion coefficient match). In addition, weight may be an issue as refractory materials such as Mo or W are alloyed or clad with pure metals to deal with the thermal stress issue. AlSi and composites of SiC with Al represent attractive new materials from an expansion match point of view while nearly meeting the thermal conductivity of the metal alloy. However, they fall short in matching diamond's thermal conductivity.

The performance improvement potential of a diamond-based package has always been realised: the issue has been cost. However, in the past 10 years significant progress [1] has been made in addressing the key issues of cost and availability of CVD diamond to make it a viable candidate for certain applications, such as high power RF amplifiers, microprocessors and laser diodes.

## B FIGURE OF MERIT: THERMAL CONDUCTIVITY

An important figure of merit for a material acting in a thermal management capacity is the thermal conductivity. Bulk diamond, natural or synthetic, has been used with limited success because of size issues. The advantage of CVD diamond over high pressure bulk diamond is two-fold: 1) its properties can be tuned to achieve the desired thermal conductivity simply by tuning the fraction of hydrocarbon in the fuel gas supply and 2) large area coverage of both planar and 3D shaped objects is possible. The thermal conductivity of CVD diamond is a function of the film microstructure, non-diamond ( $sp^2$ bonded carbon) phase and hydrogen in the lattice and grain boundaries. The TC depends predominantly on the methane gas fraction and deposition temperature. TC values as high as 2500 W/m K, by using isotopically pure methane, or as low as 100 W/m K can be obtained, depending on film thickness and process conditions. FIGURE 1 shows an example of the effect of a dominant impurity, hydrogen, entrained in the film. The decrease in TC with the hydrogen content is attributed to its function as a phonon scatterer. Similar correlations are observed for other defects.

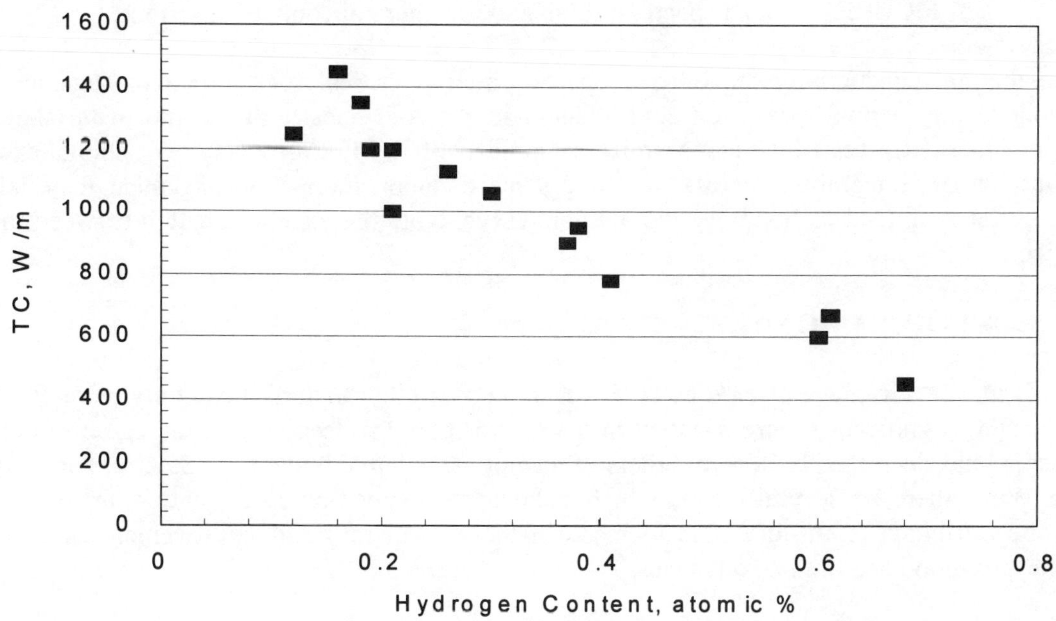


FIGURE 1 Variation of thermal conductivity with hydrogen entrained in the film.

The non-diamond carbon phase, which imparts many of the thermal, mechanical and optical properties to CVD diamond, is easily discerned using Raman spectroscopy. Typically, a broad peak is observed at 1560 $cm^{-1}$, neighbouring the diamond signature peak at 1332 $cm^{-1}$, as shown in FIGURE 2. The Raman spectra represent examples of relatively high (1200 W/m K) and low thermal conductivity (600 W/M K) materials. An additional indicator of low TC material is the background luminescence from defects, as shown by the elevated baseline.

The TC is anisotropic. In the Z-direction it is typically 20% higher than in the in-plane direction because of the stratified columnar microstructure in the growth direction. For thermal management purposes 800–1000 W/m K is sufficient to accrue the benefits of heat spreading. Higher TC values come, literally, at a higher price. In fact the correlation is quadratic.

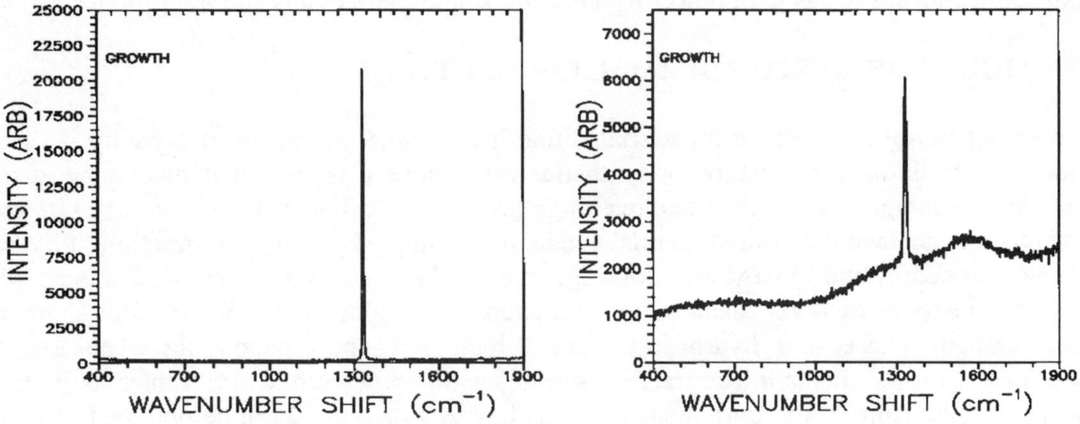


High TC Low TC

FIGURE 2 Raman spectra of high and low thermal conductivity films.

Besides thermal conductivity the other property which is critical for thermal management in IC packaging is the thermal expansion coefficient. Large mismatches with the die material result in deleterious thermal stress which compromises reliability. FIGURE 3 presents the fractional expansion as a function of temperature for diamond and some common thermal management materials. The metals, not shown, tend to lie above the alumina curve. Note the excellent match between diamond and silicon.

## C COST OF DIAMOND

The performance advantage offered by CVD diamond over other materials can only be realized if the price is right. Significant progress has been made over the past decade in reducing deposition cost. The cost [1] has decreased by several orders of magnitude, from $5000/$cm^2$ to $5/$cm^2$, during the past decade. While there are several factors which make up the deposition cost equation, the decrease can mainly be attributed to the dramatic increase in the growth rate and improvement of energy-use efficiency, as embodied in three advances:

- scaling to higher power deposition systems,
- increase in substrate capture area,
- increasing yield of useful product.

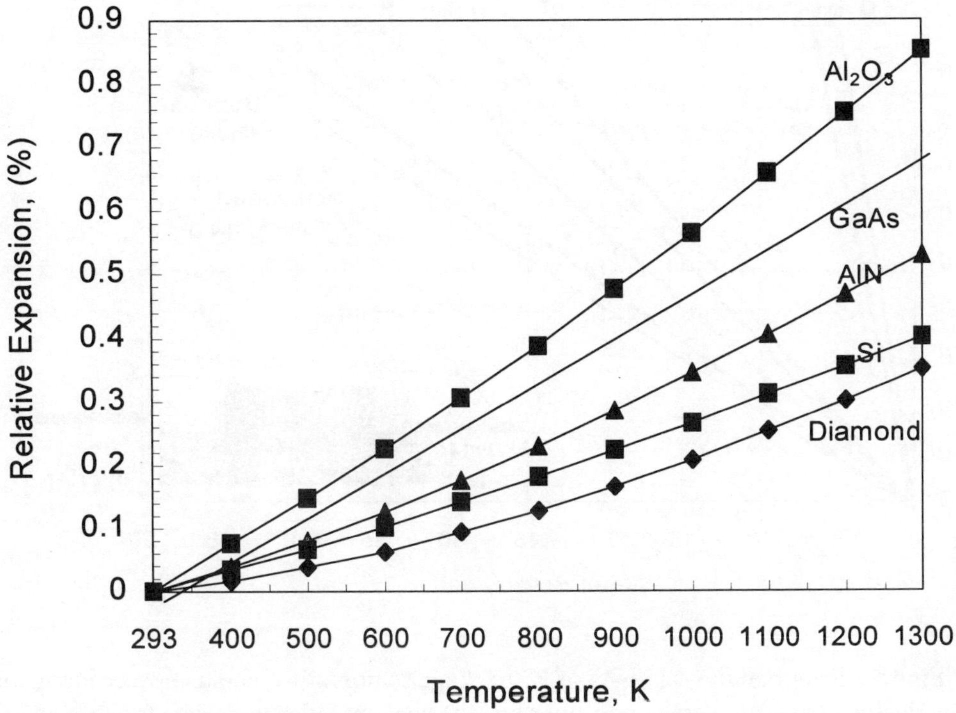


FIGURE 3 Relative expansion variation with temperature for a variety of materials used in IC packaging.

Rates have advanced from the several millicarats/hr range in late 1987 to 20 carats/hr in late 1998 while maintaining quality as the reactor power has increased from 0.5 kW to 125 kW. Rates as high as 20 carats/hr for 100–200 kW systems have been achieved. While mass growth rate is important, the more pertinent figure of merit with regard to deposition cost is the energy-use efficiency e, which is a composite term that incorporates not only deposition rate, but also atomic hydrogen generation and capture efficiency. The objective of any deposition technology should be to increase the slope of the rate-power curve so as to approach the smallest e values. CVD diamond deposition technology has reached values as low as 6.25 kW-hr/carat [1].

## D MODELLING: IC PACKAGING

Diamond is best suited in applications where it can be used to bypass a thermal bottleneck. In the case of IC packages a large thermal resistor exists between the heat spreader and the copper leadframe which serves not only as the signal carrier but also as the dominant conduit for heat dissipation to the outside world. Since diamond is electrically insulating the leadframe can be attached directly to the heat spreader, obviating the need for an insulating layer required for a metal heat spreader. This is referred to as an overlap design. The benefits of this overlap design versus a non-overlap design are quantified in a finite element model along with the effects of various heat spreader materials, and adhesion thickness for a silicon-based plastic encapsulated IC package. The results are shown in FIGURE 4 which presents the package temperature against power input to the die for a 250 μm thick heat spreader.

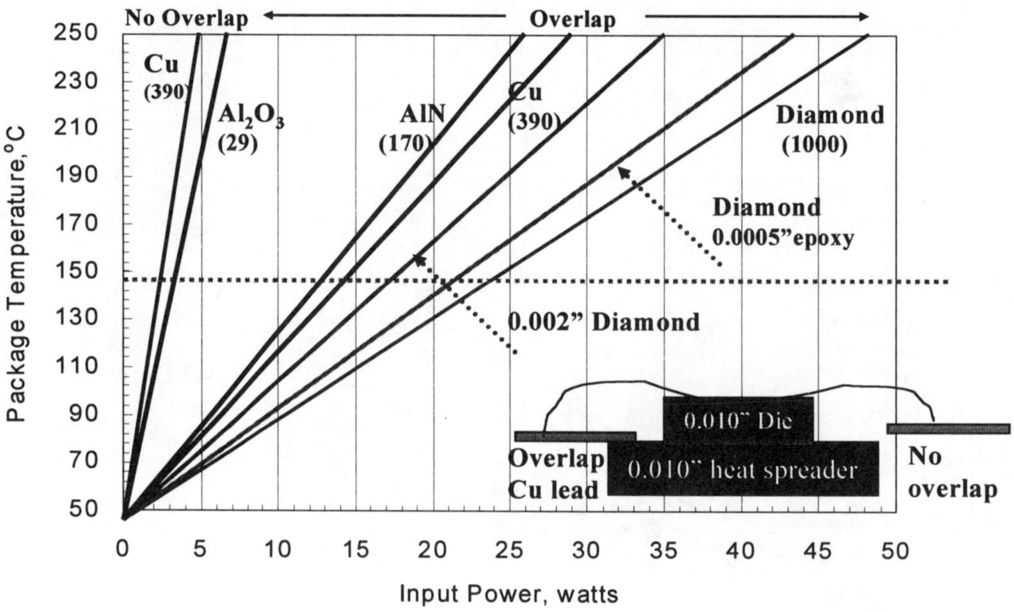


FIGURE 4 Finite element analysis (FEA) of IC package temperature versus power input for (a) various package designs (overlap versus non-overlap) (b) heat spreader materials (copper, $Al_2O_3$, AlN and diamond) and (c) heat spreader thickness (50 and 250 μm thick diamond).

The results show that the greatest benefit is achieved by package design. The overlap design increases the power handling capacity from 3 to 25 W for a copper heat spreader over the non-overlap design. Diamond's high TC provides additional performance enhancement over AlN and copper. Thinner diamond can be used to reduce cost but sacrifices performance.

## E THERMO-MECHANICAL COMPATIBILITY

Overcoming the significant thermal expansion mismatch between package components (GaAs die and Cu leadframes with CVD diamond heat spreader) while maintaining thermal performance poses a significant challenge. Finite element modelling using elastic-plastic assumptions for the GaAs/PbSnAg/CVD diamond assembly, showed interfacial peeling stresses, shear stresses, and tensile stresses to originate upon attachment. Peeling stress can cause horizontal die cracking as GaAs has a low flexural strength of about 100 MPa. By modifying the mechanical properties and thickness of the attach material(s) a 60% stress reduction in GaAs is achieved using a 100 μm Cu compliant layer between the die and the diamond to absorb the thermal stress. The GaAs/diamond package passed thermal stress cycling tests from –60 to 125°C [2].

## F THERMAL MANAGEMENT APPLICATIONS

### F1 Diamond in IC Plastic Packages

Plastic packaging has been adopted globally for low power devices because of low cost. However, as the IC power dissipation increases plastic packaging may no longer become viable with conventional heat spreaders. Diamond is an attractive solution [3–6]. A dramatic demonstration of the performance advantage of diamond has been shown by incorporating a GaAs MMIC PHEMT [2] power amplifier, soldered to a diamond substrate, and moulded into a plastic package. 20 W CW operation for 96 hr with a junction temperature of only 140°C was achieved. This represents a 10× power improvement over conventional plastic packaging. The CVD diamond enhanced packages passed the 200 hr, 110°C, 85% RH highly accelerated stress testing without electrical, mechanical or thermal degradation.

### F2 Laser Diodes

Laser diode arrays have continuous-wave output power levels of up to 100 W at room temperature [7]. Even with an efficiency of 50% these laser diodes create a tremendous heat flux. For efficient cooling of these devices it is essential to spread the narrow heat flux with a CVD diamond heat spreader. The advantage of diamond heat spreaders was illustrated on a CVD diamond heat spreader positioned between the laser diode array and a microchannel cooler. For the calculation a heat flux of 1000 W/cm$^2$ for the laser diode and a thermal conductivity of 20 W/cm K for the diamond layer were assumed. Without a diamond heat spreader the maximum temperature rise in the laser diode array amounts to < 50 K. By adding a diamond heat spreader the heat is distributed over more cooling channels producing a significantly reduced temperature rise by a factor of 3×. Even more importantly, the diamond heat spreader strongly influences the temperature gradient across the active zone in the laser diode. Elevated operation temperatures strongly affect the lifetime and efficiency of lasers and there is a wavelength shift. The diamond heat spreader reduces these deleterious effects.

## G CONCLUSION

The insertion of CVD diamond into IC packages and laser diodes can offer significant performance and reliability advantages. Utilizing surface treatment technology for high reliability metallization/die attach technology for GaAs stress management, the thermal spreading capabilities of CVD diamond have been demonstrated to elevate plastic packaging technology from the 1 W power regime to 20 W.

## REFERENCES


[1] J.E. Butler, H. Windischmann [ *MRS Bull. (USA)* vol.23 (1998) p.22-7 ]

[2] P.M. Fabis, H. Windischmann [ *Proc. Workshop on Mechanical Reliability of Polymer Materials and Plastic Packages of IC Devices* Paris, Dec. 12 1998, Ed. E. Suhir (ASME) vol.25 p.109 ]

[3] M.D. Brown, A.P. Malshe, M.H. Gordon, R.K. Ulrich, W.F. Schmidt, W.D. Brown [ *Proc. IMAPS 1997 Int. Systems Packaging Symposium* San Diego, CA, Dec. 1 1997, p.129 ]

[4] P.M. Fabis, H. Windischmann [ *Proc. INTERPACK* Maui, HI, June 12 1999 (EEP ASME, New York) vol.26-1 p.899 ]

[5] H. Windischmann, P.M. Fabis [ *ISPS '99 Proc. Int. Systems Packaging* San Diego, CA, Jan. 11 1999 (IMAPS, 1999) p.19 ]

[6] H. Windischmann [ *Proc. 1998 Int. Conf. Multichip Modules and High Density Packaging* Denver, CO, April 15 1998, p.224 ]

[7] Frauenhofer Institute [ http://www.iaf.fhg.de/diamond/index.html ]

# SUBJECT INDEX